Categorical Foundations

The book offers categorical introductions to order, topology, algebra, and sheaf theory, suitable for graduate students, teachers, and researchers of pure mathematics. Readers familiar with the most basic notions of category theory will learn about the main tools that are used in modern categorical mathematics but are not readily available in the literature. Hence, in eight independent chapters, the reader will encounter various ways of how to study "spaces": order-theoretically via their open-set lattices, as objects of a fairly abstract category via their interaction with other objects, or via their topoi of set-valued sheaves. Likewise, "algebras" are treated as both models for Lawvere's algebraic theories and Eilenberg-Moore algebras for monads, but they appear also as the objects of an abstract category with various levels of "exactness" conditions. The abstract methods are illustrated by applications that, in many cases, lead to results not yet found in more traditional presentations of the various subjects, for instance, on the exponentiability of spaces and embeddability of algebras. Suggestions for further studies and research are also given.

Maria Cristina Pedicchio is Professor of Mathematics at the University of Trieste. She is President of the Area Science Park of Trieste and Consultant for the Italian Ministry of Education.

Walter Tholen is Professor of Mathematics at York University. His research interests include category theory and its applications to algebra, topology, and computer science.

ENCYCLOPEDIA OF MATHEMATICS AND ITS APPLICATIONS

Categorical Foundations

Special Topics in Order, Topology, Algebra, and Sheaf Theory

Edited by

MARIA CRISTINA PEDICCHIO
University of Trieste

WALTER THOLEN
York University, Toronto

CAMBRIDGE
UNIVERSITY PRESS

PUBLISHED BY THE PRESS SYNDICATE OF THE UNIVERSITY OF CAMBRIDGE
The Pitt Building, Trumpington Street, Cambridge, United Kingdom

CAMBRIDGE UNIVERSITY PRESS
The Edinburgh Building, Cambridge CB2 2RU, UK
40 West 20th Street, New York, NY 10011-4211, USA
477 Williamstown Road, Port Melbourne, VIC 3207, Australia
Ruiz de Alarcón 13, 28014 Madrid, Spain
Dock House, The Waterfront, Cape Town 8001, South Africa

http://www.cambridge.org

First published 2004

Printed in the United States of America

Typeface Times New Roman PS 10/12.5 pt.   *System* LATEX 2_ε [TB]

A catalog record for this book is available from the British Library.

Library of Congress Cataloging in Publication data
Categorical foundations : special topics in order, topology, and sheaf theory / edited by
Maria Cristina Pedicchio, Walter Tholen.
p.   cm. – (Encyclopedia of mathematics and its applications)
Includes bibliographical references and index.
ISBN 0-521-83414-7
1. Categories (Mathematics)   I. Pedicchio, M. C. (Maria Cristina), 1953-   II. Tholen, W.
(Walter), 1947-   III. Series.
QA169.C345   2004
511.3 – dc21   2003055124

ISBN 0 521 83414 7 hardback

Dedicated to the memory of

Japie Vermeulen

Summary of Contents

Preface

In 1998, with the support of the European Union and the Government of Canada, mathematicians from four European and three Canadian universities embarked on a four-year program promoting, on the one hand, transatlantic student exchanges and, on the other hand, a collaborative research project on the coordinated teaching of various topics in pure mathematics, the common thread of which is the active use of categorical foundations and methods. Primarily during group meetings held in the summers of 1999 in Coimbra (Portugal), 2000 in Toronto (Canada), and 2001 in Barisciano (Italy), small teams were formed to work on a variety of themes of current interest for which an adequate modern treatment in the literature was not available. These teams included not only faculty members of the sponsoring universities, but also graduate students of these universities who were participating in the exchange program, as well as visiting researchers to the partner universities.

The result of this collaboration is documented in this book which provides a categorical introduction to some key areas of modern mathematics, readable by anyone with a modest knowledge of category theory. While the degree of required general mathematical maturity and particular categorical expertise may vary among chapters, graduate students and researchers should have no difficulty following the main thrust of the theories presented, which in many instances should stimulate further research.

We dedicate this book to the memory of an admired researcher and a great friend of many of its authors, Japie Vermeulen of the University of Cape Town. Had it not been for his untimely death in February of 2001 we would have very much liked this brilliant mind to participate in the project, over and above the stimulating advice which many of the book's authors had the privilege to receive from him even long before this project was conceived.

Among the many colleagues who offered us their valuable comments and thoughtful advice during various stages of the project we would like to thank foremost Bill Lawvere for his loyal mentorship and Francis Borceux for his willingness to read large portions of earlier versions of the text and to provide feedback. Their contributions led to innumerable improvements. In the same vein we would like to express our appreciation for the constructive criticism offered by the anonymous referees contacted by the publisher.

This project would not have come to life without the generous support and the dedication of:

- the EC/Canada Cooperation Programme of the European Commission, Directorate-General of Education and Culture,
- the International Academic Mobility Programme of the Department of Human Resources and Development Canada,
- the participating universities and the local project leaders, namely the University of L'Aquila (Anna Tozzi), the University of British Columbia (John MacDonald), the University of Coimbra (Manuela Sobral), Dalhousie University (Richard Wood), the Université Catholique de Louvain (Enrico Vitale), the University of Trieste (Maria Cristina Pedicchio), and York University (Walter Tholen),
- Isabella Fontana who helped drafting the original project proposal and coordinated the European project activities,
- the staff at the international offices of the participating universities, especially at the lead institutions in Europe (Trieste) and Canada (York),
- the national research granting agencies, namely the Natural Sciences and Engineering Council of Canada, the Italian "Ministero dell'Istruzione dell'Università e della Ricerca - Programmi di Ricerca Scientifica di rilevante interesse nazionale",
- the Centre for Mathematics at the University of Coimbra.

We thank them all, as well as Cambridge University Press for publishing this book.

Last, but not least, we express our gratitude to Isabel Clementino and Jorge Picado without whose TeXnical help and expertise we would have felt at a loss, and to our spouses Pierpaolo Ferrante and Jane Cleve whose never-ending patience, encouragement, and support meant so much to both of us.

M.C.P. and W.T.

Trieste and Toronto, February 2003.

Table of Contents

Introduction

Walter Tholen

Following the many splendid mathematical discoveries of the nineteenth century, mathematicians like Felix Hausdorff and Emmy Noether formalized the key notions and thereby the categories which enable us to pursue modern structural mathematics. Bourbaki's monumental work gives the most comprehensive testimony of this rapid development during the first half of the twentieth century. Despite the laudable trend toward emphasizing applications, for the past few decades there has been little change in the overall perception of which general notions are to be considered important, and they continue to dominate the standard courses on topology and algebra.

Soon after Samuel Eilenberg and Saunders Mac Lane coined the notions of category, functor and natural transformation in the 1940s, a very different way of doing structural mathematics emerged which was promoted especially in homological algebra. Rather than building objects made up from sets with a structure, like topological spaces and modules over a ring, one imposes additional conditions on a category which make its objects behave like the structured sets one has in mind. Hence, rather than relying on a set-theoretic foundation on which to build the structures at issue, one moves into a "fully equipped building" in which to explore the objects of interest. Indeed, the category theory of the 1950s, especially Alexander Grothendieck's work, showed that a great deal of module theory can be performed in any abelian category and which abelian categories are actually equivalent to module categories.

The exclusive standpoint that all abstract mathematics is done by studying sets with a suitable structure was shattered even more severely when in the late 1960s Bill Lawvere and Myles Tierney axiomatized categories of sheaves of sets. It soon became clear that their notion of elementary topos with its internal intuitionistic logic allows us to dramatically alter the view of what the mathematical objects of our consideration are and how to study them, starting with what seemed to underly all modern mathematics: sets themselves. However, having grown up on traditional Boolean and set-theoretic foundations, for most mathematicians it continues to feel foreign to study a mathematical object by exclusively looking at its interaction with all the other players in the same league, rather than by trying to create a "once-and-for-all" (but perhaps elusive) definition of what these objects are, in terms of their elements and the structures put onto them.

This book gives its readers categorical glimpses at a number of key areas of mathematics, showing in fact a kaleidoscope of modern approaches to them. Rather than promoting a one-fits-all theory, it presents in eight rather independent chapters, with varying categorical rigor, important mathematical subjects, letting believers of "objects are sets with structure" and champions of categorical axiomatics cohabitate peacefully and in fact interact to each others' benefit, just like module theory and abelian category theory do. Hence, in extension of this classical interaction, the reader will learn in up-to-date presentations about the models of algebraic theories and monads as well as the axioms that characterize the categories formed by them. Likewise, the topology-minded reader may study "spaces" as locales, as the objects of a suitably structured abstract category, or via their topoi of set-valued sheaves. To a large extent, rudimentary knowledge of category theory suffices to understand each chapter, and except for some "collateral" cross-references there is little interdependence between chapters. Less familiar notions are introduced in the respective chapters, normally with references to the literature for further reading. The Index for the entire book should help readers find terms unknown to them.

Here is a brief outline of the book. Chapter I presents perhaps the easiest mathematical structure of the entire book, ordered sets, but it does so under the most rigorous categorical regime, taking "set" to mean "object of an elementary topos" and thereby making the theory entirely constructive for the intuitionist. But even the reader who disregards this fact will learn to appreciate two further features which distinguish the text from any other presentation of ordered sets, regardless of what "set" may mean to him or her: the power and simplistic beauty of the use of adjunctions, and the clear isolation of the role of Booleaness and of choice principles.

Chapter II introduces the span between order and topology, giving a hands-on approach to the theory of locales. While algebraically these are just complete lattices satisfying an infinite distributive law, the reader will quickly learn how to view them as "spaces" that have open or closed parts but not necessarily points. Keeping the categorical prerequisites at a minimum, the chapter gives a detailed but compact presentation of important topological themes, including separation, regularity, and compactness.

A strictly axiomatic approach to general topology is presented in Chapter III. "Spaces" are now objects of a category with some additional structure and properties, and this category may indeed be the category of the topological spaces of a Zermelo-Fraenkel set-theorist, the category of spaces over a fixed base space as in fibred topology, or the category of locales of Chapter II. How far do you go in such generality? The answer is given in terms of separation, compactness, and exponentiabilty and contains many facts not found in traditional textbooks, but also leaves plenty of room for further exploration.

Methodologically, Chapter IV is a close relative of its predecessor: If Chapter III teaches "intrinsic topology", then here the reader learns about "intrinsic algebra". Its key notion is that of a protomodular category which, in effect, makes the validity of the Short Five Lemma of homological algebra an axiom in the category at hand. Especially in conjunction with Barr's exactness axiom, it makes its objects behave like groups, to the same extent that objects of an abelian category look like abelian groups. The reader is then exposed to those algebraic constructions which may be performed under the characteristic categorical properties carried by the main categories of algebra, such as the categories of groups, rings, modules, and algebras.

Monads, as presented in Chapter V, and Lawvere's algebraic theories, as used in Chapter VI, are syntactic categorical tools which allow for the formation of categories of fairly general algebraic structures, including those considered in "universal algebra". Chapter V gives a first impression of the ubiquity and applicability of modern monad theory, with its many occurrences throughout mathematics as well as theoretical computer science. It sheds light on very particular classical theorems of algebra, such as the Birkhoff-Witt Theorem for Lie algebras and Schreier's Theorem for free products of groups with amalgamated subgroup. The purpose of Chapter VI is to characterize various types of algebraic categories internally. Regularity and exactness of a category as treated in Chapter IV, monadicity over the category of sets as treated in Chapter V, as well as the Gabriel-Ulmer notion of locally finitely presentable category are the key ingredients to a series of finely interlocked characterization theorems.

The introduction to sheaf theory as given in Chapter VII touches upon many categorical themes and tools mentioned earlier in the book but represents nevertheless a self-contained first account of a beautiful mathematical subject. While presenting en passant a variety of powerful categorical techniques it leads the reader gently to the modern characterization of localizations of presheaf categories (i.e. of Grothendieck topoi), as the Barr-exact and Carboni-Lack-Walters-extensive categories with a generator. Finally, Chapter VIII gives a brief look at descent theory which, like sheaf theory, turns up in many different mathematical contexts. The focus is on basic-fibrational descent theory, especially on the role of effective descent morphisms in categories which fail to enjoy the good properties like exactness as assumed in many other chapters. Links with other mathematical theories, like monad, sheaf, and Galois theory are explained briefly with directions for further studies.

In general, each chapter should be readable by a graduate student with some basic knowledge of pure mathematics and modest acquaintance with category theory. In principle, chapters are readable independently of each other, irrespective of the lose interrelationships as mentioned earlier and roughly depicted by the following chart.

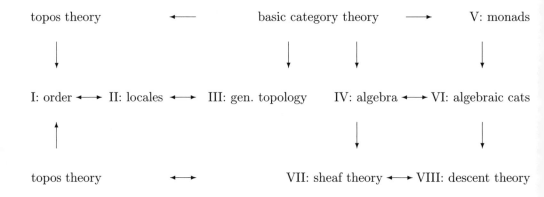

I

Ordered Sets via Adjunctions

R.J. Wood

'Sets for Mathematics', by F.W. Lawvere and R. Rosebrugh, [5] is a ground-breaking, undergraduate, set theory textbook. Categories provide the metalanguage and, for a substantial part of the book, axioms are gradually imposed on a category $\mathcal{S}$ until its objects and arrows capture the key features of sets and functions that are used in mathematical practice. To those who would say that sets and functions are themselves lurking in the definition of *category*, the rejoinder should surely be that sets and functions are present to the same extent in the metalanguage of traditional set theory texts. By the time a student starts to think critically about sets and functions in an undergraduate mathematics program, he or she has already implicitly studied several categories—continuous, differentiable, linear, order-preserving, and so on. It is to these categories, and other categories of mathematical structures, that a student turns repeatedly in the course of studying Mathematics. To see these categories as categories of sets with structure, it seems to this writer most appropriate to put the formal study of sets themselves on the same footing.

Lawvere and Rosebrugh accomplish in [5] much more than is possible in a traditional set theory book because they isolate those categorical axioms for sets and functions that allow sets to admit both variation and the intuitionistically valid constructs and theorems of the subject. A category satisfying the axioms in question is called a(n *elementary) topos*. For each topos $\mathcal{S}$ there is a category of groups in $\mathcal{S}$, of topological spaces in $\mathcal{S}$, and so on. Rather remarkably, such categories can be studied as though their objects are sets with structure, using quite conventional methodology, provided that one proceeds as an intuitionist. Part of the goal of this chapter is to make a modest incursion into the 'category' of ordered objects in a topos.

For undergraduate students, much of the material in [5] is a serious prerequisite for this chapter. At the time of this writing, [5] itself is very new but most of the material it contains has been known to the category theory community for about thirty years and certainly there are category theorists who have taught set theory classes to their undergraduates employing the ideas of [5] but without benefit of a published text. There are other texts on toposes which, strictly speaking, could serve as prerequisites for this work but they are aimed at a higher level than this chapter itself. For readers familiar with topos theory via older texts, it is hoped that the present contribution may serve as reference material.

Just as there is a category of groups in $\mathcal{S}$ ($\mathcal{S}$ a topos) so there is a category whose objects are categories in $\mathcal{S}$ and whose arrows are functors. After all, in the metalanguage of categories, wherein $\mathcal{S}$ itself appears, we speak of a category as having a 'set' of objects, a 'set' of arrows, and functions, satisfying functional identities, that prescribe domain, codomain, identity arrows and composition. If the set theory given by $\mathcal{S}$ is to be taken at all seriously then it is quite natural to speak of categories in $\mathcal{S}$. (In fact, less than the full set of axioms for a topos suffices for this purpose.) Let us write $\mathbf{cat}(\mathcal{S})$ for the category of categories and functors in $\mathcal{S}$. For $F, G : \mathbf{X} \to \mathbf{A}$ in $\mathbf{cat}(\mathcal{S})$ it is a straightforward matter to define natural transformations from F to G — just as in the metalanguage — and, with these at hand, to define *adjunction* in $\mathbf{cat}(\mathcal{S})$ — again, just as in the metalanguage.

To some, including this writer, adjunction is the most important concept in category theory. Its foundational nature is witnessed by the fact that the axioms for a topos can be given in terms of adjunctions. Yet, adjunction tends to appear rather late in most accounts of category theory and is often presented in a forbiddingly technical way. Fortunately, the concept makes sense for order-preserving functions and for these it is almost impossible to obfuscate the central ideas.

A student who has studied but a little of [5] already knows that if a category $\mathcal{C}$ of the metalanguage has the property that, for any objects X and A in $\mathcal{C}$, there is at most one arrow from X to A then $\mathcal{C}$ is an 'ordered set' and that one might as well write $X \leq A$ precisely when there is an arrow from X to A. Such a student also knows that if categories $\mathcal{C}$ and $\mathcal{D}$ are both 'ordered sets' in this sense then to give a functor $F : \mathcal{C} \to \mathcal{D}$ is to give a function from the objects of $\mathcal{C}$ to the objects of $\mathcal{D}$ with the property that $X \leq A$ implies $FX \leq FA$. We assume this familiarity in the process of defining $\mathbf{ord}(\mathcal{S})$, as the bicategory of ordered objects in $\mathcal{S}$, order-preserving arrows, and inequalities. We then immediately define adjunction in $\mathbf{ord}(\mathcal{S})$ and keep it as a central theme throughout.

We say no more about $\mathbf{cat}(\mathcal{S})$. By concentrating on adjunction in $\mathbf{ord}(\mathcal{S})$ we are able to pursue the concept without coming to terms with the issues of 'size' and 'coherence' while at the same time somewhat advancing the programme of [5] in an obvious direction. One might make the case for simply pursuing the study of ordered sets via adjunction in the metalanguage but to do so would lose sight of the valuable generality gained in [5]. Most of what we chose to include is valid in any topos. We find it convenient to write $\mathbf{set}$ for $\mathcal{S}$ and $\mathbf{ord}$ for $\mathbf{ord}(\mathcal{S})$ and we do not give any explicit examples of how our $\mathbf{set}$ might be variable because these are to be found in [5]. The student who has *carefully* studied [5] is encouraged to examine key concepts of this chapter in toposes such as $\mathbf{set}^2$ by translating them back to $\mathbf{set}$.

Of course the study of ordered sets can be taken up in many ways using traditional foundations and there are some serious limitations on what we can hope to say without introducing more pure topos theory than is already found in [5]. An important case in point concerns the issue of finiteness. In our $\mathbf{set}$ there are different, inequivalent notions of finiteness that must be employed for a careful treatment of some theorems that are traditionally quite straightforward. We do not pursue this problem. On the few occasions when we find it convenient, for expository reasons, to include material that addresses finiteness we flag such results with an asterisk.

In Section 1 we provide some preliminaries that allow us to build ordered set theory from the bicategory of relations in **set**. In fact we define adjunctions therein, and show that from these we can recover functions. In Section 2 we formally define ordered sets, the bicategory of these, adjunctions therein, and provide some important examples. We put adjunctions to work in Section 3 to describe semi-lattices, lattices and a few of their specializations. So as to discuss complete lattices in these terms in Section 5, Section 4 addresses power set Heyting algebras.

Of course power sets are traditionally considered to be Boolean algebras and here we should underscore a central point of departure with traditional approaches. We do not assume of our **set** that a subset of a set is equal to its double 'complement' — in other words we do not assume that our topos **set** is *Boolean*. (A fortiori, we do not invoke the axiom of choice, which is known to imply Booleaness — a fact first discovered in the context of topos theory by Radu Diaconescu.) Actually, independently of any wish to work in a topos, there is a certain economy when it comes to ordered sets in confronting Heyting rather than Boolean algebra from the outset. For an ordered set, in anybody's set theory, the *downsets* of the ordered set (ordered by inclusion) form a Heyting rather than a Boolean algebra. It is the downsets of an ordered set X (which we fully intend to pursue in some detail) that are relevant to the study of X in a way that the power set of the underlying set of X cannot be expected to be. After all, most sets can be ordered in more than one way, in anybody's set theory. Ordered set theory in particular — category theory in general — is intrinsically intuitionistic.

Section 6 is a brief introduction to the author's own work on completely distributive lattices with a number of coauthors, most notably Rosebrugh. It is here that the approach to ordered sets via adjunction seems to add the greatest clarity to the subject to date.

Traditionally, most theorems about completely distributive (CD) lattices have invoked the axiom of choice (AC). In fact, without (AC) there is even a paucity of examples of (CD) lattices, for one of the many classical equivalent forms of (AC) states that every power set is (CD). Via adjunction it is very natural to introduce a notion that we have called 'constructively completely distributive' (CCD) lattices for which we prove the following theorem, stated in an obvious symbolic form:

$$(AC) \iff ((CD) \iff (CCD))$$

Via adjunction it follows, almost immediately from the definition of (CCD), that every power set is *intuitionistically* (CCD) (and together with the aforementioned classical result about power sets this proves half of the assertion above).

In fact, (CCD) bears further on the issue of studying Mathematics via topos theory. Let us write (^{op}CD) for the dual of (CD). In other words, a lattice L

satisfies (^{op}CD) if and only if the dual lattice L^{op} satisfies (CD). Similarly, write (^{op}CCD) for the dual of (CCD). It is classical that

$$(\text{AC}) \implies ((\text{CD}) \iff (^{op}\text{CD})).$$

Writing (BLN) to denote that the topos of enquiry is Boolean, it was shown in [7] that

$$(\text{BLN}) \iff ((\text{CCD}) \iff (^{op}\text{CCD})).$$

We include a proof of this result and observe that it, together with other results of this section, provides a new, indirect proof of the aforementioned Diaconescu result:

$$(\text{AC}) \implies (\text{BLN}).$$

1. Preliminaries

1.1. Categorical prerequisites. We assume a basic knowledge of categorical concepts, such as *pullback*, which are defined by universal mapping properties. An excellent reference is [5]. The study of ordered sets allows one to meet new examples — simultaneously on two different levels. We will see that on the one hand, ordered sets and the relevant functions between them form a category but, on the other hand, each ordered set can be regarded as a category in its own right. Thus, the category of all ordered sets will provide a simple example of a *category of categories*.

1.2. Topos axioms. A *topos*, see [5], is a category $\mathcal{S}$ which has

    i) finite limits;
    ii) power objects.

From the axioms it follows that a topos has finite colimits, mapping objects, a subobject classifier, and many other properties. Often, ii) is expressed in the context of i) by saying of $\mathcal{S}$ that it

    ii′) is cartesian closed (has mapping objects);
    ii″) has a subobject classifier.

If Ω is a subobject classifier for cartesian closed $\mathcal{S}$ with finite limits then, for each X in $\mathcal{S}$, a power object is provided by Ω^X. Conversely, if $\mathcal{S}$ with finite limits has power objects $\mathbb{P}X$ then $\mathbb{P}1$, where 1 is the terminal object, provides a subobject classifier while power objects A^X are obtained as certain subobjects of the $\mathbb{P}(A \times X)$.

    As stated in the introduction, we will write **set** for the topos that is regarded as our category of sets and functions and, unless otherwise indicated, assume no more than that. In particular, we do not assume that the $\mathbb{P}X$ are Boolean algebras in **set**. In **set** we speak of 'power sets', and so on, rather than 'power objects'.

1.3. Specified universal objects. We will understand 'has' in connection with finite limits and power sets to include specification. For if in some category we have squares

$$
\begin{array}{ccc}
P & \xrightarrow{q} & B \\
\downarrow{\scriptstyle p} & & \downarrow{\scriptstyle g} \\
A & \xrightarrow{f} & X
\end{array}
\qquad
\begin{array}{ccc}
P' & \xrightarrow{q'} & B \\
\downarrow{\scriptstyle p'} & & \downarrow{\scriptstyle g} \\
A & \xrightarrow{f} & X
\end{array}
$$

which are pullbacks then there is an isomorphism $i : P \xrightarrow{\simeq} P'$, unique with the property that

commutes. Either square is 'a' pullback of f and g and without further comment 'the' pullback of f and g might be meaningless without making some sort of choice. In many categories however there is a specific construction which produces a specific pullback, for each pair f and g with common codomain. It is our point of view that **set** is such a category but we have no need to say what the specific construction is. When we speak of 'the' pullback of f and g this specific pullback is what we will mean. Notation such as $A \times_X B$ will be used for specified pullbacks. In particular, **set** has specified inverse images of parts (monomorphisms) too and notation such as $f^{-1}(m)$ will be used for these. We make similar conventions about *the* terminal set and *the* power set of a given set. It follows that we have derived specifications — for colimits and so on. It should be noted though that specified pullbacks and the like do not enjoy many closure properties. For example, if two pullback squares are pasted side by side then the resulting large square is again a pullback but it is usually not a specified pullback when the original pullback squares are so.

1.4. Categories of parts. For X any set, we provisionally write $\mathcal{P}X$ for the collection of all parts of X. (A *part* of X is a monomorphism with codomain X.) Here and in the sequel the word 'collection' will be used as a term in the metalanguage that we are tacitly using to describe our category **set** of sets and functions. If given collections $\mathcal{Q}$ and $\mathcal{R}$ there is a rule ϕ that associates to each Q in $\mathcal{Q}$ a unique R in $\mathcal{R}$ then we use notation such as $\phi(Q) = R$ and even $\phi : \mathcal{Q} \to \mathcal{R}$, without the suggestion that $\phi : \mathcal{Q} \to \mathcal{R}$ is in **set** or for that matter *in* any other category under formal consideration. Such notation is merely a convenient extension of the meta-language. In fact, such $\mathcal{Q}$ and $\mathcal{R}$ can be seen as discrete categories, whereupon ϕ is seen as a functor. The fact that **set** is rich enough to encode within itself many statements made in the metalanguage is quite a different matter.

We can regard $\mathcal{P}X$ as a category — by taking as objects the parts of X and declaring there to be an arrow from part R to part R' precisely if $R \subseteq R'$. For any part R, we have $R \subseteq R$; while if $R \subseteq R'$ and $R' \subseteq R''$ then $R \subseteq R''$. These two observations provide the identities and composition for the category that we have in mind. It is important to understand that this composition automatically satisfies the associativity and identity axioms for a category. See, for example, Definition 1.13 of [5]. Observe that if $R \subseteq R'$ and $R' \subseteq R$ then R and R' are isomorphic parts and we write $R \cong R'$ in this case.

For a category $\mathcal{C}$ and objects X and A therein, it is often convenient to write $\mathcal{C}(X, A)$ for the collection of arrows in $\mathcal{C}$ of the form $f : X \to A$. Many authors speak of these collections as the *hom-sets* of $\mathcal{C}$ but such terminology is initially unfortunate for us. The $\mathcal{C}(X, A)$ are creatures of the metalanguage. In these terms the very structure of the category $\mathcal{C}$ is given by

$$\mathcal{C}(X, A) \times \mathcal{C}(A, Y) \to \mathcal{C}(X, Y) \quad \text{and} \quad 1 \to \mathcal{C}(X, X).$$

Here we have used $- \times -$ to build a collection of pairs and 1 to denote any convenient collection with one member. The fact that $- \times -$ and 1 are the same symbols that we use axiomatically in **set** should cause no problems. Evidently, each $\mathcal{C}(X, X)$ carries a monoid structure in virtue of the categorical structure of $\mathcal{C}$ and these monoids act on the $\mathcal{C}(X, A)$. But it is important to understand that, for a general category $\mathcal{C}$, no further structure is assumed. This will shortly become a point of departure for this Chapter. To emphasize the notation, observe that $\mathcal{P}X(R, R')$ is a collection with one member if $R \subseteq R'$ and no members otherwise.

Every part R of X in **set** gives rise to a characteristic function $\chi_R : X \to \Omega$ in **set**, unique with the property that $\chi_R^{-1}(true) \cong R$. This tells us, amongst other things, that every part is isomorphic to a specified part, where we speak of 'specified' in the sense of 1.3. Diagrammatically, we have in the metalanguage

$$\mathcal{P}X \underset{\gamma=(-)^{-1}(true)}{\overset{\phi=\chi_-}{\rightleftarrows}} \mathbf{set}(X, \Omega)$$

where we have introduced the simpler names ϕ and γ just to make the points that,

$$(\text{for every } R \text{ in } \mathcal{P}X)(\gamma\phi(R) \cong R)$$

$$(\text{for every } \omega \text{ in } \mathbf{set}(X, \Omega))(\phi\gamma(\omega) = \omega).$$

The *equation* comes from the uniqueness clause in the definition of characteristic function.

For ω and ω' in $\mathbf{set}(X, \Omega)$ we define $\omega \leq \omega'$ to hold if and only if $\gamma(\omega) \subseteq \gamma(\omega')$. It is clear that, for any ω, we have $\omega \leq \omega$; while if $\omega \leq \omega'$ and $\omega' \leq \omega''$ then $\omega \leq \omega''$. Just as we did for $\mathcal{P}X$, we can now regard $\mathbf{set}(X, \Omega)$ as a category—the objects being the functions $\omega : X \to \Omega$ and there being an arrow from ω to ω' precisely if $\omega \leq \omega'$. But if $\omega \leq \omega'$ and $\omega' \leq \omega$, so that ω and ω' are isomorphic, $\omega \cong \omega'$, then, using the uniqueness clause for characteristic functions, $\omega = \omega'$.

Categories in which isomorphic objects are equal are called *skeletal*. The arrows in the display above can be seen as inverse *equivalence functors* — about which we do not need more at this time—and we will find it convenient to redefine $\mathcal{P}X$ to mean the subcategory of specified parts. Since isomorphic specified parts are equal, this convention makes the displayed functors inverse isomorphisms. It is then harmless to write $\mathcal{P}X = \mathbf{set}(X, \Omega)$ and treat its members either as specified parts or as characteristic functions depending on what is most convenient at the time. Without further comment we will understand that if a construction takes us from specified parts to general parts then it is tacitly followed by replacement of the general part by its unique specified isomorph. We will call a specified part a *subset*.

1.5. The category of relations. Let X and A be sets. We define a *relation R from X to A* to be a subset R of $A \times X$ and write $R : X \nrightarrow A$. A relation gives a configuration of arrows $A \leftarrow R \rightarrow X$ that is quite generally called a *span from X to A*. In the case of a relation, because the corresponding arrow $R \rightarrow A \times X$ is a monomorphism, $A \leftarrow R \rightarrow X$ is called a *monic span*. If $(a, x) \in R$ we write aRx and say that a *is R-related to x*. Our use of $A \times X$ rather than $X \times A$ is a somewhat arbitrary convention. If we compose $R \rightarrowtail A \times X$ with the switch isomorphism $A \times X \xrightarrow{\simeq} X \times A$ then we obtain a relation that we call $R^{\circ} : A \nrightarrow X$. We will find it useful to be systematic about R and R°.

In [5] the *graph* of a function $f : X \rightarrow A$ is defined to be $\langle 1_X, f \rangle : X \rightarrow X \times A$. For the projection $p : X \times A \rightarrow X$ we have $p\langle 1_X, f \rangle = 1_X$, so the graph of f is a split monomorphism and provides a relation $f^* : A \nrightarrow X$. We write $f_* : X \nrightarrow A$ for $(f^*)^{\circ}$.

Given $R : X \nrightarrow A$ and $S : A \nrightarrow Y$ we have the *composite relation SR* : $X \nrightarrow Y$ where $ySRx$ if and only $(\exists a)(ySa$ and $aRy)$. (We often abbreviate a conjunction $(ySa$ and $aRy)$ by $ySaRx$.) The existential quantifier requires care if we take seriously the possibility that the objects of $\mathbf{set}$ are variable sets. To be precise: For $y : T \rightarrow Y$ and $x : T \rightarrow X$, we have $ySRx$ (that is $\langle y, x \rangle : T \rightarrow Y \times X$ factoring through $SR \rightarrowtail Y \times X$) if and only if there exists an epimorphism $e : S \twoheadrightarrow T$ and a function $a : S \rightarrow A$ with $yeSaRxe$.

To construct $SR \rightarrowtail Y \times X$ in the spirit of [5] consider

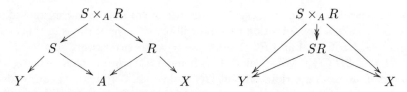

where on the left we have an evident pullback and on the right

$$S \times_A R \twoheadrightarrow SR \rightarrowtail Y \times X$$

is the image factorization of $S \times_A R \rightarrow Y \times X$. This composition of relations is associative and has, for any set X, $(1_X)_* = \Delta_X : X \nrightarrow X$, as identity on X. It follows that sets and relations form a category that we denote by $\mathbf{rel}$. Observe that if $f : X \rightarrow A$ and $g : A \rightarrow Y$ in $\mathbf{set}$ then $(gf)_* = g_* f_* : X \nrightarrow Y$ in

rel. This together with the fact that $(1_X)_*$ is the identity relation tells us that $(-)_* : \mathbf{set} \to \mathbf{rel}$ is a functor which is the identity on objects. We will write 1_X as a convenient abbreviation for Δ_X. In fact, it is usually unambiguous to write f for f_*.

1.6. The bicategory of relations. For the category **rel** we have $\mathbf{rel}(X, A) = \mathcal{P}(A \times X)$ and by 1.4 it follows that each $\mathbf{rel}(X, A)$ is a category, in fact a skeletal category, in its own right. Of course we have only considered the underlying objects of the categories $\mathbf{rel}(X, A)$ in our description of **rel** as a category. It follows easily from the definitions that if $R \subseteq R'$ for $R, R' : X \nrightarrow A$ then, for any $S : A \nrightarrow Y$, $SR \subseteq SR'$ and, for any $T : B \nrightarrow X$, $RT \subseteq R'T$. From these observations we see that composition

$$\mathbf{rel}(X, A) \times \mathbf{rel}(A, Y) \to \mathbf{rel}(X, Y)$$

is a functor and each specification of an identity $1 \to \mathbf{rel}(X, X)$ is in any event a functor, since 1 is discrete. In fact, these observations show that **rel** is a *bicategory* (a term that we do not formally define because we are not using its full generality) since its hom-collections are in general not discrete categories.

If $f, g : X \to A$ in **set** and $f_* \subseteq g_* : X \nrightarrow A$ in **rel** then $f = g : X \to A$ in **set**. For af_*x, that is $(a, x) \in f_*$, holds if and only if $a = f(x)$ and thus $f_* \subseteq g_*$ implies that $f(x) = g(x)$ for every (generalized) element x.

Exercises.

1. For relations $S : A \nrightarrow Y$ and $T : X \nrightarrow Y$, construct a relation $S \Rightarrow T : X \nrightarrow A$ with the following property: For any relation $R : X \nrightarrow A$, $SR \subseteq T$ if and only if $R \subseteq S \Rightarrow T$. (This relation can be read 'S implies T', which provides a hint.) Note that a consequence of the property is $S \cdot (S \Rightarrow T) \subseteq T$ and the latter containment is not generally an equality.
2. For relations $R : X \nrightarrow A$ and $T : X \nrightarrow Y$, construct a relation $T \Leftarrow R : A \nrightarrow Y$ with the following property: For any relation $S : A \nrightarrow Y$, $SR \subseteq T$ if and only if $S \subseteq T \Leftarrow R$. (This relation can be read 'T is implied by R').
3. Discover the relationship between the connectives $\Rightarrow$ and $\Leftarrow$.

1.7. Adjoint relations. The bicategory structure of **rel** that we have exposed in 1.6 allows us to make a key definition that we will revisit several times in this chapter. For relations $L : X \nrightarrow A$ and $R : A \nrightarrow X$, we have an *adjunction*, denoted $L \dashv R$ if $1_X \subseteq RL$ and $LR \subseteq 1_A$. We speak of L as being a *left adjoint*, of R as being a *right adjoint*, of L as having a right adjoint (R) and so on. Adjunctions can be defined in any bicategory (but what we have defined for **rel** reflects a simplification to the case at hand) and arrows which have right adjoints in a bicategory are often called *maps*. The next Theorem records that the maps in **rel** are precisely the functions. In fact this theorem can be proved using only that **set** is a regular category. (See [1].) It is convenient to state two preliminary results.

Proposition. *If $L \dashv R$ and $L \dashv R'$ in* **rel** *then $R = R'$. Similarly, if $L \dashv R$ and $L' \dashv R$ then $L = L'$.*

Proof. It suffices to prove just one of the assertions. From $1_X \subseteq R'L$ we have $R \subseteq R'LR \subseteq R'$. Similarly, $R' \subseteq R$ so $R \cong R'$ and hence $R = R'$ since $\mathbf{rel}(A, X)$ is skeletal. ☐

Lemma. *A monic span $g : A \leftarrow L \rightarrow X : h$ seen as a relation $L : X \nrightarrow A$ is of the form f_* for $f : X \rightarrow A$ a function if and only if h is an isomorphism. In this case $L = f_*$ uniquely for $f = gh^{-1}$.*

Proof. It is easy to see that $\langle g, h \rangle : L \rightarrowtail A \times X$ and $\langle gh^{-1}, 1_X \rangle : X \rightarrowtail A \times X$ are isomorphic as parts and hence equal as relations. The uniqueness statement follows from the last paragraph in 1.6. ☐

Theorem. *Every function $f : X \rightarrow A$ in $\mathbf{set}$ gives rise to an adjunction $f_* \dashv f^*$ in $\mathbf{rel}$. Moreover, every adjunction $L \dashv R : A \nrightarrow X$ in $\mathbf{rel}$ is of the form $f_* \dashv f^*$ for a, necessarily unique, function $f : X \rightarrow A$.*

Proof. Given $f : X \rightarrow A$ in $\mathbf{set}$ consider

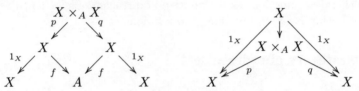

Here $p : X \leftarrow X \times_A X \rightarrow X : q$ is already a monic span and provides the composite $f^* f_*$ in $\mathbf{rel}$ while the displayed arrow $X \rightarrow X \times_A X$ on the right witnesses $1_X \subseteq f^* f_*$. In

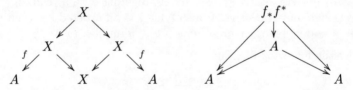

we see that the pullback stage in the construction of $f_* f^*$ is trivial so that $f_* f^*$ is just the image of $\langle f, f \rangle : X \rightarrow A \times A$, which is contained in Δ_A, as displayed on the right, and provides $f_* f^* \subseteq 1_A$. Hence $f_* \dashv f^*$.

For the second assertion write $g : A \leftarrow L \rightarrow X : h$ for a monic span that defines $L : X \nrightarrow A$ as a relation. From the preceding Proposition it suffices to show that h is an isomorphism. From $1_X \subseteq RL$ we have that for any $x : T \rightarrow X$ there exists an epimorphism $e : S \twoheadrightarrow T$ and a function $a : S \rightarrow A$ with $xeRaLxe$. Taking $1_X : X \rightarrow X$ for x gives an epimorphism $e : S \twoheadrightarrow X$ and an arrow $a : S \rightarrow A$ with

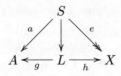

commutative. It follows that h is an epimorphism. To show that h is a monomorphism, take $l_0, l_1 : T \to L$ with $hl_0 = hl_1$. Again using $1_X \subseteq RL$, take $hl_0 = hl_1$ for x to get an epimorphism $e : S \twoheadrightarrow T$ and an $a : S \to A$ with $(hl_0 e)RaL(hl_0 e)$. Now $LR \subseteq 1_A$ is equivalent to saying that $aLxRb$ implies $a = b$. From $(gl_0 e)L(hl_0 e)Ra$ we have $gl_0 e = a$ and similarly $gl_1 e = a$. So $gl_0 e = gl_1 e$ and since e is an epimorphism $gl_0 = gl_1$. But this together with $hl_0 = hl_1$ and the fact that $\langle g, h \rangle$ is monic gives $l_0 = l_1$. $\qquad \square$

We will not continue to exercise such diligence about existential quantification in our proofs. We will usually write as if sets are 'constant' and leave the general argument to the interested reader. To contrast, it is useful to see how the second half of the proof above simplifies if **set** is truly a category of constant sets: Assume $L \dashv R$. From $1_X \subseteq RL$ it follows that $(\forall x)(\exists a)(xRaLx)$ so that L is everywhere defined. If also bLx then from $bLxRa$ and $LR \subseteq 1_A$ it follows that $b = a$ and L is single valued. In most traditional accounts of set theory, a function is *defined* to be a single valued, everywhere defined relation. It is interesting that the axiomatization of [5] which takes 'function' as a primitive nevertheless allows us to recover the usual relationship between functions and relations.

2. The bicategory of ordered sets

2.1. Ordered sets and order preservation. A relation $R : X \nrightarrow X$ is *reflexive* if $1_X \subseteq R$ (where here 1_X is the identity relation on X) and *transitive* if $RR \subseteq R$. Note that if R is both reflexive and transitive then $RR = R$.

Definition. An *ordered set* $(X, \leq)$ is a set X, together with a relation $\leq : X \nrightarrow X$, which is reflexive and transitive. If also $(A, \leq)$ is an ordered set then a function $f : X \to A$ is said to be *order-preserving* if $x \leq y$ implies $fx \leq fy$.

Consider the following diagram:

$$
\begin{array}{ccc}
X & \xrightarrow{\;f\;} & A \\
{\scriptstyle \leq}\downarrow & {\scriptstyle \subseteq} & \downarrow{\scriptstyle \leq} \\
X & \xrightarrow[\;f\;]{} & A
\end{array}
$$

We understand it to mean that in **rel**, the composites $f_* \cdot \leq$ and $\leq \cdot f_*$ satisfy

$$(f_* \cdot \leq) \subseteq (\leq \cdot f_*).$$

Again, we can write just $(f \cdot \leq) \subseteq (\leq \cdot f)$.

Proposition. *For ordered sets X and A, a function $f : X \to A$ is order-preserving if and only if it satisfies the condition prescribed by the square above.*

Proof. We have $a(f \cdot \leq)y$ if and only if $(\exists x)(a = fx$ and $x \leq y)$. On the other hand, we have $a(\leq \cdot f)y$ if and only if $(\exists b)(a \leq b$ and $b = fy)$ which is the case if and only if $a \leq fy$. Now assume that f is order-preserving and $a(f \cdot \leq)y$. So there is an x such $a = fx$ and $x \leq y$, from which we have $a = fx \leq fy$ showing that the condition of the square holds. Conversely, assume the square condition and $x \leq y$. Then $fx(f \cdot \leq)y$ holds and we conclude $fx \leq fy$. □

If $f : X \to A$ and $g : A \to Y$ are order-preserving then the composite $gf : X \to Y$ is order-preserving. Of course this is a triviality but it is worth noting that if we were to eschew completely the use of elements and take the last displayed square as a definition then we could think in terms of 'pasting' together such squares as in:

$$
\begin{array}{ccccc}
X & \xrightarrow{\ f\ } & A & \xrightarrow{\ g\ } & Y \\
\scriptstyle\leq\!\downarrow & \subseteq & \scriptstyle\leq\!\downarrow & \subseteq & \scriptstyle\leq\!\downarrow \\
X & \xrightarrow[\ f\ ]{} & A & \xrightarrow[\ g\ ]{} & Y
\end{array}
$$

The identity function 1_X is obviously order-preserving, so it is clear that ordered sets and order-preserving functions yield a category that we will call **ord**.

Like **rel**, the category **ord** has each $\mathbf{ord}(X, A)$ a category in its own right. For if $f, g : X \to A$ are order-preserving, then we can define $f \leq g$ if and only if for all (generalized) x in X, $fx \leq gx$ in A. This is equivalent to the condition that $\langle f, g \rangle : X \to A \times A$ factor through $\leq \;\rightarrowtail\; A \times A$. Just as for **rel**, we say that there is an arrow from f to g in $\mathbf{ord}(X, A)$ if and only if $f \leq g$. If $f \leq g : X \to A$ in **ord** then, for any $x : T \to X$ in **ord**, we have $fx \leq gx : T \to A$ in **ord** and, for any $h : A \to Y$ in **ord**, we have $hf \leq hg : X \to Y$ in **ord**. Thus composition

$$\mathbf{ord}(X, A) \times \mathbf{ord}(A, Y) \to \mathbf{ord}(X, Y)$$

is a functor and each specification of an identity $1 \to \mathbf{ord}(X, X)$ is trivially a functor.

2.2. Adjoint order-preserving functions. The most important definition in this Chapter is:

Definition. For order-preserving $f : X \to A$ and $u : A \to X$, f is *left adjoint* to u, written $f \dashv u$, if and only if, for all x in X, for all a in A, $fx \leq a$ if and only if $x \leq ua$.

There is a certain amount of vocabulary that surrounds *adjunctions*, by which is meant such a pair as in Definition 2.2 with $f \dashv u$. For example, one also says that u is *right adjoint* to f. In diagrams it is sometimes convenient to rotate the turnstile symbol, with the sharp end still directed towards the left adjoint.

If $X = (X, \leq)$ is an ordered set then X^{op} is defined to be $(X, \leq^\circ)$. Adjunctions for which one of the ordered sets involved bears an $(-)^{op}$ are sometimes called *Galois connections*. But since we have $X = (X^{op})^{op}$ such adjunctions are not really special at all. A bit of history may be helpful. If X and A are orders then an order-reversing function $f : X \to A$ is one for which $x \leq y$ implies $fy \leq fx$. This is exactly the same as saying that $f : X^{op} \to A$ is order-preserving but psychological rather than logical considerations often hold sway. Before proceeding, recall the fundamental theorem of Galois theory for a field extension $k \rightarrowtail K$. Now consider the constructions defining the famous pair of order-reversing bijections in the fundamental theorem but without the assumption that the field extension is finite and Galois. One no longer has an order-isomorphism of course but what remains is a Galois connection — an adjunction.

If $f \dashv u : A \rightarrow X$ is an adjunction then, for all x, $fx \leq fx$, ensures, for all x, $x \leq ufx$. In other words, $1_X \leq uf$. A similar consideration shows that $fu \leq 1_A$. In fact these inequalities characterize adjunctions.

Proposition. *For order-preserving $f : X \rightarrow A$ and $u : A \rightarrow X$, $f \dashv u$ if and only if $1_X \leq uf$ and $fu \leq 1_A$.*

Proof. Assume $1_X \leq uf$ and $fu \leq 1_A$. Assume $fx \leq a$. Then $ufx \leq ua$ since u is order-preserving and now $x \leq ufx \leq ua$. The other implication is just as easy. ☐

The Proposition suggests that we could unify the definitions of adjunction in **rel** and adjunction in **ord**. This is correct but the natural vehicle for doing so is the notion of 'bicategory' and we have chosen to not present that here. Besides, repetition has given us an excuse for emphasis that we feel is warranted.

Corollary. *For ordered sets $(X, \leq)$ and $(A, \leq)$ and functions $f : X \rightarrow A$ and $u : A \rightarrow X$, assume, for all x in X and all a in A, $fx \leq a$ if and only if $x \leq ua$. Then f and u are necessarily order-preserving and $f \dashv u$.*

Proof. The considerations preceding Proposition 2.2 above show that even the weaker hypothesis of this proposition ensures $x \leq ufx$, for all x in X, and $fua \leq a$, for all a in A. Assume $x \leq y$. Then $x \leq y \leq ufy$ and hence $fx \leq fy$. That u is order-preserving follows in the same way and hence $f \dashv u$. ☐

2.3. Antisymmetric ordered sets. A relation $R : X \nrightarrow X$ is said to be *antisymmetric* if $(xRy$ and $yRx)$ implies $x = y$. Since relations are specified parts and parts can be intersected it should be clear that we can also express antisymmetry by saying that $R \cap R^\circ \subseteq 1_X$. If an ordered set $(X, \leq)$ is antisymmetric, meaning that $\leq$ is antisymmetric, most authors call $(X, \leq)$ a *partially ordered set*. For that matter, most authors use the term *pre-ordered set* for what we have called simply an ordered set. In any ordered set $(X, \leq)$, when $x \leq y$ and $y \leq x$ we write $x \cong y$. In particular, if $x, y : T \rightarrow X$ then when $x \cong y$ we really have x isomorphic to y in the category $\mathbf{ord}((T, =), (X, \leq))$.

The relation $\cong$ is an equivalence relation. For any ordered set X, we can construct $X \twoheadrightarrow X/\cong$ and the quotient with the inherited order is clearly an anti-symmetric ordered set with $X \twoheadrightarrow X/\cong$ order-preserving. Quite generally we say that an order-preserving pair $f : X \rightleftarrows A : u$ is an equivalence if $1_X \cong uf$ and $fu \cong 1_A$. (Clearly, every order-isomorphism is an equivalence and every equivalence is an adjunction.) If $X \twoheadrightarrow X/\cong$ splits then the splitting function is necessarily order-preserving and the pair then provide an equivalence.

Existence of a splitting for $X \twoheadrightarrow X/\cong$ is guaranteed by the Axiom of Choice. For most authors, sets are 'constant' and the Axiom of Choice is assumed. For such authors every ordered set is equivalent to an antisymmetric ordered set and and when a naturally occurring ordered set X which is not antisymmetric is encountered it is routinely replaced by its antisymmetrization $X/\cong$. For those interested in possibly variable sets the matter requires a little more care. In any event, $X \twoheadrightarrow X/\cong$ is universal with respect to order-preserving arrows from X to an anytisymmetric ordered set. In other words, for all $f : X \to A$ in **ord** with A antisymmetric, there exists a unique $f' : X/\cong \to A$ such that

commutes.

It should be pointed out that an instance of $X \twoheadrightarrow X/\cong$ *might* have a canonical splitting. An interesting case is provided in the metalanguage in 1.4. The reader has no doubt observed that the category of parts of a set X is an 'ordered collection'. We provisionally called it $\mathcal{P}X$ and the original metalanguage pair

$$\phi : \mathcal{P}X \rightleftarrows \text{set}(X, \Omega) : \gamma$$

is an 'equivalence' which induces an 'order-isomorphism' $\mathcal{P}X/\cong \rightleftarrows \text{set}(X, \Omega)$. This is why we elected to redefine $\mathcal{P}X$ in terms of specified parts.

2.4. Adjunctions. An order-preserving function need have neither a right adjoint nor a left adjoint. In fact, as we will soon see, existence of adjoints imposes strong constraints. But if an order-preserving function has a right adjoint then it is essentially unique.

Lemma. *For $f : X \to A$ in **ord**, if $f \dashv u$ and $f \dashv v$ then $u \cong v$.*

Proof. For all a, $ua \le va$ since $fua \le a$. Similarly, $va \le ua$ follows from $fva \le a$. □

While $f \dashv u$ is not symmetric in f and u, study of adjoints is nevertheless simplified by duality. For X an ordered set we defined X^{op} in 2.2. For $f : X \to A$ in **ord**, the defining data prescribes equally an arrow $f^{op} : X^{op} \to A^{op}$ in **ord**. Moreover, if $f \le g : X \to A$ then $g^{op} \le f^{op} : X^{op} \to A^{op}$. (Note that $(-)^{op}$ preserves the direction of arrows but reverses the direction of inequalities.) Also $(1_X)^{op} = 1_{X^{op}}$ and $(fg)^{op} = f^{op}g^{op}$.

Proposition. *If $f \dashv u$ then $u^{op} \dashv f^{op}$.*

Proof. Assume $u^{op}(a) \le x$ in X^{op}. This means that $x \le ua$ in X and from $f \dashv u$ we have $fx \le a$ in A. This last means $a \le f^{op}(x)$ in A^{op}. □

Adjunctions compose.

Theorem. *If $f : X \to A$ and $g : A \to Y$ in **ord** with $f \dashv u$ and $g \dashv v$ then $gf \dashv uv$.*

Proof. $gfx \leq y$ iff $fx \leq vy$ iff $x \leq uvy$.   □

A few examples of adjunctions will be given below. It is convenient to begin by recalling that Ω, the subset classifier of **set**, is canonically an ordered set, arguably the single most important one. To give a $\leq : \Omega \nrightarrow \Omega$ is to give a subset of $\Omega \times \Omega$ and the canonical one to which we will always refer is that consisting of all $\langle \omega, \omega' \rangle$ such that $\omega \leq \omega'$ as in 1.4. Since global elements of Ω are subsets of 1, the terminal set, we also feel free to write $\subseteq$ for this antisymmetric order on Ω. From this order on Ω we get a pointwise ordering on Ω^X and it is a straightforward exercise to show that it brings into our axiomatic world the antisymmetric inclusion order for parts of X. We also feel free to write $\subseteq$ for this order on Ω^X and $\mathbb{P}X$ for Ω^X with this order.

(1) Let $f : X \to A$ be a function. We have $\Omega^f : \Omega^A \to \Omega^X$ and it has been seen in [5] that this function encapsulates inverse image of parts along f. On the other hand we have $f_! : \mathcal{P}X \to \mathcal{P}A$ given by taking $f_!(Y \rightarrowtail X)$ to be the subset of A determined by the image factorization of $Y \rightarrowtail X \xrightarrow{f} A$. Arguing as in [5] we get a function $\exists_f : \Omega^X \to \Omega^A$, in fact an adjunction $\exists_f \dashv \Omega^f : \mathbb{P}A \to \mathbb{P}X$, from examination of $f_!$ and $f^{-1} : \mathcal{P}A \to \mathcal{P}X$ in the metalanguage.

(2) Consider next what it would mean for f^{-1} to have also a 'right adjoint', say f_*. An element a of A would belong to $f_*(Y \rightarrowtail X)$ if and only if $\{a\} \subseteq f_*(Y \rightarrowtail X)$ and adjointness would require that this condition hold if and only if $f^{-1}(\{a\}) \subseteq (Y \rightarrowtail X)$. Thus one is led to *define* $f_*(Y \rightarrowtail X) = \{a \in A | f^{-1}(\{a\}) \subseteq (Y \rightarrowtail X)\}$. It is then possible to show that for every $Y \rightarrowtail X$ and every $B \rightarrowtail A$, $f^{-1}(B) \subseteq Y$ if and only if $B \subseteq f_*(Y)$. In other words, f_* provides a 'right adjoint' to f^{-1} in the metalanguage and because it is natural then, again as explained in [5], this suffices to construct a function $\forall_f : \Omega^X \to \Omega^A$ with $\Omega^f \dashv \forall_f : \mathbb{P}X \to \mathbb{P}A$ in **ord**.

It is desirable to use the metalanguage quite freely, without quite so many explicit comments as above. Still, sometimes it is equally desirable to work in **set** directly. For example, one might try — after the heuristics leading to f_* are understood — to construct $\forall_f : \Omega^X \to \Omega^A$ directly as a function $A \times \Omega^X \to \Omega$ and hence as a part of $A \times \Omega^X$. One then 'carves out' of $A \times \Omega^X$ the part consisting of all those pairs $\langle a, Y \rangle$ for which $\Omega^f(\{a\}) \subseteq Y$. Since we have $Z \subseteq Y$ if and only if $Z = Z \cap Y$, the containment can be seen as an equational condition so that the requisite part can be constructed as an equalizer involving $\{-\} : A \to \Omega^A$, Ω^f and $- \cap - : \Omega^X \times \Omega^X \to \Omega^X$.

The usual name for $\exists_f$ is 'direct image along f' but $\forall_f$ is not often given an analogous name. If **set** is taken to be a Boolean topos then one has the relation $\forall_f(Y) = (\exists_f(Y^c))^c$, where $(-)^c$ denotes complementation, but for a category of variable sets this equation cannot generally be assumed.

(3) Quite generally, we say that a *discrete ordered set* is one of the form $(X, =)$. The terminal set 1 gives the discrete ordered set $1 = (1, =)$. It is clear that for any X in **ord**, the unique function $!_X : X \to 1$ is order-preserving. So 1 is the terminal object of the category **ord**. *If* $!_X$ *has a right adjoint then it is a function* $t : 1 \to X$ satisfying, for all $x : T \to X$, $x \le t!_T$. Conversely the existence of such a 'top' element provides a right adjoint to $!_X$.

(4) Dually, $!_X : X \to 1$ has a left adjoint if and only if there is a $b : 1 \to X$ with $b!_T \le x$, for all $x : T \to X$ which is to say if and only if X has a 'bottom element'. Since adjoints are unique to within isomorphism, top and bottom elements are so. We will usually write 1 for t and 0 for b when such elements exist.

(5) For X and Y in **ord**, define $(x, y) \le (u, v)$ if and only if $x \le u$ and $y \le v$. It follows that this is an order on $X \times Y$, that the projection functions are order-preserving and that $X \times Y$ together with the projections is a product in **ord**. It then follows that the diagonal function $\Delta_X : X \to X \times X$ is order-preserving too. Assume that Δ_X has a right adjoint. Call it $\wedge$ and write $x \wedge y$ for $\wedge(x, y)$. The defining property is this: $z \le x \wedge y$ if and only if $z \le x$ and $z \le y$. Taking $z = x \wedge y$ we see that $x \wedge y \le x$ and $x \wedge y \le y$, so that $x \wedge y$ is a 'lower bound' for the set that one would like to write as $\{x, y\}$. The defining property says that among all such lower bounds, z, $x \wedge y$ is the 'greatest'. So $\wedge$ is a greatest lower bound operation, often called 'meet'. Any other such greatest lower bound operation is isomorphic to $\wedge$.

(6) Dually, $\Delta_X : X \to X \times X$ has a left adjoint if and only if for each pair of elements (x, y) there is prescribed an element $x \vee y$ with the property that, for all z, $x \vee y \le z$ if and only if $x \le z$ and $y \le z$. Thus $\vee$ is a least upper bound operation, often called 'join'.

Exercises.

1. For ordered sets T, X, define X^T to be the set of all order-preserving $T \to X$ ordered by $a \le b$ if and only if, for all t in T, $at \le bt$ (in X). Show that in **ord** there is a bijective correspondence between order-preserving $T \times S \to X$ and order-preserving $S \to X^T$. Note that this definition is consistent with the definition of X^T for *sets* T, X. Explain what is meant by this, regarding a set as a discrete ordered set.

2. For $f : X \to A$ in **ord**, extend the definition of X^T above to $f^T : X^T \to A^T$ by composition. Show that if $f \dashv u$ then $f^T \dashv u^T$.

3. For an order-preserving $f : X \to A$, extend the definition of S^X above to $S^f : S^A \to S^X$ by composition. (Note the reversal of direction.) Show that if $f \dashv u$ then $S^u \dashv S^f$.

3. Semilattices and lattices

3.1. The definitions via adjunction.

Definition. A *meet [join] semilattice* is an ordered set X for which the canonical

$$1 \leftarrow X \rightarrow X \times X$$

have right [left] adjoints.

Note that X is a join semilattice if and only if X^{op} is a meet semilattice.

Definition. A *lattice* is an ordered set X which is both a meet semilattice and a join semilattice.

Remark. The terminology of the last two definitions is usually only employed in the context of antisymmetric X. This is a needless restriction but nevertheless there is reason to examine the concepts in the stricter context and we will do so briefly.

Lemma. *An antisymmetric join semilattice is equivalently described as a set X (with no order assumed) together with an idempotent commutative monoid structure.*

Proof. Let $(X, \vee, 0)$ be an idempotent commutative monoid. That is, X is a monoid satisfying the further equations $x \vee x = x$ and $x \vee y = y \vee x$. Define $x \leq y$ if and only if $x \vee y = y$. Now $x \leq x$ since $x \vee x = x$ and if $x \leq y \leq z$ then $x \vee z = x \vee y \vee z = y \vee z = z$, so $x \leq z$. If $x \leq y$ and $y \leq x$ then $y = x \vee y = y \vee x = x$ shows that $(X, \leq)$ is an antisymmetric order.

From $0 \vee x = x$, for all x, we have $0 \leq x$, for all x, showing that 0 is the bottom element with respect to $\leq$. Now let x and y be any two elements of X. Then $x \vee (x \vee y) = (x \vee x) \vee y = x \vee y$ gives $x \leq x \vee y$ and, similarly, $y \leq x \vee y$ so that $x \vee y$ is an upper bound for $\{x, y\}$ with respect to $\leq$. Finally, assume that $x \leq z$ and $y \leq z$. Then $(x \vee y) \vee z = x \vee y \vee z = x \vee z = z$ so that $x \vee y \leq z$ and we have shown that $x \vee y$ is the join of x and y with respect to $\leq$.

Conversely, assume that $(X, \leq)$ is an antisymmetric join semilattice. Verify that $x \vee x = x$, $x \vee y = y \vee x$, $(x \vee y) \vee z = x \vee (y \vee z)$ and $x \vee 0 = x \ (= 0 \vee x)$ using the properties of join and bottom element. (If X is not assumed to be antisymmetric then these equations must be replaced by instances of $\cong$.)

Finally, show that these constructions are mutually inverse. $\qquad\square$

The point of this Lemma is to show that, modulo antisymmetry, semi-lattices can be regarded as purely algebraic objects. Observe too that if both X and A are idempotent commutative monoids then a homomorphism of monoids $f : X \rightarrow A$, meaning a function satisfying $f(x \vee y) = fx \vee fy$ and $f0 = 0$, is necessarily order-preserving. For given $x \leq y$ we have $fx \vee fy = f(x \vee y) = fy$. The converse is not true. However, for $f : X \rightarrow A$ order-preserving between join semilattices (not necessarily antisymmetric) we do always have $fx \vee fy \leq f(x \vee y)$ and $0 \leq f0$. The first of these 'comparison inequalities' arises from $x \leq x \vee y$ and $y \leq x \vee y$

implying $fx \leq f(x \vee y)$ and $fy \leq f(x \vee y)$ which shows that $f(x \vee y)$ is *an* upper bound for $\{fx, fy\}$. If X and A are meet semilattices then an $f : X \to A$ in **ord** has comparison inequalities, $f(x \wedge y) \leq fx \wedge fy$ and $f1 \leq 1$.

The next proposition makes the point that antisymmetric lattices can be described algebraicly too.

Proposition. *An antisymmetric lattice is equivalently described as a set X with idempotent commutative monoid structures $(X, \vee, 0)$ and $(X, \wedge, 1)$ related by the 'absorptive laws'*

$$x \wedge (x \vee y) = x \quad and \quad x \vee (x \wedge y) = x.$$

Proof. Each of the idempotent commutative monoid structures prescribes an order on X as above. The absorptive laws, which are easily seen to be valid in an antisymmetric lattice, ensure that the two orders are in fact the same. □

3.2. A test for adjunction. Lattice structure, even a fragment of it, provides a convenient test for an arrow in **ord** to admit adjoints.

Proposition. *If $f \dashv u : A \to X$ then f preserves any bottom elements and joins that exist in X and, dually, u preserves any top elements and meets that exist in A.*

Proof. Assume that X has a bottom element, 0. Now $f0 \leq a$ if and only if $0 \leq ua$ shows that $f0$ is a bottom element for A. Assume that elements x and y have a join, $x \vee y$ in X. We want to show that $f(x \vee y)$ provides a join for fx and fy in A. But $[fx \leq a$ and $fy \leq a]$ iff $[x \leq ua$ and $y \leq ua]$ iff $[x \vee y \leq ua]$ iff $[f(x \vee y) \leq a]$ shows this precisely. □

3.3. Bounds. In fact we can sharpen considerably the test provided by Proposition 3.2. Let S be a subset of an ordered set X. An *upper bound* for S is an element u in X such that, for all s in S, $s \leq u$. A *least upper bound*, otherwise known as a *supremum* or even a *sup* for S, is an upper bound l for S with the property that if u is any upper bound for S then $l \leq u$. A supremum for S, if it exists, is unique up to isomorphism. Note that if l exists then it is completely characterized, up to isomorphism, by the statement $l \leq x$ if and only if, for all s in S, $s \leq x$. When we have a definite supremum in mind we will write it as $\bigvee S$.

Check that a bottom element 0 in X is actually $\bigvee \emptyset$. Similarly check that $x \vee y = \bigvee \{x, y\}$. The notion 'dual' to supremum is *infimum* and we write $\bigwedge S$ with the duals of the preceding remarks applying.

If $f : X \to A$ is order-preserving and S is a subset of X, let us follow the simple convention of writing fS for the image of S. Suppose that $\bigvee S$ and $\bigvee fS$ exist. We have a comparison inequality $\bigvee fS \leq f \bigvee S$. To see this it suffices to show that $f \bigvee S$ is *an* upper bound for fS. But $\bigvee S$ is an upper bound for S and any arrow in **ord** takes an upper bound u for S to an upper bound fu for fS. We say that f preserves $\bigvee S$ if the comparison above is invertible.

Proposition. *If $f \dashv u : A \to X$ then f preserves any suprema that exist in X. Dually, u preserves any infima that exist in A. Moreover, for any $f : X \to A$ in* **ord**, *f has a right adjoint if and only if, for all a in A,*

(1) $\bigvee\{x | fx \leq a\}$ exists.

(2) f preserves the supremum required in (1).

Proof. For S a subset of X, assume that $\bigvee S$ exists. We want to show that $f \bigvee S$ provides a supremum for fS in A. We have $[f \bigvee S \leq a]$ iff $[\bigvee S \leq ua]$ iff $[\forall s \text{ in } S][s \leq ua]$ iff $[\forall s \text{ in } S][fs \leq a]$ iff $[\forall t \text{ in } fS][t \leq a]$ which proves the first clause. That right adjoints preserve infima follows by duality.

For the second clause let us begin by establishing the necessity of (1). Assume that f has a right adjoint, say u. We show that $ua \cong \bigvee\{x | fx \leq a\}$. This is so if and only if $ua \leq y$ precisely when for all s in $\{x | fx \leq a\}$, $s \leq y$. Assume first that $ua \leq y$ and take s with $fs \leq a$. Then $s \leq ua \leq y$. Conversely, assume that, for all s in $\{x | fx \leq a\}$, $s \leq y$. To show that $ua \leq y$ it suffices to show that ua is in $\{x | fx \leq a\}$ and this follows from $fua \leq a$.

The necessity of (2) now follows from the first clause of the proposition.

For the sufficiency of (1) and (2), define $ua = \bigvee\{x | fx \leq a\}$ and let y be an element of X. Assume $fy \leq a$. It follows that y is in $\{x | fx \leq a\}$ and hence $y \leq ua$. Conversely, assume that $y \leq ua$. Applying f to this inequality and noting the preservation property (2) we have

$$fy \leq fua = f\bigvee\{x | fx \leq a\} \cong \bigvee f\{x | fx \leq a\} = \bigvee\{fx | fx \leq a\} \leq a. \qquad \square$$

Corollary. *If every subset of X has a supremum then, for any $f : X \to A$ in* **ord**, *f has a right adjoint if and only if f preserves all suprema. (Dually, if every subset of X has an infimum then $f : X \to A$ has a left adjoint if and only if f preserves all infima).*

3.4. Distributive lattices. We defer further consideration of $\bigvee S$ for general S and return now to lattices. For elements x, y, z in a lattice we always have

$$(x \wedge y) \vee (x \wedge z) \leq x \wedge (y \vee z).$$

To see this it suffices to show $x \wedge y \leq x \wedge (y \vee z)$ and $x \wedge z \leq x \wedge (y \vee z)$. For the first of these inequalities it suffices to show $x \wedge y \leq x$, which is trivial, and $x \wedge y \leq y \vee z$ which is also trivial using $x \wedge y \leq y \leq y \vee z$. The second of the two required inequalities is established similarly. But it is an important extra property of a lattice for it to satisfy also

$$x \wedge (y \vee z) \leq (x \wedge y) \vee (x \wedge z)$$

and hence

$$x \wedge (y \vee z) \cong (x \wedge y) \vee (x \wedge z).$$

Such a lattice is said to be *distributive*. The following lattice, known as M_5, is an important example of a lattice which is not distributive.

The following simple result is somewhat surprising.

Proposition. *If X is a distributive lattice then X^{op} is also distributive.*

Proof. $(x \vee y) \wedge (x \vee z) \cong ((x \vee y) \wedge x) \vee ((x \vee y) \wedge z) \cong x \vee ((x \wedge z) \vee (y \wedge z)) \cong (x \vee (x \wedge z)) \vee (y \wedge z) \cong x \vee (y \wedge z)$. $\qquad \square$

For an element a in a lattice, one says that a *complement* of a is an element x such that $a \wedge x \cong 0$ and $a \vee x \cong 1$. So, for example, in M_5 above both b and c are complements of a.

Lemma. *For any three elements b, a, t in a distributive lattice there is, up to isomorphism, at most one x satisfying both $a \wedge x \cong b$ and $a \vee x \cong t$.*

Proof. Assume both x and y satisfy the conditions. Then $x \cong (a \vee x) \wedge x \cong t \wedge x \cong (a \vee y) \wedge x \cong (a \wedge x) \vee (y \wedge x) \cong b \vee (y \wedge x) \cong y \wedge x$. The last relation holding because $b \cong a \wedge x \cong a \wedge y$ is a lower bound for $\{x, y\}$. Similarly $y \cong y \wedge x$, so $x \cong y$. $\qquad \square$

Corollary. *In a distributive lattice complements are unique up to isomorphism when they exist.*

Proof. Take $b = 0$ and $t = 1$ above. $\qquad \square$

3.5. Boolean algebras. A *Boolean algebra* X is an antisymmetric distributive lattice equipped with an additional unary operation $(-)^c : X \to X$ such that for all a in X, a^c is a complement of a. Note that $(\ )^c$ is order reversing. It is clear that $(a^c)^c = a$. If X and A are antisymmetric lattices, a lattice homomorphism $f : X \to A$ (necessarily order-preserving) is a function that preserves both binary operations and both nullary operations. If X and A are Boolean algebras then Corollary 3.4 shows that such a homomorphism also preserves complementation. That is, for all x in X, $f(x^c) = (fx)^c$.

Exercises.

1. A Boolean ring X is a ring $(X, +, 0, \cdot, 1)$ in which multiplication $\cdot$ is idempotent. In other words, for all x in X, $xx = x$. Prove that for any Boolean ring X:
    (i) X is commutative
    (ii) For all x in X, $x + x = 0$.
    (*Hint*: Consider $x + y = (x + y)(x + y)$ and expand.)

2. Let $(X, \vee, 0, \wedge, 1)$ be a Boolean algebra. Define the 'symmetric difference' operation, written $+$, by $x+y = (x \wedge y^c) \vee (y \wedge x^c)$. Show that $(X, +, 0, \wedge, 1)$ is a Boolean ring. Now let $(A, +, 0, \cdot, 1)$ be a Boolean ring. Define an operation $\vee$ on A by $a \vee b = a + ab + b$. Show that $(A, \vee, 0, \cdot, 1)$ is a Boolean algebra. Finally, show that the constructions are inverse to each other.

3.6. Heyting lattices. For X a lattice and x an element of X, observe that $x \wedge -: X \to X$, the function whose value at y is $x \wedge y$, is order-preserving. Suppose that $x \wedge -$ has a right adjoint which we will call $x \Rightarrow -$. Then for any pair of elements y, z the definition of adjunction gives $[x \wedge y \leq z]$ if and only if $[y \leq x \Rightarrow z]$. It follows that $x \Rightarrow z$ is a largest element whose meet with x is less than or equal to z. Given an arbitrary pair x, z in an arbitrary lattice, an element with this property may or may not exist. Saying that $x \wedge -$ has a right adjoint ensures that $x \Rightarrow z$ exists, for all z.

Definition. A *Heyting lattice* is a lattice in which, for each x, the order-preserving $x \wedge -$ has a right adjoint (which we will denote by $x \Rightarrow -$). If the lattice is antisymmetric then we call a Heyting lattice a *Heyting algebra*.

Since we can define an antisymmetric lattice in terms of operations and equations without reference to a previously given order, it is natural to ask if the same holds for Heyting algebras. It does, as the next lemma, whose proof will be left as an exercise, shows.

Lemma. *For X an antisymmetric lattice with a further binary operation, $- \Rightarrow -$, (not a priori satisfying any order conditions) the resulting structure is a Heyting algebra if and only if, for all x, y, z in X we have:*

(i) $x \Rightarrow x = 1$.
(ii) $x \wedge (x \Rightarrow y) = x \wedge y$.
(iii) $y \wedge (x \Rightarrow y) = y$.
(iv) $x \Rightarrow (y \wedge z) = (x \Rightarrow y) \wedge (x \Rightarrow z)$.

One should immediately take special note of $x \Rightarrow 0$, a largest element whose meet with x is less than or equal to 0. Since 0 is less than or equal to all elements we have $x \wedge (x \Rightarrow 0) \cong 0$. This is one of the conditions for $x \Rightarrow 0$ to be a complement for x but the other condition does *not* hold in general. We will nevertheless write x^c for $x \Rightarrow 0$. It is often called the *pseudo-complement* or *negation* of x. We prefer the latter term for a general Heyting lattice and reserve the former for an important special case. Of course these observations tempt one to believe that a Boolean algebra is a Heyting algebra. That is the case and will be established shortly.

Proposition. *For any y in a Heyting lattice X, $- \Rightarrow y : X^{op} \to X$ is order-preserving and $(- \Rightarrow y)^{op} : X \to X^{op}$ is left adjoint to $- \Rightarrow y$.*

Proof. For x in X and z in X^{op} we have $[x \Rightarrow y \leq^{op} z]$ if and only if $[z \leq x \Rightarrow y]$ if and only if $[x \wedge z \leq y]$ if and only if $[z \wedge x \leq y]$ if and only if $[x \leq z \Rightarrow y]$. $\square$

Corollary. $((-)^c)^{op} \dashv (-)^c$. *(In other words, $[x \leq y^c]$ if and only if $[y \leq x^c]$.)*

3.7. Properties of negation.

Proposition. *For a Heyting lattice X, for any x, y in X, and for any subset S of X for which $\bigvee S$ exists,*

  (i) $0^c \cong 1$
  (ii) $(x \vee y)^c \cong x^c \wedge y^c$
  (iii) $(\bigvee S)^c \cong \bigwedge\{s^c | s \text{ in } S\}$
  (iv) $x \leq (x^c)^c$.

Also,

  (v) $1^c \cong 0$

and for any x, y in X, and for any subset S of X for which $\bigwedge S$ exists,

  (vi) $x^c \vee y^c \leq (x \wedge y)^c$
  (vii) $\bigvee\{s^c | s \text{ in } S\} \leq (\bigwedge S)^c$.

Proof. The right adjoint $(-)^c$ preserves top elements, meets and infima. In X^{op} these are, respectively, bottom elements, joins and suprema, so that we have statements (i) to (iii). (Note that the information obtained from $((-)^c)^{op}$ being a left adjoint merely repeats the statements (i) to (iii) and does *not* provide their duals.) Statement (iv) is merely the statement provided by both of the adjunction inequalities.

In any meet semilattice we have $x \cong 1 \wedge x$, for all x, hence $1^c \cong 1 \wedge 1^c \cong 0$ which is (v). Statements (vi) and (vii) just express the comparison inequalities for joins and suprema for $(-)^c : X^{op} \to X$. $\square$

When the inequality (vi) above is an isomorphism we say that *de Morgan's law* holds. In the study of Boolean algebras the corresponding equality and also that corresponding to (ii) are usually referred to as de Morgan's laws but in the context of Heyting algebras only the inequality $(x \wedge y)^c \leq x^c \vee y^c$, which does not hold in general, is singled out.

3.8. Boolean implies Heyting.

Proposition. *If X is a Boolean algebra then X is a Heyting algebra.*

Proof. We will write here x^c for the 'complement' of x in X, as defined after Proposition 3.4. In particular we have $x^c \vee x = 1$. Assume that in X, $x \wedge y \leq z$. It follows that $x^c \vee (x \wedge y) \leq x^c \vee z$ from which we have

$$y \leq x^c \vee y = 1 \wedge (x^c \vee y) = (x^c \vee x) \wedge (x^c \vee y) = x^c \vee (x \wedge y) \leq x^c \vee z.$$

Similarly, if we start with $y \leq x^c \vee z$ and meet both sides of the inequality with x then a similar calculation shows that $x \wedge y \leq z$. Thus for a Boolean algebra the defining property of $x \Rightarrow z$ is provided by $x^c \vee z$. $\square$

It follows that in a Boolean algebra $x \Rightarrow 0$ is given by $x^c \vee 0 = x^c$ so that our terminology in the proof of the proposition is unambiguous. Said otherwise, the complements in a Boolean algebra are Heyting negations. We do *not* have $x \Rightarrow z = (x \Rightarrow 0) \vee z$, for all x and z in a general Heyting algebra.

3.9. Heyting implies distributive.

Proposition. *If X is a Heyting lattice then X is distributive.*

Proof. Since each $x \wedge -$ has a right adjoint they preserve binary joins which is precisely the statement of distributivity. □

In a Heyting lattice, since each $x \wedge -$ preserves any suprema that exist we also have $x \wedge (\bigvee S) \cong \bigvee \{x \wedge s | s \text{ in } S\}$ whenever $\bigvee S$ exists.

3.10. 'Finite' distributive implies Heyting.

Proposition. * *A distributive lattice of the form $X = (\{x_1, ..., x_n\}, \leq)$ is a Heyting lattice.*

Proof. According to Proposition 3.3, the order-preserving $x \wedge -$ has a right adjoint if and only if, for all z, $\bigvee \{y | x \wedge y \leq z\}$ exists and such suprema are preserved by $x \wedge -$. For X of the form above, the required suprema can be calculated as iterated binary joins and the required preservation property follows from distributivity. □

3.11. When Heyting is Boolean.

Proposition. *For a Heyting algebra X, the following are equivalent:*
  (i) $((-)^c)^{op} : X \to X^{op}$ is also right adjoint to $(-)^c : X^{op} \to X$.
  (ii) For all x, $(x^c)^c = x$.
  (iii) X is a Boolean algebra.

Proof. The extra condition imposed by (i) is just $(x^c)^c \leq x$ so that (i) implies (ii) is clear. For (ii) implies (iii) we have already seen that X is distributive as required for a Boolean algebra and here we have $(-)^c : X^{op} \to X$ an isomorphism of lattices from which de Morgan's law follows. Thus to see that x^c is actually a complement here we apply $(-)^c$ to $x \wedge x^c = 0$ and get $x \vee x^c = (x^c)^c \vee x^c = (x^c \wedge x)^c = 0^c = 1$. Finally, for (iii) implies (i), it is clear that in a Boolean algebra $x = (x^c)^c$ and hence $(x^c)^c \leq x$. □

We could say that a Heyting lattice X satisfies the 'infinite' de Morgan law if $(\ )^c : X^{op} \to X$ has a right adjoint. This implies that for any subset S of X, for which $\bigwedge S$ exists, we have $(\bigwedge S)^c \cong \bigvee \{s^c | s \text{ in } S\}$. This condition should be carefully compared with (i) in Proposition 3.11. Consider the ordered set **3** which we might display as $0 \leq 1 \leq 2$. It is clear that it is a Heyting lattice and that negation is given by

$$\begin{pmatrix} 0 & 1 & 2 \\ 2 & 0 & 0 \end{pmatrix}$$

A direct calculation shows that negation has a right adjoint given by

$$\begin{pmatrix} 0 & 1 & 2 \\ 1 & 1 & 0 \end{pmatrix}$$

which is different from negation. So **3** is a Heyting algebra which satisfies the infinite de Morgan law — which quite generally implies the usual de Morgan law — but of course **3** is not a Boolean algebra.

4. Power set Heyting algebras

4.1. The lattice $\mathbb{P}1$**.** It will be useful to review the lattice structure of $\mathbb{P}1$. We have $true : 1 \to \Omega$, the generic monomorphism, and $false : 1 \to \Omega$ is the characteristic function of $0 \rightarrowtail 1$. We can *define* $\wedge : \Omega \times \Omega \to \Omega$ to be the characteristic function of $\langle true, true \rangle : 1 \rightarrowtail \Omega \times \Omega$ while $\vee : \Omega \times \Omega \to \Omega$ is the characteristic function of the image of

$$\begin{pmatrix} 1 & t \\ t & 1 \end{pmatrix} : \Omega + \Omega \to \Omega \times \Omega$$

where the 1's are 1_Ω's and t is $\Omega \to 1 \stackrel{true}{\to} \Omega$. (An arrow into a product is an ordered pair of arrows and an arrow out of a sum is an ordered pair of arrows — that can be written as a column vector — whence the 2×2 matrix of arrows.)

It is possible to show that $(\Omega, \vee, false, \wedge, true)$ is a lattice by deriving the equations of Lemma 3.1 and Proposition 3.1 from the properties of characteristic functions. The order relation for this lattice is the subobject of $\Omega \times \Omega$ given by the equalizer of the pair $p, \wedge : \Omega \times \Omega \to \Omega$, where p is the first product projection, and this can be shown to be the order on Ω that we introduced prior to the examples in 2.4. This subobject $\leq \; \rightarrowtail \; \Omega \times \Omega$ itself has a characteristic function, call it $\Rightarrow : \Omega \times \Omega \to \Omega$. Again, using just the properties of characteristic functions, it can be shown that this operation satisfies the equations of Lemma 3.6 so that Ω with the above structure is a Heyting algebra in **set**, our elementary topos of sets and functions. For Ω we refer to $(-)^c : \Omega^{op} \to \Omega$ as pseudo-complementation. We remark that even if **set** is assumed to be *Boolean*, meaning that Ω is Boolean, it does not follow that the topos **set²** of functions and commutative squares is Boolean in this sense.

However, Ω enjoys properties not shared by other Heyting algebras. The next two lemmas and subsequent results are somewhat surprising.

Lemma. *(Dennis Higgs) If* $f : \Omega \to \Omega$ *is a monomorphism in* **set** *(hence not necessarily order-preserving) then* $f^2 = 1_\Omega$.

Proof. See [4], page 44, Exercise 3. $\qquad\qquad\qquad\qquad\qquad\qquad\qquad\qquad\square$

Lemma. *(Jean Bénabou) If* $f : \Omega \to \Omega$ *is a function in* **set** *which satisfies* $f \leq 1_\Omega$ *then, for every (generalized) element* ω *of* Ω, $f(\omega) = \omega \wedge f(true)$. *Thus, if also* $f(true) = true$ *then* $f = 1_\Omega$.

Proof. Let $m : U \to \Omega$ be the subset of Ω classified by f so that

$$\begin{array}{ccc} U & \xrightarrow{\;!_U\;} & 1 \\ {\scriptstyle m}\downarrow & & \downarrow{\scriptstyle true} \\ \Omega & \xrightarrow[\;f\;]{} & \Omega \end{array}$$

is a pullback. Since 1_Ω classifies $true : 1 \to \Omega$ and $f \le 1_\Omega$ we have

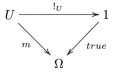

What we must show is that

$$f = (\Omega \xrightarrow{(1_\Omega, f.true.!_\Omega)} \Omega \times \Omega \xrightarrow{\wedge} \Omega).$$

Since the characteristic function of $(true, true) : 1 \to \Omega \times \Omega$ is $\wedge$, it suffices to show that

$$
\begin{array}{ccc}
U & \xrightarrow{\ !_U\ } & 1 \\
{\scriptstyle m}\downarrow & & \downarrow{\scriptstyle (true,true)} \\
\Omega & \xrightarrow[(1_\Omega, f.true.!_\Omega)]{} & \Omega \times \Omega
\end{array}
$$

is a pullback. To see that the square actually commutes, consider 'coordinates'. We have $1_\Omega.m = m = true.!_U$ using the triangle and

$$f.true.!_\Omega.m = f.true.!_U = f.m = true.!_U$$

using both the triangle and the first square. The pullback property now follows easily from that of the first square. $\qquad\square$

We gave a diagrammatic proof because in some respects this Lemma is mysterious. If Ω is Boolean and 2-valued then there are only two possibilities for an $f \le 1_\Omega$ and a verification by cases is almost a triviality — but this is of little help for an intuitionistically valid argument. Nevertheless, we can argue with generalized elements as follows: First, for any $g, f : \Omega \to \Omega$, to show that $g \le f$ it suffices to show that if $true = g(\omega)$ then $true = f(\omega)$. For *any* $f : \Omega \to \Omega$, now define $g : \Omega \to \Omega$ by $g(\omega) = \omega \wedge f(true)$. If $true = g(\omega)$ then $true = \omega \wedge f(true)$ gives $true = \omega$ and $true = f(true)$ which gives $true = f(\omega)$ so that $g \le f$ always holds. But when $f \le 1_\Omega$ the assumption $true = f(\omega)$ gives $true = \omega$ which substituted into the assumption gives $true = f(true)$ so that we have $true = \omega \wedge f(true)$ which is $true = g(\omega)$ and verifies $f \le g$.

In sharp contrast to the example we gave at the end of Section 3, Bénabou's Lemma leads to:

Proposition. *If Ω satisfies the infinite de Morgan law then Ω is Boolean.*

Proof. If pseudo-complementation $(-)^c : \Omega^{op} \to \Omega$ has a right adjoint r then we have $(-)^c.r \le 1_\Omega$. Since r is a right adjoint, $r(true) = false$ and since by (i) of Proposition 3.7 we have $false^c = true$ it follows from the previous Bénabou Lemma that $(-)^c.r = 1_\Omega$. Now from 3.6 and the assumption we have $((-)^c)^{op} \dashv$

$(-)^c \dashv r$. Take left adjoints of both sides of the last equation to get, using Lemma 2.4 and Theorem 2.4, $(-)^c.((-)^c)^{op} = 1_\Omega$. □

Theorem. *If Ω^{op} is Heyting then Ω is Boolean.*

Proof. If Ω^{op} is Heyting then it has a negation $(-)^n : \Omega = (\Omega^{op})^{op} \to \Omega^{op}$ which is a priori distinct from $((-)^c)^{op} : \Omega \to \Omega^{op}$. From Corollary 3.6 we have $((-)^n)^{op}(-)^n \leq 1_\Omega$ and $((-)^n)^{op}(-)^n(true) = true$ by Proposition 3.7. Thus $((-)^n)^{op}(-)^n : \Omega \to \Omega$ satisfies the hypotheses of Lemma 4.1 and is therefore 1_Ω. By Proposition 3.11, Ω^{op} is Boolean and hence Ω is Boolean. □

The reader is urged to note that this Theorem is *not* true for arbitrary Heyting algebras, even those that have all suprema (and infima). For example, both **3** and **3**^{op} are Heyting and have all suprema.

4.2. The lattices $\mathbb{P}X$. Each powerset Ω^X becomes a Heyting algebra using pointwise operations and, as noted before, the order on Ω^X will be written $\subseteq$ and read as 'containment', giving $\mathbb{P}X = (\Omega^X, \subseteq)$ in **ord**. For these Heyting algebras we also refer to $(-)^c : (\mathbb{P}X)^{op} \to \mathbb{P}X$ as the pseudo-complement. If $f : X \to A$ is a function then, as pointed out in the examples in 2.4, $\Omega^f : \Omega^A \to \Omega^X$ is order-preserving with adjoints $\exists_f \dashv \Omega^f \dashv \forall_f : \mathbb{P}X \to \mathbb{P}A$.

If A is an ordered set, it will sometimes be convenient to explicitly write $A = (|A|, \leq_A)$ or $A = (|A|, \leq)$ so that $|A|$ is the underlying set of A. If X is a set then we will also find $dX = (X, =)$, the *discrete* order on X, to be useful. Order-preserving $dX \to A$ are in one to one correspondence with functions $X \to |A|$. This one to one correspondence lifts to an isomorphism $|A^{dX}| \cong |A|^X$ which in turn lifts to an order-preserving $A^{dX} \cong A^X$, where $A^X = (|A|^X, (\leq_A)^X)$. We will dispense with the d in dX if the context is clear. In particular we will write $k : |A| \to A$ for the order-preserving $d|A| \to A$ (which corresponds to the identity function $|A| \to |A|$). Of course the underlying function of $k : |A| \to A$ is the identity function. Note too that for any set X, we have $\mathbb{P}X \cong \mathbb{P}1^X = (\mathbb{P}1)^X$ in **ord**.

For A an ordered set we must distinguish between $\mathbb{P}1^A$ and $\mathbb{P}1^{|A|}$. The latter is isomorphic to $\mathbb{P}|A|$ but the former depends fully on A. Recall now Exercise 3. in 2.4. For $f : X \to A$ in **ord**, we have as a special case the order-preserving $\mathbb{P}1^f : \mathbb{P}1^A \to \mathbb{P}1^X$. On the other hand, if X and A are discrete we have already described $\Omega^f : \mathbb{P}A \to \mathbb{P}X$ as inverse image and often find it convenient to write $\mathbb{P}f$ for Ω^f.

Proposition. *For sets X and A and a function $f : X \to A$, the following diagram in **ord** commutes.*

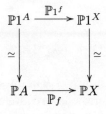

The Proposition identifies, in this context, composition of order-preserving functions (the effect of $\mathbb{P}1^f$) and pulling back subsets along f (the effect of $\mathbb{P}f$). Observe that the top arrow of the diagram makes sense without the proviso of discreteness. as in the paragraph preceding the Proposition. In particular the order-preserving $k : |A| \to A^{op}$ gives rise to $\mathbb{P}1^k : \mathbb{P}1^{A^{op}} \to \mathbb{P}1^{|A|}$. Since composing with k is at the level of functions just composing with an identity, it should be clear that $\mathbb{P}1^k$ is just the inclusion of order-preserving functions in all functions and the following question is provoked:

$$
\begin{array}{ccc}
\mathbb{P}1^{A^{op}} & \xrightarrow{\ \mathbb{P}1^k\ } & \mathbb{P}1^{|A|} \\[2pt]
\Big\downarrow{\scriptstyle\simeq} & & \Big\downarrow{\scriptstyle\simeq} \\[4pt]
? & \xrightarrow[\ ?\ ]{} & \mathbb{P}|A|
\end{array}
$$

In other words, which subsets of $|A|$ correspond to the *order-preserving* functions from A^{op} to $\mathbb{P}1$?

4.3. Downsets.

Proposition. *The subsets of $|A|$ which correspond to order-preserving functions from A^{op} to $\mathbb{P}1$ via $\mathbb{P}|A| \cong \mathbb{P}1^{|A|}$ are precisely those subsets S with the property that $a \leq b$ and b in S implies a in S.*

Definition. The subsets described in Proposition 4.3 are called *downsets* of A. (We write A rather than $|A|$ because they depend on A and not just $|A|$.) The collection of all downsets of A together with the order induced from $\mathbb{P}|A|$ ('containment') is denoted $\mathbb{D}A$.

Note that if A is discrete then the condition in Proposition 4.3 is trivial and for discrete A we have $\mathbb{D}A = \mathbb{P}A$. In particular, for any ordered A we have $\mathbb{D}|A| = \mathbb{P}|A|$ which suggests that the last diagram be filled in as:

$$
\begin{array}{ccc}
\mathbb{P}1^{A^{op}} & \xrightarrow{\ \mathbb{P}1^k\ } & \mathbb{P}1^{|A|} \\[2pt]
\Big\downarrow{\scriptstyle\simeq} & & \Big\downarrow{\scriptstyle\simeq} \\[4pt]
\mathbb{D}A & \xrightarrow[\ \mathbb{D}k\ ]{} & \mathbb{D}|A|
\end{array}
$$

where in the bottom row the instance of k in question is $k : |A| \to A$.

For $f : X \to A$ an arbitrary arrow in **ord**, we can define $\mathbb{D}f : \mathbb{D}A \to \mathbb{D}X$:

Lemma. *If $f : X \to A$ is in* **ord** *and $|f|$ is the underlying function, then for any downset S in A, $\mathbb{P}|f|(S)$ is a downset in X and we have*

$$
\begin{array}{ccc}
\mathbb{D}A & \xrightarrow{\ \mathbb{D}f\ } & \mathbb{D}X \\
\downarrow & & \downarrow \\
\mathbb{P}|A| & \xrightarrow[\ \mathbb{P}|f|\ ]{} & \mathbb{P}|X|
\end{array}
$$

where $\mathbb{D}f$ is the restriction. Moreover this definition of $\mathbb{D}f$ agrees with that given above for the special case $f = k$.

4.4. Adjoints for the $\mathbb{D}f$. We find it convenient to write $i : \mathbb{D}A \rightarrowtail \mathbb{P}|A|$ for the inclusion, although it already has the name $\mathbb{D}k$. Our immediate goal is to show that all $\mathbb{D}f$ have both left and right adjoints. We treat the special case of the $i = \mathbb{D}k$ first.

Lemma. *The arrow $i : \mathbb{D}A \rightarrowtail \mathbb{P}|A|$ has a left adjoint, called* down-closure, *whose value at S is $S^{\blacktriangledown} = \{a \in A | (\exists b)(a \leq b \text{ and } b \in S)\}$ and a right adjoint (called* down-interior*) whose value at S is $S^{\,\substack{\blacktriangledown \\ \circ}} = \{a \in A | (\forall b)(b \leq a \text{ implies } b \in S)\}$.*

For $f : X \to A$ in **ord**, f is monic if and only if $|f|$ is monic in **set**. In particular, the $k : |A| \to A$ are monics. A more useful notion than 'monic' when dealing with ordered sets is 'fully faithful'. Say that $i : X \to A$ is *fully faithful* when, for all x, y in X, $ix \leq iy$ if and only if $x \leq y$. It follows that if i is fully faithful then $ix \cong iy$ if and only if $x \cong y$. In particular, for antisymmetric X, i fully faithful implies i monic. That the converse is not true is seen by considering $k : |A| \to A$ for non-discrete A. If X is a subset of an ordered set A and X is given the induced ordering, then the inclusion of X in A is fully faithful. In particular, $i : \mathbb{D}A \rightarrowtail \mathbb{P}|A|$ is fully faithful for each A.

Theorem. *For $f : X \to A$ in* **ord**, *$\mathbb{D}f : \mathbb{D}A \to \mathbb{D}X$ has a left adjoint, $f_!$, which is described as the composite in the following diagram:*

$$
\begin{array}{ccc}
\mathbb{P}|X| & \xrightarrow{\ \exists_{|f|}\ } & \mathbb{P}|A| \\
i \uparrow & & \downarrow (-)^{\blacktriangledown} \\
\mathbb{D}X & \xrightarrow[\ f_!\ ]{} & \mathbb{D}A
\end{array}
$$

where $\exists_{|f|}$ is the left adjoint of $\mathbb{P}|f|$. Also, $\mathbb{D}f : \mathbb{D}A \to \mathbb{D}X$ has a right adjoint, f_*, which is described as the composite in the following diagram:

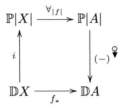

where $\forall_{|f|}$ is the right adjoint of $\mathbb{P}|f|$.

Proof. For Y in $\mathbb{D}X$ and B in $\mathbb{D}A$, we have in Gentzen-style calculus:

$$
\begin{aligned}
(\exists_{|f|}(iY))^{\blacktriangledown} &\subseteq B \quad [\text{ in } \mathbb{D}A] \\
\exists_{|f|}(iY) &\subseteq iB \quad [\text{ in } \mathbb{P}|A|] \\
iY &\subseteq \mathbb{P}|f|(iB) \quad [\text{ in } \mathbb{P}|X|] \\
iY &\subseteq i\mathbb{D}f(B) \quad [\text{ in } \mathbb{P}|X|] \\
Y &\subseteq \mathbb{D}f(B) \quad [\text{ in } \mathbb{D}X]
\end{aligned}
$$

which establishes the first claim. The proof of the second claim is similar. □

4.5. Full suborders. If $i : X \to A$ in **ord** has $|i| : |X| \to |A|$ a subset inclusion, we will say that X is a *full suborder* of A if $i : X \to A$ is fully faithful. It will be no real loss of generality to assume that X is *replete*, meaning that, for x in X, $x \cong y$ in A implies y in X.

Proposition. *For X a full suborder of A, if A is a meet (respectively join) semilattice and X is closed with respect to 1 (respectively 0) and $- \wedge -$ (respectively $- \vee -$), then X is a meet (respectively join) semilattice and $i : X \to A$ preserves 1 (respectively 0) and $- \wedge -$ (respectively $- \vee -$).*

It is important to realize that, for X a full suborder of A and A a meet (respectively join) semilattice, X may well be a meet (respectively join) semilattice without enjoying the closure properties. For example, let V be a vector space with underlying set $|V|$ and consider the full suborder $\mathbb{L}V$ of $\mathbb{P}|V|$ consisting of the subspaces of V. The empty set is not in $\mathbb{L}V$ and $\mathbb{L}V$ is not closed with respect to unions but $\mathbb{L}V$ is a lattice.

4.6. Reflectors and coreflectors. A full suborder X of A is said to be *reflective [coreflective]* if the inclusion $i : X \to A$ has a left [right] adjoint. The left [right] adjoint is called the *reflector [coreflector]*. Observe that for any ordered set X, $(-)^{\blacktriangledown} : \mathbb{P}|X| \to \mathbb{D}X$ provides a reflector for the inclusion $i : \mathbb{D}X \to \mathbb{P}|X|$, while $(-)^{\blacktriangledown} : \mathbb{P}|X| \to \mathbb{D}X$ provides a coreflector.

Lemma. *If X is a full, reflective, replete suborder of A with reflector l then, for a in A, a is in X if and only if the adjunction inequality $a \leq ila$ is an isomorphism.*

Proof. If $a \cong ila$, then a, being isomorphic to an element of X, namely la, is in X. On the other hand, if x is in X then, as for any adjunction in **ord**, we have $ix \cong ilix$ and the result follows since $x = ix$. $\square$

(Of course the dual result holds for coreflectives.) Note too that for x in X, the adjunction inequality $lix \leq x$ is necessarily an isomorphism.

Proposition. *Full reflective suborders of meet semilattices are meet semilattices and full reflective suborders of lattices are lattices. In either case, the inclusion preserves the meet structure.*

Proof. For the first clause assume that A is a meet semilattice and that we have $i : X \to A$ full with reflector l. We have $1 \leq il1$ so $1 \cong il1$. For x and y in X, consider $il(x \wedge y) \leq ilx \leq x$. Similarly $il(x \wedge y) \leq y$. So $il(x \wedge y) \leq x \wedge y$ and $x \wedge y \cong il(x \wedge y)$. By Proposition 4.5 and Lemma 4.6, X is a meet semilattice and the inclusion preserves this structure.

For the second clause, assume in addition that A is a lattice and consider first $l(0)$. For x in X we have $l(0) \leq x$ since $0 \leq ix$ and it follows that $l(0)$ is a bottom element for X. For x and y in X, consider $l(ix \vee iy)$. For any z in X we have

$$
\begin{array}{rcll}
l(ix \vee iy) & \leq & z & \text{in } X \\
ix \vee iy & \leq & iz & \text{in } A \\
\langle ix, iy \rangle & \leq & \langle iz, iz \rangle & \text{in } A \times A \\
\langle x, y \rangle & \leq & \Delta z & \text{in } X \times X.
\end{array}
$$

Thus x and y have a join in X given by $l(ix \vee iy)$. $\square$

Note that 0 and joins in X are typically different from their counterparts in A. Consider again the example $i : \mathbb{L}V \to \mathbb{P}|V|$ mentioned above. It is reflective with reflector given by 'the subspace generated by $-$' and illustrates Proposition 4.6. Remember that left adjoints preserve bottom elements and joins but typically do not preserve top elements and meets. For $i : X \to A$ full reflective with reflector l, one should observe that nevertheless l preserves top elements. It is an important extra property however for l to preserve meets.

4.7. Meet preserving reflectors. To illustrate the last remark consider now $i : \mathbb{D}X \to \mathbb{P}|X|$. In terms of our current vocabulary, $\mathbb{D}X$ is both reflective and coreflective in a power set. For two different reasons $\mathbb{D}X$ is a lattice. *Here* we have both meets and joins in $\mathbb{D}X$ as in the larger order because the inclusion has both left and right adjoints. Suppose that in X we have distinct x and y with a lower bound b. Now $(\{x\} \cap \{y\})^{\blacktriangledown} = \emptyset^{\blacktriangledown} = \emptyset$ but b is in $\{x\}^{\blacktriangledown} \cap \{y\}^{\blacktriangledown}$ so that the binary meet comparison inequality for $(-)^{\blacktriangledown}$ is strict. Note that a consequence of this observation is that $(-)^{\blacktriangledown}$ cannot have a left adjoint unless X is fairly trivial. The inclusion $i : \mathbb{L}V \to \mathbb{P}|V|$ also has a reflector that fails to preserve $- \wedge -$.

This is a good place to point out that $\{x\}^{\blacktriangledown}$ will be abbreviated by $\downarrow x$. Read 'down-seg x' for $\downarrow x$. Note that for any downset S in X that we have x in S if and only if $\downarrow x \subseteq S$.

Proposition. *If $i : X \to A$ is reflective with left adjoint l and A is a Heyting lattice then, for all x in X and a in A, $a \Rightarrow x$ is in X if and only if l preserves $- \wedge -$.*

Proof. Assume that l preserves $- \wedge -$ and consider $il(a \Rightarrow x)$. To show it is at most $a \Rightarrow x$ is precisely to show that $a \wedge il(a \Rightarrow x) \le x$. Now il preserves $-\wedge-$ and for any meet-preserving f, we have a comparison inequality $f(a \Rightarrow b) \le fa \Rightarrow fb$. So $a \wedge il(a \Rightarrow x) \le a \wedge (ila \Rightarrow ilx) \le a \wedge (a \Rightarrow ilx) \le ilx \cong x$. (In the second inequality we used $a \le ila$ and the fact that $- \Rightarrow b$ is order reversing.)

For the converse it is convenient to supress instances of i and use the following Gentzen-style deduction, where we leave justification of the steps to the reader:

$$
\begin{aligned}
a \wedge b &\le l(a \wedge b) \\
a &\le b \Rightarrow l(a \wedge b) \\
la &\le b \Rightarrow l(a \wedge b) \\
b &\le la \Rightarrow l(a \wedge b) \\
lb &\le la \Rightarrow l(a \wedge b) \\
la \wedge lb &\le l(a \wedge b).
\end{aligned}
$$

$\square$

Corollary. *If $i : X \to A$ is reflective with meet-preserving reflector and A is a Heyting lattice then X is a Heyting lattice.*

4.8. Meet preserving coreflectors. Now in fact $\mathbb{D}X$ is a Heyting lattice for any ordered X but it does not follow from Corollary 4.7 and the inclusion of $\mathbb{D}X$ in $\mathbb{P}|X|$ because, as we have seen, the reflector $(-)^{\blacktriangledown}$ does not preserve meets. In fact, more generally, for $X = (|X|, \mathbb{O}X)$ a topological space we have $\mathbb{O}X$ a Heyting lattice as will be seen from the next Proposition. For X a topological space we have $i : \mathbb{O}X \to \mathbb{P}|X|$ coreflective, with the coreflector given by open-interior, and i preserves finite meets.

Proposition. *If $i : X \to A$ is full coreflective with coreflector r and A is Heyting, then $r(ix \Rightarrow iy)$ provides a Heyting operation for X if and only if i preserves $-\wedge-$.*

Proof. Assume that i preserves $- \wedge -$ and x, y, z in X.

$$
\begin{aligned}
x \wedge z &\le y & \text{in } X \\
i(x \wedge z) &\le iy & \text{in } A \\
ix \wedge iz &\le iy & \text{in } A \\
iz &\le ix \Rightarrow iy & \text{in } A \\
z &\le r(ix \Rightarrow iy) & \text{in } X.
\end{aligned}
$$

Conversely, assume that $r(i- \Rightarrow i-)$ provides a Heyting operation for X.

$$\begin{aligned} ix \wedge iy &\leq i(x \wedge y) \\ iy &\leq ix \Rightarrow i(x \wedge y) \\ y &\leq r(ix \Rightarrow i(x \wedge y)) \\ x \wedge y &\leq x \wedge y. \end{aligned}$$

$\square$

Recall that a topological space can be presented as a set $|X|$ together with an interior operator $(-)^o : \mathbb{P}|X| \rightarrow \mathbb{P}|X|$ that preserves the top element and meets. For an ordered set X, down-interior followed by the inclusion provides such an interior operator on $\mathbb{P}|X|$, with the extra property that it preserves *arbitrary* infima. The downsets are the opens for the topology. This topology, $\mathbb{D}X$, is called the *Alexandroff topology* on $|X|$.

Corollary. *For any topological space X, $\mathbb{O}X$ is a Heyting lattice and, in particular, for any ordered set X, $\mathbb{D}X$ is a Heyting lattice.*

4.9. The down-segment embedding. For X any ordered set $\downarrow : X \rightarrow \mathbb{D}X$ is fully faithful. It is monic precisely if X is antisymmetic but fully faithfulness is the important property. We have spoken of 'suborders' in this section to avoid circumlocution but it is easy to rephrase the results of this section for fully faithful arrows in **ord**.

It is nevertheless useful to think of X as being contained in $\mathbb{D}X$ and it is time now to think about adjoints for $\downarrow : X \rightarrow \mathbb{D}X$. A right adjoint r would provide, for each downset S of X, a largest element x with the property that $\downarrow x \subseteq S$. Since this must apply in particular to $S = \emptyset$, the empty subset, and the $\downarrow x$ are not empty, it follows that $\downarrow$ never has a right adjoint. It should be pointed out that if $\downarrow_X$ is regarded as being an arrow $X \rightarrow \mathbb{P}1^{X^{op}}$ then the corresponding $X^{op} \times X \rightarrow \mathbb{P}1$ has as its underlying function the characteristic function of the order relation for X.

The possibility of a *left* adjoint for $\downarrow : X \rightarrow \mathbb{D}X$ is a quite different matter and important. Recall (the dual of) Proposition 3.3. Let S be a subset of X and assume that $\bigwedge S$ exists. It is characterized (up to isomorphism) by the requirement $x \leq \bigwedge S$ if and only if, for all s in S, $x \leq s$. We have the comparison inequality $\downarrow (\bigwedge S) \leq \bigwedge \{\downarrow s | s \in S\}$, provided the right side exists, merely because $\downarrow$ is order-preserving.

Lemma. *For any subset S of X, $\bigwedge \{\downarrow s | s \in S\}$ exists in $\mathbb{D}X$ and if $\bigwedge S$ exists in X then the comparison inequality $\downarrow (\bigwedge S) \leq \bigwedge \{\downarrow s | s \in S\}$ is an isomorphism (necessarily an equality).*

Proof. Infima in $\mathbb{D}X$ are necessarily as they are in $\mathbb{P}|X|$ — intersections. We have x in $\bigwedge \{\downarrow s | s$ in $S\}$ if and only if $\downarrow x \subseteq \bigwedge \{\downarrow s | s$ in $S\}$. By the characterizing property of an infimum this inequality holds if and only if $\downarrow x \subseteq \downarrow s$, for all s in S, which is the case if and only if $x \leq s$, for all s in S. Assuming that $\bigwedge S$ exists, this last is the case if and only if $x \leq \bigwedge S$ if and only if x is in $\downarrow \bigwedge S$.   □

Said otherwise $\downarrow$, the down-segment embedding preserves any infima that exist. Sometimes $\downarrow$ is called a Yoneda embedding, after an early worker in Category Theory. In a Heyting lattice, the elements $x \Rightarrow y$ share some properties with infima. As we will see, the down-segment embedding also preserves any instances of $\Rightarrow$ that exist in a meet semilattice. More precisely, if an arrow $f : X \to A$ in **ord** preserves $- \wedge -$, as for example down-segment does, then there is a comparison inequality $f(x \Rightarrow y) \leq fx \Rightarrow fy$ and to say that f *preserves implications* is to say that these comparisons are isomorphisms.

Proposition. *The down-segment embedding* $\downarrow: X \to \mathbb{D}X$ *preserves any implications that exist in* X.

Proof. Assume that $x \Rightarrow y$ exists in X. For any z in X, $z \in (\downarrow x \Rightarrow \downarrow y)$ if and only if $\downarrow z \subseteq \downarrow x \Rightarrow \downarrow y$ if and only if $\downarrow x \cap \downarrow z \subseteq \downarrow y$ if and only if $\downarrow (x \wedge z) \subseteq \downarrow y$ if and only if $x \wedge z \leq y$ if and only if $z \leq x \Rightarrow y$ if and only if $z \in \downarrow (x \Rightarrow y)$.   □

Note that a general right adjoint need not preserve implications, — although, for example, a fully faithful right adjoint whose left adjoint is meet-preserving does.

5. Completeness

5.1. A key definition. From Proposition 3.3 we see that existence of a left adjoint for $\downarrow_X$ is entirely dependent upon the existence in X of all infima of the form $\bigwedge\{x | S \subseteq \downarrow x\}$ for S in $\mathbb{D}X$.

Definition. An ordered set X is said to be *complete* if $\downarrow_X : X \to \mathbb{D}X$ has a left adjoint.

Theorem. *For an ordered set* X, *the following are equivalent:*
   (1) X is complete.
   (2) For every downset S of X, $\bigwedge\{x | S \subseteq \downarrow x\}$ exists.
   (3) For every downset S of X, $\bigvee S$ exists.
   (4) For every subset S of X, $\bigvee S$ exists.

Proof. The equivalence of (1) and (2) is what has just been discussed. To say that $\downarrow_X$ has a left adjoint, l, is to say that, for every downset S, there is an element lS in X with the property that for every x in X, $lS \leq x$ if and only if $S \subseteq \downarrow x$. But to say that $S \subseteq \downarrow x$ is precisely to say that x is an upper bound for S so it is clear that lS is a supremum for S which establishes the equivalence of (1) and (3). Obviously (4) implies (3). To see that (3) implies (4), compose the adjunction $\bigvee \dashv \downarrow: X \to \mathbb{D}X$ with the adjunction $(-)^{\blacktriangledown} \dashv i : \mathbb{D}X \to \mathbb{P}|X|$ thus providing a left adjoint to the composite $X \to \mathbb{P}|X|$. For S an arbitrary subset of X, we nevertheless have $S \subseteq i(\downarrow x)$ if and only if x is an upper bound for S so the left adjoint to the composite provides a supremum function defined on arbitrary subsets. □

The apparent asymmetry with respect to existence of infs and sups is illusory. Existence of all suprema is equivalent to the existence of all infima. The clue is provided by (2) above: The supremum of S is the infimum of the set of all upper bounds of S. Dually then, the infimum of a subset T should be the supremum of the set of lower bounds for T. Simple though it is, a detailed explanation in terms of adjoints is worthwhile.

5.2. Upsets. In addition to the down-segment embedding $\downarrow_X \colon X \to \mathbb{D}X$ we also have $\downarrow_{X^{op}} \colon X^{op} \to \mathbb{D}X^{op}$. Now $\downarrow_{X^{op}}(x) = \{y \in X \mid x \leq y\}$ which is an 'upset' of X. In fact *upsets* of X can be defined quite generally to be downsets of X^{op}. We define $\mathbb{U}X$ to be the set of upsets of X ordered by reverse inclusion. That is, for upsets T and U we take $T \leq U$ in $\mathbb{U}X$ if and only if $T \supseteq U$, meaning that $U \subseteq T$ in $\mathbb{P}|X|$. It follows that $\mathbb{U}X = (\mathbb{D}(X^{op}))^{op}$. We define $\uparrow_X = (\downarrow_{X^{op}})^{op}$ so that $\uparrow_X \colon X \to \mathbb{U}X$. Slavishly following the notation, X^{op} is complete if and only if $\downarrow_{X^{op}} \colon X^{op} \to \mathbb{D}X^{op}$ has a left adjoint but by Proposition 2.4 this is so if and only if $\uparrow_X \colon X \to \mathbb{U}X$ has a right adjoint. For S in $\mathbb{D}X$ we write S^+ for the set of upper bounds of S. It is clearly an upset. If $S \subseteq R$ in $\mathbb{D}X$ then $R^+ \subseteq S^+$ so that $(-)^+$ defines $(-)^+ \colon \mathbb{D}X \to \mathbb{U}X$. Similarly, for T in $\mathbb{U}X$ we define T^- to be the downset of lower bounds of T.

Proposition. *For any ordered set* X, $(-)^+ \dashv (-)^- \colon \mathbb{U}X \to \mathbb{D}X$ *and the following diagram commutes:*

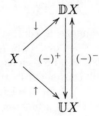

Theorem. *For an ordered set* X, X *is complete if and only if* X^{op} *is complete, in which case the following diagram commutes:*

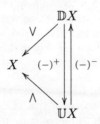

In terms of what we have seen of adjoints thus far, Theorem 5.2 is somewhat strange. The right adjoint $\bigwedge$ is computed as a right adjoint followed by a left adjoint. A dual remark applies to $\bigvee$. The matter is further explained in [9]. The adjunction $(-)^+ \dashv (-)^-$ deserves to be known as an *Isbell conjugation adjunction*.

There is another adjunction connecting $\mathbb{D}X$ and $\mathbb{U}X$, quite distinct from $(-)^+ \dashv (-)^-$. For any Heyting lattice H, recall that we have $((-)^c)^{op} \dashv (\ )^c : H^{op} \rightarrow H$, where $x^c = x \Rightarrow 0$ is the negation of x. We know that for any set X, $\mathbb{P}X$ is a Heyting algebra. Here, for S an element of $\mathbb{P}X$, that is a subset of X, S^c is what is often called the pseudo-complement of S. For X an ordered set, the adjunction

$$((-)^c)^{op} \dashv (-)^c : \mathbb{P}|X|^{op} \rightarrow \mathbb{P}|X|$$

restricts to give $((-)^c)^{op} \dashv (-)^c : \mathbb{U}X \rightarrow \mathbb{D}X$. Said more readably, the pseudo-complement of an upset is a downset and vice versa. (These arrows are not inverse to each other in general.)

Exercises.

1. Suppose that $\mathcal{B}$ and $\mathcal{C}$ are categories which, like **ord**, carry the extra structure of ordered hom collections and functorial composition. The relevant functors between such categories are those $F : \mathcal{B} \rightarrow \mathcal{C}$ for which each effect on hom collections $F_{X,A} : \mathcal{B}(X, A) \rightarrow \mathcal{C}(FX, FA)$ is also a functor. In other words, we require that if $f \leq g : X \rightarrow A$ in $\mathcal{B}$ then $Ff \leq Fg : FX \rightarrow FA$ in $\mathcal{C}$. Such an F is a *2-functor* (but what we have said does not exhaust the generality of that term). Define *adjunction* in such $\mathcal{B}$ and $\mathcal{C}$ as are under consideration and show that if $F : \mathcal{B} \rightarrow \mathcal{C}$ is a 2-functor and $f \dashv u$ in $\mathcal{B}$ then $Ff \dashv Fu$ in $\mathcal{C}$.

2. For such $\mathcal{B}$ as are under consideration, in addition to the usual dual $\mathcal{B}^{op}$, with $\mathcal{B}^{op}(X, A) = \mathcal{B}(A, X)$ we have also $\mathcal{B}^{co}$ with $\mathcal{B}^{co}(X, A) = \mathcal{B}(X, A)^{op}$ and $\mathcal{B}^{coop} = (\mathcal{B}^{co})^{op}$, which is equally $(\mathcal{B}^{op})^{co}$. Show that if $f \dashv u : A \rightarrow X$ in $\mathcal{B}$ then $f \dashv u : X \rightarrow A$ in $\mathcal{B}^{coop}$.

3. In [5] it is shown that inverse image is contravariantly functorial. Show that $\mathbb{D}$ provides a 2-functor $\mathbb{D} : \mathbf{ord}^{coop} \rightarrow \mathbf{ord}$ (which can also be seen as a 2-functor $\mathbf{ord} \rightarrow \mathbf{ord}^{coop}$). Conclude that $\mathbb{U}$ can be seen as a 2-functor $\mathbb{U} : \mathbf{ord} \rightarrow \mathbf{ord}^{coop}$.

4. For $f : X \rightarrow A$ in **ord** it will be convenient in this exercise, and those that follow below, to write $Df = f_! : \mathbb{D}X \rightarrow \mathbb{D}A$ for the left adjoint of $\mathbb{D}f$. Use properties of adjunctions to show that D provides a 2-functor $D : \mathbf{ord} \rightarrow \mathbf{ord}$, given on objects by $\mathbb{D}$.

5. For $f : X \rightarrow A$ in **ord** it follows from the definition of $\mathbb{U}$ and Theorem 4.4 that $\mathbb{U}f$ has both a left adjoint and a right adjoint. Write Uf for the *right* adjoint of $\mathbb{U}f$ and verify that it is constructed with the help of $\exists_{|f|}$. Use adjoints to show that U provides a 2-functor $U : \mathbf{ord} \rightarrow \mathbf{ord}$, given on objects by $\mathbb{U}$.

6. Show that the composite 2-functors $\mathbb{D}U : \mathbf{ord} \rightarrow \mathbf{ord}$ and $D\mathbb{U} : \mathbf{ord} \rightarrow \mathbf{ord}$ are equal by showing that for any $f : X \rightarrow A$ in **ord** both $\mathbb{D}Uf$ and $D\mathbb{U}f$ have the same right adjoint.

The equality $\mathbb{D}U = D\mathbb{U}$ of the last exercise provides the starting point for an investigation that is beyond the scope of this chapter. The interested reader who is familiar with monads may wish to follow it in [6]. In primitive terms it says that for $f : X \rightarrow A$ in **ord** and $\mathcal{S}$ a downset of upsets of X,

$$(f^{-1})^{-1}(\mathcal{S}) = \{\{fs|s \in S\}^{\blacktriangle}|S \in \mathcal{S}\}^{\overline{\blacktriangledown}}$$

and is tedious to verify directly, without the help of adjoints, in anybody's set theory. The reader is urged to do this. Of course a mere function $f : X \to A$ in **set** can be regarded as an arrow in **ord**, between discrete orders, but observe that even in this case the down-closure in the display is non-trivial. In other words, it is *not* true that $\mathbb{P}\mathbb{P}f$ is given by $\exists_{\exists_f}$.

5.3. Complete ordered sets are lattices. If an ordered set X is complete, it has $\bigvee S$ and $\bigwedge S$ for every $S \subseteq |X|$ but we should exercise some initial care about existence of $x \wedge y$ etcetera. Given a pair (x, y), an element of $X \times X$, one expects that $x \wedge y$ should be $\bigwedge\{x, y\}$ but given our meagre axioms for **set** one might ask critically about the construction of $\{x, y\}$ and about its properties. We have used $\{x, y\}$ somewhat informally already but, fortunately, we can avoid it here.

Proposition. *If X is a complete ordered set then X is a lattice.*

Proof. It suffices by duality to show that X is a meet semilattice. Thus we require right adjoints for $1 \xleftarrow{!} X \xrightarrow{\Delta} X \times X$. By Proposition 3.3, it suffices for a complete X to show that these preserve the requisite suprema and this we leave as an (easy) exercise. □

In view of the Proposition we often say 'X is a complete lattice' rather than 'X is a complete ordered set'. Some caution is in order. A lattice need not be distributive, as we have seen. A complete lattice need not be distributive either. We also saw that meets distribute over joins if and only if joins distribute over meets but a 'distributive law' $x \wedge (\bigvee S) \cong \bigvee\{x \wedge s | s \text{ in } S\}$ relating the binary meet operation and general suprema operation says nothing about the situation with respect to binary join and general infima. We will have more to say about distributive laws later.

5.4. Completeness of the $\mathbb{D}X$. We need examples of complete lattices. Naively, a 'finite' lattice is a complete lattice but again, using the axioms of **set**, finiteness is a complicated issue. There is no single definition, that we know of, which suffices for all purposes in the study of ordered sets.

Our next theorem is fundamental for the study of complete lattices.

Lemma. *For any ordered set X, $\downarrow_{\mathbb{D}X} = (\downarrow_X)_* : \mathbb{D}X \to \mathbb{D}\mathbb{D}X$.*

Proof. Starting with $\downarrow_X : X \to \mathbb{D}X$, we obtain $(\downarrow_X)_* : \mathbb{D}X \to \mathbb{D}\mathbb{D}X$ as the right adjoint to $\mathbb{D}(\downarrow_X) : \mathbb{D}X \leftarrow \mathbb{D}\mathbb{D}X$, by Theorem 4.4. For T, S in $\mathbb{D}X$ we have

$$
\begin{array}{rcl}
T & \in & (\downarrow_X)_*(S) \\
\downarrow_{\mathbb{D}X}(T) & \subseteq & (\downarrow_X)_*(S) \\
\mathbb{D}(\downarrow_X)(\downarrow_{\mathbb{D}X}(T)) & \subseteq & S \\
\{x \in X | \downarrow_X(x) \in \downarrow_{\mathbb{D}X}(T)\} & \subseteq & S \\
\{x \in X | \downarrow_X(x) \subseteq T\} & \subseteq & S \\
T & \subseteq & S \\
T & \in & \downarrow_{\mathbb{D}X}(S)
\end{array}
$$

which shows that $(\downarrow_X)_* = \downarrow_{\mathbb{D}X}$ as claimed. $\qquad\square$

The Lemma above, in much greater generality, is due to Street and Walters and can be found in their [10]. It is extremely basic but in some sense is something that one would not even look for without the guiding discipline of searching for adjoints, as will be evident in the next theorem.

Theorem. *For any ordered set X, $\mathbb{D}X$ is a complete lattice.*

Proof. We have $\mathbb{D}(\downarrow_X) \dashv (\downarrow_X)_* = \downarrow_{\mathbb{D}X} : \mathbb{D}X \to \mathbb{D}\mathbb{D}X$ by the Lemma above. $\qquad\square$

The proof of the Theorem actually shows more than the statement. We have also $(\downarrow_X)_! \dashv \mathbb{D}(\downarrow_X)$, again by Theorem 4.4 and a *left* adjoint to supremum, which is here $\mathbb{D}(\downarrow_X)$, is certainly not a feature of general complete lattices.

Corollary. *For any set X, $\mathbb{P}X$ is a complete lattice.*

Proof. For X a set, $\mathbb{P}X = \mathbb{D}(dX)$ where $dX = (X, =)$ is the discrete order on X. $\qquad\square$

Of course we know that $\bigvee$ for $\mathbb{P}X$, X a set, is 'union'. It is useful to calculate $\mathbb{D}(\downarrow_X)(\mathcal{S})$, for $\mathcal{S}$ a (containment-)downset of subsets of X. In other words for $\mathcal{S}$ an element of $\mathbb{D}\mathbb{P}X$.

We have x in $\mathbb{D}(\downarrow_X)(\mathcal{S})$ if and only if $\downarrow x$ is in $\mathcal{S}$. Since x is in $\downarrow x$ we certainly have $\mathbb{D}(\downarrow_X)(\mathcal{S})$ contained in the union of $\mathcal{S}$. But if x is any element of the union of $\mathcal{S}$ then there exists S with x in S and S in $\mathcal{S}$. Necessarily we have $\downarrow x \subseteq S \in \mathcal{S}$ and hence $\downarrow x \in \mathcal{S}$ (because $\mathcal{S}$ is a downset) and hence x in $\mathbb{D}(\downarrow_X)(\mathcal{S})$. Observe that the argument works exactly as stated for any ordered X, not necessarily discrete. In fact the only simplification that follows for discrete X is $\downarrow x = \{x\}$.

5.5. Examples of complete lattices. We now generate many examples of complete lattices with some simple observations about reflections and coreflections. It seems useful to isolate some of the background in lemmata.

Lemma. *If $f \dashv u : A \to X$ then f is fully faithful if and only if $1_X \le uf$ is an isomorphism.*

Lemma. *If $f : X \to A$ is fully faithful then $f_! : \mathbb{D}X \to \mathbb{D}A$ is fully faithful.*

Lemma. *For any $f : X \to A$ in **ord**, the following diagram commutes:*

$$
\begin{array}{ccc}
X & \xrightarrow{\ f\ } & A \\
\downarrow{\scriptstyle \downarrow_X} & & \downarrow{\scriptstyle \downarrow_A} \\
\mathbb{D}X & \xrightarrow[\ f_!\ ]{} & \mathbb{D}A
\end{array}
$$

Proposition. *Full reflective suborders of complete lattices are complete lattices.*

Proof. For $i : X \to A$ with reflector l and A complete, it is now an easy exercise to show that $\bigvee_X : \mathbb{D}X \to X$ is given by the composite $l \cdot \bigvee_A \cdot i_!$. $\qquad\square$

Corollary. *If A is an 'algebra' in the sense of universal algebra then $\mathbb{S}A$, the set of subalgebras of A ordered by inclusion, is a complete lattice.*

Proof. Without expanding on what is meant by 'universal algebra', the point is simply this: we have $i : \mathbb{S}A \to \mathbb{P}|A|$ a full inclusion, where here $|A|$ denotes the underlying set of A, and a reflector is provided by the construction 'subalgebra generated by -'. $\qquad\square$

Corollary. *If V is a vector space then $\mathbb{L}V$ is a complete lattice.*

Proposition. *Full coreflective suborders of complete lattices are complete lattices.*

Proof. For $i : X \to A$ with coreflector r and A complete we have $i^{op} : X^{op} \to A^{op}$ with reflector r^{op} and A^{op} complete. It follows that X^{op} and hence X is complete. $\qquad\square$

Corollary. *For X any topological space, $\mathbb{O}X$ is complete.*

Exercises.

1. Show, in any elementary topos, that $\downarrow : \mathbb{P}1 \to \mathbb{D}\mathbb{P}1$ has a left adjoint, which has a left adjoint, which has a left adjoint, which has a left adjoint. (*Hint:* Note that $\mathbb{P}1$ is $\mathbb{D}1$ and that 1 is $\mathbb{D}\emptyset$.)

2. For a topos of constant sets, describe explicitly, in terms of elements, each of the five arrows between Ω and $\mathbb{D}\Omega$ that are prescribed by the preceding exercise. For each of the three from Ω to $\mathbb{D}\Omega$, describe explicitly, in terms of elements, the corresponding relation, $\Omega \twoheadrightarrow \Omega$.

3. For X a complete lattice, define a new relation, $\ll$, on $|X|$ by $x \ll y$ if and only if for every downset S of X, $y \leq \bigvee S$ implies $x \in S$. (This relation has been called the 'totally below' relation so that $x \ll y$ is read 'x is totally below y'.) Prove the following:
   (i) $w \leq x \ll y$ implies $w \ll y$.
   (ii) $x \ll y \leq z$ implies $x \ll z$.
   (iii) $x \ll y$ implies $x \leq y$.
   (iv) $x \ll y \ll z$ implies $x \ll z$.

4. For the lattice M_5 (in constant sets) draw a 'Hasse' type diagram to explicitly illustrate the resulting $\ll$. Since $\ll$ is not reflexive in general, instances of $x \ll x$ must be shown explicitly.

5. Describe $\ll$ explicitly for $([0,1]), \leq)$, the closed unit interval with its usual order.

5.6. Frames and locales. For X complete, x in X and S in $\mathbb{P}|X|$, consider the inequality $\bigvee\{x \wedge s | s \in S\} \leq x \wedge \bigvee S$ which is easily established and easily recognized as the $\bigvee$-comparison inequality for the arrow $x \wedge - : X \to X$ in **ord**. Let us say that X satisfies condition (L) if, for each x and each S, the inequality is an isomorphism so that we have

$$x \wedge \bigvee S \cong \bigvee\{x \wedge s | s \in S\} \qquad (L)$$

which manifestly says that the binary meet distributes over suprema. The arrow $x \wedge - : X \to X$ preserves all suprema if and only if it has a right adjoint.

Definition. A complete lattice that satisfies condition (L) is called a *complete Heyting lattice* or a *frame* or a *locale*.

Just for emphasis let us note:

Proposition. *For X complete, X is a complete Heyting lattice if and only if X is a Heyting lattice.*

Proof. X satisfies (L) if and only if each $x \wedge -$ has a right adjoint $(x \Rightarrow -)$.   □

The term 'frame' tends to be employed when we focus on condition (L), rather than the operation $- \Rightarrow -$ and we consider arrows $f : X \to A$ between frames that preserve $- \wedge -$ and 1 and suprema. Such an arrow f necessarily has a right adjoint that is often written f_*. Any order-preserving function between meet semilattices that preserves binary meets and top $(- \wedge -$ and $1)$ is said to be 'left exact'. The term 'locale' tends to be used when we focus on condition (L) and we consider arrows $f : A \to X$ between locales which have left exact left adjoints. In this context a left adjoint for f is written f^*.

The lattices $\mathbb{P}X$, for X a set, $\mathbb{D}X$, for X an ordered set and $\mathbb{O}X$, for X a topological space are all examples of complete Heyting lattices. We have proved that they are complete and Heyting. It is clear that an $(\mathbb{O}X)^{op}$ can fail to be Heyting. For constant sets, all $(\mathbb{P}X)^{op}$ and all $(\mathbb{D}X)^{op}$ are Heyting *but* in the elementary topos $\mathbf{set}_0^2$, where $\mathbf{set}_0$ denotes the category of constant finite sets, even $(\mathbb{P}1)^{op}$ fails to be Heyting. Thus, in the generality in which we wish to work, we cannot assume that the $(\mathbb{P}X)^{op}$ are Heyting and, a fortiori, we cannot assume that the $(\mathbb{D}X)^{op}$ are Heyting.

Suppose that X and A are topological spaces and that $f : X \to A$ is a continuous function. Consider the following commutative square,

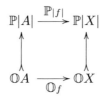

where $\mathbb{O}f$ is the restriction of inverse image. Note that it is the very definition of 'continuous' that tells us that we have such a restriction. Note too that $\mathbb{O}f$ is left exact because binary meet and top in $\mathbb{O}-$ are as in $\mathbb{P}| - |$ and $\mathbb{P}|f|$ preserves them. But unlike $\mathbb{P}|f|$, $\mathbb{O}f$ cannot have a left adjoint in general, because $\mathbb{O}f$ does not generally preserve *all* infima. (Note that infima in $\mathbb{O}-$ are given by interior applied to intersections.) On the other hand, $\mathbb{O}f$ preserves suprema, for these are as in $\mathbb{P}| - |$, so that each $\mathbb{O}f$ has a right adjoint. It should be clear that such right adjoints are given by inclusion in the power set followed by $\forall|f|$ followed by application of the interior operator. The point is that any continuous $f : X \to A$ gives rise to the kind of arrow $\mathbb{O}X \to \mathbb{O}A$ that we said above is considered for locales. The 'spatial' sounding word 'locale' was chosen for this reason. In fact the obvious categories of locales and topological spaces are very closely related. The subject of locales is also known as 'pointless topology'. The text [3] by Peter Johnstone, deals with this subject in great detail and is strongly recommended.

We leave this interesting area before even getting involved. The next Lemma is crucial for study of further distributive laws.

Lemma. *For X a meet semilattice and S and T in $\mathbb{D}X$,*

$$S \cap T = \{s \wedge t | \langle s, t \rangle \in S \times T\}.$$

Proof. Given u in $S \cap T$ and writing $u = u \wedge u$ establishes $S \cap T \subseteq \{s \bigwedge t | s \in S$ and $t \in T\}$. If $s \in S$ and $t \in T$, then $s \wedge t \leq s \in S$ and $s \wedge t \leq t \in T$ shows that $s \bigwedge t$ is in $S \cap T$. □

For any complete X, $\bigvee : \mathbb{D}X \to X$ preserves the top element.

Theorem. *For X complete, X is a complete Heyting lattice if and only if $\bigvee : \mathbb{D}X \to X$ preserves binary meets.*

Proof. To say that $\bigvee$ preserves binary meets is to say that for all S and T in $\mathbb{D}X$, $\bigvee(S \cap T) \cong (\bigvee S) \wedge (\bigvee T)$. It is easy to show that this is equivalent to condition (L) above by using the Lemma above. □

Since $\bigvee : \mathbb{D}X \to X$ preserves 1 in any event, $\bigvee$ is left exact for a complete Heyting lattice. Here the use of $\mathbb{D}X$ rather than $\mathbb{P}|X|$ is crucial. In neither Lemma 5.6 nor Theorem 5.6 can $\mathbb{D}X$ be replaced by $\mathbb{P}|X|$. In an important sense that becomes clear after a careful reading of [10], $\mathbb{D}-$ is the correct domain for $\bigvee$. It makes sense to ask for other preservation properties of $\bigvee : \mathbb{D}X \to X$.

6. Complete distributivity

6.1. The definitions. Classically, a complete lattice X is said to be *completely distributive* (CD) if

$$(\forall \mathcal{S} \subseteq \mathbb{P}|X|)(\bigwedge \{\bigvee S | S \in \mathcal{S}\} \cong \bigvee \{\bigwedge \{T(S) | S \in \mathcal{S}\} | T \in \Pi\mathcal{S}\}).$$

In [2] the following definition was introduced. It has been pursued a great deal, especially in [8]. Further references will be found in [9].

Definition. A complete lattice X is said to be *constructively completely distributive* (CCD) if $X \leftarrow \mathbb{D}X : \bigvee$ has a left adjoint.

Since right adjoints preserve all infima, it follows immediately from Theorem 5.6 that:

Proposition. *If X is a (CCD) lattice then X is a complete Heyting lattice.*

Moreover, an immediate supply of (CCD) lattices is provided by the adjunction $(\downarrow_X)_! \dashv \mathbb{D}(\downarrow_X) = \bigvee_{\mathbb{D}X}$, noted immediately after Theorem 5.4:

Theorem. *For all X in* **ord**, *the lattice of downsets $\mathbb{D}X$ is (CCD).*

Since each set X has $\mathbb{P}X = \mathbb{D}(dX)$, we have:

Corollary. *For all X in* **set**, *the power set $\mathbb{P}X$ is (CCD). In particular, Ω is (CCD).*

The Corollary, which holds assuming only the topos axioms for **set**, stands in stark contrast to the classical result:

$$(\text{AC}) \iff (\text{for every set } X)(\mathbb{P}X \text{ is (CD)})$$

where (AC) denotes the Axiom of Choice, and we address this point next.

6.2. (CD) versus (CCD) We begin by extending Lemma 5.6.

Lemma. *For X a complete lattice and $\mathcal{S} \subseteq \mathbb{D}X$,*

$$\bigcap \mathcal{S} = \{\bigwedge \{T(S) | S \in \mathcal{S}\} | T \in \Pi\mathcal{S}\}.$$

Proof. For $x \in \bigcap \mathcal{S}$, define $\hat{x} \in \Pi\mathcal{S}$ by $\hat{x}(S) = x$, for all $S \in \mathcal{S}$. Since $\bigwedge \{\hat{x}(S) | S \in \mathcal{S}\} = \bigwedge \{x\} = x$, we have $\bigcap \mathcal{S} \subseteq \{\bigwedge \{T(S) | S \in \mathcal{S}\} | T \in \Pi\mathcal{S}\}$. For all $S \in \mathcal{S}$, we have $\bigwedge \{T(S) | S \in \mathcal{S}\} \leq T(S) \in S$ and hence by the downset property, for all $S \in \mathcal{S}$, $\bigwedge \{T(S) | S \in \mathcal{S}\} \in S$. Thus $\{\bigwedge \{T(S) | S \in \mathcal{S}\} | T \in \Pi\mathcal{S}\} \subseteq \bigcap \mathcal{S}$. $\square$

Proposition. *For X a complete lattice, X is (CCD) if and only if*

$$(\forall \mathcal{S} \subseteq \mathbb{D}X)(\bigwedge \{\bigvee S | S \in \mathcal{S}\} \cong \bigvee \{\bigwedge \{T(S) | S \in \mathcal{S}\} | T \in \Pi\mathcal{S}\})$$

and hence, (CD) implies (CCD).

Proof. Since infima in $\mathbb{D}X$ are given by intersection, it follows from Corollary 3.3 that X is (CCD) if and only if, for all $\mathcal{S} \subseteq \mathbb{D}X$,

$$\bigvee \bigcap \mathcal{S} \cong \bigwedge \{\bigvee S | S \in \mathcal{S}\}$$

so that the statement of the proposition follows immediately from the preceding Lemma. $\square$

Theorem.
$$(\text{AC}) \iff ((\text{CD}) \iff (\text{CCD})).$$

Proof. Assume (AC) and that X is (CCD). For $\mathcal{S} \subseteq \mathbb{P}|X|$ we have

$$\bigwedge \{\bigvee S | S \in \mathcal{S}\} = \bigwedge \{\bigvee S^{\triangledown} | S \in \mathcal{S}\} \cong \bigvee \bigcap \{S^{\triangledown} | S \in \mathcal{S}\}.$$

The equality follows from the fact that we may take the supremum of a general subset of X to be the supremum of its down closure. The isomorphism is an instance of the isomorphism displayed in the proof of the last Proposition. To show that X is (CD) it suffices to show that

$$\bigcap \{S^{\triangledown} | S \in \mathcal{S}\} = \{\bigwedge \{T(S) | S \in \mathcal{S}\} | T \in \Pi\mathcal{S}\}^{\triangledown}.$$

For $T \in \Pi\mathcal{S}$, we have $\bigwedge\{T(S)|S \in \mathcal{S}\} \leq T(S) \in S \subseteq S^{\blacktriangledown}$, for all $S \in \mathcal{S}$. Since the intersection is a downset, $\{\bigwedge\{T(S)|S \in \mathcal{S}\}|T \in \Pi\mathcal{S}\}^{\blacktriangledown} \subseteq \bigcap\{S^{\blacktriangledown}|S \in \mathcal{S}\}$. For $x \in \bigcap\{S^{\blacktriangledown}|S \in \mathcal{S}\}$ we have $(\forall S \in \mathcal{S})(x \in S^{\blacktriangledown})$, which is to say $(\forall S \in \mathcal{S})(\exists y \in S)(x \leq y)$. In the presence of (AC) this last is equivalent to $(\exists T \in \Pi\mathcal{S})(\forall S \in \mathcal{S})(x \leq T(S)$ which gives $(\exists T \in \Pi\mathcal{S})(x \leq \bigwedge\{T(S)|S \in \mathcal{S}\})$ and hence $x \in \{\bigwedge\{T(S)|S \in \mathcal{S}\}|T \in \Pi\mathcal{S}\}^{\blacktriangledown}$.

On the other hand, if we assume that (CD) is equivalent to (CCD) then, by Corollary 6.1, every power set is (CD) and by the displayed classical result of 6.1 we have (AC). □

6.3. Two closure properties of (CCD). The reader is encouraged to supply proofs that employ only the calculus of adjunctions for the statements of the next Proposition, where we have written $\rightarrowtail\!\!\!\rightarrow$ to indicate a fully faithful arrow in **ord**.

Proposition. *If A is (CCD) and*

$$X \xrightarrow[]{} A$$

*in **ord** then X is (CCD).*
  If A is (CCD) and

$$X \xrightarrow[]{} A$$

*in **ord** then X is (CCD).*

In the first diagram of the Proposition observe that the arrow from X to A preserves all infima and all suprema. Such an arrow, between (antisymmetric) complete lattices, fully faithful or not, is sometimes called a complete homomorphism. When A in the first diagram is of the form $\mathbb{P}S$, for S a set, and X is literally a subset of A, earlier terminology referred to X as a *complete ring of sets*. In the second diagram of the Proposition the arrow from A to X is surjective (for antisymmetric X and A) and in this case X is sometimes called a complete homomorphic image of A.

Exercises.

1. Show that a complete ring of sets $X \rightarrowtail\!\!\!\rightarrow \mathbb{P}S$ necessarily has $X = \mathbb{D}(S, \leq)$ for a unique order on S.
2. For an adjoint sequence $f \dashv u \dashv g$ in **ord**, show that f is fully faithful if and only if g is fully faithful.
3. For an adjoint sequence $f \dashv u \dashv g$ in **ord**, with f, equivalently g, fully faithful, show that $f \leq g$.

6.4. The Raney-Büchi theorem. The following theorem, stated for (CD) lattices, was discovered independently by Raney and Büchi. Their classical result follows immediately from the following constructive version using Theorem 6.2.

Theorem. *An (antisymmetric) complete lattice is (CCD) if and only if it is a complete homomorphic image of a complete ring of sets.*

Proof. If X is (CCD) then by definition we have

$$X \xleftarrow{\perp} \mathbb{D}X \xrightarrow{\perp} \mathbb{P}|X|$$

which provides a presentation of the kind required. (The left adjoint to $X \leftarrow \mathbb{D}X$: $\bigvee$ is necessarily fully faithful by Exercise 2 of 6.3.)

Conversely, if we have

$$X \xleftarrow{\perp} R \xrightarrow{\perp} \mathbb{P}S$$

for a set S then $\mathbb{P}S$ is (CCD) by Corollary 6.1, whereupon R is (CCD) by the first clause of Proposition 6.3 and finally X is (CCD) by the second clause of Proposition 6.3   □

6.5. The dual of (CCD). We say that X is (^{op}CCD) if X^{op} is (CCD). We use (^{op}CD) similarly and recall the classical result:

$$(\text{AC}) \implies ((^{op}\text{CD}) \iff (\text{CD})).$$

We write (BLN) to indicate that **set** is further assumed to satisfy the Boolean axiom.

Proposition. *If (BLN) holds and X is (CCD) then X is $(^{op}\,CCD)$.*

Proof. Apply $(-)^{op}$ (which provides a 2-functor $(-)^{op}$: $\mathbf{ord}^{co} \to \mathbf{ord}$, see exercises 1 and 2 of 5.2) to the diagram

$$X \xleftarrow{\perp} \mathbb{D}X$$

and obtain by Proposition 2.4 and the first Lemma in 5.5

$$X^{op} \xleftarrow{\perp} (\mathbb{D}X)^{op} \; .$$

Now $(\mathbb{D}X)^{op} = \mathbb{U}X^{op}$ and as pointed out immediately prior to the exercises in 5.2 we always have pseudo-complementation $(-)^c : \mathbb{U}X \to \mathbb{D}X$. In the case at hand, (BLN) gives an isomorphism $(-)^c : (\mathbb{D}X)^{op} \xrightarrow{\sim} \mathbb{D}X^{op}$. Now $\mathbb{D}X^{op}$ is in any event (CCD) and thus (BLN) implies that $(\mathbb{D}X)^{op}$ is (CCD), whereupon the second display above and the second clause of Proposition 6.3 show that X^{op} is (CCD).   □

    Since $(-)^{op} : \mathbf{ord}^{co} \to \mathbf{ord}$ is in any event an involutive isomorphism it is clear that if (CCD) $\implies$ (^{op}CCD) then (CCD) $\iff$ (^{op}CCD). Using the very special property of Ω given in Theorem 4.1 we can expand on the last Proposition as follows:

Theorem.

$$(BLN) \iff ((CCD) \iff (^{op}\,CCD)).$$

Proof. It only remains to be shown that (BLN) follows from (CCD) $\implies$ (^{op}CCD). Since $\Omega \cong \mathbb{P}1$ is in any event (CCD) the assumption ensures that Ω^{op} is (CCD) and by Proposition 6.1 is thus Heyting. But we have seen in Theorem 4.1 that when Ω^{op} is Heyting Ω is Boolean. $\qquad\Box$

Just as (AC) ensures the equivalence of (CD) and (CCD) so it also provides the equivalence of (^{op}CD) and (^{op}CCD). Combining these equivalences with the classical result that (AC) implies the equivalence of (CD) and (^{op}CD), the Theorem above provides a very indirect proof of Diaconescu's Theorem:

Corollary.
$$(AC) \implies (BLN).$$

References

[1] A. Carboni, S. Kasangian, and R. Street, Bicategories of spans and relations, *J. Pure Appl. Algebra* 33 (1984) 259–267.

[2] B. Fawcett and R. J. Wood, Constructive complete distributivity I, *Math. Proc. Cam. Phil. Soc.* 107 (1990) 81–89.

[3] P. T. Johnstone, *Stone Spaces*, Cambridge University Press, 1982.

[4] P. T. Johnstone, *Topos Theory*, Academic Press, 1977.

[5] F.W. Lawvere and R. Rosebrugh, *Sets for Mathematics*, Cambridge University Press, 2001.

[6] F. Marmolejo, R. Rosebrugh and R. J. Wood, A Basic Distributive Law, *J. Pure Appl. Algebra*, to appear.

[7] R. Rosebrugh and R. J. Wood, Constructive complete distributivity II, *Math. Proc. Cam. Phil. Soc.* 110 (1991) 245–249.

[8] R. Rosebrugh and R. J. Wood, Constructive complete distributivity IV, *Applied Categorical Structures* 2 (1994) 119–144.

[9] R. Rosebrugh and R. J. Wood, Boundedness and complete distributivity, *Applied Categorical Structures*, to appear.

[10] R. Street and R. F. C. Walters, Yoneda structures on 2-categories, *Journal of Algebra* 50 (1978) 350–379.

Department of Mathematics and Statistics,
Dalhousie University,
Halifax NS B3H 3J5, Canada
E-mail: rjwood@mathstat.dal.ca

II

Locales

Jorge Picado, Aleš Pultr, and Anna Tozzi

This chapter is an introduction to the basic concepts, constructions, and results concerning locales. Locales (frames) are the object of study of the so called point-free topology. They sufficiently resemble the lattices of open sets of topological spaces to allow the treatment of many topological questions. One motivation for the theory of locales is building topology on the intuition of "places of non-trivial extent" rather than on points. Not the only one; hence it is not surprising that the theory has developed beyond the purely geometric scope. Still, we can think of a locale as of a kind of space, more general than the classical one, allowing us to see topological phenomena in a new perspective. Other aspects are, for instance, connections with domain theory [53, 52], continuous lattices [5, 31], logic [65, 20] and topos theory [42, 20].

Modern topology originates, in principle, from Hausdorff's "*Mengenlehre*" [30] in 1914. One year earlier there was a paper by Caratheodory [23] containing the idea of a point as an entity localized by a special system of diminishing sets; this is also of relevance for the modern point-free thinking. In the twenties and thirties the importance of (the lattice of) open sets (which are, typically, "places of non-trivial extent") became gradually more and more apparent (see e.g. Alexandroff [1] or Sierpinski [54]). In [57] and [58], Stone presented his famous duality theorem from which it followed that compact zero-dimensional spaces and continuous maps are well represented by the Boolean algebras of closed open sets and lattice homomorphisms. This was certainly an encouragement for those who endeavoured to treat topology other than as a structure on a given system of points (Wallman [66] in 1938, Menger [45] in 1940, McKinsey and Tarski [44] in 1944). In the Ehresmann seminar of the late fifties [28, 19, 46], we encounter the theory of frames (introduced as "local lattices") already in the form that we know today (it should be noted that almost at the same time there appeared independently two important papers, by Bruns [22] and Thron [59], on homeomorphism of spaces with isomorphic lattices of open sets, under weak separation axioms). Then many authors got interested (C.H. Dowker, D. Papert (Strauss), J. Isbell, B. Banaschewski, etc.) and the field started to develop rapidly. The pioneering paper by Isbell [32], which opened several topics and placed specific emphasis on the dual of the category of frames, introducing the term "locale", merits particular mentioning. In 1982, Johnstone published his monograph [34] which is still a primary source of reference.

The notion of a locale can be viewed as an extension of the notion of a (topological) space. Extending or generalizing a notion calls for justification (as Johnstone says in [35], "*there remains the question: why study locales at all?*"). Several questions naturally arise: When abandoning points, do we not lose too much information? Is the broader range of "spaces" we now have desirable at all? That is, is the theory in this context, in whatever sense, more satisfactory? And is it not so that the new techniques obscure the geometric contents? Here are some answers:

1. Starting with very low separation axioms (sobriety, T_D) the point-free representation contains all the information of the original space.

2. The class of Hausdorff locally compact spaces is represented equivalently by distributive continuous lattices, also a very satisfactory fact.

3. The broader context does yield, in some areas, better theory. For instance, the passage from sober spaces to locales is a full embedding of categories which, in general, does not preserve products. This discrepancy between the two products is, however, beneficial in some respects, as it was firstly observed by Isbell in [32]. If one recalls how badly the notion of paracompactness (which is very important in applications of topology) behaves under constructions in the classical context, one starts to be quite happy about the product being changed sometimes: products of paracompact and metric spaces are not necessarily paracompact but this is not the case in the point-free context (the category of paracompact locales is reflective in the category of locales).

   Subobjects in this broader context also behave differently, again with advantages for locales. This is clear from the fact that the intersection of any family of dense sublocales of a given locale is again dense. In the words of Johnstone [35],

   > "...*the single most important fact which distinguishes locales from spaces: the fact that every locale has a smallest dense sublocale. If you want to 'sell' locale theory to a classical topologist, it's a good idea to begin by asking him to imagine a world in which any intersection of dense subspaces would always be dense; once he has contemplated some of the wonderful consequences that would flow from this result, you can tell him that that world is exactly the category of locales.*
   >
   > (...) *It is certainly clear that in order to achieve such a world, we have to abandon the idea of a space as a set of points equipped with some kind of structure; for there will be examples in any category of this type of pairs of dense subspaces of a nontrivial space having no points in common.*"

4. The techniques are sometimes less intuitive than the classical ones; but it can be argued that they are very often simpler. And, perhaps surprisingly, they often yield constructive results where the classical counterparts cannot. For instance, the Tychonoff product theorem is fully constructive (meaning: no choice principle and no excluded middle, see [33, 6]).

   So, locales have characteristics that go beyond the interest they may deserve as generalized topological spaces. In many situations, certain spaces are non-trivial only by virtue of some choice principle, whereas their lattices of open sets already have previous existence, before such assumptions. This means that, in some sense, we always see the lattice of opens, but to see their points may require some additional tool in the form of some choice principle. This idea was nicely expressed by Banaschewski, with the following slogan [7]:

$$\frac{\begin{array}{c} \textit{choice-free localic argument} \\ + \\ \textit{suitable choice principle} \end{array}}{\textit{classical result on spaces}}$$

5. In many of the localic constructions that precede certain familiar spaces, in more general contexts than classical set theory, the aforementioned spaces are not sufficient and the corresponding lattices of open sets take over their place. A typical example is the Gelfand Duality: classically, it is a duality between the category of commutative C^*-algebras and the category of compact Hausdorff spaces; within the constructive context of a Grothendieck topos, the Gelfand Duality takes the form of a dual equivalence between the category of commutative C^*-algebras and the category of compact completely regular locales [15, 16]. This is the real version of Gelfand's Duality, its classical version being an accidental consequence because of the special assumptions assumed: the category of compact regular locales is, in the presence of the Boolean Ultrafilter Theorem, equivalent to the category of compact Hausdorff spaces.

   Thus, locales play an important role in a constructive approach to topology, allowing us to develop topology in an arbitrary topos and other non-classical contexts (see e.g. [14, 16, 61]).

For more information on the history and development of point-free topology we advise the reader to see the excellent survey in Johnstone's [35] or [36] (see also the introduction to [34]).

We assume that the reader is acquainted with basic categorical notions such as adjunctions, limits, colimits, and factorization systems. We also assume that the reader is familiar with the basics of lattice theory. A general reference for categorical concepts is [20] and for lattice theoretical concepts we refer to Chapter I of this volume.

1. Spaces, frames, and locales

1.1. From topological spaces to frames. Take a topological space X viewed as a set $|X|$ endowed with a system $\mathcal{O}X \subseteq \{M \mid M \subseteq |X|\}$ of *open sets*. The system $(\mathcal{O}X, \subseteq)$ is a complete lattice since the union of any family of open sets is again open; evidently the infinite distributive law

$$U \wedge \bigvee_{i \in I} V_i = \bigvee_{i \in I} (U \wedge V_i)$$

is valid in $\mathcal{O}X$, since finite meet $\wedge$ and supremum $\bigvee$ in $\mathcal{O}X$ are given by the usual set theoretical operations of intersection $\cap$ and union $\bigcup$, respectively (note that arbitrary meet $\bigwedge_{i \in I} V_i$ is given by the interior, $int(\bigcap_{i \in I} V_i)$, of the intersection $\bigcap_{i \in I} V_i$).

The question naturally arises how much information of the space X is contained in this lattice considered as an algebraic object (that is, ignoring the fact that and how the elements $U \in \mathcal{O}X$ consist of the points $x \in |X|$).

On the other hand, a continuous mapping $f : X \to Y$ induces a lattice homomorphism $h : \mathcal{O}Y \to \mathcal{O}X$, defined by $h(U) = f^{-1}[U]$, which clearly preserves arbitrary joins and finite meets; again, one may naturally ask how accurately the homomorphism h represents the original continuous map f.

Suppose the answers to these questions are satisfactory. Then we might treat topology, or a considerable part of it, as a part of algebra (perhaps with some advantages of a handy calculus).

Now the answers are indeed fairly pleasing. For instance, if the spaces in question are sober (a separation axiom considerably weaker than the Hausdorff one, see 1.3 below), they can be reconstructed from the lattices $\mathcal{O}X$, and the continuous map f can be reconstructed from the homomorphism h above. (And it turns out that the ensuing calculus does bring surprising advantages.)

Abstracting the properties of the lattices $\mathcal{O}X$ and the mappings $(U \mapsto f^{-1}[U]) : \mathcal{O}Y \to \mathcal{O}X$, one defines a *frame* A as a complete lattice satisfying the infinite distributive law

$$a \wedge \bigvee S = \bigvee \{a \wedge s \mid s \in S\} \tag{1.1.1}$$

for all $a \in A$ and $S \subseteq A$, and a *frame homomorphism* $h : A \to B$ as a mapping preserving all joins, including the least element 0, and finite meets, including the largest element 1. The resulting category will be denoted by

$$\mathsf{Frm}.$$

Besides the lattices $\mathcal{O}X$ of open subsets of a topological space, obvious examples of frames are the finite distributive lattices, the complete Boolean algebras and the complete chains.

Remarks. (1) Recall from Chapter I that (1.1.1) makes a frame a Heyting algebra. In fact, the frame distributivity law says that for each $a \in A$ the map

$$- \wedge a : \quad A \quad \to \quad A$$
$$c \quad \mapsto \quad c \wedge a$$

preserves suprema. Consequently, by a standard fact on posetal categories (Corollary I.3.3), it has a right adjoint; denoting it by $a \to -$ we see that a frame is a complete Heyting algebra, that is, a complete lattice endowed with an extra operation $\to$ satisfying

$$c \wedge a \leq b \quad \text{if and only if} \quad c \leq a \to b.$$

Using the distributivity rule in the frame we immediately see that $a \to b = \bigvee\{x \in A \mid x \wedge a \leq b\}$. In particular,

$$a^c = \bigvee\{x \in A \mid x \wedge a = 0\}$$

is the *pseudo-complement* $a \to 0$ of a (also called *negation* in Chapter I), that is, one has

$$x \wedge a = 0 \quad \text{if and only if} \quad x \leq a^c. \tag{1.1.2}$$

The following are the standard basic properties of pseudo-complements:

(C1) $a \mapsto a^c$ is antitone, $0^c = 1$ and $1^c = 0$.
(C2) $a \leq a^{cc}$.
(C3) $a^c = a^{ccc}$.
(C4) $a \wedge b = 0$ iff $a^{cc} \wedge b = 0$.
(C5) $(\bigvee a_i)^c = \bigwedge(a_i^c)$.
(C6) $(a \wedge b)^{cc} = a^{cc} \wedge b^{cc}$.

(2) It should be noted right away that not every frame A is isomorphic to an $\mathcal{O}X$ (see 1.8 and 2.14 below for such examples). Thus, viewing frames as representations of spaces, we have "more spaces than before". Such an extension of the scope of the objects considered as spaces may be seen as becoming, and again, it may not. During the development of point-free topology it has turned out that it is unequivocally of advantage.

Exercises.

1. Prove formulas (C1)-(C6) of Remark 1 (cf. Proposition I.3.7).
2. Show that in $\mathcal{O}X$ we have:
   (a) $U \to V = int((X \setminus U) \cup V)$;
   (b) $U^{cc} = int(\overline{U})$, the interior of the closure of U.
   (Compare this with classical logic, where $p \to q \Leftrightarrow \neg p \vee q$ and $\neg\neg p \Leftrightarrow p$.)
3. Show that in a frame, $a \wedge a^c = 0$. Find an example in the locale $\mathcal{O}\mathbb{R}$ of open subsets of the real line, provided with its usual topology, for which $a \vee a^c \neq 1$.
4. Show that the locale $\mathcal{O}\mathbb{R}$ does not satisfy the de Morgan Law $(a \wedge b)^c = a^c \vee b^c$.
5. Show that the lattice $\mathcal{O}X$ of open sets of a topological space X need not satisfy the dual of the infinite distributive law (1.1.1).

6. Give examples to show that the frame homomorphism $f^{-1} : \mathcal{O}Y \to \mathcal{O}X$ need not preserve infinite meets, nor the Heyting implication $\to$, and not even the pseudo-complements U^c.

7. Prove that if $f : X \to Y$ is any open continuous map, then $f^{-1} : \mathcal{O}Y \to \mathcal{O}X$ preserves infinite meets and the Heyting implication.

1.2. Locales. Let Top denote the category of topological spaces and continuous maps. Sending X to $\mathcal{O}X$ and f to f^{-1} yields a contravariant functor Top $\to$ Frm. To obtain a category which is extending (in a way) that of spaces (we will see shortly that we can think at least of generalized sober spaces) we consider the dual category of Frm. It will be denoted by

$$\mathsf{Loc}$$

and called the category of *locales*. The functor Top $\to$ Frm mentioned above becomes now a covariant functor

$$\mathsf{Lc} : \mathsf{Top} \to \mathsf{Loc}.$$

Thus, a locale X (that is, an object of Loc thought of as a space) is the same thing as a frame, but if we wish to emphasize the algebraic (lattice) aspects we often write $\mathcal{O}X$ for the same object (as if it would be the "lattice of open sets" of the "generalized space" X). A morphism of locales (*localic map*) is thought of as represented by a frame homomorphism in the opposite direction. If $f : X \to Y$ is a localic map, the corresponding frame homomorphism will be indicated by $f^* : \mathcal{O}Y \to \mathcal{O}X$, as if it were a left Galois adjoint of f (see I.2.2)—which point of view has a certain substantiation, see 1.5 below. This notation will make it clear whether we wish to think of a given object as sitting in Loc or in Frm. Notice that since f^* preserves arbitrary joins it has a right adjoint. This right adjoint is denoted by f_* and is given by the formula

$$
\begin{aligned}
f_* : \quad \mathcal{O}X \quad &\to \quad \mathcal{O}Y \\
a \quad &\mapsto \quad \bigvee \{b \mid f^*(b) \le a\}.
\end{aligned}
$$

Exercise. Let $f : X \to Y$ be a continuous map between topological spaces and let $f_* : \mathcal{O}X \to \mathcal{O}Y$ be the right adjoint of $f^{-1} : \mathcal{O}Y \to \mathcal{O}X$.

(a) Show that $f_*(U) = Y \setminus \overline{f(X \setminus U)}$, for every $U \in \mathcal{O}X$.

(b) Give an example to show that f_* need not preserve joins.

1.3. Sober spaces. An element $a \neq 1$ of a lattice is *meet-irreducible* if $x \wedge y \le a$ implies $x \le a$ or $y \le a$ (or, equivalently, if $x \wedge y = a$ implies $x = a$ or $y = a$). A topological space X is *sober* if it is T_0 and if there are no other meet-irreducibles $U \in \mathcal{O}X$ but the $X \setminus \overline{\{x\}}$.

Exercises.

1. Let X be a topological space, $x \in X$. Show that:

(a) $X \setminus \overline{\{x\}}$ is a meet-irreducible of $\mathcal{O}X$;

(b) X is sober if and only if each meet-irreducible $U \neq X$ in $\mathcal{O}X$ is $X \setminus \overline{\{x\}}$ for a unique $x \in X$.

2. Prove that each Hausdorff space is sober.
3. Give examples to show that not every T_1-space is sober, nor is every sober space T_1.
4. Prove that:
   (a) an open subset of a sober space is a sober space;
   (b) a closed subset of a sober space is a sober space.
5. Construct a sober space with a non-sober subspace. (Hint: consider $\overline{\mathbb{N}} = \mathbb{N} \cup \{\infty\}$ with $[n, \infty]$ as open subsets.)

Lemma. *Let X, Y be topological spaces, Y sober. Then, for each frame homomorphism $h : \mathcal{O}Y \to \mathcal{O}X$, there is exactly one continuous map $f : X \to Y$ such that $h = f^{-1}$.*

Proof. For $x \in X$ consider $\mathcal{F}_x = \{U \in \mathcal{O}Y \mid x \notin h(U)\}$. Since h preserves unions, $F_x = \bigcup \mathcal{F}_x \in \mathcal{F}_x$, hence it is the largest element in $\mathcal{F}_x$ and we have

$$U \in \mathcal{F}_x \text{ if and only if } U \subseteq F_x. \tag{1.3.1}$$

Each F_x is meet-irreducible: if $F_x = U \cap V$, we have $h(F_x) = h(U) \cap h(V)$; but then, say $x \notin h(U)$ and $U \subseteq F_x \subseteq U$. Thus, $F_x = Y \setminus \overline{\{y\}}$ with uniquely defined y (since Y is T_0). Denoting this y by $f(x)$ we obtain, from (1.3.1),

$$
\begin{aligned}
x \in h(U) &\Leftrightarrow U \nsubseteq Y \setminus \overline{\{f(x)\}} \\
&\Leftrightarrow f(x) \in U \\
&\Leftrightarrow x \in f^{-1}[U].
\end{aligned}
$$

Thus, the mapping $f : X \to Y$ is continuous ($f^{-1}[U] = h(U) \in \mathcal{O}X$) and we have $h = f^{-1}$. $\qquad \square$

Denote by Sob the full subcategory of Top defined by sober spaces. Then:

Corollary. *The restriction* Lc : Sob $\to$ Loc *is a full embedding.*

1.4. The points of a locale. Using Lemma 1.3 we can easily reconstruct a sober space X from the lattice $A = \mathcal{O}X$. The one-point space is the terminal object T in Top, and a point of X can be viewed as a continuous map $T \to X$. Since the two-element Boolean algebra $\mathbf{2} = \{0, 1\}, 0 \neq 1$, is (isomorphic to) $\mathcal{O}T$, we have by Lemma 1.3 the points of X in a natural one-one correspondence with the frame homomorphisms $A \to \mathbf{2}$; the open sets U from the original X are then seen as the $\widetilde{U} \subseteq \{h \mid h : A \to \mathbf{2} \text{ in Frm}\}$ defined by $h \in \widetilde{U}$ if and only if $h(U) = 1$ (such a h is f_x^{-1} for f_x sending the sole point of T to x, and $x \in U$ if and only if $f_x^{-1}[U]$ is non-void).

This construction may be generalized as follows. For any locale X define a *point* of X to be a localic map $T \to X$ from the terminal object of Loc to X (cf. Definition 1.3 of [41]). Equivalently, in frame terms, set

$$\text{Pt}(X) = \left(\{h \mid h : \mathcal{O}X \to \mathbf{2} \text{ in Frm}\}, \{\Sigma_a \mid a \in \mathcal{O}X\} \right)$$

where $\Sigma_a = \{h \mid h(a) = 1\}$. The following is straightforward:

Remark. $\Sigma_0 = \emptyset$, $\Sigma_1 = \mathsf{Pt}(X)$, $\Sigma_{a \wedge b} = \Sigma_a \cap \Sigma_b$ and $\Sigma_{\bigvee a_i} = \bigcup \Sigma_{a_i}$. Thus, in particular, $\{\Sigma_a \mid a \in \mathcal{O}X\}$ is a topology on $\{h \mid h : \mathcal{O}X \to \mathbf{2}\}$.

The space $\mathsf{Pt}(X)$ is often referred to as the *spectrum* of the locale X.

For a localic map $f : X \to Y$ define $\mathsf{Pt}(f) : \mathsf{Pt}(X) \to \mathsf{Pt}(Y)$ by setting $\mathsf{Pt}(f)(h) = h \cdot f^*$.

Lemma. *We have* $(\mathsf{Pt}(f))^{-1}[\Sigma_a] = \Sigma_{f^*(a)}$. *Thus* $\mathsf{Pt}(f)$ *is a continuous mapping.*

Proof. $(\mathsf{Pt}(f))^{-1}[\Sigma_a] = \{h \mid h \cdot f^* \in \Sigma_a\} = \{h \mid h(f^*(a)) = 1\}$. $\quad\square$

Obviously, $\mathsf{Pt}(1_X) = 1_{\mathsf{Pt}(X)}$ and $\mathsf{Pt}(f \cdot g) = \mathsf{Pt}(f) \cdot \mathsf{Pt}(g)$. Therefore we have a functor

$$\mathsf{Pt} : \mathsf{Loc} \to \mathsf{Top}.$$

Exercise. Let B be a complete Boolean algebra. Show that frame homomorphisms $B \to \mathbf{2}$ correspond bijectively to atoms in B.

1.5. Two alternative representations of the spectrum Pt.

(1) Recall that a *completely prime filter* (briefly, *complete filter*) in a lattice A is a proper filter $F \subseteq A$ such that for any system $\{a_i \mid i \in I\}$,

$$\bigvee_{i \in I} a_i \in F \Rightarrow \exists i, a_i \in F.$$

Complete filters are in a obvious one-one correspondence with frame homomorphisms $h : A \to \mathbf{2}$ ($h(a) = 1$ if and only if $a \in F$) and thus we can represent $\mathsf{Pt}(X)$ as

$$\Big(\{F \mid F \text{ complete filter in } \mathcal{O}X\}, \{\Sigma_a \mid a \in \mathcal{O}X\}\Big),$$

where $F \in \Sigma_a$ if and only if $a \in F$. The mapping $\mathsf{Pt}(f)$ sends F to $(f^*)^{-1}(F)$.

(2) A slightly less obvious translation of the Pt construction is based on the following result:

Proposition. *The formulas*

$$h \mapsto a = \bigvee \{x \mid h(x) = 0\}$$

$$a \mapsto h, \text{ where } h(x) = 1 \text{ iff } x \not\leq a$$

constitute a one-one correspondence between the set of all frame homomorphisms $h : A \to \mathbf{2}$ *and the set* $\mathrm{irr}(A)$ *of all meet-irreducible elements in* A.

Proof. Checking that $\bigvee\{x \mid h(x) = 0\}$ is meet-irreducible and that the above defined h is a homomorphism is immediate. If h is sent to a and a is sent to k as above we have $k(y) = 1$ iff $y \not\leq \bigvee\{x \mid h(x) = 0\}$ iff $h(y) = 1$. Finally if $a \mapsto h \mapsto b$ we have $b = \bigvee\{x \mid x \leq a\} = a$. $\quad\square$

Now we can represent $\mathsf{Pt}(X)$ as

$$\Big(\mathrm{irr}(\mathcal{O}X), \{\Sigma_a \mid a \in \mathcal{O}X\}\Big),$$

with $\Sigma_a = \{x \in \mathrm{irr}(\mathcal{O}X) \mid a \not\leq x\}$.

Lemma. *Let* $h : A \to B$ *be a frame homomorphism. Then* $h_*[\text{irr}(B)] \subseteq \text{irr}(A)$.

Proof. If $a \in \text{irr}(B)$ and $x \wedge y \leq h_*(a)$ then $h(x) \wedge h(y) \leq a$ and hence, say, $h(x) \leq a$ and $x \leq h_*(a)$. $\square$

In the original description of Pt we had $\text{Pt}(f)(h) = h \cdot f^*$. Represent $a \in \text{irr}(\mathcal{O}X)$ by the h_* above. Then the resulting $\text{Pt}(f)(h)$ corresponds to

$$\bigvee\{x \mid h(f^*(x)) = 0\} = \bigvee\{x \mid f^*(x) \leq a\} = (f^*)_*(a).$$

Thus, if we think, just for the moment, of the localic morphisms as of the right adjoints of f^*, we have now $\text{Pt}(f)$ represented simply as the restriction of f. Note that this is one of the reasons for denoting the algebraic (frame) correspondent of the localic map f by f^*.

1.6. Pt is right adjoint to Lc.

Proposition. *Each* $\text{Pt}(X)$ *is sober.*

Proof. We will use the representation from 1.5(2). First, note that we have, since $\Sigma_a = \{x \mid a \not\leq x\}$,

$$x \in \overline{\{y\}} \Leftrightarrow (a \not\leq x \Rightarrow a \not\leq y) \Leftrightarrow y \leq x. \tag{1.6.1}$$

Now let Σ_a be meet-irreducible in $\mathcal{O}(\text{Pt}(X))$. Set $b = \bigvee\{c \mid \Sigma_c \subseteq \Sigma_a\}$. Then, obviously, $\Sigma_b = \Sigma_a$, and b is meet-irreducible; indeed, if $x \wedge y \leq b$ then $\Sigma_x \cap \Sigma_y = \Sigma_{x \wedge y} \subseteq \Sigma_b = \Sigma_a$ and hence, say, $\Sigma_x \subseteq \Sigma_a$ and $x \leq b$. We have

$$x \in \Sigma_b \Leftrightarrow b \not\leq x \Leftrightarrow x \notin \overline{\{b\}} \Leftrightarrow x \in \text{Pt}(X) \setminus \overline{\{b\}},$$

thus $\Sigma_a = \Sigma_b = \text{Pt}(X) \setminus \overline{\{b\}}$. By (1.6.1), $\text{Pt}(X)$ is T_0. $\square$

Define morphisms $\eta_X : X \to \text{PtLc}(X)$ and $\varepsilon_Y : \text{LcPt}(Y) \to Y$ by setting $\eta_X(x)(U) = 1$ if and only if $x \in U$ and $\varepsilon_Y^*(a) = \Sigma_a$, respectively. It is easy to check that each $\eta_X(x)$ is indeed a homomorphism $\text{Lc}(X) \to \mathbf{2}$. Since

$$\eta_X^{-1}[\Sigma_U] = \{x \mid \eta_X(x)(U) = 1\} = U, \tag{1.6.2}$$

η_X is a continuous mapping. On the other hand, ε_Y is a homomorphism by 1.5.

Lemma. *The systems* $\eta = (\eta_X)_{X \in \text{Top}}$ *and* $\varepsilon = (\varepsilon_Y)_{Y \in \text{Loc}}$ *are natural transformations* $\eta : 1_{\text{Top}} \overset{\cdot}{\to} \text{PtLc}$ *and* $\varepsilon : \text{LcPt} \overset{\cdot}{\to} 1_{\text{Loc}}$, *respectively.*

Proof. If $f : X \to Y$ is a continuous mapping we have

$$(\text{PtLc}(f) \cdot \eta_X)(x)(U) = (\eta_X(x) \cdot \text{Lc}(f))(U) = \eta_X(x)(\text{Lc}(f)(U)).$$

Therefore

$$\begin{aligned}
1 = (\text{PtLc}(f) \cdot \eta_X)(x)(U) &\Leftrightarrow x \in \text{Lc}(f)(U) \\
&\Leftrightarrow f(x) \in U \\
&\Leftrightarrow (\eta_Y \cdot f)(x)(U) = 1.
\end{aligned}$$

If $f : X \to Y$ is a localic map we have, by 1.5,

$$
\begin{aligned}
(\varepsilon_X \cdot \mathsf{LcPt}(f))^*(a) &= (\mathsf{LcPt}(f))^*(\varepsilon_X^*(a)) = (\mathsf{Pt}(f))^{-1}[\Sigma_a] \\
&= \Sigma_{f^*(a)} = \varepsilon_Y^*(f^*(a)) = (f \cdot \varepsilon_Y)^*(a). \qquad \square
\end{aligned}
$$

Theorem. $\mathsf{Pt} : \mathsf{Loc} \to \mathsf{Top}$ *is right adjoint to* $\mathsf{Lc} : \mathsf{Top} \to \mathsf{Loc}$, *with unit* η *and co-unit* ε *as above.*

Proof. Consider the composition

$$
\mathsf{Lc}(X) \xrightarrow{\mathsf{Lc}(\eta_X)} \mathsf{LcPtLc}(X) \xrightarrow{\varepsilon_{\mathsf{Lc}(X)}} \mathsf{Lc}(X).
$$

We have

$$
(\varepsilon_{\mathsf{Lc}(X)} \cdot \mathsf{Lc}(\eta_X))^*(U) = \eta_X^{-1}[\varepsilon_{\mathsf{Lc}(X)}^{-1}(U)] = \eta_X^{-1}[\Sigma_U] = \{x \mid \eta_X(x)(U) = 1\} = U
$$

so that

$$
\varepsilon_{\mathsf{Lc}(X)} \cdot \mathsf{Lc}(\eta_X) = 1_{\mathsf{Lc}(X)}. \tag{1.6.3}
$$

Consider the composition

$$
\mathsf{Pt}(Y) \xrightarrow{\eta_{\mathsf{Pt}(Y)}} \mathsf{PtLcPt}(Y) \xrightarrow{\mathsf{Pt}(\varepsilon_Y)} \mathsf{Pt}(Y).
$$

We have

$$
((\mathsf{Pt}(\varepsilon_Y) \cdot \eta_{\mathsf{Pt}(Y)})(h))(U) = (\eta_{\mathsf{Pt}(Y)}(h) \cdot \varepsilon_Y^*)(U) = \eta_{\mathsf{Pt}(Y)}(h)[\Sigma_U],
$$

thus

$$
1 = ((\mathsf{Pt}(\varepsilon_Y) \cdot \eta_{\mathsf{Pt}(Y)})(h))(U) \Leftrightarrow h \in \Sigma_U \Leftrightarrow h(U) = 1,
$$

and again $\mathsf{Pt}(\varepsilon_Y) \cdot \eta_{\mathsf{Pt}(Y)} = 1_{\mathsf{Pt}(Y)}$. $\qquad \square$

1.7. A reflection of Top onto Sob. The natural transformation η yields a reflection of the category Top onto the subcategory Sob:

Theorem. *The following statements on a space X are equivalent:*

(i) X *is sober;*
(ii) η_X *is one-one and onto;*
(iii) η_X *is a homeomorphism.*

Proof. (i)$\Rightarrow$(ii) is an immediate consequence of Lemma 1.3, (ii)$\Rightarrow$(iii) follows from (1.6.2) (which yields, for an invertible η_X, $\eta_X[U] = \Sigma_U$), and (iii)$\Rightarrow$(i) follows from Proposition 1.6. $\qquad \square$

1.8. Spatial locales. A locale X is said to be *spatial* if it is (isomorphic to) $\mathsf{Lc}(Y)$ for some space Y. Here is an easy criterion of spatiality:

Theorem. *The following statements on a locale X are equivalent:*

(i) X *is spatial;*
(ii) ε_X^* *is one-one;*
(iii) ε_X^* *is an isomorphism;*
(iv) *each $a \in \mathcal{O}X$ is a meet of meet-irreducible elements.*

Proof. (i)⇒(ii) since, by (1.6.3),

$$(\mathsf{Lc}(\eta_Y))^* \cdot \varepsilon^*_{\mathsf{Lc}(Y)} = (\varepsilon_{\mathsf{Lc}(Y)} \cdot \mathsf{Lc}(\eta_Y))^* = 1_{\mathsf{Lc}(Y)}$$

for $X = \mathsf{Lc}(Y)$ and therefore $\varepsilon^*_X = \varepsilon^*_{\mathsf{Lc}(Y)}$ is one-one. (ii)⇒(iii) since each ε^*_X is onto, and (iii)⇒(i) is trivial. Assertion (iv) is just a reformulation of (ii) in the representation from 1.5(2). □

All finite distributive lattices and all complete chains are spatial but, by way of contrast, not every Boolean algebra is spatial, showing that locales "considerably transcend topology". In fact the intersection of the class of spatial locales and that of Boolean algebras are only the discrete spaces. We have:

Proposition. *Each meet-irreducible element of a Boolean algebra is a co-atom. Consequently, each spatial Boolean algebra is atomic.*

Proof. Let a be meet-irreducible and let $a < x$. Since $a = (a \vee x) \wedge (a \vee x^c)$, where x^c is the complement of x, we have to have $a = a \vee x^c$ and hence $x^c \leq a < x$. Then $x = x^{cc} \geq a^c$ and hence $1 = a \vee a^c \leq x$ and $x = 1$. □

1.9. The "maximal" equivalence induced by the adjunction Lc ⊣ Pt. Every adjunction induces a "maximal" equivalence between a pair of full subcategories. Here the adjunction Lc ⊣ Pt gives:

Theorem. *The category of spatial locales is equivalent to the category of sober topological spaces.*

Proof. If X is a topological space, the locale $\mathsf{Lc}(X)$ is spatial. Therefore, by Proposition 1.6, the adjunction of Theorem 1.6 between Top and Loc restricts to the full subcategories of sober spaces and spatial locales. By definition of spatial locale, ε_X is an isomorphism. By Theorem 1.7, if a space X is sober, η_X is an isomorphism. Therefore we get the required equivalence. □

The results in this section show that the category of locales is an appropriate environment in which to develop topology (for more motivation consult [11, 34, 35, 36]). From now on we develop locale theory in this perspective. The topological intuition will be apparent from the use of topological adjectives to describe localic concepts.

2. Sublocales

2.1. Epimorphisms. Consider the frame $\mathbb{S} = \{0 < s < 1\}$. Obviously, for every frame A and every $a \in A$ the mapping $\sigma_a : \mathbb{S} \to A$, sending 0 to 0, 1 to 1 and s to a, is a frame homomorphism. Consequently, if a frame homomorphism $h : A \to B$ is a monomorphism, it must be one-one (if $h(a) = h(b)$ then $h \cdot \sigma_a = h \cdot \sigma_b$). Thus:

Proposition. *The epimorphisms in Loc are exactly the f such that f^* is one-one.*

Usually epimorphisms in Loc are called *surjections*.

2.2. Extremal monomorphisms. Recall that a monomorphism μ in a category is said to be *extremal* if, for each factorization $\mu = \nu \cdot \varepsilon$ with ε an epimorphism, ε is an isomorphism.

Proposition. *The extremal monomorphisms in* Loc *are precisely the f such that f^* is onto.*

Proof. In other words, we should prove that the extremal epimorphisms in Frm are exactly the homomorphisms $h : A \to B$ that are onto. If $h : A \to B$ is an extremal epimorphism and we factor it through its image $h[A]$, the embedding $h[A] \hookrightarrow B$ must be an isomorphism, and h is onto. On the other hand, every onto homomorphism $h : A \to B$ is obviously an epimorphism, and if we have a factorization $h = m \cdot h'$ with a monomorphism m, then m is one-one and onto and hence an isomorphism in Frm.   □

The structure of general monomorphisms in Loc is by far not so transparent (see [34, 43]). For instance, there is a locale X such that for each cardinal number α there is a monomorphism $f : Y \to X$ with $|Y| \geq \alpha$ (see 3.9 below).

2.3. Decomposition of morphisms. Every localic map $f : X \to Y$ decomposes as

with $\mathcal{O}Z = f^*[\mathcal{O}Y]$, $e^* : \mathcal{O}Z \hookrightarrow \mathcal{O}X$ and $m^* = (y \mapsto f^*(y)) : \mathcal{O}Y \twoheadrightarrow \mathcal{O}Z$. Thus, every localic map can be factored as an epimorphism followed by an extremal monomorphism.

We point out that, furthermore, it can be proved that the classes $\mathcal{E}$ of epimorphisms and $\mathcal{M}$ of extremal monomorphisms constitute a factorization system (see III.1.2) in Loc.

2.4. Sublocales. In many everyday life categories (like that of topological spaces, graphs, posets, or general relational systems), extremal monomorphisms represent well the subobjects (as opposed to plain monomorphisms $m : A \to B$ that may not — like for instance the one-one continuous maps — relate the structure of A closely enough to that of B). This point of view is also adopted in Loc and a *sublocale* $j : Y \rightarrowtail X$ of X is defined as a localic map such that the corresponding frame homomorphism $j^* : \mathcal{O}X \to \mathcal{O}Y$ is onto (recall 2.2). It should be noted that in many other categories (like that of Hausdorff spaces, rings, small categories, etc.) choosing extremal monomorphisms for subobjects would be too restrictive. But our situation is closer to that of general spaces, and the definition is also supported by the notion of a subspace Y of a topological space X, where the topology $\mathcal{O}Y$ is defined as $\{U \cap Y \mid U \in \mathcal{O}X\}$, making $j^{-1} : \mathcal{O}X \to \mathcal{O}Y$, for $j : Y \hookrightarrow X$, an onto homomorphism.

On the class of sublocales of X we have the natural preorder $j_1 \sqsubseteq j_2$ if and only if there is a j such that $j_2 \cdot j = j_1$ (note that this j is necessarily again a sublocale). Sublocales $j_i : Y_i \rightarrowtail X$ $(i = 1, 2)$ are *equivalent* if $j_1 \sqsubseteq j_2$ and $j_2 \sqsubseteq j_1$ or, equivalently, if there is an isomorphism $j : Y_1 \to Y_2$ such that $j_2 \cdot j = j_1$. The ensuing partially ordered set will be denoted by $\mathcal{S}(X)$.

We will use the symbol $\sqsubseteq$ also for the corresponding frame homomorphisms. Thus, for frame homomorphisms $h_i : A \to B_i$ $(i = 1, 2)$, $h_1 \sqsubseteq h_2$ if there is an h such that $h \cdot h_1 = h_2$, and h_1 and h_2 are considered equivalent if $h_1 \sqsubseteq h_2$ and $h_2 \sqsubseteq h_1$. So, for $j_1, j_2 \in \mathcal{S}(X)$,

$$j_1 \sqsubseteq j_2 \quad \text{iff} \quad j_2^* \sqsubseteq j_1^*. \tag{2.4.1}$$

2.5. Frame congruences. We have the natural correspondence between surjective frame homomorphisms $h : A \twoheadrightarrow B$ and congruences (with respect to finite meets and general joins) on A. More precisely, the formulas

$$h \mapsto C_h = \{(a, b) \mid h(a) = h(b)\}, \quad C \mapsto h_C = \{a \mapsto aC\} : A \to A/C \tag{2.5.1}$$

constitute a one-one correspondence between the set of (the equivalence classes of) the onto homomorphisms $h : A \twoheadrightarrow B$ and the set $\mathcal{C}(A)$ of all congruences on A.

The following is an immediate observation:

In the correspondence (2.5.1) above, we have $h_1 \sqsubseteq h_2$ iff $C_{h_1} \subseteq C_{h_2}$. (2.5.2)

Since any intersection of congruences is a congruence, $\mathcal{C}(A)$ is a complete lattice. Therefore, by (2.4.1) and (2.5.2), $\mathcal{S}(X)$ is also a complete lattice, isomorphic to $\mathcal{C}(\mathcal{O}X)^{op}$. The meets and joins in $\mathcal{S}(X)$ will be denoted by

$$j \sqcap k, \quad \bigsqcap_{i \in I} j_i, \text{ etc., resp. } j \sqcup k, \quad \bigsqcup_{i \in I} j_i, \text{ etc.}$$

These symbols will be also used when dealing with the associated frame homomorphisms. Note that $j \sqcap k$ is represented by the pullback

in Loc.

The initial object in Loc will be denoted by 0; this is the locale that corresponds to the frame $\{0 = 1\}$ (that we denote by $\mathbf{1}$ in analogy with the definition of $\mathbf{2}$). The unique localic map $0_X : 0 \rightarrowtail X$ is the bottom of $\mathcal{S}(X)$. The top element of $\mathcal{S}(X)$ is the (equivalence class of the) identity morphism $1_X : X \to X$ and will be denoted by 1_X or simply by X.

2.6. Open and closed sublocales. We consider now some simple examples of sublocales which resemble open and closed subspaces of a topological space. We have the sublocales given by the frame homomorphisms

$$\hat{a} : \quad \mathcal{O}X \quad \longrightarrow \quad \downarrow a := \{x \in \mathcal{O}X \mid x \le a\}$$
$$x \quad \longmapsto \quad x \wedge a, \tag{2.6.1}$$

for every $a \in \mathcal{O}X$, and the sublocales given by the frame homomorphisms

$$\begin{aligned} \breve{a}: \quad \mathcal{O}X \quad &\longrightarrow \quad {\uparrow}a := \{x \in \mathcal{O}X \mid x \geq a\} \\ x \quad &\longmapsto \quad x \vee a, \end{aligned} \tag{2.6.2}$$

for every $a \in \mathcal{O}X$. The former will be referred to as *open* sublocales and the latter as *closed* ones. The $\hat{a}$ (resp. $\breve{a}$) will also be referred to as open (resp. closed), and similarly one speaks of the corresponding congruences

$$\Delta_a := \{(x,y) \mid x \wedge a = y \wedge b\} \text{ and } \nabla_a := \{(x,y) \mid x \vee a = y \vee b\}.$$

We write X_a for the locale given by the frame ${\downarrow}a$ and $X\text{-}X_a$ for the locale given by ${\uparrow}a$. Then (2.6.1) describes a sublocale

$$X_a \rightarrowtail X$$

and (2.6.2) describes a sublocale

$$X\text{-}X_a \rightarrowtail X.$$

Spatially, when we write $X\text{-}X_a \rightarrowtail X$ we are thinking of the closed subspace corresponding to the set theoretic complement of the open a.

Exercises.
1. Prove that a localic map $f : X \to Y$ factors through the open sublocale $Y_a \rightarrowtail Y$ generated by $a \in \mathcal{O}Y$ if and only if $f^*(a) = 1$.
2. Prove that a localic map $f : X \to Y$ factors through the closed sublocale $Y\text{-}Y_a \rightarrowtail Y$ if and only if $f^*(a) = 0$.

We list some properties of open and closed congruences:

Proposition.
  (1) $a \leq b \Leftrightarrow \Delta_b \subseteq \Delta_a \Leftrightarrow \nabla_a \subseteq \nabla_b$.
  (2) $\nabla_0 = 0 = \Delta_1$ *and* $\nabla_1 = 1 = \Delta_0$.
  (3) $\nabla_a \cap \nabla_b = \nabla_{a \wedge b}$.
  (4) $\bigvee \nabla_{a_i} = \nabla_{\bigvee a_i}$.
  (5) $\Delta_a \vee \Delta_b = \Delta_{a \wedge b}$.
  (6) $\bigcap \Delta_{a_i} = \Delta_{\bigvee a_i}$.

Proof. (1) If $\nabla_a \subseteq \nabla_b$ we have, in particular, $(a,0) \in \nabla_b$ and hence $a \vee b = b$. Similarly, if $\Delta_b \subseteq \Delta_a$ then $(b,1) \in \Delta_a$ and $a \wedge b = a$. Obviously, if $a \leq b$ then $x \vee a = y \vee a$ implies $x \vee b = y \vee b$, and similarly with the meet.
(2), (3) and (6) are immediate; we show (4) and (5):
   By (1), $\Delta_{a \wedge b} \supseteq \Delta_a, \Delta_b$ and $\nabla_{\bigvee a_i} \supseteq \nabla_{a_i}$ for all i. If C is a congruence such that $C \supseteq \Delta_a, \Delta_b$ and $x \wedge a \wedge b = y \wedge a \wedge b$, then $(x \wedge a, y \wedge a) \in C$, since $C \supseteq \Delta_b$, and $xCx \wedge aCy \wedge aCy$, since $C \supseteq \Delta_a$. Thus, $C \supseteq \Delta_{a \wedge b}$. If $C \supseteq \nabla_{a_i}$ for each i, we have $(0, a_i) \in C$ for each i, hence $(0, \bigvee a_i) \in C$. Thus, if $x \vee \bigvee a_i = y \vee \bigvee a_i$, we have $(x,y) = (x \vee 0, y \vee \bigvee a_i) = (x \vee 0, x \vee \bigvee a_i) \in C$. □

As a consequence, we have:
  (1) $a \leq b \Leftrightarrow X_a \sqsubseteq X_b \Leftrightarrow X\text{-}X_b \sqsubseteq X\text{-}X_a$;
  (2) $X\text{-}X_0 = X_1 = 1_X$ and $X\text{-}X_1 = X_0 = 0_X$;

(3) $X\text{-}X_a \sqcup X\text{-}X_b = X\text{-}X_{a\wedge b}$;

(4) $\bigsqcap X\text{-}X_{a_i} = X\text{-}X_{\bigvee a_i}$;

(5) $X_a \sqcap X_b = X_{a\wedge b}$;

(6) $\bigsqcup X_{a_i} = X_{\bigvee a_i}$.

2.7. Open and closed sublocales are complemented in $\mathcal{S}(X)$. Open and closed sublocales, corresponding to the same element $a \in \mathcal{O}X$, are complements in $\mathcal{S}(X)$:

Proposition. $X_a \sqcup X\text{-}X_a = 1_X$ and $X_a \sqcap X\text{-}X_a = 0_X$.

Proof. The proof will be done in terms of congruences.

If $a \wedge x = a \wedge y$ and $a \vee x = a \vee y$, then $x = x \wedge (a \vee x) = x \wedge (a \vee y) = (x \wedge a) \vee (x \wedge y) = (y \wedge a) \vee (y \wedge x) = y \wedge (a \vee x) = y \wedge (a \vee y) = y$. Thus, $\Delta_a \cap \nabla_a = \{(x, x) \mid x \in \mathcal{O}X\}$, the bottom of $\mathcal{C}(\mathcal{O}X)$. If $C \supseteq \nabla_a, \Delta_a$ we have $(a, 0) \in C$ and $(a, 1) \in C$, thus $(0, 1) \in C$ and hence $x = x \wedge 1 C x \wedge 0 = y \wedge 0 C y \wedge 1 = y$ for every (x, y). Thus, $C = \mathcal{O}X \times \mathcal{O}X$. $\square$

Remark. Unlike subspaces of spaces, however, not every sublocale is complemented in $\mathcal{S}(X)$ (not even the sublocales of spaces that are subspaces).

2.8. A representation of a general sublocale.

Proposition. *Let $j : Y \rightarrowtail X$ be a sublocale of X. Then*

$$j = \bigsqcap \{X_a \sqcup X\text{-}X_b \mid j^*(a) = j^*(b)\}.$$

In the language of frame congruences: for any congruence C on $\mathcal{O}X$ we have

$$C = \bigvee \{\Delta_a \cap \nabla_b \mid (a, b) \in C\}.$$

Proof. The proof will be done for the congruences.

Let $(a, b) \in C$. Then $\Delta_a \cap \nabla_b \subseteq C$; indeed, if $(x, y) \in \Delta_a \cap \nabla_b$, that is, if $x \wedge a = y \wedge a$ and $x \vee b = y \vee b$, we have $x = x \wedge (y \vee b) C x \wedge (y \vee a) = (x \wedge y) \vee (x \wedge a) = (x \wedge y) \vee (y \wedge a) = y \wedge (x \vee a) C y \wedge (x \vee b) = y$; thus, $(x, y) \in C$.

On the other hand, let $D \supseteq \Delta_a \cap \nabla_b$ for all $(a, b) \in C$. In particular, we have $(a, a \vee b) \in \Delta_a \cap \nabla_b$ and $(b, a \vee b) \in \Delta_a \cap \nabla_b$, hence $a D a \vee b D b$ and $(a, b) \in D$. $\square$

2.9. Closure. For a localic map $f : Y \to X$ set

$$\mathsf{c}_f = \bigvee \{a \in \mathcal{O}X \mid f^*(a) = 0\}.$$

Proposition. *Let $j : Y \rightarrowtail X$ be a sublocale. Then the sublocale $X\text{-}X_{\mathsf{c}_j} \rightarrowtail X$ is the smallest closed sublocale k such that $j \sqsubseteq k$.*

Proof. Define $h : {\uparrow}\mathsf{c}_j \to \mathcal{O}Y$ by setting $h(x) = j^*(x)$. Since $j^*(\mathsf{c}_j) = 0$, h is a frame homomorphism (preserving non-void joins and finite meets being trivial), and we have $(h \cdot \check{\mathsf{c}}_j)(x) = j^*(\mathsf{c}_j \vee x) = j^*(x)$. Thus, $\check{\mathsf{c}}_j \sqsubseteq j^*$, that is, $j \sqsubseteq X\text{-}X_{\mathsf{c}_j}$. If $j \sqsubseteq X\text{-}X_a$ there is a $\varphi : {\uparrow}a \to \mathcal{O}Y$ such that $j^*(x) = \varphi(x \vee a)$ and hence, in particular, $j^*(a) = \varphi(a) = 0$ and $a \leq c_j$ so that $X\text{-}X_{c_j} \sqsubseteq X\text{-}X_a$ by property (1) of 2.6. □

The sublocale map $X\text{-}X_{c_j} \rightarrowtail X$ will be denoted by $\bar{j}$ and called the *closure* of the sublocale j.

2.10. Closure behaves like in spaces ... We list some properties of the closure:

Proposition.

    (1) $\overline{0_X} = 0_X$.
    (2) $j \sqsubseteq k \Rightarrow \bar{j} \sqsubseteq \bar{k}$.
    (3) $\bar{\bar{j}} = \bar{j}$.
    (4) $\overline{j \sqcup k} = \bar{j} \sqcup \bar{k}$.

Proof. (1) $0_X = (1_X)^{\text{`}}$. (2) and (3) follow immediately from Proposition 2.9. By (2), $\overline{j \sqcup k} \sqsupseteq \bar{j} \sqcup \bar{k}$. By Proposition 2.6, the join of two closed sublocales is closed and hence, by Proposition 2.9, $\overline{j \sqcup k} \sqsubseteq \bar{j} \sqcup \bar{k}$ and (4) follows. □

2.11. ... but not in all respects. In spaces, the topology is determined by the closures of subsets. Here we have:

Proposition. *Let $\mathcal{O}X$ be a Boolean algebra. Then each sublocale of X is closed.*

Proof. Let $j : Y \rightarrowtail X$ be a sublocale, c_j as in 2.9. We can define $h : X\text{-}X_{c_j} \rightarrow Y$, by setting $h^*(j^*(a)) = c_j \vee a$, since if $j^*(a) = j^*(b)$ then $c_j \vee a = c_j \vee b$ (indeed, $j^*(a) = j^*(b)$ implies $j^*(a \wedge b^c) = 0$, hence $a \wedge b^c \leq c_j$ making $c_j \vee a \leq c_j \vee b$, and $c_j \vee b \leq c_j \vee a$ by symmetry). This shows that $\bar{j} = j$. □

Now T_0-spaces with Boolean topology are necessarily discrete. We have, however, rather non-trivial locales X such that $\mathcal{O}X$ is Boolean (see 2.13 below).

Exercise. Prove that, for each locale X:

    (a) $\mathcal{O}X$ is a Boolean algebra if and only if every sublocale of X is closed;
    (b) $\mathcal{O}X$ is a Boolean algebra if and only if every sublocale of X is open.

2.12. Density. A sublocale $j : Y \rightarrowtail X$ is said to be *dense* if $\bar{j} = X$. The corresponding j^* will also be referred to as dense.

Proposition. *A sublocale $j : Y \rightarrowtail X$ is dense if and only if $j^*(a) = 0$ implies $a = 0$.*

Proof. $\bar{j} = X$ iff $0 = \bigvee\{a \in \mathcal{O}X \mid j^*(a) = 0\}$ iff $a = 0$ whenever $j^*(a) = 0$. □

More generally, a localic map $f : Y \rightarrow X$ is said to be *dense* if $f^*(a) = 0$ implies $a = 0$.

For any $f : Y \to X$, $f^*(c_f) = 0$ and c_f is the largest such element. Since $\uparrow c_f \to \mathcal{O}Y$, given by $x \mapsto f^*(x)$, is clearly a frame homomorphism we have the factorization

$$
\begin{array}{ccc}
& X\text{-}X_{c_f} & \\
{\scriptstyle g}\nearrow & & \searrow{\scriptstyle h} \\
Y & \xrightarrow{\quad f \quad} & X
\end{array}
$$

where $g^*(x) = f^*(x \vee c) = f^*(x)$. Evidently, g is a dense map. This gives the so called *dense factorization*, where each f is factorized as a dense map followed by a closed sublocale. Note that denseness is a condition weaker than injectivity.

2.13. Booleanization and Isbell's Density Theorem. Recall, from 1.1, properties (C1)-(C6) of pseudo-complements a^c.

Lemma. *For each locale X, $\{a \in \mathcal{O}X \mid a^{cc} = a\}$ is a complete Boolean algebra.*

Proof. First, it is easy to see that the formula $\bigvee' a_i = (\bigvee a_i)^{cc}$ yields a join in $\{a \in \mathcal{O}X \mid a^{cc} = a\}$; hence, we have a complete lattice. By property (C6), $\wedge$ gives finite meets. Finally, we have $a \wedge a^c = 0$ and $a \vee' a^c = 1$ since $(a \vee a^c)^c = 0$; indeed, if $x \wedge (a \vee a^c) = 0$ we have both $x \wedge a = 0$ (and hence $x \leq a^c$) and $a \wedge a^c = 0$ (and hence $x \leq a^{cc}$) so that $x \leq a^c \wedge a^{cc} = 0$. $\qquad\square$

Moreover, using properties (C5) and (C6), it is also easy to see that the map

$$\mathcal{O}X \to \{a \in \mathcal{O}X \mid a^{cc} = a\}$$

defined by $a \mapsto a^{cc}$ is a frame homomorphism, obviously onto and dense.

Therefore, defining $\mathcal{B}X$ by $\mathcal{O}(\mathcal{B}X) = \{a \in \mathcal{O}X \mid a^{cc} = a\}$ we have a dense sublocale $\beta_X : \mathcal{B}X \rightarrowtail X$ given by $\beta_X^*(a) = a^{cc}$. This sublocale is called the *Booleanization* of X [18].

Proposition. *β_X is the least dense sublocale of X.*

Proof. Let $j : Y \rightarrowtail X$ be dense. If $j^*(a) = j^*(b)$ we have $j^*(a \wedge b^c) = j^*(a) \wedge j^*(b^c) \leq j^*(a) \wedge j^*(b)^c = 0$ (by (1.1.2), $h(x^c) \leq h(x)^c$ for any homomorphism). Thus, $a \wedge b^c = 0$ and $a \leq b^{cc}$. By symmetry also $b \leq a^{cc}$ and we see that

$$j^*(a) = j^*(b) \Rightarrow a^{cc} = b^{cc}$$

and that we can define a localic $\varphi : \mathcal{B}X \to Y$, by putting $\varphi^*(j^*(a)) = a^{cc}$, to obtain $\beta_X \sqsubseteq j$. $\qquad\square$

This is a new feature of locale theory: it is not the case that all topological spaces have least dense subspaces.

2.14. Sublocales and subspaces. Now let X be a topological space and let $S \subseteq X$ be a subspace. We have the obvious representation of S as a sublocale $\widetilde{S} \rightarrowtail \mathsf{Lc}(X)$, determined by the congruence C_S:

$$(U, V) \in C_S \Leftrightarrow U \cap S = V \cap S \text{ (that is, } j^{-1}[U] = j^{-1}[V] \text{ for the } j : S \hookrightarrow X).$$

This representation is, however, not always satisfactory. We say that a topological space X satisfies the *axiom T_D* if, for every $x \in X$, there is an open $U \ni x$ such that $U \setminus \{x\}$ is also open.

Axiom T_D is stronger than T_0 and (much) weaker than T_1. It is incomparable with sobriety.

Exercises.

1. Prove that $T_1 \Rightarrow T_D \Rightarrow T_0$.
2. Show that a space X is T_D if and only if there is a neighborhood V such that $V \cap \overline{\{x\}} = \{x\}$.

Proposition. *In a spatial locale $\mathsf{Lc}(X)$ we have, for subspaces S, T of the space X, the implication $(\widetilde{S} = \widetilde{T} \Rightarrow S = T)$ if and only if X satisfies T_D.*

Proof. Let X satisfy T_D and let, for any open sets U and V,

$$U \cap S = V \cap S \text{ iff } U \cap T = V \cap T. \tag{2.14.1}$$

Let $S \not\subseteq T$. Choose $x \in S \setminus T$ and U open, $U \ni x$, such that $V = U \setminus \{x\}$ is open. Then $U \cap S \neq V \cap S$ while $U \cap T = V \cap T$, contradicting (2.14.1).

On the other hand, let T_D not hold and let x be a point such that, for U open, $U \ni x$, $V = U \setminus \{x\}$ is never open. Then, for $S = X \setminus \{x\}$, one has $U \cap S = V \cap S$, that is, $U \setminus \{x\} = V \setminus \{x\}$, only if $U = V$, and hence $\widetilde{S} = \widetilde{X}$. $\qquad \square$

For a topological space X, $\mathcal{B}X$ (more exactly, $\mathcal{B}(\mathsf{Lc}(X))$) is the Boolean algebra of the *regular open sets* U of X, that is, the U that are equal to $int(\overline{U})$. Thus (recall Proposition 1.8) they are typically not spatial. Hence, sublocales of a space (or, spatial locale) are not necessarily spatial ("we have more sublocales than subspaces").

The Booleanization $\beta : \mathcal{B}X \rightarrowtail X$ also illustrates the fact that the intersection of sublocales need not agree with the intersection of spaces (in the notation above, $\widetilde{S} \cap \widetilde{T}$ is not necessarily the same as $\widetilde{S \cap T}$). For instance, if X is the space of reals, the intersection of the subspaces of the rationals and of the irrationals is void; however, by 2.13, the intersection of the respective sublocales contains at least $\mathcal{B}X$, which is a rather large Boolean algebra.

On the other hand, the unions do not bring any surprise. We have $\widetilde{\bigcup_{i \in I} S_i} = \bigcup_{i \in I} \widetilde{S_i}$ since, obviously, $U \cap \bigcup_{i \in I} S_i = V \cap \bigcup_{i \in I} S_i$ if and only if $U \cap S_i = V \cap S_i$ for all i.

2.15. Factorization Theorem. We will now be concerned with a technique of producing sublocales by means of extending a relation (identifying elements according to the needs of a construction) to a congruence. During this paragraph, by convenience, we will consistently use the frame language.

Let $R \subseteq A \times A$ be an arbitrary binary relation on a frame A. An element $s \in A$ is *saturated* (more precisely, *R-saturated*) if

$$\forall a, b, c \; aRb \Rightarrow (a \wedge c \leq s \text{ iff } b \wedge c \leq s). \tag{2.15.1}$$

In case R is *meet-stable*, that is, if there is a subset $M \subseteq A$ such that

(1) $1 \in M$ and $a = \bigvee\{x \in M \mid x \leq a\}$ for every $a \in A$,
(2) $\forall a, b \in A \; \forall x \in M, \; aRb \Rightarrow a \wedge xRb \wedge x$,

then $s \in A$ is saturated if and only if

$$\forall a, b \;\; aRb \Rightarrow (a \leq s \text{ iff } b \leq s). \tag{2.15.2}$$

In fact, aRb if and only if $a \wedge xRb \wedge x$ for every $x \in M$, and $a \wedge c \leq s$ if and only if $a \vee x \leq s$ for every $x \leq c$.

Obviously, any meet of saturated elements is saturated. Consequently, we have the saturated

$$\nu(a) = \nu_R(a) = \bigwedge\{s \mid s \text{ saturated }, a \leq s\}.$$

Recall Remark 1 of 1.1. For the Heyting implication $\rightarrow$ we have:

Lemma. *Let s be saturated. Then each $x \rightarrow s$ is saturated.*

Proof. $a \wedge c \leq x \rightarrow s$ iff $a \wedge (c \wedge x) \leq s$ iff $b \wedge (c \wedge x) \leq s$ iff $b \wedge c \leq x \rightarrow s$. $\square$

We show some properties of mapping ν:

Proposition.

(1) $\nu : A \rightarrow A$ *is monotone.*
(2) $a \leq \nu(a)$.
(3) $\nu\nu(a) = \nu(a)$.
(4) $\nu(a \wedge b) = \nu(a) \wedge \nu(b)$.

Proof. (1), (2) and (3) are obvious.
(4) By the monotonicity, $\nu(a \wedge b) \leq \nu(a) \wedge \nu(b)$. Now since $a \wedge b \leq \nu(a \wedge b)$ we have $a \leq b \rightarrow \nu(a \wedge b)$ and, by the Lemma, $\nu(a) \leq b \rightarrow \nu(a \wedge b)$. Thus, $\nu(a) \wedge b \leq \nu(a \wedge b)$ and repeating the procedure we obtain $\nu(a) \wedge \nu(b) \leq \nu(a \wedge b)$. $\square$

Mappings $\nu : A \rightarrow A$ satisfying properties (1)-(4) from the Proposition are called *nuclei*. Denote by $\mathcal{N}(A)$ the system of all nuclei on A, endowed with the natural order.

Note that we have already encountered a nucleus, namely the $\beta_X^* : a \mapsto a^{cc}$, if viewed as a map $\mathcal{O}X \rightarrow \mathcal{O}X$.

Exercises.

1. For each nucleus ν on A let $A_\nu = \{a \in A \mid \nu(a) = a\}$. Show that:
   (a) A_ν is a frame;
   (b) $\nu : A \rightarrow A_\nu$ is a surjective frame homomorphism whose right adjoint is the inclusion $A_\nu \subseteq A$;
   (c) Conclude that, for any locale X, the partially ordered sets $\mathcal{S}(X)$ and $\mathcal{N}(\mathcal{O}X)^{op}$ are isomorphic.
2. Prove that a subset S of a frame A is equal to some A_ν if and only if
   - S is closed under arbitrary meets in A, and
   - $s \in S$, $a \in A$ implies $a \rightarrow s \in S$.

3. Define a subset S of a frame A as a *sublocale set* if it satisfies the conditions of the preceding exercise, and denote the system of all sublocale sets of A, ordered by inclusion, by $\mathcal{S}'(A)$. Prove that the correspondences $\nu \mapsto A_\nu$ and $S \mapsto \nu_S$, where $\nu_S(a) = \bigwedge\{s \in S \mid a \leq s\}$, constitute an isomorphism $\mathcal{S}'(A) \cong \mathcal{N}(A)^{op}$.

Denote the set of all R-saturated elements by A/R and view the mapping ν as restricted to $\nu : A \to A/R$.

Theorem.

(1) A/R *is a frame and the restriction* $\nu : A \to A/R$ *is a frame surjection. If aRb then $\nu(a) = \nu(b)$, and for every frame homomorphism $h : A \to B$ such that $h(a) = h(b)$ whenever aRb, there is a frame homomorphism $\overline{h} : A/R \to B$ such that $\overline{h} \cdot \nu = h$. Moreover, $\overline{h}(a) = h(a)$ for all $a \in A/R$.*

(2) *If R is meet-stable then, moreover, for every join-preserving $f : A \to B$, there is a join-preserving $\overline{f} : A/R \to B$ such that $\overline{f} \cdot \nu = f$.*

Proof. We have suprema in A/R given by $\bigvee' a_i = \nu(\bigvee a_i)$: if $a = \nu(a)$ and $a \geq a_i$ for all i then $a \geq \bigvee a_i$ and $a = \nu(a) \geq \nu(\bigvee a_i)$. We have, for general $a_i \in A$, $\nu(\bigvee a_i) \leq \nu(\bigvee \nu(a_i)) = \bigvee' \nu(a_i)$ and $\bigvee' \nu(a_i) \leq \nu(\bigvee a_i)$, by monotonicity. Preservation of finite meets follows from Proposition ($\nu(1) = 1$ by (2)). Thus we have a complete lattice A/R and $\nu : A \to A/R$ preserving all joins and finite meets. Since it is onto, A/R satisfies the distributivity requirement for frames.

If aRb then $b \leq \nu(a)$, since $a \leq \nu(a)$ and $\nu(a)$ is saturated. Hence $\nu(b) \leq \nu(a)$, and by symmetry $\nu(b) = \nu(a)$.

Let $h : A \to B$ be such that $h(x) = h(y)$ whenever xRy. For $a \in A$ set $\tau(a) = \bigvee\{a' \in A \mid h(a') \leq h(a)\}$. Then

$$a \leq \tau(a) \text{ and } h\tau(a) = h(a). \tag{2.15.3}$$

Let xRy and $x \wedge z \leq \tau(a)$. Then $h(y \wedge z) = h(x \wedge z) \leq h\tau(a) = h(a)$ and hence $y \wedge z \leq \tau(a)$. Thus, τ is saturated. Note that if R is meet-stable we can use (2.15.2) and do not need h to preserve the meet. Using (2.15.3) we see that $a \leq \nu(a) \leq \tau(a)$ and hence $h(a) \leq h\nu(a) \leq h\tau(a) = h(a)$. Thus, we can define $\overline{h} : A/R \to B$ by $\overline{h}(a) = h(a)$ to obtain $\overline{h} \cdot \nu = h$. $\qquad\square$

2.16. The coframe structure of $\mathcal{S}(X)$. So far we met three equivalent ways of representing sublocales of a locale X, given by three different complete lattices that are isomorphic to $\mathcal{S}(X)$: $\mathcal{C}(\mathcal{O}X)^{op}$, in 2.5, and $\mathcal{N}(\mathcal{O}X)^{op}$ and $\mathcal{S}'(\mathcal{O}X)$, in 2.15. The lattice structure of these partially ordered sets is particularly transparent in $\mathcal{S}'(\mathcal{O}X)$: meets are simply intersections (since any intersection of sublocale sets is a sublocale set) and joins are given by

$$\bigvee_{i \in I} S_i = \left\{ \bigwedge T \mid T \subseteq \bigcup_{i \in I} S_i \right\}$$

(indeed, a sublocale set containing all S_i necessarily contains $\{\bigwedge T \mid T \subseteq \bigcup_{i \in I} S_i\}$; on the other hand, this set is clearly closed under meets, and for any $a \in \mathcal{O}X$ and $T \subseteq \bigcup_{i \in I} S_i$, $a \to \bigwedge T = \bigwedge\{a \to t \mid t \in T\}$ and each $a \to t$ is in $\bigcup_{i \in I} S_i$).

Furthermore,

$$\bigcap_{i \in I}(S_i \vee T) \subseteq \left(\bigcap_{i \in I} S_i\right) \vee T :$$

We may assume $I \neq \emptyset$. If $a \in \bigcap_{i \in I}(S_i \vee T)$ then, for each i, $a = s_i \wedge t_i$ for some $s_i \in S_i$ and $t_i \in T$. Let $t = \bigwedge_{i \in I} t_i \in T$; then $a = s_i \wedge t = (t \to s_i) \wedge t$ for every i. On the other hand, $t \wedge s_i = t \wedge s_j$ for all $i, j \in I$ means that $t \to s_i = t \to s_j$. Therefore $t \to s_i$ does not depend on i; denote it by s (s belongs to $\bigcap_{i \in I} S_i$ since, for each i, $s = t \to s_i \in S_i$). Thus $a = s \wedge t$ with $s \in \bigcap_{i \in I} S_i$ and $t \in T$.

Since the reverse inclusion $(\bigcap_{i \in I} S_i) \vee T \subseteq \bigcap_{i \in I}(S_i \vee T)$ is trivial, we have just proved that $\mathcal{S}'(\mathcal{O}X)$ is a coframe. Consequently, $\mathcal{S}(X)$ is also a coframe and $\mathcal{N}(\mathcal{O}X)$ and $\mathcal{C}(\mathcal{O}X)$ are frames.

2.17. Images. Let $f : X \to Y$ be a localic map. The *image* of a sublocale $j : X' \rightarrowtail X$ under f, denoted by $f[j]$ (or, sometimes, by abuse of notation, by $f[X']$), is the unique m in the $(\mathcal{E}, \mathcal{M})$-factorization of $f \cdot j$ (recall 2.3)

$$
\begin{array}{ccc}
X' & \xrightarrow{\;e\;} & Y' \\
{\scriptstyle j}\downarrow & & \downarrow{\scriptstyle m} \\
X & \xrightarrow{\;f\;} & Y
\end{array}
$$

Technically, of course, one uses the congruence on $\mathcal{O}Y$ defined by

$$a \sim b \text{ if and only if } j^*(f^*(a)) = j^*(f^*(b)).$$

The definition of the *preimage* needs some more knowledge of the category Loc and is postponed to the next section.

3. Limits and colimits

3.1. Equalizers, products, and coequalizers in Frm. If $h_1, h_2 : A \to B$ are frame homomorphisms then

$$k : \{a \in A \mid h_1(a) = h_2(a)\} \hookrightarrow A$$

is obviously the equalizer of h_1, h_2 in Frm.

If A_i, $i \in I$, is any system of frames, the projections of the (cartesian) product

$$p_j : \prod_{i \in I} A_i \to A_j$$

(with the structure in $\prod A_i$ defined coordinatewise) obviously constitute the product in Frm. Thus,

the category Loc *is cocomplete (the category* Frm *is complete).*

Also, *coequalizers* in Frm (*equalizers* in Loc) are easy. If $h_1, h_2 : A \to B$ are frame homomorphisms consider the relation $R = \{(h_1(a), h_2(a)) \mid a \in A\}$ and the homomorphism $\nu : B \to B/R$ from 2.15. We immediately see that ν is the coequalizer of h_1, h_2 in Frm.

The following immediate consequence of Theorem 2.15(2) will be useful.

Proposition. *Let $h_1, h_2 : A \to B$ be frame homomorphisms and let $g : B \to C$ be their coequalizer. Let $\varphi : B \to D$ be a join-preserving map such that $\varphi(h_1(a) \wedge b) = \varphi(h_2(a) \wedge b)$ for all $a \in A$ and $b \in M$, where M join-generates B. Then there is a join-preserving $\overline{\varphi} : C \to D$ such that $\varphi = \overline{\varphi} \cdot g$.* $\qquad\square$

Concerning limits and colimits in Loc, the only problem is the product (co-product in Frm). The existence of products (and, more generally, of all limits and colimits, as well as the exactness of the category Frm) is known at once by the fact that Frm is monadic over the category Set of sets (see V.2.5(5)). But here, we need to know their structure. The major part of this section will be devoted to construct them. For technical reasons, that will be apparent shortly, we will use the frame language.

3.2. Useful facts about the category of semilattices. From Chapter I, paragraph 5.23, recall the downset functor $\mathfrak{D}$ sending a meet-semilattice S to the frame

$$\mathfrak{D}S = \Big(\{U \subseteq S \mid U = {\downarrow}U\}, \subseteq \Big),$$

where ${\downarrow}U = \{a \in S \mid \exists b(a \leq b \text{ and } b \in U)\}$ is the *down-closure* of U (cf. I.4.3). For our purposes, it will be handier to use the modification

$$\mathfrak{D}_0 : \mathsf{SLat}_0 \to \mathsf{Frm}$$

where SLat_0 is the category of *bounded* meet-semilattices (that is, semilattices with bottom 0 and top 1), and (bounded) semilattice homomorphisms (that is, mappings preserving the meet, including the top 1, and 0),

$$\mathfrak{D}_0 S = \Big(\{U \subseteq S \mid \emptyset \neq U = {\downarrow}U\}, \subseteq \Big)$$

and

$$\mathfrak{D}_0 h(U) = {\downarrow}h[U].$$

Note that the joins in $\mathfrak{D}_0 S$ are, again, the unions, with one exception: the join of the void system (the bottom of $\mathfrak{D}_0 S$) is $\{0\}$, not the union (which is $\emptyset$ and is not an element of $\mathfrak{D}_0 S$). Moreover, it is obvious that $\mathfrak{D}_0 h$ preserves the bottom, the top and all unions; it also preserves the meet since

$$
\begin{aligned}
{\downarrow}h[U] \cap {\downarrow}h[V] &= \{x \mid \exists a \in U, b \in V, \ x \leq h(a) \wedge h(b)\} \\
&\subseteq \{x \mid \exists c \in U \cap V, \ x \leq h(c)\} \\
&= {\downarrow}h[U \cap V] \\
&\subseteq {\downarrow}h[U] \cap {\downarrow}h[V]
\end{aligned}
$$

(the last inclusion because of the monotonicity).

Consider the maps $\lambda_S : S \to \mathfrak{D}_0 S$ given by $\lambda_S(a) = \downarrow a$. It is an important fact that the λ_S are the universal bounded semilattice homomorphisms, analogously as in the well known situation with the $\mathfrak{D}$ *including the empty set:*

Proposition. *The mapping λ_S is a morphism in* SLat$_0$ *and, for each frame A and each $h : S \to A$ in* SLat$_0$, *there is exactly one frame homomorphism $\overline{h} : \mathfrak{D}_0 S \to A$ such that $\overline{h} \cdot \lambda_S = h$.*

Proof. The first statement is obvious, since $\downarrow(a \wedge b) = \downarrow a \cap \downarrow b$, by the definition of $a \wedge b$. If $\overline{h} \cdot \lambda_S = h$ then

$$\overline{h}(U) = \overline{h}(\bigcup \{\downarrow a \mid a \in U\}) = \bigvee \{\overline{h}(\downarrow a) \mid a \in U\} = \bigvee \{h(a) \mid a \in U\}$$

and hence $\overline{h}$ is uniquely determined. On the other hand, the formula $\overline{h}(U) = \bigvee \{h(a) \mid a \in U\}$ determines a frame homomorphism. $\square$

In SLat$_0$ the coproducts are obtained as follows. Set

$$\prod{}_{i \in I}' S_i = \Big\{(a_i)_{i \in I} \in \prod_{i \in I} S_i \mid a_i = 1 \text{ for all but finitely many } i\Big\} \cup \Big\{(0)_{i \in I}\Big\}$$

and define

$$\gamma_j : S_j \to \prod{}_{i \in I}' S_i \quad \text{by setting} \quad (\gamma_j(a))_i = \begin{cases} a & \text{for } i = j, \\ 1 & \text{otherwise.} \end{cases}$$

Obviously, if $h_j : S_j \to T$ are morphisms in SLat$_0$, we have a uniquely defined $h : \prod' S_i \to T$ such that $h \cdot \gamma_j = h_j$, namely that given by $h((a_i)_{i \in I}) = \bigwedge_{i \in I} h_i(a_i)$ — the meet is finite, all but finitely many $h_i(a_i)$ being 1.

3.3. Coproducts of frames. Let now A_i, $i \in I$, be frames. View them, for a moment, as objects of SLat$_0$, and take $S = \prod_{i \in I}' A_i$. On the frame $\mathfrak{D}_0(\prod_{i \in I}' A_i)$ consider the relation

$$R = \Big\{\Big(\lambda_S \gamma_j(\bigvee_{m \in M} a_m), \bigvee_{m \in M} \lambda_S \gamma_j(a_m)\Big) \mid j \in I, \ M \text{ any set}, \ a_m \in A_j\Big\},$$

and set

$$\bigoplus_{i \in I} A_i = \mathfrak{D}_0(\prod{}_{i \in I}' A_i)/R.$$

Let $\nu : \mathfrak{D}_0(\prod_{i \in I}' A_i) \to \bigoplus_{i \in I} A_i$ be the homomorphism from 2.15.

Remark. The $\iota_j = \nu \cdot \lambda \cdot \gamma_j$ are frame homomorphisms. Indeed, 0,1 and $\wedge$ are preserved trivially and

$$\Big(\lambda \gamma_j(\bigvee_{m \in M} a_m), \bigvee_{m \in M} \lambda \gamma_j(a_m)\Big) \in R$$

and hence $\iota_j(\bigvee_{m \in M} a_m) = \bigvee_{m \in M} \iota_j(a_m)$.

Proposition. *The system $(\gamma_j : A_j \to \bigoplus_{i \in I} A_i)_{j \in I}$ is a coproduct in* Frm.

Proof. Consider the diagram

$$
\begin{array}{ccccccc}
A_j & \xrightarrow{\gamma_j} & \prod_{i\in I}' A_i & \xrightarrow{\lambda} & \mathfrak{D}_0(\prod_{i\in I}' A_i) & \xrightarrow{\nu} & \bigoplus_{i\in I} A_i \\
\downarrow{h_j} & & \downarrow{h'} & & \downarrow{h''} & & \downarrow{h} \\
B & = & B & = & B & = & B
\end{array}
$$

where h_j are some frame homomorphisms, h' is the coproduct morphism in SLat_0 and h'' is the frame homomorphism from Proposition 3.2. We have

$$
h''(\bigvee_M \lambda\gamma_j(a_m)) = \bigvee \{h'((b_i)_{i\in I}) \mid (b_i)_{i\in I} \le \gamma_j(a_m) \text{ for some } m \in M\} =
$$

$$
= \bigvee_M h'\gamma_j(a_m) = \bigvee_M h_j(a_m) = h_j(\bigvee_M a_m) = h'\gamma_j(\bigvee_M a_m) = h''\lambda\gamma_j(\bigvee_M a_m).
$$

Hence, by Theorem 2.15, there is a frame homomorphism h such that $h \cdot \nu = h''$. Thus, $h \cdot \iota_j = h \cdot \nu \cdot \lambda \cdot \gamma_j = h'' \cdot \lambda \cdot \gamma_j = h' \cdot \gamma_j = h_j$. The unicity follows from the obvious fact that all the elements of $\mathfrak{D}_0(\prod_{i\in I}'A_i)$ are joins of finite meets of the $\lambda\gamma_j(a)$ and hence all the elements of $\bigoplus_{i\in I} A_i$ are joins of finite meets of the $\iota_j(a)$ $(j \in I, a \in A_j)$. $\qquad\square$

Recalling the equalizers from 3.1 we conclude that

the category Loc *is complete (the category* Frm *is cocomplete).*

For finite systems we write $A \oplus B$, $A_1 \oplus A_2 \oplus A_3$ etc. to denote frame coproducts.

3.4. More about the coproduct structure. Recalling that the join $\bigvee_{m\in M} U_m$ in $\mathfrak{D}_0 A$ is equal to the union $\bigcup_{m\in M} U_m$ if $M \ne \emptyset$, and to the set $\{0\}$ if the index set M is void, we see that, in particular,

$$
\left(\downarrow\gamma_j(0), \{(0)_{i\in I}\}\right) \in R \quad \text{for all } j.
$$

Set

$$
\mathbf{0} = \left\{ (a_i)_{i\in I} \in \prod_{i\in I}' A_i \mid \exists i, a_i = 0 \right\}.
$$

Obviously, $\bigcup_{m\in M} \downarrow\gamma_j(a_m) = \downarrow\gamma_j(\bigcup_{m\in M} a_m)$ for $M \ne \emptyset$. Then, since $\mathfrak{D}_0(\prod_{i\in I}' A_i)$ is generated by the $\downarrow(b_i)_{i\in I}$, we easily infer (recall 2.15) that

> $U \in \mathfrak{D}_0(\prod' L_j)$ *is saturated if and only if*
> (1) $\mathbf{0} \subseteq U$, *and*
> (2) *for* $M \ne \emptyset$, *whenever* $x_{im} = x_i$ *for* $i \ne j$, $x_j = \bigvee_{m\in M} x_{jm}$ *and* $(x_{im})_{i\in I} \in U$ *for all* m *then* $(x_i)_{i\in I} \in U$.

Lemma. *For any* $(a_i)_{i\in I} \in \prod_{i\in I}' A_i$, *the set* $\bigoplus_{i\in I} a_i := \downarrow(a_i)_{i\in I} \cup \mathbf{0}$ *is saturated.*

Proof. Let $(x_{im})_{i\in I}$ and x_j be as in the condition above. If $x_i = 0$ for some i then $(x_i)_{i\in I} \in \bigoplus_{i\in I} a_i$. Otherwise all $x_{im} \ne 0$ for $i \ne j$ and $\bigvee_{i\in I} x_{jm} \ne 0$. Hence $x_{jn} \ne 0$ for some n, $(x_{in})_{i\in I}$ is not in $\mathbf{0}$ and therefore $x_i \le a_i$ for all $i \ne j$; but then also all $x_{jm} \le a_j$ and $x_j = \bigvee_{i\in I} x_{jm} \le a_j$. $\qquad\square$

Corollary. *If $\oplus_{i\in I}a_i \leq \oplus_{i\in I}b_i$ and $a_i \neq 0$ for all i, then $a_i \leq b_i$ for all i.* □

For finite index sets we write

$$a \oplus b, \; a_1 \oplus a_2 \oplus a_3 \;\; \text{etc.}$$

Note that, for $(a_i)_{i\in I} \in \prod'_{i\in I} A_i$, we have $\bigwedge_{i\in I} \iota_i(a_i) = \oplus_{i\in I}a_i$. Thus the set of the elements of the form $\oplus_{i\in I}a_i$ generates $\bigoplus_{i\in I} A_i$ by joins and we have, for each $u \in \bigoplus_{i\in I} A_i$,

$$u = \bigvee\{\oplus_{i\in I}a_i \mid (a_i)_{i\in I} \in u\} = \bigvee\{\oplus_{i\in I}a_i \mid \oplus_{i\in I}a_i \leq u\}.$$

Exercises.

1. Show that $\mathcal{P}X \oplus \mathcal{P}Y \cong \mathcal{P}(X \times Y)$, and hence that the product of discrete spatial locales is a discrete spatial locale.
2. For the usual topology on $\mathbb{Q}$, prove that the frame $\mathcal{O}(\mathbb{Q} \times \mathbb{Q})$ of open subsets of $\mathbb{Q} \times \mathbb{Q}$ is not isomorphic to the frame $\mathcal{O}(\mathbb{Q}) \oplus \mathcal{O}(\mathbb{Q})$.

The following technical statement will be used later.

Proposition. *Let A_i, $i = 1, 2$, be frames and $a_i \in A_i$. Then we have $\downarrow a_1 \oplus \downarrow a_2 = \downarrow(a_1 \oplus a_2)$. More precisely, if $\iota_i : A_i \to A_1 \oplus A_2$ are the coproduct injections, then*

$$\iota'_i : \quad \downarrow a_i \quad \to \quad \downarrow(a_1 \oplus a_2)$$
$$x \quad \mapsto \quad \iota_i(x) \wedge (a_1 \oplus a_2)$$

constitute the coproduct of frames $\downarrow a_1$ and $\downarrow a_2$.

Proof. Let $h_i : \downarrow a_i \to B$ be frame homomorphisms. Consider the $g : A_1 \oplus A_2 \to B$ such that $g \cdot \iota_i = h_i \cdot \hat{a}_i$. We have $g(x_1 \oplus x_2) = g(\iota_1(x_1) \wedge \iota_2(x_2)) = h_1(x_1 \wedge a_1) \wedge h_2(x_2 \wedge a_2)$. Hence, if $(x_1 \oplus x_2) \wedge (a_1 \oplus a_2) = (y_1 \oplus y_2) \wedge (a_1 \oplus a_2)$, then $g(x_1 \oplus x_2) = g(y_1 \oplus y_2)$. Thus there is a frame homomorphism $h : \downarrow(a_1 \oplus a_2) \to B$ such that $h \cdot (a_1 \oplus a_2)\hat{} = g$. For $x \in \downarrow a_i$ we have $h(\iota'_i(x)) = h(\iota_i(x) \wedge (a_1 \oplus a_2)) = g(\iota_i(x)) = h_i(x)$. The unicity of such an h is obvious. □

For any frame homomorphisms $h_i : A_i \to B_i$, $i = 1, 2$, we write $h_1 \oplus h_2$ for the unique frame homomorphism $A_1 \oplus A_2 \to B_1 \oplus B_2$ that makes the following diagram commutative

$$
\begin{array}{ccccc}
A_1 & \xrightarrow{\iota_1} & A_1 \oplus A_2 & \xleftarrow{\iota_2} & A_2 \\
{\scriptstyle h_1}\downarrow & & \downarrow {\scriptstyle h_1 \oplus h_2} & & \downarrow {\scriptstyle h_2} \\
B_1 & \xrightarrow{j_1} & B_1 \oplus B_2 & \xleftarrow{j_2} & B_2
\end{array}
$$

Obviously,

$$(h_1 \oplus h_2)\left(\bigvee_{i\in I}(a_i^1 \oplus a_i^2)\right) = \bigvee_{i\in I}(h_1(a_i^1) \oplus h_2(a_i^2)). \qquad (3.4.1)$$

3.5. Coproducts and join-preserving maps. Making the relation R in 3.3 meet-stable (recall 2.15) is easy but it would obscure the notation. For the coproduct of two frames, it is transparent enough. We can replace R by the relation R' consisting of all

$$\left(\downarrow(\bigvee_{i\in I} a_i, b), \bigvee_{i\in I} \downarrow(a_i, b)\right) \text{ and } \left(\downarrow(a, \bigvee_{i\in I} b_i), \bigvee_{i\in I} \downarrow(a, b_i)\right).$$

Using the second part of Theorem 2.15 we can then easily deduce that:

Proposition. *Let A_1, A_2, B be frames and let mappings $\varphi_i : A_i \to B$ preserve all joins. Then there is (exactly one) $\varphi : A_1 \oplus A_2 \to B$ preserving all joins such that $\varphi(a_1 \oplus a_2) = \varphi_1(a_1) \wedge \varphi_2(a_2)$.* □

3.6. Preimages. Now, knowing that the category Loc is complete and cocomplete, we can add to the definition of image from 2.17 the definition of preimage. It is a general categorical fact that

(∗) *in a category with pullbacks and pushouts, extremal monomorphisms are stable under pullbacks.*

Hence we can define the *preimage* of the sublocale $j : Y' \rightarrowtail Y$ under $f : X \to Y$, that we denote by $f^{-1}[j]$, (or, sometimes, by abuse of notation, by $f^{-1}[Y']$) as the sublocale $m : X' \rightarrowtail X$ from the pullback

$$\begin{array}{ccc} X' & \xrightarrow{f'} & Y' \\ {\scriptstyle m}\downarrow & & \downarrow{\scriptstyle j} \\ X & \xrightarrow{f} & Y \end{array} \qquad (3.6.1)$$

Images and preimages satisfy the inequalities

$$j \sqsubseteq f^{-1}[f[j]] \quad \text{and} \quad f[f^{-1}[j]] \sqsubseteq j \qquad (3.6.2)$$

(see III.1.6 for more details).

For the sake of completeness, let us finish this paragraph by proving the statement (∗):

In the pullback (3.6.1), let $m = n \cdot e$ with e an epimorphism. Consider the pushout

$$\begin{array}{ccc} X' & \xrightarrow{f'} & Y' \\ {\scriptstyle e}\downarrow & & \downarrow{\scriptstyle e'} \\ X'' & \xrightarrow{f''} & Y'' \end{array}$$

Since $(f \cdot n) \cdot e = s \cdot f'$, we have an n' such that $n' \cdot e' = s$ and $n' \cdot f'' = f \cdot n$. Since e' is an epimorphism (epimorphisms are stable under pushouts in any category) and since s is extremal, e' is an isomorphism and we can assume that $e' = 1$. Hence $n' = s$ and $s \cdot f'' = f \cdot n$. From the last equation and the original pullback we obtain a k such that $m \cdot k = n$. Hence $m \cdot k \cdot e = n \cdot e = m$, and since m is a monomorphism, $k \cdot e = 1$. Finally, since e is an epimorphism, we may conclude, using $e \cdot k \cdot e = e$, that also $e \cdot k = 1$. □

3.7. Preimages of open and closed sublocales.

Proposition. *Let $f : X \to Y$ be a localic map and let $a \in \mathcal{O}Y$. Then:*

(1) *The preimage of an open sublocale is open. More precisely,*

$$f^{-1}[Y_a] = X_{f^*(a)};$$

(2) *The preimage of a closed sublocale is closed. More precisely,*

$$f^{-1}[Y\text{-}Y_a] = X\text{-}X_{f^*(a)}.$$

Proof. (1) Let us check that

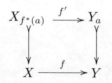

is a pullback in Loc, where $(f')^*(a) = f^*(a)$. It is a commutative square by Exercise 1 of 2.6. Given localic morphisms $g : Z \to X$ and $h : Z \to Y_a$ such that the outer part of the diagram

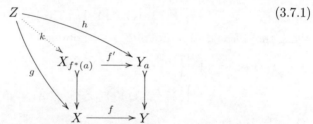

(3.7.1)

is commutative, define $k : Z \to X_{f^*(a)}$ by $k^*(x) = g^*(x)$ for each $x \le f^*(a)$ in $\mathcal{O}X$. This is a localic map; indeed k^* preserves binary meets and arbitrary joins, since g^* does, and moreover $k^*(f^*(a)) = g^*(f^*(a)) = h^*(a \wedge a) = h^*(a) = 1$. Furthermore, given $b \in \mathcal{O}X$ and $a' \in \mathcal{O}Y$ with $a \le a'$,

$$k^*(b \wedge f^*(a)) = g^*(b \wedge f^*(a)) = g^*(b) \wedge g^*(f^*(a)) = g^*(b) \wedge h^*(a) = g^*(b) \wedge 1 = g^*(b)$$

and $k^*(f^*(a')) = g^*(f^*(a')) = h^*(a \wedge a') = h^*(a')$. Thus k is a factorization in diagram (3.7.1). This factorization is unique since $X_{f^*(a)} \rightarrowtail X$ is a monomorphism. (2) The argument for closed sublocales is similar. $\square$

3.8. Preimage as a (co)frame homomorphism.
Let $f : X \to Y$ be a localic map. By (3.6.2), $f^{-1} : \mathcal{S}(Y) \to \mathcal{S}(X)$ has a left Galois adjoint $f[-] : \mathcal{S}(X) \to \mathcal{S}(Y)$. Thus,

$$f^{-1} : \mathcal{S}(Y) \to \mathcal{S}(X) \text{ preserves all meets.} \qquad (3.8.1)$$

Lemma. $f^{-1}[Y_a \sqcup Y\text{-}Y_b] = X_{f^*(a)} \sqcup X\text{-}X_{f^*(b)}.$

Proof. Since f^{-1} preserves meets, we have

$$f^{-1}[Y_a \sqcup Y\text{-}Y_b] \sqcap f^{-1}[Y\text{-}Y_a \sqcap Y_b] = f^{-1}[(Y_a \sqcup Y\text{-}Y_b) \sqcap (Y\text{-}Y_a \sqcap Y_b)]$$
$$= f^{-1}[0_Y] = 0_X,$$

the zero of $\mathcal{S}(X)$ by 2.5. On the other hand,

$$f^{-1}[Y_a \sqcup Y\text{-}Y_b] \sqcup f^{-1}[Y\text{-}Y_a \sqcap Y_b]$$
$$= f^{-1}[Y_a \sqcup Y\text{-}Y_b] \sqcup (f^{-1}[Y\text{-}Y_a] \sqcap f^{-1}[Y_b])$$
$$\sqsupseteq f^{-1}[Y_a] \sqcup f^{-1}[Y\text{-}Y_b] \sqcup (f^{-1}[Y\text{-}Y_a] \sqcap f^{-1}[Y_b])$$
$$= X_{f^*(a)} \sqcup X\text{-}X_{f^*(b)} \sqcup (X\text{-}X_{f^*(a)} \sqcap X_{f^*(b)})$$
$$= X,$$

the top element of $\mathcal{S}(X)$ by (2.5). Thus, $f^{-1}[Y_a \sqcup Y\text{-}Y_b] = (f^{-1}[Y\text{-}Y_a \sqcap Y_b])^c = (f^{-1}[Y\text{-}Y_a] \sqcap f^{-1}[Y_b])^c = (X\text{-}X_{f^*(a)} \sqcap X_{f^*(b)})^c = X_{f^*(a)} \sqcup X\text{-}X_{f^*(b)}$. □

Proposition. $f^{-1} : \mathcal{S}(Y) \to \mathcal{S}(X)$ *is a coframe homomorphism.*

Proof. By (3.8.1) f^{-1} preserves all meets and by Proposition 3.7 $f^{-1}[0_Y] = 0_X$. Thus, it remains to prove that f^{-1} preserves the join $\sqcup$.

By Proposition 2.8 we have, for each sublocale j of Y,

$$j = \bigsqcap \{Y_a \sqcup Y\text{-}Y_b \mid j^*(a) = j^*(b)\}.$$

Using the coframe structure of $\mathcal{S}(Y)$ we obtain

$$j_1 \sqcup j_2 = \bigsqcap \{Y_{a_1} \sqcup Y\text{-}Y_{b_1} \sqcup Y_{a_2} \sqcup Y\text{-}Y_{b_2} \mid j_i^*(a_i) = j_i^*(b_i), i = 1, 2\}$$
$$= \bigsqcap \{Y_{a_1 \vee a_2} \sqcup Y\text{-}Y_{b_1 \wedge b_2} \mid j_i^*(a_i) = j_i^*(b_i), i = 1, 2\}.$$

Then, by the Lemma, $f^{-1}[j_1 \sqcup j_2]$ is equal to

$$\bigsqcap \{X_{f^*(a_1 \vee a_2)} \sqcup X\text{-}X_{f^*(b_1 \wedge b_2)} \mid j_i^*(a_i) = j_i^*(b_i), i = 1, 2\}$$
$$= \bigsqcap \{X_{f^*(a_1) \vee f^*(a_2)} \sqcup X\text{-}X_{f^*(b_1) \wedge f^*(b_2)} \mid j_i^*(a_i) = j_i^*(b_i), i = 1, 2\}$$
$$= \bigsqcap \{X_{f^*(a_1)} \sqcup X_{f^*(a_2)} \sqcup X\text{-}X_{f^*(b_1)} \sqcup X\text{-}X_{f^*(b_2)} \mid j_i^*(a_i) = j_i^*(b_i), i = 1, 2\}$$
$$= \bigsqcap \{f^{-1}[Y_{a_1} \sqcup Y\text{-}Y_{b_1}] \sqcup f^{-1}[Y_{a_2} \sqcup Y\text{-}Y_{b_2}] \mid j_i^*(a_i) = j_i^*(b_i), i = 1, 2\}$$
$$= f^{-1}[j_1] \sqcup f^{-1}[j_2],$$

since $f^{-1}[j_i] = \bigsqcap \{f^{-1}[Y_{a_i} \sqcup Y\text{-}Y_{b_i}] \mid j_i^*(a_i) = j_i^*(b_i)\}$. □

Corollary. *The preimage f^{-1} preserves complementarity of sublocales.* □

3.9. $\mathcal{S}$ as a functor Loc $\to$ Loc. Peculiar monomorphisms. The fact from 3.8 allows us to extend the construction $\mathcal{S}(X)$ from 2.4 (see also 2.16) to a functor

$$\mathsf{S} : \mathsf{Loc} \to \mathsf{Loc},$$

by defining $S(X) = \mathcal{S}(X)^{op}$ and $S(f) : \mathcal{S}(X)^{op} \to \mathcal{S}(Y)^{op}$ by $(S(f))^*(j) = f^{-1}[j]$ for every $j \in \mathcal{S}(Y)^{op}$. In terms of frames, this means that the construction $\mathcal{C}(A)$ from 2.5 gives a functor $\mathcal{C} : \mathsf{Frm} \to \mathsf{Frm}$.

Moreover, we have a natural transformation

$$s : S \overset{.}{\to} 1_{\mathsf{Loc}},$$

defined by $s_X^*(a) = X\text{-}X_a$. In fact we have, for each $f : X \to Y$, $(S(f))^*(s_Y^*(a)) = f^{-1}(Y_a) = X_{f^*(a)} = s_X^*(f^*(a))$.

Remark. The homomorphisms s_X are monomorphisms in Loc. Indeed, let $f, g : Y \to S(X)$ such that $s_X \cdot f = s_X \cdot g$. Since a complement, if it exists, is uniquely determined, and since each frame homomorphism preserves complements, homomorphisms $f^*, g^* : \mathcal{O}(S(X)) \to \mathcal{O}Y$ coinciding on all elements of the form $X\text{-}X_a$ have to coincide on all elements of the form X_a as well, by 2.7. But then, by 2.8, they coincide on all $j \in S(X)$.

Functor S can be iterated by setting $S^0 = 1_{\mathsf{Loc}}$, $S^{\alpha+1}(X) = S(S^\alpha(X))$, for non-limit ordinals $\alpha + 1$, and $s_X^{\alpha+1} = s_X^\alpha \cdot s_{S^\alpha(X)}$, and taking the limit of the obvious diagram in the limit ordinals. More precisely, set $\delta_{\beta\beta} = 1_{S^\beta(X)} : S^\beta(X) \to S^\beta(X)$; if $\delta_{\beta\gamma} : S^\gamma(X) \to S^\beta(X)$ are already defined, set $\delta_{\beta,\gamma+1} = \delta_{\beta\gamma} \cdot s_{S^\gamma(X)}$; for a limit ordinal, if $S^\beta(X)$ for $\beta < \alpha$ and $\delta_{\beta\gamma}$ for $\beta, \gamma < \alpha$ are already defined, take the limit

$$\left(\delta_{\beta\alpha} : S^\alpha(X) \to S^\beta(X) \right)_{\beta < \alpha}$$

of the diagram

$$\left(\delta_{\beta\gamma} : S^\gamma(X) \to S^\beta(X) \right)_{\beta,\gamma < \alpha};$$

finally, set $s_X^\alpha = \delta_{0\alpha}$.

Then

all the $s_X^\alpha : S^\alpha(X) \to X$ are monomorphisms.

This shows that the structure of monomorphisms in Loc is rather complex. There are locales X for which the iteration $S^\alpha(X)$ never stops increasing in size ([34], 2.10). Consequently,

there is a locale X such that, for any cardinality α, there exists a locale Y with $|Y| \geq \alpha$, and a monomorphism $f : Y \to X$.

Exercise. If the locale $S(X)$ is a Boolean algebra, prove that it is the reflection of the locale X in the full subcategory of Boolean locales.

4. Some subcategories of locales

4.1. A very weak separation axiom: subfitness. A locale X is said to be *subfit* ([32], *conjunctive* in [56]) if, for every $a, b \in \mathcal{O}X$,

$$a \not\leq b \quad \Rightarrow \quad \exists c \in \mathcal{O}X, \, a \vee c = 1 \neq b \vee c.$$

Exercises.

1. Let X be a topological space. Prove: if X is T_1 then $\mathsf{Lc}(X)$ is subfit, but not conversely.
2. Show that, for T_D-spaces, subfitness coincides with T_1.

4.2. Relations $\prec$ and $\prec\!\prec$. Define also $a \prec b$ if $a^c \vee b = 1$.

Exercises.

1. Prove that $a \prec b$ if and only if there is a c such that $a \wedge c = 0$ and $b \vee c = 1$.
2. Verify that, if X is a space and $U, V \in \mathcal{O}X$, $U \prec V$ if and only if $\overline{U} \subseteq V$.

Lemma. *The relation $\prec$ has the following properties:*

(1) $a \prec b \Rightarrow a \leq b$;
(2) $a \leq b \prec c \leq d \Rightarrow a \prec d$;
(3) $a_1, a_2 \prec b \Rightarrow a_1 \vee a_2 \prec b$;
(4) $a \prec b_1, b_2 \Rightarrow a \prec b_1 \wedge b_2$;
(5) $a \prec b \Rightarrow b^c \prec a^c$;
(6) *If $f : X \to Y$ is a localic map and $a \prec b$ in $\mathcal{O}Y$, then $f^*(a) \prec f^*(b)$.*

Proof. (1) If $a^c \vee b = 1$ then $a \wedge b = a \wedge (a^c \vee b) = b$.
(2) It follows from the fact that $a \leq b$ implies $b^c \leq a^c$.
(3) Let $a_i \wedge c_i = 0$ and $a_i \vee b = 1$. Set $c = c_1 \wedge c_2$. Then $c \vee b = (c_1 \vee b) \wedge (c_2 \vee b) = 1$ and $(a_1 \vee a_2) \wedge c = 0$.
(4) Similarly, let $a \wedge c_i = 0$ and $c_i \vee b_i = 1$. Set $c = c_1 \vee c_2$. Then $c \vee (b_1 \wedge b_2) = (c \vee b_1) \wedge (c \vee b_2) = 1$ and $a \wedge c = 0$.
(5) If $a^c \vee b = 1$ then $b^{cc} \vee a^c = 1$.
(6) Since $(f^*(a))^c \geq f^*(a^c)$ we have $(f^*(a))^c \vee f^*(b) \geq f^*(a^c \vee b) = 1$. $\square$

A transitive relation R is *interpolative* if whenever aRb there is a c such that $aRcRb$. It is easy to check that, for each transitive R, there is the largest (transitive) interpolative $\tilde{R} \subseteq R$, namely the following one. Denote by D the set of dyadic rationals in the unit interval; then

   $a\tilde{R}b$ *iff there are a_d, $d \in D$, such that $a = a_0$, $b = a_1$, and $c < d \Rightarrow$*
   $a_c R a_d$.

The relation $\tilde{\prec}$ is usually denoted by $\prec\!\prec$. It is easy to see that

   Lemma 4.2 holds with $\prec$ replaced by $\prec\!\prec$.

4.3. Regular and completely regular locales. Let X be a locale. For each $a \in \mathcal{O}X$ set

   $\sigma_X(a) = \{x \in \mathcal{O}X \mid x \prec a\}$ and $\rho_X(a) = \{x \in \mathcal{O}X \mid x \prec\!\prec a\}$.

X is said to be *regular* (resp. *completely regular*) if

$$\forall a \in \mathcal{O}X, \ a = \bigvee \sigma(a) \quad (\text{resp. } a = \bigvee \rho(a)).$$

The category of regular (resp. completely regular) locales, and the localic maps between them, will be denoted by

$$\mathsf{RegLoc} \quad (\text{resp. } \ \mathsf{CRegLoc}).$$

Proposition. RegLoc *and* CRegLoc *are reflective subcategories of the category* Loc. *Consequently, these categories are complete and cocomplete.*

Proof. For a locale X set

$$R_1(X) = \{a \in \mathcal{O}X \mid a = \bigvee \sigma(a)\}.$$

Obviously, $R_1(X)$ is a subframe of $\mathcal{O}X$. For ordinals α set

$$R_{\alpha+1}(X) = R_1(R_\alpha(X)), \text{ and if } \alpha \text{ is a limit one, } R_\alpha(X) = \bigcap_{\beta<\alpha} R_\beta(X).$$

Now if $f : X \to Y$ is a localic map and Y is regular, $f^*[\mathcal{O}Y] \subseteq R(X)$, by Lemma 4.2(6). Thus, if we set $R_\infty(X) = \bigcap_\alpha R_\alpha(X)$, the epimorphisms $e_X : X \twoheadrightarrow R_\infty(X)$, given by the frame inclusions $e_X^* : R_\infty(X) \hookrightarrow X$, constitute a reflection of Loc onto RegLoc.

Similarly for complete regularity (the situation is in fact simpler: here, the procedure stops after the first step). $\qquad\square$

Exercises.

1. Let X be a topological space. Prove that:
   (a) X is regular in the classical sense if and only if the locale $\mathsf{Lc}(X)$ is regular in the sense just defined (recall Exercise 1 of 4.2);
   (b) X is completely regular in the classical sense if and only if the locale $\mathsf{Lc}(X)$ is completely regular in the sense just defined. (Hint: use the procedure from the standard proof of the Urysohn Lemma.)
2. Conclude from Lemma 4.2(6) that a sublocale of a (completely) regular locale is (completely) regular.
3. Prove that a product of regular locales is regular.

4.4. Normality. A locale X is said to be *normal* if for any $a, b \in \mathcal{O}X$ such that $a \vee b = 1$ there are $u, v \in \mathcal{O}X$ such that

$$u \vee b = 1, \quad a \vee v = 1 \quad \text{and} \quad u \wedge v = 1.$$

This is an immediate translation of the homonymous property of spaces; thus, trivially, a space X is normal in the classical sense if and only if $\mathsf{Lc}(X)$ is normal in the sense just defined.

Lemma. *In a normal locale the relation* $\prec$ *interpolates (and hence coincides with* $\prec\!\prec$).

Proof. Let $a \prec b$, that is, $a^c \vee b = 1$. Then there are u, v such that $a^c \vee v = 1$, $u \vee b = 1$ and $u \wedge v = 0$ (and hence $u \leq v^c$). Thus, $a \prec v \prec b$. $\qquad\square$

Proposition. *A subfit normal locale is completely regular.*

Proof. By the Lemma it suffices to prove that it is regular. Let $c \vee a = 1$. We shall prove that then $c \vee \bigvee \sigma(a) = 1$, so that, by subfitness, $a = \bigvee \sigma(a)$. If $c \vee a = 1$ we have u, v such that $u \vee a = 1$, $c \vee v = 1$ and $u \leq v^c$. Thus, $v \leq \bigvee \sigma(a)$ and $c \vee \bigvee \sigma(a) = 1$. $\qquad\square$

4.5. Hausdorff locales. Recall that a topological space is Hausdorff if and only if the diagonal $\{(x, x) \mid x \in X\}$ is closed in $X \times X$. In analogy with this fact, a locale X is called *Hausdorff* (*strongly Hausdorff* in [32]) if the diagonal

$$\Delta_X : X \to X \times X$$

is a closed sublocale (or, in frame terms, if the codiagonal

$$\nabla_{\mathcal{O}X} = \Delta_X^* : \mathcal{O}(X \times X) = \mathcal{O}X \oplus \mathcal{O}X \to \mathcal{O}X$$

is a closed surjection).

Since, in Frm, $\nabla_{\mathcal{O}X}(a \oplus b) = \nabla_{\mathcal{O}X}(\iota_1(a) \wedge \iota_2(b)) = a \wedge b$, we immediately infer that $\check{d}_{\mathcal{O}X} : \mathcal{O}X \oplus \mathcal{O}X \to {\uparrow}d_{\mathcal{O}X}$, where

$$d_{\mathcal{O}X} = \bigvee\{a \oplus b \mid a \wedge b = 0\},$$

is the closure of $\nabla_{\mathcal{O}X}$. Hence the Hausdorff condition amounts to the existence of a frame homomorphism $\alpha : \mathcal{O}X \to {\uparrow}d_{\mathcal{O}X}$ such that

$$\alpha \cdot \nabla_{\mathcal{O}X} = \check{d}_{\mathcal{O}X}. \tag{4.5.1}$$

Remark. Unlike the previous separation axioms, this Hausdorff condition is only an analogy of the classical one. The functor $\mathsf{Lc} : \mathsf{Top} \to \mathsf{Loc}$ does not, in general, preserve products, and a Hausdorff topological space X need not have a closed localic diagonal, that is, it does not necessarily yield a Hausdorff $\mathsf{Lc}(X)$. There are other analogues of the Hausdorff axiom in the literature [26, 40], useful in various contexts. Note that the Hausdorff type axioms presented there are weaker than the Hausdorff property discussed here. From the point of view of categorical topology, the latter (considered first by Isbell in [32]) seems to be, for obvious reasons, of a particular importance.

The category of Hausdorff locales will be denoted by $\mathsf{HausLoc}$.

Lemma. *A locale X is Hausdorff if and only if, for any $a, b \in \mathcal{O}X$,*

$$a \oplus b \leq \check{d}_{\mathcal{O}X}\Big((a \wedge b) \oplus (a \wedge b)\Big).$$

Proof. For the α from (4.5.1) we have

$(a \oplus b) \vee d_{\mathcal{O}X} = \alpha(\nabla_{\mathcal{O}X}(a \oplus b)) = \alpha(\nabla_{\mathcal{O}X}((a \wedge b) \oplus (a \wedge b))) = ((a \wedge b) \oplus (a \wedge b)) \vee d_{\mathcal{O}X}$.

On the other hand, if the condition is satisfied set $\alpha(x) = (x \oplus x) \vee d_{\mathcal{O}X}$. As $x_i \oplus x_j \leq ((x_i \wedge x_j) \oplus (x_i \wedge x_j)) \vee d_{\mathcal{O}X}$, we have $(x_i \oplus x_j) \vee d_{\mathcal{O}X} \leq (x_i \oplus x_i) \vee d_{\mathcal{O}X}$. Hence $\alpha(\bigvee_{i \in I} x_i) = (\bigvee_{i \in I} x_i \oplus \bigvee_{i \in I} x_i) \vee d_{\mathcal{O}X} = \bigvee_{i,j \in I}(x_i \oplus x_j) \vee d_{\mathcal{O}X} = \bigvee_{i \in I}(x_i \oplus x_i) \vee d_{\mathcal{O}X} = \bigvee_{i \in I} \alpha(x_i)$. Trivially, α preserves finite meets and hence $\alpha : \mathcal{O}X \to {\uparrow}d_{\mathcal{O}X}$ is a frame homomorphism. Since

$$\check{d}_{\mathcal{O}X}(a \oplus b) = (a \oplus b) \vee d_{\mathcal{O}X} = ((a \wedge b) \oplus (a \wedge b)) \vee d_{\mathcal{O}X} = \alpha(\nabla_{\mathcal{O}X}(a \oplus b)),$$

and since the $a \oplus b$ generate $\mathcal{O}X \oplus \mathcal{O}X$, we have $\check{d}_{\mathcal{O}X} = \alpha \cdot \nabla_{\mathcal{O}X}$. $\quad\square$

Proposition. *Each regular locale X is Hausdorff.*

Proof. If $x \prec a$ and $y \prec b$ then

$$
\begin{aligned}
x \oplus y &= (x \wedge (y^c \vee b)) \oplus (y \wedge (x^c \vee a)) \\
&= ((x \wedge b) \vee (x \wedge y^c)) \oplus ((a \wedge y) \vee (x^c \wedge y)) \\
&\leq ((a \wedge b) \oplus (a \wedge b)) \vee (x \oplus x^c) \vee (y^c \oplus y) \\
&\leq ((a \wedge b) \oplus (a \wedge b)) \vee d_{\mathcal{O}X}.
\end{aligned}
$$

If X is regular we have

$$
\begin{aligned}
a \oplus b &= \bigvee \{x \in \mathcal{O}X \mid x \prec a\} \oplus \bigvee \{y \in \mathcal{O}X \mid y \prec b\} \\
&= \bigvee \{x \oplus y \mid x \prec a, y \prec b\} \\
&\leq ((a \wedge b) \oplus (a \wedge b)) \vee c
\end{aligned}
$$

and, by the Lemma, X is Hausdorff. $\square$

4.6. Some special properties of HausLoc. For the equalizers in HausLoc (and hence in RegLoc and CRegLoc) one has a very simple formula, obtained first by Banaschewski for the regular case in [8]. The Hausdorff case appeared in Chen's thesis [24].

Theorem. *Equalizers in HausLoc are closed sublocales. More precisely, if X is Hausdorff and $f_1, f_2 : Y \to X$ are localic maps, then the equalizer of f_1, f_2 is given by*

$$
Y\text{-}Y_c \rightarrowtail Y,
$$

where c stands for $\bigvee \{f_1^*(a) \wedge f_2^*(b) \mid a \wedge b = 0\}$.

Proof. Recall (3.4.1). Obviously, $c = \nabla_{\mathcal{O}X}((f_1^* \oplus f_2^*)(d_{\mathcal{O}X}))$. For the α from (4.5.1) we have

$$
(x \oplus 1) \vee d_{\mathcal{O}X} = \alpha(\nabla_{\mathcal{O}X}(x \oplus 1)) = \alpha(x) = \alpha(\nabla_{\mathcal{O}X}(1 \oplus x)) = (1 \oplus x) \vee d_{\mathcal{O}X}.
$$

Therefore

$$
\begin{aligned}
\check{c}(f_1^*(x)) &= f_1^*(x) \vee c \\
&= \nabla_{\mathcal{O}X}((f_1^* \oplus f_2^*)((x \oplus 1) \vee d_{\mathcal{O}X})) \\
&= \nabla_{\mathcal{O}X}((f_1^* \oplus f_2^*)((1 \oplus x) \vee d_{\mathcal{O}X})) \\
&= f_2^*(x) \vee c = \check{c}(f_2^*(x)).
\end{aligned}
$$

On the other hand, if $f_1 \cdot \varphi = f_2 \cdot \varphi = f$ for a $\varphi : Z \to Y$, we have

$$
\varphi^*(c) = \bigvee \{f^*(x \wedge y) \mid x \wedge y = 0\} = 0
$$

and hence we can define $\overline{\varphi} : Z \to Y\text{-}Y_c$ by $\overline{\varphi}^*(x) = \varphi^*(x)$ to obtain $Y\text{-}Y_c \cdot \overline{\varphi} = \varphi$. $\square$

Proposition. *In HausLoc (and in RegLoc and CRegLoc) each dense morphism is an epimorphism.*

Proof. Let f be dense and let $f_1 \cdot f = f_2 \cdot f$. Then, for the c from Theorem 4.6, we have $f^*(c) = 0$ and hence, by density, $c = 0$. Thus, $f_1^*(a) = f_1^*(a) \vee c = f_2^*(a) \vee c = f_2^*(a)$. $\qquad\square$

Remark. Conversely, each epimorphism in HausLoc is dense (the proof is easy but space consuming). Therefore epimorphisms in HausLoc are precisely the dense morphisms.

4.7. Some special properties of RegLoc. In regular locales, congruences are completely described by the congruence class of the top element 1. Indeed:

Proposition. *Let X be regular and let C_1, C_2 be two congruences on $\mathcal{O}X$ such that the congruence classes $C_1[1], C_2[1]$ coincide. Then $C_1 = C_2$.*

Proof. Let $(a, b) \in C_1$ and let $x \prec a$. Then $x^c \vee a = 1$ and therefore $(x^c \vee b, 1) \in C_2$. Consequently, $(x \wedge b, x) = (x \wedge (x^c \vee a), x) \in C_2$ and $(a \wedge b, a) = ((\bigvee \sigma(a)) \wedge b, a) = (\bigvee \{x \wedge b \mid x \prec a\}, \bigvee \{x \mid x \prec a\}) \in C_2$. Similarly, $(a \wedge b, b) \in C_2$. This shows that $(a, b) \in C_2$. $\qquad\square$

A localic map $f : Y \to X$ is said to be *codense* if $f^*(a) = 1$ implies $a = 1$. From the proposition above we immediately obtain:

Corollary. *If X is regular then every codense $f : Y \to X$ is an epimorphism.*

It should be noted that the statement of the Proposition holds, more generally, for *fit* locales, that is, locales X in which

$$a \not\leq b \;\Rightarrow\; \exists c, \; a \vee c = 1 \text{ and } c \to b \neq b,$$

(in fact it characterizes fit locales) and the Corollary holds already for the subfit ones. The relation between fit and subfit is in the following fact:

A locale is fit iff each of its sublocales is subfit.

Exercises.

1. Show that each regular frame is fit.
2. Show that fitness is hereditary, that is, if X is fit and $j : Y \rightarrowtail X$ is a sublocale then Y is fit.
3. Prove that the following statements are equivalent for a locale X:
   (i) X is subfit;
   (ii) $C[1] = \{1\}$ for a congruence C implies that C is trivial;
   (iii) each open sublocale of X is a join of closed sublocales (more exactly, $X_a = \bigsqcup \{X\text{-}X_b \mid b \vee a = 1\}$).
4. Prove that the following statements are equivalent for a locale X:
   (i) X is fit;
   (ii) for any two congruences C_1 and C_2, $C_1[1] = C_2[1]$ implies $C_1 = C_2$;
   (iii) each sublocale of X is a meet of open sublocales (more exactly, $j = \bigsqcap \{X_a \mid j^*(a) = 1\}$);
   (iv) each sublocale of X is subfit.

4.8. First notes on compact locales. Compact locales will have a special section. Here we will just mention a few facts connected with regularity.

A *cover* of a locale X is a subset $A \subseteq \mathcal{O}X$ such that $\bigvee A = 1$, and a locale X is *compact* if each cover contains a finite subcover.

Proposition.

    (1) *Each compact regular locale is normal (and hence, by Lemma 4.4, completely regular).*

    (2) *Let $f : Y \to X$ be a dense localic map, with Y compact and X regular. Then f is an epimorphism.*

    (3) *A compact sublocale of a regular locale is closed.*

Proof. (1) Let $a \vee b = 1$. Thus, $\sigma(a) \cup \sigma(b)$ is a cover and hence there are

$$x_1, \ldots, x_n \prec a \text{ and } y_1, \ldots, y_m \prec b$$

such that $\bigvee_{i=1}^n x_i \vee \bigvee_{i=1}^m y_i = 1$. Set $x = \bigvee_{i=1}^n x_i$ and $y = \bigvee_{i=1}^m y_i$. Then, by Lemma 4.2, $x^c \vee a = 1$ and $y^c \vee b = 1$. As $x \vee y = 1$ we have $x \vee b = 1 = a \vee y$. Set $u = x \wedge y^c$ and $v = x^c \wedge y$. Then $u \vee b = (x \vee b) \wedge (y^c \vee b) = 1$ and, similarly, $a \vee v = 1$. Trivially, $u \wedge v = 0$.

(2) By Corollary 4.7 it suffices to verify that f is codense. Suppose $f^*(a) = 1$. Consequently $\{f^*(x) \mid x \prec a\}$ is a cover of Y and hence there are $x_1, \ldots, x_n \prec a$ such that $\bigvee_{i=1}^n f^*(x_i) = 1$. Set $x = \bigvee_{i=1}^n x_i$. By Lemma 4.2, $x \prec a$. Thus, we have $f^*(x) = 1$ and $x^c \vee a = 1$. Since $f^*(x^c) \leq f^*(x)^c = 1^c = 0$, $x^c = 0$ and finally $a = 1$.

(3) Let $f : Y \rightarrowtail X$ be a compact sublocale of X. For the

$$c = \bigvee\{x \in \mathcal{O}X \mid f^*(x) = 0\}$$

from the closure we have a dense sublocale $f : Y \rightarrowtail X_c$. By (2), f is an epimorphism and hence an isomorphism. $\qquad\square$

More generally, one can prove the normality for the regular Lindelöf locales, by the standard procedure imitating the classical proof.

Exercise. Prove that a closed sublocale of a compact locale is compact.

5. Open and closed maps

5.1. Open maps of locales. The Heyting structure (recall Remark 1 of 1.1) will play a crucial role in this section.

A localic map $f : X \to Y$ is said to be *open* (resp. *closed*) if the image of each open (resp. closed) sublocale under f is open (resp. closed).

Proposition. *Let $f : X \to Y$ be a localic map. The following conditions are equivalent:*

    (i) *f is open;*

(ii) f^* is a complete Heyting homomorphism (that is, it preserves all suprema and infima as well as the Heyting operation);

(iii) f^* admits a left adjoint $f_!$ that satisfies the "Frobenius Identity"

$$f_!(a \wedge f^*(b)) = f_!(a) \wedge b,$$

for every $a \in \mathcal{O}X, b \in \mathcal{O}Y$.

Proof. (i)$\Rightarrow$(ii): By the definition of image, f open means that, for each $a \in \mathcal{O}X$, there exists $f_!(a) \in \mathcal{O}Y$ such that

$$b \wedge f_!(a) = c \wedge f_!(a) \quad \text{iff} \quad f^*(b) \wedge a = f^*(c) \wedge a, \tag{5.1.1}$$

or, equivalently,

$$b \wedge f_!(a) \le c \quad \text{iff} \quad f^*(b) \wedge a \le f^*(c). \tag{5.1.2}$$

In particular, for $b = 1$, we obtain $f_!(a) \le c$ if and only if $a \le f^*(c)$. Thus, $f_!$ is a left adjoint of f^* and we see that f^* preserves all meets. Returning to (5.1.2) and using the Heyting formula we obtain

$$
\begin{aligned}
a \le f^*(b \to c) \quad &\Leftrightarrow \quad f_!(a) \le b \to c \\
&\Leftrightarrow \quad b \wedge f_!(a) \le c \\
&\Leftrightarrow \quad f^*(b) \wedge a \le f^*(c) \\
&\Leftrightarrow \quad a \le f^*(b) \to f^*(c).
\end{aligned}
$$

Thus, $f^*(b \to c) = f^*(b) \to f^*(c)$, that is, f^* is a Heyting homomorphism.

(ii)$\Rightarrow$(iii): If f^* is a complete Heyting homomorphism, it admits a left adjoint $f_!$. Obviously $f_!(a \wedge f^*(b)) \le f_!(a) \wedge b$ since $f_! \cdot f^* \le 1$. Moreover,

$$
\begin{aligned}
a \wedge f^*(b) \le f^*(f_!(a \wedge f^*(b))) \quad &\Leftrightarrow \quad a \le f^*(b) \to f^*(f_!(a \wedge f^*(b))) \\
&\Leftrightarrow \quad a \le f^*(b \to f_!(a \wedge f^*(b))) \\
&\Leftrightarrow \quad f_!(a) \le b \to f_!(a \wedge f^*(b)) \\
&\Leftrightarrow \quad b \wedge f_!(a) \le f_!(a \wedge f^*(b)).
\end{aligned}
$$

(iii)$\Rightarrow$(i): If $f_! \dashv f^*$ and $f_!$ satisfies Frobenius Identity we have

$$b \wedge f_!(a) \le c \Leftrightarrow f_!(a \wedge f^*(b)) \le c \Leftrightarrow a \wedge f^*(b) \le f^*(c),$$

so that (5.1.2) is satisfied. $\qquad\square$

Remarks. (1) Recall from 4.7 that a congruence on a regular locale is determined by the congruence class of the top element. Thus, in the regular case, the formula (5.1.1) above is equivalent to

$$f_!(a) = c \wedge f_!(a) \quad \text{iff} \quad a = f^*(c) \wedge a,$$

that is, $f_!(a) \le c$ if and only if $a \le f^*(c)$. Hence

if Y is regular, the open localic maps $f : X \to Y$ are those for which f^* is a complete lattice homomorphism.

Note that, hence, each complete lattice homomorphism $h : L \to M$ between frames, with L regular, is automatically Heyting.

(2) If X and Y are spaces and $f : X \to Y$ is an open continuous map, then f is open also as a localic map. The converse is not generally true. It holds for the Y that are T_D; in fact, the coincidence of classical open and localic open maps characterizes property T_D (see [17, 51]).

Exercises.

1. Show that a sublocale $Y \rightarrowtail X$ is open if and only if it is open as a localic map.
2. Let Y be a fit locale. Prove that a localic map $f : X \to Y$ is open if and only if $f^* : \mathcal{O}Y \to \mathcal{O}X$ is a complete lattice homomorphism. (By Exercise 4 of 4.7, this generalizes Remark 1 above.)

5.2. Pullback stability. A straightforward application of the Frobenius Identity yields a characterization of surjections among open maps:

Lemma. *An open $f : X \to Y$ is a surjection if and only if $f_!(1) = 1$.*

Proof. Since f^* is one-one, $f_! \cdot f^* = 1_{\mathcal{O}Y}$. Hence $f_!(1) = f_!(f^*(1)) = 1$.

Conversely, if $f_!(1) = 1$ we have, by the Frobenius Identity, $a = f_!(1) \wedge a = f_!(f^*(a))$, which shows that f^* is one-one. $\square$

We may now check pullback stability for open maps and open surjections.

Theorem. *Consider the pullback square*

$$
\begin{array}{ccc}
P & \xrightarrow{q} & Z \\
\scriptstyle p \downarrow & & \downarrow \scriptstyle g \\
X & \xrightarrow{f} & Y
\end{array}
$$

in Loc, *where f is open. Then:*

(1) *q is open;*

(2) *for each $a \in \mathcal{O}X$, $g^*(f_!(a)) = q_!(p^*(a))$;*

(3) *q is a surjection whenever f is a surjection.*

Proof. (1) Consider the diagram

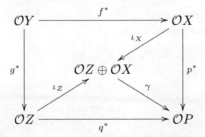

in Frm, where the outer diagram is a pushout, constructed in the standard way from the coproduct and the coequalizer of $\iota_Z \cdot g^*$ and $\iota_X \cdot f^*$. By Proposition

5.1, if f is open, f^* has a left adjoint $f_!$ satisfying Frobenius Identity. Define $\varphi : \mathcal{O}Z \oplus \mathcal{O}X \to \mathcal{O}Z$ by $\varphi(z \oplus x) = z \wedge g^*(f_!(x))$ (recall 3.4). Then

$$
\begin{aligned}
\varphi(\iota_Z(g^*(a)) \wedge (z \oplus x)) &= \varphi((g^*(a) \wedge z) \oplus x) \\
&= z \wedge g^*(a \wedge f_!(x)) \\
&= z \wedge g^*(f_!(f^*(a) \wedge x)) \\
&= \varphi(z \oplus (f^*(a) \wedge x)) \\
&= \varphi(\iota_X(f^*(a)) \wedge (z \oplus x)),
\end{aligned}
$$

and since φ preserves joins, we have a $\overline{\varphi} : \mathcal{O}P \to \mathcal{O}Z$ such that $\overline{\varphi} \cdot \gamma = \varphi$ (recall Proposition 3.1). Then

$$
\overline{\varphi}(q^*(z)) = \varphi(\iota_Z(z)) = \varphi(z \oplus 1) = z \wedge g^*(f_!(1)) \leq z
$$

and

$$
q^*(\overline{\varphi}(\gamma(z \oplus x))) = q^*(z \wedge g^*(f_!(x))) = q^*(z) \wedge p^*(f(f_!(x))) = \gamma(z \oplus f f_!(x)) \geq \gamma(z \oplus x).
$$

Thus, $\overline{\varphi} = q_!$, the left Galois adjoint of q^*. Since we have

$$
\begin{aligned}
q_!(q^*(a) \wedge \gamma(z \oplus x)) &= q_!(\gamma(\iota_Z(a) \wedge (z \oplus x))) \\
&= \varphi((a \wedge z) \oplus x) \\
&= a \wedge z \wedge g^*(f_!(x)) \\
&= a \wedge q_!(\gamma(z \oplus x)),
\end{aligned}
$$

q is open.

(2) For $a \in \mathcal{O}X$ we obtain $q_!(p^*(a)) = \varphi(1 \oplus a) = g^*(f_!(a))$.

(3) If f is a surjection we have $f_!(1) = 1$ by the Lemma. Then, by (2), $q_!(1) = g^*(1) = 1$ and hence q is surjective as well. $\square$

Remark. As is well known, surjections are not stable under pullback in Loc, as the following example [47] shows. Let $\overline{\mathbb{N}}$ with topology

$$
\mathcal{O}\overline{\mathbb{N}} = \{ U \subseteq \overline{\mathbb{N}} \mid U = \emptyset \text{ or } \overline{\mathbb{N}} \setminus U \text{ is a finite subset of } \mathbb{N} \}
$$

and let $\mathbb{N}$ with the discrete topology $\mathcal{P}\mathbb{N}$. The frame homomorphism $i^{-1} : \mathcal{O}\overline{\mathbb{N}} \to \mathcal{P}\mathbb{N}$ given by the continuous map $i : \mathbb{N} \rightarrowtail \overline{\mathbb{N}}$ is a monomorphism but the pushout of i^{-1} along the monomorphism $\mathcal{O}\overline{\mathbb{N}} \rightarrowtail \mathcal{P}\overline{\mathbb{N}}$ is

$$
\begin{array}{ccc}
\mathcal{O}\overline{\mathbb{N}} & \longrightarrow & \mathcal{P}\overline{\mathbb{N}} \\
\big\downarrow{\scriptstyle i^{-1}} & & \big\downarrow{\scriptstyle i^{-1}} \\
\mathcal{P}\mathbb{N} & \xrightarrow{\ 1\ } & \mathcal{P}\mathbb{N}
\end{array}
$$

and $i^{-1} : \mathcal{P}\overline{\mathbb{N}} \to \mathcal{P}\mathbb{N}$ is not a monomorphism.

In [8] Banaschewski characterizes the locales Y for which pullback along every $g : Z \to Y$ preserves surjections: precisely the ones such that $\mathcal{S}(Y)$ is Boolean.

Corollary. *Let $f : X \to Y$ be an open localic map. For every sublocale $j : Y' \rightarrowtail Y$,*

$$f^{-1}[\bar{j}] = \overline{f^{-1}[j]}. \tag{5.2.1}$$

Proof. We know, by 2.9, that $\bar{j}$ is the sublocale $Y\text{-}Y_{c_j} \rightarrowtail Y$ and $\overline{f^{-1}[j]}$ is the sublocale $X\text{-}X_{c_{f^{-1}(j)}} \rightarrowtail X$, where

$$c(j) = \bigvee \{b \in \mathcal{O}Y \mid j^*(b) = 0\}$$

and

$$c_{f^{-1}(j)} = \bigvee \{a \in \mathcal{O}X \mid (f^{-1}(j))^*(a) = 0\}.$$

Moreover, by Proposition 3.7, $f^{-1}(Y\text{-}Y_{c_j}) = X\text{-}X_{f^*(c(j))}$. Therefore it suffices to verify that $f^*(c(j)) = c_{f^{-1}(j)}$, that is,

$$\bigvee \{f^*(b) \mid j^*(b) = 0\} = \bigvee \{a \mid (f^{-1}(j))^*(a) = 0\}.$$

The inequality $\bigvee \{f^*(b) \mid j^*(b) = 0\} \leq \bigvee \{a \mid (f^{-1}(j))^*(a)\} = 0$ is obvious. Conversely, for each a such that $(f^{-1}(j))^*(a) = 0$ take $b = f_!(a)$. Then $a \leq f^*(b)$ and, by condition (2) of the Theorem, $j^*(b) = 0$. $\square$

The converse is not true [37], as the following example due to P. Johnstone shows. Take, for f, the dense embedding $\beta_X : \mathcal{B}X \rightarrowtail X$ from 2.13. Then the condition is trivially satisfied since, in $\mathcal{B}X$, every sublocale is closed, by Proposition 2.11. The localic map β_X is not open, though.

This contrasts with classical topology, where the formula

$$f^{-1}(\overline{A}) = \overline{f^{-1}(A)}, \text{ for any } A \subseteq Y$$

is equivalent to f being open. However, if a localic map f stably has the property (5.2.1) above (that is, if all pullbacks of f satisfy it), then f is necessarily open and we do get a characterization of localic openness (see [37] for a proof; see also III.7.3).

Exercise. For X the unit interval, find an example of a system $\{U_i \in \mathcal{O}X \mid i \in I\}$ such that $\bigwedge_{i \in I} U_i = \emptyset$ and $\bigwedge_{i \in I} \beta_X(U_i) = I$.

5.3. Closed and proper maps of locales. We end this section with a characterization of closed maps via a "co-Frobenius identity".

Proposition. *A localic map $f : X \to Y$ is closed if and only if f_* satisfies*

$$f_*(a \vee f^*(b)) = f_*(a) \vee b.$$

Proof. Similarly like in Proposition 5.1 we see that the closedness amounts to the existence of a map φ such that

$$a \vee f^*(b) = a \vee f^*(c) \quad \text{iff} \quad \varphi(a) \vee b = \varphi(a) \vee c,$$

which is easily seen to be equivalent to

$$f^*(c) \leq a \vee f^*(b) \quad \text{iff} \quad c \leq \varphi(a) \vee b.$$

Setting $b = 0$ we see that $\varphi = f_*$. Further, the first inequality is equivalent to $c \leq f_*(a \vee f^*(b))$ so that we finally transform the condition into the form

$$c \leq f_*(a \vee f^*(b)) \quad \text{iff} \quad c \leq f_*(a) \vee b,$$

yielding the desired equation.   □

Exercise 1 of 5.1 shows that open maps generalize open sublocales and Theorem 5.2 asserts that open maps and open surjections are stable under pullback. In order to have similar results for closed maps one has to restrict the class of closed maps: a localic map $f : X \to Y$ is said to be *proper* if it is closed and f_* preserves directed joins (see [63] and [64] for some alternative descriptions).

Exercise. Observe that proper maps generalize closed sublocales, by proving that a sublocale $Y \rightarrowtail X$ is closed if and only if it is proper as a localic map.

The classes of proper maps and proper surjections are stable under pullback (see [63] for a proof). In particular, if this property is weakened to mention only pullbacks along product projections $Z \times Y \to Y$ one concludes that, for proper $f : X \to Y$, $1_Z \times f : Z \times X \to Z \times Y$ is closed for all locales Z, which corresponds to one of the standard definitions of properness for continuous maps between topological spaces [21]. The converse is also true [64] and so the condition

$$1_Z \times f : Z \times X \to Z \times Y \text{ is closed for all locales } Z$$

characterizes the properness property of a map $f : X \to Y$ of locales (similarly to topological spaces).

For more information on open, closed and proper maps consult [38, 60, 63]. In [60] the results about open and proper maps are proved side by side with "parallel proofs for parallel results", showing the similarities between the two classes. For instance the proof that proper maps are stable under pullback is really just a repetition of the proof that open maps are stable under pullback but with "has a left adjoint which is a sup-lattice homomorphism" being replaced with "has a right adjoint which is a preframe homomorphism".

6. Compact locales and compactifications

6.1. Some machinery. The converse of Proposition 4.5 is valid for compact locales and it was first proved constructively by Vermeulen [62]. In order to present it we need a few results of a technical nature, involving binary products of locales.

Let X_1 and X_2 be locales. The maps $\pi_1, \pi_2 : \mathcal{D}(\mathcal{O}X_1 \times \mathcal{O}X_2) \to \mathcal{D}(\mathcal{O}X_1 \times \mathcal{O}X_2)$ defined by

$$\pi_1(U) = \{(\bigvee S, y) \mid S \times \{y\} \subseteq U\}$$

and

$$\pi_2(U) = \{(x, \bigvee S) \mid \{x\} \times S \subseteq U\}$$

are nuclei on the frame $\mathcal{D}(\mathcal{O}X_1 \times \mathcal{O}X_2)$. The map

$$\pi_0 : \begin{array}{ccc} \mathcal{D}(\mathcal{O}X_1 \times \mathcal{O}X_2) & \rightarrow & \mathcal{D}(\mathcal{O}X_1 \times \mathcal{O}X_2) \\ U & \mapsto & \pi_1(U) \cup \pi_2(U) \end{array}$$

is a *prenucleus*, that is, for all $U, V \in \mathcal{D}(\mathcal{O}X_1 \times \mathcal{O}X_2)$, $U \subseteq \pi_0(U)$, $\pi_0(U) \cap V \subseteq \pi_0(U \cap V)$ and $\pi_0(U) \subseteq \pi_0(V)$ whenever $U \subseteq V$. But for each prenucleus π_0 there is a unique nucleus π which has the same fixed points as π_0, which is given by

$$\pi(a) = \bigwedge \{b \mid a \leq b, \pi_0(b) = b\} \ [6].$$

In this case, since

$$Fix(\pi_0) = \{U \in \mathcal{D}(\mathcal{O}X_1 \times \mathcal{O}X_2) \mid \pi_0(U) = U\} = \mathcal{O}X_1 \oplus \mathcal{O}X_2,$$

the associated nucleus π is given by

$$\pi(U) = \bigcap \{V \in \mathcal{O}X_1 \oplus \mathcal{O}X_2 \mid U \subseteq V\}.$$

Furthermore define, for any $U \in \mathcal{D}(\mathcal{O}X_1 \times \mathcal{O}X_2)$,

$$\sigma_0(U) = \{\bigvee D \mid \text{ directed } D \subseteq U\}.$$

This defines a prenucleus. Let σ denote the associated nucleus. Note that $\sigma \leq \pi$, since $\sigma_0(U) \subseteq \pi(U)$ for every U. Indeed, for every directed set $D = \{(c_i, d_i) \mid i \in I\} \subseteq U$, we have $(\bigvee_{i \in F} c_i, \bigvee_{i \in F} d_i) \in U$ for every finite $F \subseteq I$, which implies $(c_i, d_j) \in U$ for every $i, j \in I$. Consequently, $(c_i, \bigvee_{i \in I} d_i) \in \pi(U)$ and finally $\bigvee D = (\bigvee_{i \in I} c_i, \bigvee_{i \in I} d_i) \in \pi(U)$.

As a consequence of this inclusion we have:

Lemma. $\pi = \sigma \cdot \pi_2 \cdot \pi_1$. □

A constructive proof of this result was first provided by Banaschewski [10].

6.2. Compact elements. An element a of a complete lattice A is said to be *compact* (*finite* in [34]) if $a \leq \bigvee S$ for some $S \subseteq A$ implies $a \leq \bigvee F$ for some finite $F \subseteq S$. Clearly a locale X is compact if and only if 1 is a compact element of $\mathcal{O}X$.

Exercises.

1. Let A be a complete lattice, $a \in A$. Prove that a is compact if and only if for every directed subset $D \subseteq A$ with $a \leq \bigvee D$, there exists $d \in D$ with $a \leq d$.
2. Let A be a frame, $a \in A$. Prove that a is compact if and only if for every $S \subseteq A$ with $a = \bigvee S$, there exists a finite $F \subseteq S$ with $\bigvee F = a$.

Lemma. *Let* $U \in \mathcal{D}(\mathcal{O}X_1 \times \mathcal{O}X_2)$ *and let* $a \in \mathcal{O}X_1$ *be a compact element. If* $a \oplus b \leq \pi(U)$ *then* $(a, b) \in \pi_2(\pi_1(U))$.

Proof. Let $S = \pi_2(\pi_1(U)) \in \mathcal{D}(\mathcal{O}X_1 \times \mathcal{O}X_2)$ and

$$W = \left\{V \in \mathcal{D}(\mathcal{O}X_1 \times \mathcal{O}X_2) \mid S \subseteq V \subseteq \pi(S), (a, b) \in V \Rightarrow (a, b) \in S\right\}.$$

Then:

- $S \in W$;

- W is σ_0-stable, that is, $V \in W$ implies $\sigma_0(V) \in W$. Indeed, take $V \in W$ and consider $(a, b) = \bigvee D$, for directed $D \subseteq V$. Since a is compact there exists $(c_0, d_0) \in D$ such that $a \leq c_0$. For any $(c, d) \in D$ with $(c_0, d_0) \leq (c, d)$ we have $(a, d) \in V$, thus $(a, d) \in S$. Since $b = \bigvee\{d \mid (c, d) \in D, (c_0, d_0) \leq (c, d)\}$, we get $(a, b) \in \pi_2(S) = S$;
- trivially, $W := \bigcup W \in W$.

Thus $\sigma_0(W) \in W$ and $\sigma_0(W) = W$, which implies $\sigma(W) = W \in W$. Hence $\sigma(S) \subseteq \sigma(W) = W$ and $\sigma(S) \in W$. On the other hand, by Lemma 6.1, $\pi(U) \subseteq \sigma(S)$ so $\pi(U) \in W$. $\qquad\square$

6.3. A technical lemma. The following notation will be convenient in the sequel. For locales X_1, X_2 and $I \in \mathcal{O}X_1 \oplus \mathcal{O}X_2$ let

$$I_1[b] = \bigvee\{a \in \mathcal{O}X_1 \mid (a, b) \in I\} \qquad (b \in \mathcal{O}X_2)$$

and

$$I_2[a] = \bigvee\{b \in \mathcal{O}X_2 \mid (a, b) \in I\} \qquad (a \in \mathcal{O}X_1).$$

Note that $(I_1[b], b) \in I$ and $(a, I_2[a]) \in I$ for every $a \in \mathcal{O}X_1$ and $b \in \mathcal{O}X_2$.

Lemma. *Consider locales X_1, X_2 and the projections $p_i : X_1 \times X_2 \to X_i$ ($i = 1, 2$). If X_1 is compact then, for every $a \in \mathcal{O}X_1$ and $I \in \mathcal{O}X_1 \oplus \mathcal{O}X_2$, we have*

$$p_{2_*}(p_1^*(a) \vee I) = \bigvee\{b \in \mathcal{O}X_2 \mid a \vee I_1[b] = 1\}.$$

Proof. Let $b \in \mathcal{O}X_2$ with $a \vee I_1[b] = 1$. In order to show that $b \leq p_{2_*}(p_1^*(a) \vee I)$ it suffices to check that $(1, b) \in p_1^*(a) \vee I$, which is true because $(a, b) \in p_1^*(a)$ and $(I_1[b], b) \in I$. So

$$p_{2_*}(p_1^*(a) \vee I) \geq \bigvee\{b \in \mathcal{O}X_2 \mid a \vee I_1[b] = 1\}.$$

Now, for $U = p_1^*(a) \cup I \in \mathcal{D}(\mathcal{O}X_1 \times \mathcal{O}X_2)$ and $u = p_{2_*}(p_1^*(a) \vee I)$, let us prove that $u \leq \bigvee\{b \mid a \vee I_1[b] = 1\}$. Since $p_2^*(u) \leq p_1^*(a) \vee I$, we have that $(1, u) \in p_1^*(a) \vee I = \pi(U)$. Hence, by Lemma 6.2, $(1, u) \in \pi_2(\pi_1(U))$. But

$$\pi_2(\pi_1(U)) = \left\{(x, y) \mid y \leq \bigvee\{b \mid x \leq a \vee I_1[b]\}\right\}. \tag{6.3.1}$$

Indeed, if $(x, y) \in \pi_2(\pi_1(U))$ then $y = \bigvee S$ with $(x, s) \in \pi_1(U)$ for every $s \in S$. Then $(x, s) = (\bigvee R_s, s)$ where $R_s \times \{s\} \subseteq U$, from which it follows that, for each $s \in S$ and $r \in R_s$, $r \leq a \vee I_1[s]$. Therefore $x = \bigvee R_s \leq a \vee I_1[s]$ for every $s \in S$, which means that $\bigvee\{b \mid x \leq a \vee I_1[b]\} \geq \bigvee S = y$. On the other hand, let (x, y) be such that $y \leq \bigvee\{b \mid x \leq a \vee I_1[b]\}$. The conclusion that $(x, y) \in \pi_2(\pi_1(U))$ follows immediately from the fact that, for each such b, $(a, b) \in p_1^*(a)$ and $(I_1[b], b) \in I$ so $(a \vee I_1[b], b) \in \pi_1(U)$ and then $(x, b) \in \pi_1(U)$.

Finally, it follows from (6.3.1) that $u \leq \bigvee\{b \mid a \vee I_1[b] = 1\}$. $\qquad\square$

6.4. Hausdorffness and regularity. We may now prove, at last, the converse of Proposition 4.5 for compact locales.

Theorem. *Each compact Hausdorff locale is regular.*

Proof. Applying Lemma 6.3 in the case $X = X_1 = X_2$ and $I = d_{\mathcal{O}X} = \bigvee\{a \oplus b \mid a \wedge b = 0\}$ we get

$$p_{2_*}(p_1^*(a) \vee d_{\mathcal{O}X}) = \bigvee\{b \in \mathcal{O}X \mid a \vee b^c = 1\} = \bigvee\{b \in \mathcal{O}X \mid b \prec a\}.$$

Since X is Hausdorff, we may conclude, from Theorem 4.6, that the sublocale

$$X \times X\text{-}(X \times X)_{d_{\mathcal{O}X}} \rightarrowtail X$$

is the equalizer of p_1 and p_2, since

$$\bigvee\{p_1^*(a) \wedge p_2^*(b) \mid a \wedge b = 0\} = \bigvee\{a \oplus b \mid a \wedge b = 0\} = d_{\mathcal{O}X}.$$

Therefore $p_{2_*}(p_1^*(a) \vee d_{\mathcal{O}X}) = p_{2_*}(p_2^*(a) \vee d_{\mathcal{O}X}) \geq p_{2_*}(p_2^*(a)) \geq a$, which shows that $\bigvee\{b \in \mathcal{O}X \mid b \prec a\} = a$ for every $a \in \mathcal{O}X$. $\square$

6.5. The Kuratowski-Mrówka Theorem for locales. The Kuratowski-Mrówka Theorem characterizes compact spaces K by the fact that for each X the projection $K \times X \to X$ is closed (see, for example, [29]). Its counterpart for locales is also valid and was first obtained by Pultr and Tozzi [50]. By applying the results on binary coproducts of Sections 6.1 and 6.2 we can present a constructive proof [25].

Theorem. *A locale K is compact if and only if $p_2 : K \times X \to X$ is closed for any locale X.*

Proof. Let K be a compact locale. By Proposition 5.3, p_2 is closed if and only if it satisfies the co-Frobenius Identity

$$p_{2_*}(I \vee p_2^*(x)) = p_{2_*}(I) \vee x \tag{6.5.1}$$

for all $I \in \mathcal{O}(K \times X)$ and $x \in \mathcal{O}X$. Since

$$p_{2_*}(I) = \bigvee\{a \in \mathcal{O}X \mid p_2^*(a) \leq I\} = \bigvee\{a \in \mathcal{O}X \mid (1, a) \in I\},$$

(6.5.1) holds if and only if $(I \vee p_2^*(x))_2[1] = I_2[1] \vee x$ or, equivalently, $(I \vee p_2^*(x))_2[1] \leq I_2[1] \vee x$, that is,

$$(1, y) \in I \vee p_2^*(x) \Rightarrow y \leq I_2[1] \vee x.$$

Let $U = I \cup p_2^*(x) \in \mathcal{D}(\mathcal{O}K \times \mathcal{O}X)$. Then $\pi_2(U) = \{(a, b) \mid b \leq x \vee I_2[a]\}$, and $(1, y) \in \pi_2(U)$ if and only if $y \leq x \vee I_2[1]$. Thus (6.5.1) holds if and only if

$$(1, y) \in \pi(U) \Rightarrow (1, y) \in \pi_2(U).$$

This is true by Lemma 6.1 and the fact that $\pi_1(U) = U$.

Conversely, suppose U is a directed cover of $\mathcal{O}K$. We shall prove that $1 \in U$. In the set $\mathcal{O}K$ define

$$\mathcal{T}(\mathcal{O}K) = \Big\{S \subseteq \mathcal{O}K \mid 1 \in S \Rightarrow \uparrow u \subseteq S \text{ for some } u \in U\Big\}.$$

This is a topology on $\mathcal{O}K$. Let X be the corresponding locale. By hypothesis, $p_2 : K \times X \to X$ is closed, that is, $p_{2*}((1 \oplus a) \vee I) = a \vee p_{2*}(I)$ for all $a \in \mathcal{O}X$, $I \in \mathcal{O}(K \times X)$. Consider $a = K \setminus \{1\} \in \mathcal{O}X$ and $I = \bigvee\{u \oplus \uparrow\!u \mid u \in U\}$. Then

$$
\begin{aligned}
(1 \oplus a) \vee I &= \bigvee\{u \oplus a \mid u \in U\} \vee \bigvee\{u \oplus \uparrow\!u \mid u \in U\} \\
&= \bigvee\{u \oplus (a \cup \uparrow\!u) \mid u \in U\} \\
&= \{u \oplus 1_{\mathcal{O}X} \mid u \in U\} \\
&= 1_{\mathcal{O}(K \times X)}.
\end{aligned}
$$

Hence $a \vee p_{2*}(I) = p_{2*}((1 \oplus a) \vee I) = p_{2*}(1) = 1_{\mathcal{O}X} = \mathcal{O}K$, which implies $1 \in p_{2*}(I)$. By definition of $\mathcal{T}(\mathcal{O}K)$ this means that $\uparrow\!v \subseteq p_{2*}(I)$ for some $v \in U$, that is, $p_{2*}(\uparrow\!v) \subseteq I$. Then $(1 \oplus \uparrow\!v) \wedge (1 \oplus \downarrow\!v) \leq I \wedge (1 \oplus \downarrow\!v)$ is equivalent to $1 \oplus \{v\} \leq \bigvee\{u \oplus [u, v] \mid u \in U\}$, where $[u, v] = \{k \in \mathcal{O}K \mid u \leq k \leq v\}$. For $W = \{\bigvee V \mid V \subseteq U\}$ let

$$
S = \downarrow\!\{(w, [w, v]) \mid w \in W\} \in \mathcal{D}(\mathcal{O}K \times \mathcal{O}X).
$$

It can be easily checked that $\pi_1(S) = S$ and $1 \oplus \{v\} \subseteq \pi(S)$. Since $\{v\}$ is an atom of $\mathcal{O}X$, it is a compact element, and we may apply Lemma 6.2 to conclude that $(1, \{v\}) \in \pi_1(\pi_2(S))$. Again by $\{v\}$ being an atom of $\mathcal{O}X$, this implies that $(1, \{v\}) \in \pi_1(S) = S$. Finally, $(1, \{v\}) \in S$ means that $1 = w$ and $\{v\} = [w, v]$ for some $w \in W$, that is, $1 = w = v \in U$, as required. $\square$

6.6. Regular ideals. In the sequel we will present an easy construction of compact locales starting from general ones. This will give us the compactification of completely regular locales due to Banaschewski and Mulvey [12].

An *ideal* in a frame A is a non-void subset $J \subseteq A$ such that

   (1) $a, b \in J \implies a \vee b \in J$, and
   (2) $a \in J \ \& \ b \leq a \implies b \in J$.

It is said to be *regular* if, moreover

   (3) for each $a \in J$ there is a $b \in J$ such that $a \prec\!\!\prec b$.

Examples. The sets $\downarrow\!a$, or the $\sigma(a)$ from 4.3, are ideals; $\rho(a)$ is a regular ideal.

The collection of all ideals (resp. all regular ideals) in A will be denoted by

$$
\mathfrak{J}A \quad (\text{resp. } \mathfrak{R}A).
$$

Proposition. $\mathfrak{J}A$ *and* $\mathfrak{R}A$ *are compact frames with bottom* $\{0\}$, *top* A, *intersection for meet and the join defined by*

$$
\bigvee_{i \in I} J_i = \left\{ \bigvee F \mid F \text{ finite}, \ F \subseteq \bigcup_{i \in I} J_i \right\}.
$$

Proof. $\{0\}$ and A are regular ideals and a finite intersection of a system of (regular) ideals is obviously a (regular) ideal. Also, obviously, $\{\bigvee F \mid F \text{ finite}, \ F \subseteq \bigcup_{i \in I} J_i\}$ is an ideal containing all J_i, and if K is an ideal containing all J_i, it has to contain all finite joins of the elements of $\bigcup_{i \in I} J_i$. Thus, $\{\bigvee F \mid F \text{ finite}, F \subseteq \bigcup_{i \in I} J_i\}$ is the supremum of the system J_i in $\mathfrak{J}A$. If all the J_i are regular and $\{a_1, \dots, a_n\} \subseteq \bigcup_{i \in I} J_i$ choose b_i, $a_i \prec\!\prec b_i$ in $\bigcup_{i \in I} J_i$; then $a_1 \vee \cdots \vee a_n \prec\!\prec b_1 \vee \cdots \vee b_i$ by Lemma 4.2 and we see that $\bigvee_{i \in I} J_i$ is regular as well. Trivially, $(\bigvee_{i \in I} J_i) \cap K \supseteq \bigvee_{i \in I}(J_i \cap K)$. If $a \in (\bigvee_{i \in I} J_i) \cap K$ we have $a = a_1 \vee \cdots \vee a_n \in K$, $a_j \in \bigcup_{i \in I} J_i$. Since K is an ideal, all the a_i are in K and hence $\{a_1, \dots, a_n\} \subseteq \bigcup_{i \in I}(J_i \cap K)$ and $a \in \bigvee_{i \in I}(J_i \cap K)$. Thus, $\mathfrak{J}A$ is a frame and $\mathfrak{R}A$ one of its subframes.

Finally, let $\{J_i \mid i \in I\}$ be a cover of $\mathfrak{J}A$. Then $\bigvee_{i \in I} J_i = A \ni 1$ and hence there are $a_1, \dots, a_n \in \bigcup_{i \in I} J_i$ such that $a_1 \vee \cdots \vee a_n = 1$. Choose J_{i_j} containing a_j. Then $\bigvee_{j=1}^{n} J_{i_j} \ni 1$ and hence $\bigvee_{j=1}^{n} J_{i_j} = A$ by the down-closedness condition (2) of ideals. $\qquad\square$

6.7. The Stone-Čech compactification of locales. The constructions $\mathfrak{J}A$ and $\mathfrak{R}A$ can be extended to functors by setting

$$\mathfrak{J}h(J) \ (\text{resp. } \mathfrak{R}(h)(J)) = \downarrow h[J].$$

(Checking that $\downarrow h[J]$ is an ideal — a regular one if J is regular — is straightforward and so is the preserving of 0, 1 and joins. Also preserving the meets is an easy exercise.)

Proposition. *If A is completely regular then $\mathfrak{R}A$ is regular. Thus, $\mathfrak{R}$ can be viewed as a functor from* CRegLoc *to the category* KRegLoc *of compact regular locales.*

Proof. For a regular ideal J we obviously have $J = \bigvee\{\rho(a) \mid a \in J\} \ (= \bigcup\{\rho(a) \mid a \in J\})$, and for each a, $\rho(a) = \bigvee\{\rho(b) \mid b \prec\!\prec a\}$. Thus, it suffices to prove that

$$b \prec\!\prec a \text{ in } A \quad\Rightarrow\quad \rho(b) \prec \rho(a) \text{ in } \mathfrak{R}A.$$

Interpolate $b \prec\!\prec x \prec\!\prec y \prec\!\prec a$. Obviously, $\rho(b^c) \cap \rho(b) = \{0\}$ and consequently $\rho(b^c) \subseteq \rho(b)^c$. Then, by 4.2, $x^c \subseteq \rho(b)^c$. Thus, $1 = x^c \vee y \in \rho(b)^c \vee \rho(a)$ and hence $\rho(b)^c \vee \rho(a) = J$. $\qquad\square$

For a completely regular locale X define $\upsilon_X : X \to \mathfrak{R}X$ by setting $\upsilon_X^*(J) = \bigvee J$.

Lemma. υ_X *is a dense sublocale.*

Proof. We obviously have

$$\upsilon^*(\rho(a)) = a \quad\text{and}\quad \rho(\upsilon^*(J)) \supseteq J. \qquad (6.7.1)$$

Thus, υ^* is a left adjoint and preserves all joins. Furthermore, $\upsilon^*(\mathcal{O}X) = 1$ and

$$\upsilon^*(J) \wedge \upsilon^*(K) = \bigvee J \wedge \bigvee K = \bigvee\{a \wedge b \mid a \in J, b \in K\} \leq$$

$$\leq \bigvee\{c \mid c \in J \cap K\} = \upsilon^*(J \cap K) \leq \upsilon^*(J) \wedge \upsilon^*(K),$$

the last inequality being trivial. Obviously υ^* is onto, and if $\upsilon^*(J) = \bigvee J \neq 0$ then $J \neq \{0\}$. $\qquad\square$

The parallelism with the classical situation is now apparent; we have a reflection of CRegLoc onto KRegLoc, called by obvious reasons the *Stone-Čech compactification* for locales [12]:

Theorem. *The functor $\mathfrak{R}$ and the system of mappings v_X constitute a reflection of* CRegLoc *onto* KRegLoc.

Proof. Checking that, for a localic map $f : X \to Y$, $\mathfrak{R}(f) \cdot v_X = v_Y \cdot f$ is immediate. Thus, it remains to be proved that if X is compact, v_X is an isomorphism. According to formulas (6.7.1), it suffices to prove that $\rho(v^*(J)) \subseteq J$. Let $a \in \rho(\bigvee J)$. Then $a^c \vee \bigvee J = 1$ and if X is compact there are $x_1, \ldots, x_n \in J$ such that $a^c \vee x_1 \vee \cdots \vee x_n = 1$. Then $x = x_1 \vee \cdots \vee x_n \in J$ and $a \leq x$ so that $a \in J$. $\qquad\square$

Remarks. (1) More generally, a *compactification* of a locale X is a dense extremal monomorphism $f : X \to Y$ with compact regular codomain. A locale which has a compactification is called *compactifiable*. For a comprehensive view of compactifications of locales consult [9].

(2) Note that the entire procedure was constructive (no choice principle or the law of the excluded middle was used). Banaschewski and Mulvey presented this construction in [12]. In [13] they presented an alternative construction. Realizing that the fact that a reflective subcategory is closed under limits can be also proved constructively, we can conclude that products of compact regular locales (by Proposition 4.5 and Theorem 6.4 this is the same as compact Hausdorff locales) are compact and this fact does not need non-constructive principles. This is even true in the non-regular case (where a reflection here does not exist), that is, Tychonoff's Theorem is choice-free for locales (the proof is much more difficult, though — see [33] or [6]). This came as a remarkable surprise when Johnstone [33] was able to prove it within Zermelo-Fraenkel axiomatic without choice ($\mathbb{ZF}$). Indeed, this contrasts with the classical case where Tychonoff's Theorem is equivalent to the axiom of choice (in the Hausdorff setting, with the Boolean Ultrafilter Theorem). It turns out that in fact the non-constructive principle is needed for products having enough points rather than for preserving compactness (see III.11.4).

Later refinements of this important result were given by Kříž [39], who uses fewer axioms of $\mathbb{ZF}$ (namely, Kříž's proof does not depend on the non-constructive axiom of replacement as Johnstone's did), and by Vermeulen [61], whose proof is constructively valid in the sense of topos theory (meaning: valid in an arbitrary topos).

7. Locally compact locales

7.1. The "way below" relation and continuous lattices. Recall that in a complete lattice A, a is *well below* b, written

$$a \ll b,$$

if for each directed $D \subseteq A$

$$b \leq \bigvee D \quad \Rightarrow \quad \exists d \in D, \ a \leq d.$$

By Exercise 1 of 6.2, an element a of a complete lattice is compact precisely when $a \ll a$.

A *continuous lattice* [53] is a complete lattice A such that, for every $a \in A$,

$$a = \bigvee \{b \in A \mid b \ll a\}.$$

Lemma. *The relation $\ll$ satisfies the following properties:*

(1) $0 \ll a$ *for all* a;

(2) *if* $x \leq a \ll b \leq y$ *then* $x \ll y$;

(3) *if* $a_1, a_2 \ll b$ *then* $a_1 \vee a_2 \ll b$. *Thus, the set* $\{x \mid x \ll a\}$ *is always directed*;

(4) *in any frame,* $a \prec b \ll 1$ *implies* $a \ll b$;

(5) *in any regular (resp. completely regular) frame,*

$$a \ll b \ \Rightarrow \ a \prec b \ (\textit{resp. } a \ll b \ \Rightarrow \ a \prec\!\prec b);$$

(6) *in any continuous lattice, the relation $\ll$ interpolates.*

Proof. (1), (2), and (3) are immediate.

(4) Let $b \leq \bigvee D$ for a directed D. Then $1 \leq a^c \vee \bigvee D$ and hence there is a $d \in D$ such that $b \leq a^c \vee d$. Then $a = a \wedge b \leq a \wedge d$, that is, $a \leq d$.

(5) For regular frames we have $b = \bigvee \{x \mid x \prec b\}$ and since the join is directed, there is an $x \prec b$ such that $a \leq b$. For completely regular frames the proof is analogous.

(6) We have $b = \bigvee \{\bigvee \{y \mid y \ll x\} \mid x \ll b\} = \bigvee \{y \mid \exists x, \ y \ll x \ll b\}$ and the join is obviously directed. Thus, if $a \ll b$ there are x, y such that $a \leq y \ll x \ll b$. $\square$

Exercise. Show that, for open subsets U, V of a space X, if there is a compact K such that $U \subseteq K \subseteq V$, then $U \ll V$. Conclude that, for every locally compact space X, the frame $\mathcal{O}X$ is continuous.

7.2. Locally compact locales. Exercise 7.1 above suggests that the continuity may be a good description of local compactness in the localic setting: a locale X is said to be *locally compact* if the frame $\mathcal{O}X$ is a continuous lattice. In fact we will see that it is even better than that (see Theorem 7.4 below).

A general compact locale is not necessarily locally compact. But we have:

Proposition.

(1) *A compact Hausdorff locale is locally compact. More generally, any open sublocale of a compact Hausdorff locale is locally compact. Moreover, in such a case the relations $\ll$, $\prec$ and $\prec\!\prec$ coincide.*

(2) *A completely regular locale X is locally compact if and only if it is open in its Stone-Čech compactification $v_X : X \to \mathfrak{R}X$.*

Proof. (1) Let X be a compact Hausdorff locale. Compactness means $1 \ll 1$ and hence, by Lemma 7.1(4), $a \prec b$ implies $a \ll b$ in $\mathcal{O}X$. By Theorem 6.4, X is regular (and hence, by 4.8, completely regular). Therefore, by Lemma 7.1(5), $a \ll b$ implies $a \prec\!\prec b$.

(2) By (1) it suffices to prove that, in case X is locally compact, there is a regular ideal J in $\mathcal{O}X$ such that the congruence associated with v_X^* is the open Δ_J, that is, for any regular ideals J_1 and J_2,

$$v_X(J_1) = v_X(J_2) \text{ if and only if } J_1 \cap J = J_2 \cap J.$$

Consider $J = \{x \in \mathcal{O}X \mid x \ll 1\}$. By Lemma 7.1 it is a regular ideal. By continuity, $v_X^*(J) = \bigvee J = 1$ and hence $J_1 \cap J = J_2 \cap J$ implies $v_X^*(J_1) = v_X^*(J_2)$. On the other hand, let $\bigvee J_1 = \bigvee J_2$, let $a \in J_1 \cap J$ and choose $b \in J_1 \cap J$ such that $a \prec\!\prec b$ (and hence $a \ll b$). Since ideals are directed, we may conclude from $b \leq \bigvee J_2 = \bigvee J_1$ that there is an $x \in J_2$ with $a \leq x$, and hence $a \in J_2$. □

Furthermore, as for compact locales, the converse of Proposition 4.5 is valid for locally compact locales [62] and Hausdorff also means regular. In ([62], Proposition 4.7) it is also proved that, for Hausdorff locales, local compactness can be defined in terms of compact neighborhoods.

7.3. Scott topology. Let A be a lattice. A subset $U \subseteq A$ is said to be *Scott open* if $U = \uparrow U$ and, whenever $D \subseteq A$ is directed and $\bigvee D \in U$, $U \cap D \neq \emptyset$. Obviously the Scott open subsets constitute a topology (the so-called *Scott topology* on A).

Consider now a locale X and represent the spectrum $\mathsf{Pt}(X)$ by complete filters $P \subseteq \mathcal{O}X$ (as in 1.5). One can characterize compact subsets of $\mathsf{Pt}(X)$ in terms of Scott opens:

Lemma. *A subset $K \subseteq \mathsf{Pt}(X)$ is compact if and only if $\bigcap\{P \mid P \in K\}$ is Scott open.*

Proof. Let $\bigcap\{P \mid P \in K\}$ be Scott open and let $K \subseteq \bigcup\{\Sigma_a \mid a \in A\}$. Then $\bigvee A \in \bigcap\{P \mid P \in K\}$ since for each $P \in K$ there is an $a \in A$ such that $a \in P$, and hence $\bigvee A \in P$. Thus, there are $a_1, \ldots, a_n \in A$ with $a_1 \vee \cdots \vee a_n \in \bigcap\{P \mid P \in K\}$ and hence $K \subseteq \Sigma_{a_1 \vee \cdots \vee a_n} = \bigcup_{i=1}^n \Sigma_{a_i}$. If K is compact and $\bigvee A \in \bigcap\{P \mid P \in K\}$ then $K \subseteq \Sigma_{\bigvee A} = \bigcup\{\Sigma_a \mid a \in A\}$ and there are $a_1, \ldots, a_n \in A$ such that $K \subseteq \Sigma_{a_1 \vee \cdots \vee a_n} = \bigcup_{i=1}^n \Sigma_{a_i}$. Finally $a_1 \vee \cdots \vee a_n \in \bigcap\{P \mid P \in K\}$. □

7.4. The Hofmann-Lawson Duality. On the other hand, prime Scott open filters of $\mathcal{O}X$ give the points of the spectrum $\mathsf{Pt}(X)$:

Lemma. *The elements $P \in \mathsf{Pt}(X)$ are precisely the prime Scott open filters.*

Proof. A completely prime P is obviously Scott open. Now let P be Scott open and prime, and let $\bigvee_{i \in I} a_i \in P$. Since P is open there are $a_{i_1}, \ldots, a_{i_n}$ such that $a_{i_1} \vee \cdots \vee a_{i_n} \in P$. Since it is prime, $a_{i_j} \in P$ for some j. □

Proposition. *Let F be a Scott open filter of $\mathcal{O}X$ such that $a \in F$ and $b \notin F$. Then there is a complete filter $P \supseteq F$ such that $a \in P$ and $b \notin P$. Consequently, each Scott open filter is an intersection of complete filters.*

Proof. This is just the famous Birkhoff's Theorem with the openness added. Using Zorn's Lemma in the standard way (taking into account that unions of open sets are open), we obtain an open filter $P \supseteq F$ maximal with respect to the condition $b \notin P \ni a$. We will prove that it is prime (and hence, by the Lemma, complete). Suppose it is not; then there are $u, v \notin P$ such that $u \vee v \in P$. Set $G = \{x \mid x \vee v \in P\}$. Then G is obviously a Scott open filter and, because of the u, $P \subset G$. Thus, $b \in G$, $b \vee v \in P$ and we can repeat the procedure with $v, b \notin P$, $v \vee b \in P$ and $H = \{x \mid x \vee b \in P\}$ to obtain the contradiction $b = b \vee b \in P$. $\qquad\square$

If X is locally compact and $c \ll a$, interpolate inductively

$$a \gg x_1 \gg x_2 \gg \cdots \gg x_n \gg \cdots c,$$

choose the x_n fixedly for each such couple a, c, and set

$$F(a, c) = \{x \mid x \geq x_k \text{ for some } k\}.$$

Then $F(a, c)$ is obviously a Scott open filter. These filters are useful to prove the following theorem, that justifies the definition of locally compact locale.

Theorem. *Each locally compact locale is spatial, and functors* Lc *and* Pt *restrict to an equivalence between the category of sober locally compact spaces and the category of locally compact locales.*

Proof. Let X be a locally compact locale and consider $a, b \in \mathcal{O}X$ with $a \nleq b$. Then there is a $c \ll a$ such that $c \nleq b$. Hence $b \notin F(a, c) \ni a$ and, by the Proposition, there exists a complete P such that $b \notin P \ni a$.

Thus, since we already know (Exercise 7.1) that $\mathsf{Lc}(X)$, with locally compact space X, is locally compact, it suffices to show that each $\mathsf{Pt}(X)$, with locally compact locale X, is locally compact.

Let $P \in \Sigma_a$, that is, $a \in P$. Since $a = \bigvee\{x \mid x \ll a\}$ and P is open there is a $c \ll a$, $c \in P$. Set $K = \{Q \in \mathsf{Pt}(X) \mid F(a, c) \subseteq Q\}$. By the Proposition, $\bigcap K = F(a, c)$, and, by Lemma 7.3, K is compact. If $Q \in \Sigma_c$, that is, $c \in Q$, we have $F(a, c) \subseteq Q$, and if $F(a, c) \subseteq Q$ then $a \in Q$. Thus, $P \in \Sigma_c \subseteq K \subseteq \Sigma_a$. $\qquad\square$

By Proposition 7.2(1) the equivalence above restricts to an equivalence between the category of compact Hausdorff spaces and $\mathsf{KRegLoc}$. The contravariant version of Theorem 7.4 (in terms of continuous frames) is the well-known Hofmann-Lawson Duality ([31], see also [4]).

7.5. Preservation of products by Lc. The functor Lc is a left adjoint and generally does not preserve products. But we have:

Proposition. *The functor* Lc : Top $\to$ Loc *preserves finite products of sober completely regular locally compact spaces.*

Proof. In view of Theorem 7.4 it suffices to show that finite products in the smaller categories coincide with those in Top resp. Loc. This is obvious for locally compact spaces. Using Proposition 7.2 we infer from 3.4 that a product of two completely regular locally compact locales is locally compact (and completely regular, by Proposition 4.3). □

References

[1] P.S. Alexandroff, Zur Begründung der N-dimensionalen mengentheoretischen Topologie, *Math. Annalen* 94 (1925) 296-308.

[2] R.N. Ball and A.W. Hager, On the localic Yoshida representation of an archimedean lattice ordered group with weak unit, *J. Pure Appl. Algebra* 70 (1991) 17-43.

[3] B. Banaschewski, Frames and compactifications, in: *Extension Theory of Topological Structures and its Applications*, Deutscher Verlag der Wissenschaften (1969) 29-33.

[4] B. Banaschewski, The duality of distributive continuous lattices, *Canad. J. Math.* 32 (1980) 385-394.

[5] B. Banaschewski, Coherent frames, in: *Continuous Lattices*, Springer Lecture Notes in Math. 871 (1981) 1-11.

[6] B. Banaschewski, Another look at the localic Tychonoff theorem, *Comment. Math. Univ. Carolinae* 26 (1985) 619-630.

[7] B. Banaschewski, On proving the Tychonoff Product Theorem, *Kyungpook Math. J.* 30 (1990) 65-73.

[8] B. Banaschewski, On pushing out frames, *Comment. Math. Univ. Carolinae* 31 (1990) 13-21.

[9] B. Banaschewski, Compactification of frames, *Math. Nachr.* 149 (1990) 105-116.

[10] B. Banaschewski, Bourbaki's fixpoint lemma reconsidered, *Comment. Math. Univ. Carolinae* 33 (1992) 303-309.

[11] B. Banaschewski, Recent results in pointfree topology, in: *Papers on General Topology and Applications* (Brookville, NY, 1990), Ann. New York Acad. Sci. 659 (1992) 29-41.

[12] B. Banaschewski and C.J. Mulvey, Stone-Čech compactification of locales I, *Houston J. Math.* 6 (1980) 301-312.

[13] B. Banaschewski and C.J. Mulvey, Stone-Čech compactification of locales II, *J. Pure Appl. Algebra* 33 (1984) 107-122.

[14] B. Banaschewski and C.J. Mulvey, A constructive proof of the Stone-Weierstrass theorem, *J. Pure Appl. Algebra* 116 (1997) 25-40.

[15] B. Banaschewski and C.J. Mulvey, The spectral theory of commutative C^*-algebras: the constructive spectrum, *Quaest. Math.* 23 (2000) 425-464.

[16] B. Banaschewski and C.J. Mulvey, The spectral theory of commutative C*-algebras: the constructive Gelfand-Mazur theorem, *Quaest. Math.* 23 (2000) 465-488.

[17] B. Banaschewski and A. Pultr, Variants of openness, *Appl. Categ. Structures* 1 (1993) 181-190.

[18] B. Banaschewski and A. Pultr, Booleanization, *Cahiers Topologie Géom. Différentielle Catég.* 37 (1996) 41-60.

[19] J. Bénabou, Treillis locaux et paratopologies, *Séminaire Ehresmann* (1re année, exposé 2, Paris 1958).

[20] F. Borceux, *Handbook of Categorical Algebra*, Encyclopedia of Mathematics and its Applications 52 (Cambridge University Press, Cambridge 1994).

[21] N. Bourbaki, *Éléments de Mathématique: Topologie Générale* (Hermann, Paris, 1966).

[22] G. Bruns, Darstellungen und Erweiterungen geordneter Mengen II, *J. für Math.* 210 (1962) 1-23.

[23] C. Caratheodory, Über die Begrenzung einfach zusamenhängender Gebiete, *Math. Annalen* 73 (1913).

[24] X. Chen, *Closed Frame Homomorphisms*, Doctoral Dissertation (McMaster University, 1991).

[25] X. Chen, On binary coproducts of frames, *Comment. Math. Univ. Carolinae* 33 (1992) 699-712.

[26] C.H. Dowker and D. Strauss, Separation axioms for frames, *Colloq. Mat. Soc. J. Bolyai* 8 (1972) 223-240.

[27] C.H. Dowker and D. Strauss, Sums in the category of frames, *Houston J. Math.* 3 (1977) 7-15.

[28] C. Ehresmann, Gattungen von lokalen Strukturen, *Jber. Deutsch. Math. Verein* 60 (1957) 59-77.

[29] R. Engelking, *General Topology*, Sigma Series in Pure Mathematics 6 (Heldermann Verlag, Berlin, 1989).

[30] F. Hausdorff, *Grundzüge der Mengenlehre* (Veit & Co., Leipzig, 1914).

[31] K.H. Hofmann and J.D. Lawson, The spectral theory of distributive continuous lattices, *Trans. Amer. Math. Soc.* 246 (1978) 285-310.

[32] J. Isbell, Atomless parts of spaces, *Math. Scand.* 31 (1972) 5-32.

[33] P.T. Johnstone, Tychonoff Theorem without the axiom of choice, *Fund. Math.* 113 (1981) 21-35.

[34] P.T. Johnstone, *Stone Spaces*, Cambridge Studies in Advanced Mathematics 3 (Cambridge University Press, Cambridge 1982).

[35] P.T. Johnstone, The point of pointless topology, *Bull. Amer. Math. Soc. (N.S.)* 8 (1983) 41-53.

[36] P.T. Johnstone, The Art of Pointless Thinking: a Student's Guide to the Category of Locales, in: *Category Theory at Work* (Proc. Workshop Bremen 1990, edited by H. Herrlich and H.-E. Porst), Research and Exposition in Math. 18, Heldermann Verlag, Berlin (1991) 85-107.

[37] P.T. Johnstone, Complemented sublocales and open maps, preprint (Cambridge University, 2002).

[38] A. Joyal and M. Tierney, *An extension of the Galois Theory of Grothendieck*, Mem. Amer. Math. Soc. 309, 1984.

[39] I. Kříž, A constructive proof of the Tychonoff's theorem for locales, *Comment. Math. Univ. Carolinae* 26 (1985) 619-630.

[40] I. Kříž and A. Pultr, A spatiality criterion and an example of a quasitopology which is not a topology, *Houston J. Math.* 15 (1989) 215-234.

[41] F.W. Lawvere and R. Rosebrugh, *Sets for Mathematics* (Cambridge University Press, Cambridge, 2001).

[42] S. MacLane and I. Moerdijk, *Sheaves in Geometry and Logic: A First Introduction to Topos Theory* (Springer, Berlin, 1992).

[43] J.J. Madden and A. Molitor, Epimorphisms of frames, *J. Pure Appl. Algebra* 70 (1991) 129-132.

[44] J.C.C. McKinsey and A. Tarski, The algebra of topology, *Ann. Math.* 45 (1944) 141-191.

[45] K. Menger, Topology without points, *Rice Institute pamphlet* 27 (1940) 80-107.

[46] D. Papert and S. Papert, Sur les treillis des ouverts et paratopologies, *Séminaire Ehresmann* (1re année, exposé 1, Paris 1958).

[47] A. Pitts, Amalgamation and interpolation in the category of Heyting algebras, *J. Pure Appl. Algebra* 29 (1983) 155-165.

[48] T. Plewe, Localic products of spaces, *Proc. London Math. Soc.* (3) 73 (1996) 642-678.

[49] T. Plewe, A. Pultr and A. Tozzi, Regular monomorphisms of Hausdorff frames, *Appl. Categ. Structures* 9 (2001) 15-33.

[50] A. Pultr and A. Tozzi, Notes on Kuratowski-Mrówka theorems in point-free context, *Cahiers Topologie Géom. Différentielle Catég.* 33 (1992) 3-14.

[51] A. Pultr and A. Tozzi, Separation axioms and frame representation of some topological facts, *Appl. Categ. Structures* 2 (1994) 107-118.

[52] G. Sambin, Formal topology and domains, *Electron. Notes Theor. Comput. Sci.* 35 (2000) 14 pp.

[53] D. Scott, Continuous lattices, in: *Toposes, Geometry and Logic*, Springer Lecture Notes in Math. 247 (1972) 97-136.

[54] W. Sierpinski, La notion de derivée come base d'une theorie des ensembles abstraits, *Math. Annalen* 27 (1927) 321-337.

[55] H. Simmons, A framework for topology, in: *Logic Colloq. '77*, North-Holland (1978) 239-251.

[56] H. Simmons, The lattice theoretic part of topological separation properties, *Proc. Edinburgh Math. Soc.* (2) 21 (1978) 41-48.

[57] M.H. Stone, Boolean algebras and their application in topology, *Proc. Nat. Acad. Sci. USA* 20 (1934) 197-202.

[58] M.H. Stone, The theory of representations for Boolean algebras, *Trans. Amer. Math. Soc.* 40 (1936) 37-111.

[59] W.J. Thron, Lattice-equivalence of topological spaces, *Duke Math. J.* 29 (1962) 671-679.

[60] C. Townsend, *Preframe Techniques in Constructive Locale Theory*, Doctoral Dissertation (University of London, Imperial College 1996).

[61] J.J.C. Vermeulen, *Constructive techniques in functional analysis*, Doctoral Dissertation (University of Sussex 1987).

[62] J.J.C. Vermeulen, Some constructive results related to compactness and the (strong) Hausdorff property for locales, in: *Category Theory* (Proceedings, Como 1990), Springer Lecture Notes in Math. 1488 (1991) 401-409.

[63] J.J.C. Vermeulen, Proper maps of locales, *J. Pure Appl. Algebra* 92 (1994) 79-107.

[64] J.J.C. Vermeulen, A note on stably closed maps of locales, *J. Pure Appl. Algebra* 157 (2001) 335-339.

[65] S. Vickers, *Topology via Logic*, Cambridge Tracts in Theoretical Computer Science 5 (Cambridge University Press, Cambridge 1989).

[66] H. Wallman, Lattices of topological spaces, *Ann. Math.* 39 (1938) 112-126.

Department of Mathematics,
University of Coimbra,
3001-454 Coimbra, Portugal
E-mail: picado@mat.uc.pt

KAM and Institute of Theoretical Computer Science (ITI)
MFF, Charles University,
Prague, Czech Republic
E-mail: pultr@kam.ms.mff.cuni.cz

Department of Pure and Applied Mathematics
University of L'Aquila
67100 L'Aquila, Italy
E-mail: tozzi@univaq.it

III

A Functional Approach to General Topology

Maria Manuel Clementino, Eraldo Giuli, and Walter Tholen

In this chapter we wish to present a categorical approach to fundamental concepts of General Topology, by providing a category $\mathcal{X}$ with an additional structure which allows us to display more directly the geometric properties of the objects of $\mathcal{X}$ regarded as *spaces*. Hence, we study topological properties for them, such as Hausdorff separation, compactness, and local compactness, and we describe important topological constructions, such as the compact-open topology for function spaces and the Stone-Čech compactification. Of course, in a categorical setting, spaces are not investigated "directly" in terms of their points and neighborhoods, as in the traditional set-theoretic setting; rather, one exploits the fact that the relations of points and parts inside a space become categorically special cases of the relation of the space to other objects in its category. It turns out that many stability properties and constructions are established more economically in the categorical rather than the set-theoretic setting, leave alone the much greater level of generality and applicability.

The idea of providing a category with some kind of topological structure is certainly not new. So-called Grothendieck topologies (see Chapter VII) and, more generally, Lawvere-Tierney topologies are fundamental for the geometrically inspired construction of topoi. Specifically, these structures provide a notion of closure and thereby a notion of closed subobject, for every object in the category, such that all morphisms become "continuous". The notion of Dikranjan-Giuli *closure operator* [17] axiomatizes this idea and can be used to study topological properties categorically (see, for example, [9, 12]).

Here we go one step further and follow the approach first outlined in [48]. Hence, we provide the given category with a factorization structure and a special class $\mathcal{F}$ of morphisms of which we think as of the *closed morphisms*, satisfying three basic axioms; however, there is no a-priori provision of "closure" of subobjects. Depending on the parameter $\mathcal{F}$, we introduce and study basic topological properties as mentioned previously, but encounter also more advanced topics, such as exponentiability. The following features distinguish our presentation from the treatment of the same topological themes in existing topology books:

1. We emphasize the object-morphism interplay: every object notion corresponds to a morphism notion, and vice versa. For example, compact spaces *are* proper

maps, and conversely. Consequently, every theorem on compact spaces *is* a theorem on proper maps, and conversely.

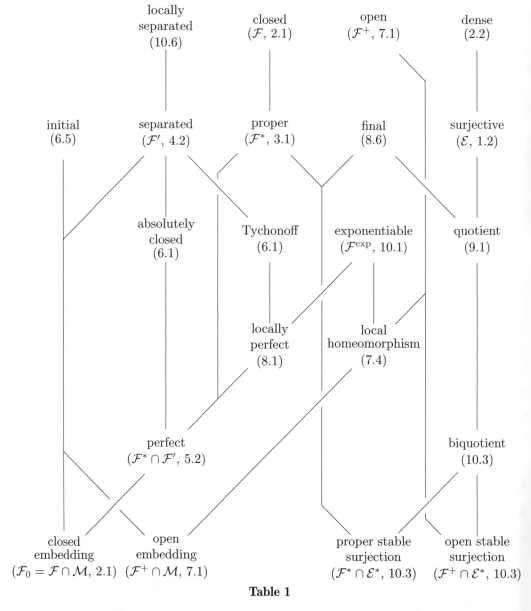

Table 1

2. We reach a wide array of applications not only by considering various categories, such as the categories of topological spaces, of locales (see Chapter II), etc., but also by varying the notion of closed morphism within the same category.

For example, in our setting, compact spaces behave just like discrete spaces, and perfect maps just like local homeomorphisms, because these notions arise from a common generalization.

3. Our categorical treatment is entirely constructive. In particular, we avoid the use of the Axiom of Choice (also when we interpret the categorical theory in the category of topological spaces), and we restrict ourselves to finitary properties (with the exception of some general remarks on the Tychonoff Theorem and the existence of the Stone-Čech compactification at the end of the chapter).

We expect the Reader to be familiar with basic categorical notions, such as adjoint functor and limit, specifically pullback, and we also assume familiarity with basic topological notions, such as topological space, neighborhood, closure of subsets, continuous map. The Reader will find throughout the chapter mostly easy exercises; those marked with ∗ are deemed to be more demanding.

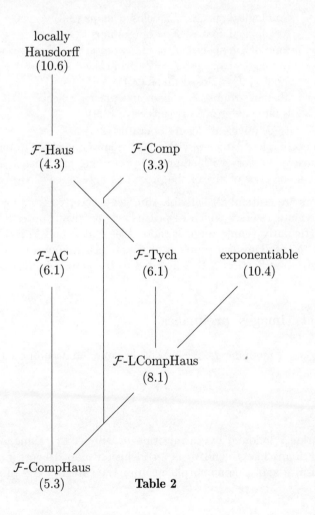

Table 2

Our basic hypotheses on the category we are working with are formulated as *axioms*

(F0)-(F2)	in 1.2	(axioms for factorization systems),
(F3)-(F5)	in 2.1	(axioms for closed maps),
(F6)-(F8)	in 11.1	(axioms giving closures),
(F9)	in 11.2	(infinite product axiom).

The morphism notions discussed in this chapter are summarized in Table 1, which also indicates the relevant subsections and the abbreviations used. Upward-directed lines indicate implications; some of these may need extra (technical) hypotheses.

Table 2 replicates part of Table 1 at the object level.

Many of these notions get discussed throughout the chapter in standard examples, as introduced in Section 2. We list some of them here in summary form, for the Reader's convenience:

$\mathcal{T}op$:	topological spaces, $\mathcal{F}$ = closed maps (2.3),
$\mathcal{T}op_{\text{open}}$:	topological spaces, $\mathcal{F}$ = open maps (2.8),
$\mathcal{T}op_{\text{clopen}}$:	topological spaces, $\mathcal{F}$ = maps preserving clopen sets (2.6),
$\mathcal{T}op_{\text{Zariski}}$:	topological spaces, $\mathcal{F}$ = Zariski-closed sets (2.7),
$\mathcal{L}oc$:	locales, $\mathcal{F}$ = closed maps (2.4),
$\mathcal{A}bGrp$:	abelian groups, $\mathcal{F}$ = homomorphisms whose restriction to the torsion subgroups is surjective (2.9),

any topos with a universal closure operator (2.5),
any finitely complete category with a Dikranjan-Giuli closure operator (2.5),
any lextensive category with summand-preserving morphisms (2.6),
any comma-category of any of the preceding categories (2.10).

The authors are grateful for valuable comments received from many colleagues and the anonymous referees which helped us getting this chapter into its current form. We particularly thank Jorge Picado, Aleš Pultr, and Peter Johnstone for their interest in resolving the problem of characterizing open maps of locales in terms of closure (see Section 7.3), which led to the comprehensive solution given in [31].

1. Subobjects, images, preimages

1.1. Motivation. A mapping $f : X \to Y$ of sets may be decomposed as

$$\begin{array}{ccc} & Z & \\ {\scriptstyle e}\nearrow & & \searrow{\scriptstyle m} \\ X & \xrightarrow{\quad f \quad} & Y \end{array} \tag{1}$$

with a surjection e followed by an injection m, simply by taking $Z = f[X]$, the image of f; e maps like f, and m is an inclusion map. Hence, $f = m \cdot e$ with an epimorphism e and a monomorphism m of the category $\mathcal{S}et$ of sets. If f is a homomorphism of groups X, Y, then Z is a subgroup of Y, and e and m become epi- and monomorphisms in the category $\mathcal{G}rp$ of groups, respectively. If f is a continuous mapping of topological spaces X, Y, then Z may be endowed with the subspace topology inherited from Y, and e and m are now epi- and monomorphisms in the category $\mathcal{T}op$ of topological spaces, respectively. However, the situation in $\mathcal{T}op$ is rather different from that in $\mathcal{S}et$ and $\mathcal{G}rp$: applying the decomposition (1) to a *mono*morphism f, in $\mathcal{S}et$ and $\mathcal{G}rp$ we obtain an isomorphism e, but not so in $\mathcal{T}op$, simply because the $\mathcal{S}et$-inverse of the continuous bijective mapping e may fail to be continuous.

In what follows we therefore consider a (potentially quite special) class $\mathcal{M}$ of monomorphisms and, symmetrically, a (potentially quite special) class $\mathcal{E}$ of epimorphisms in a category $\mathcal{X}$ and assume the existence of $(\mathcal{E}, \mathcal{M})$-decompositions for all morphisms, as follows.

1.2. Axioms for factorization systems. Throughout this chapter we work in a finitely complete category $\mathcal{X}$ with two distinguished classes of morphisms $\mathcal{E}$ and $\mathcal{M}$ such that

(F0) $\mathcal{M}$ is a class of monomorphisms and $\mathcal{E}$ is a class of epimorphisms in $\mathcal{X}$, and both are closed under composition with isomorphisms;

(F1) every morphism f decomposes as $f = m \cdot e$ with $m \in \mathcal{M}$, $e \in \mathcal{E}$;

(F2) every $e \in \mathcal{E}$ is *orthogonal* to every $m \in \mathcal{M}$ (written as $e \perp m$), that is: given any morphisms u, v with $m \cdot u = v \cdot e$, then there is a uniquely determined morphism w making the following diagram commutative:

$$\begin{array}{ccc} \cdot & \xrightarrow{\ u\ } & \cdot \\ {\scriptstyle e}\downarrow & {\scriptstyle w}\nearrow & \downarrow {\scriptstyle m} \\ \cdot & \xrightarrow{\ v\ } & \cdot \end{array} \qquad (2)$$

Any pair $(\mathcal{E}, \mathcal{M})$ satisfying conditions (F0)-(F2) is referred to as a *proper* (*orthogonal*) *factorization system* of $\mathcal{X}$; "proper" gets dropped if one allows $\mathcal{M}$ and $\mathcal{E}$ to be arbitrary morphism classes, i.e., if one drops the "mono" and "epi" condition in (F0). In that case the unicity requirement for w in (F2) becomes essential; however it is redundant in our situation.

Exercises.

1. Show that every category $\mathcal{X}$ has two *trivial* but generally non-proper factorization systems: (Iso, All), (All, Iso).
2. Show that (Epi, Mono) is a proper factorization system in $\mathcal{S}et$ and $\mathcal{G}rp$, but not in $\mathcal{T}op$.
3. Find a class $\mathcal{M}$ such that (Epi, $\mathcal{M}$) is a proper factorization system in $\mathcal{T}op$.
4. Show that $f : X \to Y$ is an epimorphism in the full subcategory $\mathcal{H}aus$ of Hausdorff spaces in $\mathcal{T}op$ if and only if f is dense (i.e., its image meets every non-empty open set in Y). Find a class $\mathcal{M}$ such that (Epi, $\mathcal{M}$) is a proper factorization system in $\mathcal{H}aus$. Compare this with $\mathcal{T}op$.

(Here Iso, Epi, Mono denotes the class of all iso-, epi-, monomorphisms in the respective category, and All the class of all morphisms.)

1.3. One parameter suffices. Having introduced factorization systems with two parameters we hasten to point out that one determines the other:

Lemma.
$$\mathcal{E} = \{f \in \operatorname{mor}\mathcal{X} \mid \forall m \in \mathcal{M} \ : \ f \perp m\},$$
$$\mathcal{M} = \{f \in \operatorname{mor}\mathcal{X} \mid \forall e \in \mathcal{E} \ : \ e \perp f\}.$$

Proof. By (F2), every $f \in \mathcal{E}$ satisfies $f \perp m$ for all $m \in \mathcal{M}$. Conversely, if $f \in \operatorname{mor}\mathcal{X}$ has this property, we may write $f = m \cdot e$ with $e \in \mathcal{E}$ and $m \in \mathcal{M}$, by (F1), and then apply the hypothesis on f as follows:

$$\tag{3}$$

The monomorphism m with $m \cdot w = 1$ is now recognized as an isomorphism, and we obtain $f = m \cdot e \in \mathcal{E}$ with (F0). This shows the first identity, the second follows dually. □

Exercises.
1. Show that the Lemma holds true without assuming properness of the factorization system.
2. Show that $(\mathcal{E}, \mathcal{M})$ is a (proper) factorization system of $\mathcal{X}$ if and only if $(\mathcal{M}, \mathcal{E})$ is a (proper) factorization system of $\mathcal{X}^{\mathrm{op}}$. Conclude that the second identity of the Lemma *follows* from the first.
3. Using properness of $(\mathcal{E}, \mathcal{M})$, show that if $g \cdot f = 1$, then $f \in \mathcal{M}$ and $g \in \mathcal{E}$. More generally: every *extremal monomorphism* (so that $m = h \cdot e$ with e epic only if e iso) lies in $\mathcal{M}$; dually, every extremal epimorphism lies in $\mathcal{E}$.

1.4. Properties of $\mathcal{M}$. We list some stability properties of $\mathcal{E}$ and $\mathcal{M}$:

Proposition.
(1) $\mathcal{E}$ and $\mathcal{M}$ are both closed under composition, and $\mathcal{E} \cap \mathcal{M} = \mathrm{Iso}$.
(2) If $n \cdot m \in \mathcal{M}$, then $m \in \mathcal{M}$.
(3) $\mathcal{M}$ is stable under pullback; hence, for every pullback diagram

$$\tag{4}$$

$n \in \mathcal{M}$ implies $n' \in \mathcal{M}$.

(4) $\mathcal{M}$ *is* stable under intersection (=multiple pullback)*; hence, if in the commutative diagram*

$$\begin{array}{ccc} & M_i & \\ {\scriptstyle j_i}\nearrow & & \searrow {\scriptstyle m_i} \\ M & \xrightarrow{\ m\ } & X \end{array} \qquad (5)$$

$(M, m, j_i)_{i \in I}$ *is the limit of the diagram given by the morphisms* $m_i \in \mathcal{M}$ $(i \in I)$, *then also* $m \in \mathcal{M}$.

(5) *With all* $m_i : X_i \to Y_i$ $(i \in I)$ *in* $\mathcal{M}$, *also* $\prod_{i \in I} m_i : \prod_{i \in I} X_i \to \prod_{i \in I} Y_i$ *is in* $\mathcal{M}$ *(if the products exist)*.

Proof. We show (1), (2) and leave the rest as an exercise (see below). Clearly, if we factor an isomorphism $f = m \cdot e$ as in (F1), both e, m are isomorphisms as well, hence $f \in \mathcal{E} \cap \mathcal{M}$ with (F0). Conversely, $f \perp f$ for $f \in \mathcal{E} \cap \mathcal{M}$ shows that f must be an isomorphism.

Given a composite morphism $n \cdot m$ in $\mathcal{M}$, we may factor $m = m' \cdot e$ with $e \in \mathcal{E}$, $m' \in \mathcal{M}$ and obtain from $e \perp n \cdot m$ a morphism w with $w \cdot e = 1$. Hence, the epimorphism e is an isomorphism, and $m = m' \cdot e \in \mathcal{M}$. This shows (2).

Having again a composite $f = n \cdot m$ with both $n, m \in \mathcal{M}$, we may now factor $f = m^* \cdot e^*$ with $e^* \in \mathcal{E}$, $m^* \in \mathcal{M}$. Then $e^* \perp n$ gives a morphism w with $w \cdot e^* = m \in \mathcal{M}$, hence $e^* \in \mathcal{E} \cap \mathcal{M} = \mathrm{Iso}$ by what already has been established, and $f \in \mathcal{M}$ follows.

The assertion on $\mathcal{E}$ follows dually. $\qquad\square$

Exercises.

1. Complete the proof of the Proposition, using the same factorization technique as in the part already established.
2. Formulate and prove the dual assertions for $\mathcal{E}$; for example: if $e \cdot d \in \mathcal{E}$, then $e \in \mathcal{E}$.
3. For any class $\mathcal{E}$ of morphisms, let $\mathcal{M} := \{f \in \mathrm{mor}\mathcal{X} \,|\, \forall e \in \mathcal{E} \;:\; e \perp f\}$. Show that $\mathcal{M}$ is *closed* under composition, and *under arbitrary limits*; that is: for functors $H, K : \mathcal{D} \to \mathcal{X}$ and any natural transformation $\mu : H \to K$ with $\mu_d \in \mathcal{M}$ for all $d \in \mathcal{D}$, also

   $$\lim \mu : \lim H \longrightarrow \lim K$$

   is in $\mathcal{M}$ (if the limits exist). Dualize this statement.
4. Prove that *any* pullback-stable class $\mathcal{M}$ satisfies: if $n \cdot m \in \mathcal{M}$ with n monic, then $m \in \mathcal{M}$.
5. * Prove that *any* class $\mathcal{M}$ closed under arbitrary limits (see 3) satisfies assertions (3)-(5) of the Proposition.

1.5. Subobjects. We usually refer to morphisms m in $\mathcal{M}$ as *embeddings* and call those with codomain X *subobjects* of X. The class of all subobjects of X is denoted by

$$\mathrm{sub}X.$$

It is preordered by

$$m \le m' \iff \exists j \ : \ m' \cdot j = m;$$

$$
\begin{array}{ccc}
M & \xrightarrow{\;j\;} & M' \\
& \searrow{\scriptstyle m} \quad \swarrow{\scriptstyle m'} & \\
& X &
\end{array}
\qquad (6)
$$

Such j is uniquely determined and necessarily belongs to $\mathcal{M}$. Furthermore, it is easy to see that

$$m \le m' \ \& \ m' \le m \iff \exists \text{ isomorphism } j \ : \ m' \cdot j = m;$$

we write $m \cong m'$ in this case and think of m and m' as *representing the same subobject of* X.

Exercises.

1. Let $A \subseteq X$ be sets. Prove that all injective mappings $m : M \to X$ with $m[M] = A$ represent the same subobject of X in $\mathcal{S}et$ (with $\mathcal{M} = $ Mono).
2. Prove that $m : M \to X$ in $\mathcal{M}$ is an isomorphism if and only if $m \cong 1_X$.

As a consequence of Proposition 1.4 we note:

Corollary. $(\mathcal{E}, \mathcal{M})$-*factorizations are essentially unique. Hence, if $f = m \cdot e = m' \cdot e'$ with $e, e' \in \mathcal{E}$, $m, m' \in \mathcal{M}$, then there is a unique isomorphism j with $j \cdot e = e'$, $m' \cdot j = m$; in particular, $m \cong m'$.*

Proof. Since $e \perp m'$ there is j with $j \cdot e = e'$, $m' \cdot j = m$, and with Prop. 1.4(1),(2),(2)^{op} we have $j \in \mathcal{E} \cap \mathcal{M} = $ Iso. $\qquad\square$

1.6. Image and preimage. For $f : X \to Y$ in $\mathcal{X}$ and $m \in \mathrm{sub}X$, one defines the *image* $f[m] \in \mathrm{sub}Y$ *of m under f* by an $(\mathcal{E}, \mathcal{M})$-factorization of $f \cdot m$, as in

$$
\begin{array}{ccc}
M & \xrightarrow{\;e\;} & f[M] \\
{\scriptstyle m}\downarrow & & \downarrow{\scriptstyle f[m]} \\
X & \xrightarrow{\;f\;} & Y
\end{array}
\qquad (7)
$$

The *preimage* (or *inverse image*) $f^{-1}[n] \in \mathrm{sub}X$ *of $n \in \mathrm{sub}Y$ under f* is given by the pullback diagram (cp. 1.4(3))

$$
\begin{array}{ccc}
f^{-1}[N] & \xrightarrow{\;f'\;} & N \\
{\scriptstyle f^{-1}[n]}\downarrow & & \downarrow{\scriptstyle n} \\
X & \xrightarrow{\;f\;} & Y
\end{array}
\qquad (8)
$$

The pullback f' of f along n is also called a *restriction of f*. Both, image and preimage are uniquely defined, up to isomorphism, and the constructions are related, as follows:

Lemma. *For $f : X \to Y$ and $m \in \mathrm{sub}X$, $n \in \mathrm{sub}Y$ one has:*

$$f[m] \leq n \Leftrightarrow m \leq f^{-1}[n].$$

Proof. Consider diagrams (7) and (8). Then, having $j : f[M] \to N$ with $n \cdot j = f[m]$ gives $k : M \to f^{-1}[N]$ with $f^{-1}[n] \cdot k = m$, by the pullback property of (8). Viceversa, the existence of k gives the existence of j since $e \perp n$. □

Exercises.

1. Conclude from the Lemma:
   (a) $m \leq f^{-1}[f[m]]$, $f[f^{-1}[n]] \leq n$;
   (b) $m \leq m' \Rightarrow f[m] \leq f[m']$, $n \leq n' \Rightarrow f^{-1}[n] \leq f^{-1}[n']$.
2. Show: $f^{-1}[1_Y] \cong 1_X$, and $(f \in \mathcal{E} \Leftrightarrow f[1_X] \cong 1_Y)$.

1.7. Image is left adjoint to preimage. Lemma 1.6 (with the subsequent Exercise) gives:

Proposition. *For every $f : X \to Y$ there is a pair of adjoint functors*

$$f[-] \dashv f^{-1}[-] : \mathrm{sub}Y \longrightarrow \mathrm{sub}X.$$

Consequently, $f^{-1}[-]$ preserves all infima and $f[-]$ preserves all suprema.

Exercises.

1. Complete the proof of the Proposition.
2. For $g : Y \to Z$, establish natural isomorphisms

   $$(g \cdot f)[-] \cong g[-] \cdot f[-], \quad (g \cdot f)^{-1}[-] \cong f^{-1}[-] \cdot g^{-1}[-],$$

   $$1_X[-] \cong \mathrm{id}_{\mathrm{sub}X}, \quad 1_X^{-1}[-] \cong \mathrm{id}_{\mathrm{sub}X}.$$

3. Show that the $\cong$-classes of elements in $\mathrm{sub}X$ carry the structure of a (possibly large) meet-semilattice, with the largest element represented by 1_X, and with the infimum $m \wedge n$ represented by $m \cdot (m^{-1}[n]) \cong n \cdot (n^{-1}[m])$.

1.8. Pullback stability. We give sufficient conditions for the image-preimage functors to be partially inverse to each other:

Proposition. *Let $f : X \to Y$ be a morphism in $\mathcal{X}$.*

(1) *If $f \in \mathcal{M}$, then $f^{-1}[f[m]] \cong m$ for all $m \in \mathrm{sub}X$.*
(2) *If every pullback of f along a morphism in $\mathcal{M}$ lies in $\mathcal{E}$ (so that in every pullback diagram (4) with $n \in \mathcal{M}$ one has $f' \in \mathcal{E}$), then $f[f^{-1}[n]] \cong n$ for all $n \in \mathrm{sub}Y$.*

Proof. (1) If $f \in \mathcal{M}$, in (7) we may take $f[m] = f \cdot m$ and $e = 1_M$, and since f is a monomorphism, (7) is now a pullback diagram, which implies the assertion.
(2) Under the given hypothesis, in (8) the morphism f' lies in $\mathcal{E}$, hence n is the image of $f^{-1}[n]$ under f. □

Exercises.

1. Prove that the sufficient condition given in (2) is also necessary.
2. Check that for the factorization systems of $\mathcal{S}et$, $\mathcal{G}rp$, $\mathcal{T}op$ given in Exercises 2 and 3 of 1.2, the class $\mathcal{E} = \text{Epi}$ is stable under pullback, but not so for the factorization system of $\mathcal{H}aus$ described in Exercise 4 of 1.2.
3. With the factorization system $(\text{Epi}, \mathcal{M})$ of $\mathcal{T}op$, find a morphism $f : X \to Y$ with $f^{-1}[f[M]] = M$ for all $M \subseteq X$ which fails to belong to $\mathcal{M}$.

2. Closed maps, dense maps, standard examples

2.1. Axioms for closed maps. A topology on a set X is traditionally defined by giving a system of open subsets of X which is stable under arbitrary joins (unions) and finite meets (intersections); equivalently, one may start with a system of closed subsets of X which is stable under arbitrary meets and finite joins, with the correspondence between the two approaches given by complementation. Continuity of $f : X \to Y$ is then characterized by preservation of closed subsets under inverse image. Among the continuous maps, these are those for which closedness of subsets is also preserved by image, called *closed maps*. Since closedness of subobjects is transitive, closed subobjects are precisely those subobjects for which the representing morphism is a closed map.

In what follows we assume that in our finitely complete category $\mathcal{X}$ with its proper $(\mathcal{E}, \mathcal{M})$-factorization system (satisfying (F0)-(F2)) we are given a special class $\mathcal{F}$ of morphisms of which we think as of the closed maps, satisfying the following conditions:

(F3) $\mathcal{F}$ contains all isomorphisms and is closed under composition;
(F4) $\mathcal{F} \cap \mathcal{M}$ is stable under pullback;
(F5) whenever $g \cdot f \in \mathcal{F}$ with $f \in \mathcal{E}$, then $g \in \mathcal{F}$.

We often refer to the morphisms in $\mathcal{F}$ as $\mathcal{F}$-*closed maps*, and to those of

$$\mathcal{F}_0 := \mathcal{F} \cap \mathcal{M}$$

as $\mathcal{F}$-*closed embeddings* . Having fixed $\mathcal{F}$ does not prevent us from considering further classes with the same properties; hence, we call any class $\mathcal{H}$ of morphisms in $\mathcal{X}$ $(\mathcal{E}, \mathcal{M})$-*closed* if $\mathcal{H}$ satisfies (F3)-(F5) in lieu of $\mathcal{F}$.

Exercises.

1. Show that the class of closed maps in $\mathcal{T}op$ (with the factorization structure of Exercise 3 of 1.2) satisfies (F3)-(F5).
2. A continuous map of topological spaces is *open* if images of open sets are open. Prove that the class of open morphisms in $\mathcal{T}op$ is (like the system of closed maps) $(\text{Epi}, \mathcal{M})$-closed.
3. Prove that each of the following classes in $\mathcal{X}$ is $(\mathcal{E}, \mathcal{M})$-closed: Iso, $\mathcal{E}$, $\mathcal{M}$, $\mathcal{F}_0$.

4. Given a pullback-stable class $\mathcal{H}_0 \subseteq \mathcal{M}$ containing all isomorphisms and being closed under composition, define the morphism class $\mathcal{H}$ by

$$(f : X \to Y) \in \mathcal{H} \Leftrightarrow \forall m \in \text{sub}X, \ m \in \mathcal{H}_0 \ : \ f[m] \in \mathcal{H}_0.$$

Show: $\mathcal{H} \cap \mathcal{M} = \mathcal{H}_0$, and $\mathcal{H}$ is $(\mathcal{E}, \mathcal{M})$-closed if every pullback of a morphism in $\mathcal{E}$ along a morphism in $\mathcal{H}_0$ lies in $\mathcal{E}$ (see Proposition 1.8(2)). Furthermore, $\mathcal{H}$ satisfies the dual of (F5): whenever $g \cdot f \in \mathcal{H}$ with $g \in \mathcal{M}$, then $f \in \mathcal{H}$.

5. Find an $(\mathcal{E}, \mathcal{M})$-closed system $\mathcal{H}$ which does not arise from a class $\mathcal{H}_0$ as described in Exercise 4.

The following statements follow immediately from the axioms (see also Exercise 3 of 1.7).

Proposition.

(1) *The $\mathcal{F}$-closed subobjects of an object X form (a possibly large) subsemilattice of the meet-semilattice* $\text{sub}X$.

(2) *Every morphism $f : X \to Y$ in $\mathcal{X}$ is $\mathcal{F}$-continuous, that is: $f^{-1}[-]$ preserves $\mathcal{F}$-closedness of subobjects.*

(3) *For every $\mathcal{F}$-closed morphism $f : X \to Y$ in $\mathcal{X}$, $f[-]$ preserves $\mathcal{F}$-closedness of subobjects* (but not necessarily viceversa, see Exercise 5).

□

2.2. Dense maps. A morphism $d : X \to Y$ of $\mathcal{X}$ is $\mathcal{F}$-*dense* if in any factorization $d = m \cdot h$ with an $\mathcal{F}$-closed subobject m one necessarily has $m \in \text{Iso}$.

Lemma. *d is $\mathcal{F}$-dense if and only if $d \perp m$ for all $m \in \mathcal{F}_0$.*

Proof. "if": From $d = m \cdot h$ with $m \in \mathcal{F}_0$ and $d \perp m$ one obtains m iso as in diagram (3). "only if": If $v \cdot d = m \cdot u$, one obtains h with $d = n \cdot h$, where $n = v^{-1}[m] \in \mathcal{F}_0$ by Prop. 2.1. Consequently, n is iso, and we may put $w = v' \cdot n^{-1}$ with the pullback v' of v along m. Then $m \cdot w = m \cdot v' \cdot n^{-1} = v \cdot n \cdot n^{-1} = v$, which implies $w \cdot u = d$, as desired. □

Corollary.

(1) *Any morphism in $\mathcal{E}$ is $\mathcal{F}$-dense.*

(2) *An $\mathcal{F}$-dense and $\mathcal{F}$-closed subobject is an isomorphism.*

(3) *The class of all $\mathcal{F}$-dense morphisms is closed under composition and satisfies the duals of properties (2)-(5) of Proposition* 1.4.

Proof. See Lemma 1.3 and Exercises 1.4. □

Exercises.

1. Prove that $f : X \to Y$ is an $\mathcal{F}$-dense morphism if and only if $f[1_X] : f[X] \to Y$ is an $\mathcal{F}$-dense embedding.

2. Show that in $\mathcal{T}op$ (with $\mathcal{F}$ the class of closed maps, as in Exercise 1 of 2.1) a morphism $d : X \to Y$ is $\mathcal{F}$-*dense* if and only if every non-empty open set in Y meets $d[X]$.

3. Verify that the class of $\mathcal{F}$-dense maps in $\mathcal{T}op$ is not (Epi, $\mathcal{M}$)-closed.

In what follows we shall discuss some important examples which we shall refer to later on as *standard examples*, by just mentioning the respective category $\mathcal{X}$; the classes $\mathcal{E}$, $\mathcal{M}$, $\mathcal{F}$ are understood to be chosen as set out below.

2.3. Topological spaces with closed maps. The category $\mathcal{T}op$ will always be considered with its (Epi, $\mathcal{M}$)-factorization structure; here $\mathcal{M}$ is the class of *embeddings*, i.e., of those monomorphisms $m : M \to X$ for which every closed set of M has the form $m^{-1}[F]$ for a closed set F of X. (The fact that $\mathcal{M}$ is precisely the class of regular monomorphisms of the category $\mathcal{T}op$ is not being used in this chapter.) We emphasize again that here $\mathcal{E} =$ Epi is stable under pullback (see Exercise 2 of 1.8). In the standard situation, $\mathcal{F}$ is always the class of closed maps. $\mathcal{F}$-closedness and $\mathcal{F}$-density for subobjects take on the usual meaning.

2.4. Locales with closed maps. The category $\mathcal{L}oc$ of *locales* has been introduced as the dual of the category $\mathcal{F}rm$ of *frames* in Chapter II. The (Epi, $\mathcal{M}$)-factorization structure of $\mathcal{L}oc$ coincides with the ($\mathcal{E}$, Mono)-factorization structure of $\mathcal{F}rm$, where $\mathcal{E}$ is the class of surjective frame homomorphisms; following the notation of Chapter II, this means:

$$m : M \to X \text{ in } \mathcal{M} \;\Leftrightarrow\; m^* : \mathcal{O}X \to \mathcal{O}M \text{ in } \mathcal{E}.$$

The subobject m is *closed* if there is $a \in \mathcal{O}X$ and an isomorphism $h : \mathcal{O}M \to\uparrow a$ of frames such that $h \cdot m^* = \breve{a}$, with $\breve{a} : \mathcal{O}X \to\uparrow a$ given by $\breve{a}(x) = a \vee x$. The class $\mathcal{F}$ of closed maps in $\mathcal{L}oc$ can now be defined just as in $\mathcal{T}op$, by preservation of closed subobjects under image (see II.5.1).

We point out a major difference to the situation in $\mathcal{T}op$: the class Epi in $\mathcal{L}oc$ fails to be stable under pullback (see II.5.2). However, pullbacks of epimorphisms along complemented sublocales (which include both closed sublocales and open sublocales – see II.2.7) are again epic in $\mathcal{L}oc$, since for any morphism f in $\mathcal{L}oc$ the inverse image functor $f^{-1}[-]$ preserves finite suprema, and, hence, also complements (see II.3.8).

Next we mention three (schemes of) examples of a more general nature.

2.5. A topos with closed maps with respect to a universal closure operator. Every elementary topos $\mathcal{S}$ has an (Epi, Mono)-factorization system, and the class Epi is stable under pullback (see [28], 1.5). A *universal closure operator* $j = (j_X)_{X \in \text{ob}\mathcal{S}}$ is given by functions $j_X : \text{sub}X \to \text{sub}X$ satisfying the conditions 1. $m \le j_X(m)$, 2. $m \le m' \Rightarrow j_X(m) \le j_X(m')$, 3. $j_X(j_X(m)) \cong j_X(m)$, 4. $f^{-1}[j_Y(n)] \cong j_X(f^{-1}[n])$, for all $f : X \to Y$, $m, m' \in \text{sub}X$, $n \in \text{sub}Y$ (see [28], 3.13, and [36], V.1). Taking for $\mathcal{F}$ now the class of those morphisms f for which not only $f^{-1}[-]$ but also $f[-]$ commutes with the closure operator, one obtains an (Epi, Mono)-closed system $\mathcal{F}$ for $\mathcal{S}$; its closed maps are also described by the fact that image preserves closedness of subobjects (characterized by $m \cong j_X(m)$). A very particular feature of this structure is that the class of $\mathcal{F}$-dense maps is stable under pullback; see Chapter VII.

Although we shall not discuss this aspect any further in this chapter, we mention that the notion of universal closure operator may be generalized dramatically if we are just interested in the generation of a closed system. In fact, for any finitely complete category $\mathcal{X}$ with a proper $(\mathcal{E}, \mathcal{M})$-factorization system and a so-called Dikranjan-Giuli closure operator one may take for $\mathcal{F}$ those morphisms for which image commutes with closure (see [18, 9] for details, as well as Section 11 below).

2.6. An extensive category with summand-preserving morphisms. A finitely complete category $\mathcal{X}$ with binary coproducts (denoted by $X + Y$) is *(finitely) extensive* if the functor

$$\mathrm{pb}_{X,Y} : \mathcal{X}/(X + Y) \longrightarrow \mathcal{X}/X \times \mathcal{X}/Y$$

(given by pullback along the coproduct injections) is an equivalence of categories, for all objects X and Y; equivalently, one may ask its left adjoint (given by co-product) to be an equivalence, see [6, 5, 7]. (Note that in Chapter VII *extensive* is used in the *infinitary* sense where the binary coproducts are replaced by arbitrary ones, see VII.4.1; however, in this chapter extensive always means finitely extensive.) In order to accommodate our setting we also assume that $\mathcal{X}$ has a proper $(\mathcal{E}, \mathcal{M})$-factorization structure such that every coproduct injection lies in $\mathcal{M}$. Taking for $\mathcal{F}_0$ now those morphisms $m : M \to X$ for which there is some morphism $n : N \to X$ such that $X \cong M + N$ with coproduct injections m, n, one lets $\mathcal{F}$ contain those morphisms for which image preserves the class $\mathcal{F}_0$: see Exercise 4 of 2.1.

For example, $\mathcal{X} = \mathcal{T}op$ is extensive, and the $\mathcal{F}$-closed morphisms are precisely those maps which map clopen (=closed and open) sets onto clopen sets. Whenever we refer to $\mathcal{T}op$ with this structure $\mathcal{F}$ and its (Epi, $\mathcal{M}$)-factorization system, we write $\mathcal{T}op_{\mathrm{clopen}}$ instead of $\mathcal{T}op$. Another prominent example of an extensive category is the dual of the category $\mathcal{CR}ng$ of commutative rings (where $\mathcal{CR}ng$ is considered with its $(\mathcal{E}, \mathrm{Mono})$-factorization system); for details we refer to [5].

Exercises.

1. Prove that the finitely complete category $\mathcal{X}$ with binary coproducts is extensive if and only if in every commutative diagram

$$\begin{array}{ccc} A & \longrightarrow & C & \longleftarrow & B \\ \downarrow & & \downarrow & & \downarrow \\ X & \xrightarrow{\ i_1\ } & X + Y & \xleftarrow{\ i_2\ } & Y \end{array} \qquad (9)$$

the top row represents a coproduct (so that $C \cong A + B$) precisely when the two squares are pullback diagrams.
2. Prove that it is enough to require the existence of the coproduct $1 + 1$ and that $\mathrm{pb}_{1,1}$ is an equivalence to obtain extensivity, including the existence of all binary coproducts.
3. Prove that in an extensive category coproduct injections are monomorphisms, and that they are *disjoint*, i.e. the pullback of i_1, i_2 is an initial object. In particular, $\mathcal{X}$ has an initial object.
4. Prove that an extensive category is *distributive*, so that the canonical morphism $(X \times Y_1) + (X \times Y_2) \to X \times (Y_1 + Y_2)$ is an isomorphism. In addition, $X \times 0 \cong 0$, with 0 denoting the initial object (see Ex. 3).

5. Verify the extensivity of $\mathcal{T}op$ and $\mathcal{CR}ng^{\mathrm{op}}$, and of every topos.

2.7. Algebras with Zariski-closed maps. For a fixed set K let $\Omega = \{\omega_i : K^{n_i} \to K \mid i \in I\}$ be a given class of operations on K of arbitrary arities; hence, the n_i are arbitrary cardinal numbers, and there is no condition on the size of the indexing system I. The category Ω-$\mathcal{S}et$ has as objects sets X which come equipped with an Ω-subalgebra $A(X)$ of K^X; here K^X carries the pointwise structure of an Ω-algebra:

$$(K^X)^{n_i} \cong (K^{n_i})^X \to K^X.$$

A morphism $f : X \to Y$ in Ω-$\mathcal{S}et$ must satisfy $K^f[A(Y)] \subseteq A(X)$, that is: $f \cdot \beta \in A(X)$ for all $\beta \in A(Y)$. Like $\mathcal{T}op$, Ω-$\mathcal{S}et$ has a proper (Epi, $\mathcal{M}$)-factorization structure, with the morphisms in $\mathcal{M}$ represented by embeddings $M \hookrightarrow X$, so that $A(M) = \{\alpha|_M \mid \alpha \in A(X)\}$. Such a subobject M is called *Zariski-closed* if any $x \in X$ with

$$\forall \alpha, \beta \in A(X) \, (\alpha|_M = \beta|_M \Rightarrow \alpha(x) = \beta(x))$$

already lies in M. The class $\mathcal{F}$ of closed maps contains precisely those maps for which image preserves Zariski-closedness (see [15, 22]).

We mention in particular two special cases. First, take $K = 2 = \{0, 1\}$, with the operations given by the frame structure of $\{0 \le 1\}$, that is: by arbitrary joins and finite meets. Equipping a set X with an Ω-subalgebra of 2^X is putting a topology (of open sets) on X, and we have Ω-$\mathcal{S}et = \mathcal{T}op$. The Zariski-closed sets on X define a new topology on X with respect to which a basic neighborhood of a point $x \in X$ has the form $U \cap \overline{\{x\}}$, where U is a neighborhood of x and $\overline{\{x\}}$ the closure of $\{x\}$ in the original topology. Whenever we refer to $\mathcal{T}op$ with $\mathcal{F}$ given as above, we shall write $\mathcal{T}op_{\mathrm{Zariski}}$.

Secondly, in the "classical" situation, one considers the ground field K in the category of commutative K-algebras, so the operations in Ω are

$$0, 1 : K^0 \to K, \quad a(-) : K \to K, \quad +, \cdot : K^2 \to K,$$

where $a(-)$ is (left) multiplication by a, for every $a \in K$. Hence, an Ω-set X comes with a subalgebra $A(X)$ of K^X, and $M \subseteq X$ is Zariski-closed if

$$M = \{x \in X \mid \forall \alpha \in A(X) \, (\alpha|_M = 0 \Rightarrow \alpha(x) = 0)\}.$$

Putting $J = \{\alpha \in A(X) \mid \alpha|_M = 0\}$ one sees that the Zariski-closed sets are all of the form

$$Z(J) = \{x \in X \mid \forall \alpha \in J \, : \, \alpha(x) = 0\}$$

for some ideal J of $A(X)$. For $X = K^n$ and $A(X)$ the algebra of polynomial functions $K^n \to K$ we have exactly the classical Zariski-closed sets considered as in commutative algebra and algebraic geometry (see [15]).

Exercises.

1. Let X be the cartesian product of the sets X_ν, with projections p_ν ($\nu \in N$). If each X_ν is an Ω-set, let $A(X)$ be the subalgebra of K^X generated by $\{\alpha \cdot p_\nu \mid \nu \in N, \alpha \in A(X_\nu)\}$. Conclude that Ω-$\mathcal{S}et$ has products. More generally, show that Ω-$\mathcal{S}et$ is small-complete.

2. Show that in $\mathcal{T}op_{\text{Zariski}}$ every Zariski-closed subset of a subspace Y of X is the intersection of a Zariski-closed set in X with Y. Generalize this statement to $\Omega\text{-}\mathcal{S}et$.

3. Consider $K = 2 = \{0, 1\}$ and $\Omega = \emptyset$. Show that $\Omega\text{-}\mathcal{S}et$ is the category $\mathcal{C}hu_2^{\text{ext}}$ of extensive Boolean Chu spaces (see [42]), whose objects are sets X with a "generalized topology" $\tau \subseteq 2^X$ (no further condition on τ), and whose morphisms are "continuous maps". Prove that a subspace M of X is Zariski-closed if for every $x \in X \setminus M$ there are ("open sets") $U, V \in \tau$ with $U \cap M = V \cap M$ and $x \in U \setminus V$.

2.8. Topological spaces with open maps. Here we mention a somewhat counter-intuitive but nevertheless important example. Again, we consider $\mathcal{T}op$ with its (Epi, $\mathcal{M}$)-factorization system, but now take $\mathcal{F}$ to be the class of open maps (see Exercise 2 of 2.1). Here the $\mathcal{F}$-dense maps $f : X \to Y$ are characterized by the property that the closure $\overline{\{y\}}$ of every point $y \in Y$ meets the image $f[X]$. When equipping $\mathcal{T}op$ with this structure, we refer to $\mathcal{T}op_{\text{open}}$ rather than to $\mathcal{T}op$ (see 2.3).

2.9. Abelian groups with torsion-preserving maps. The category $\mathcal{A}b\mathcal{G}rp$ of abelian groups has, like $\mathcal{G}rp$, an (Epi, Mono)-factorization system. Here we consider for $\mathcal{F}$ the class of homomorphisms $f : A \to B$ which map the torsion subgroup of A onto the torsion subgroup of B: $f[\text{Tor}A] = \text{Tor}B = \{b \in B \mid nb = 0 \text{ for some } n \geq 1\}$. (More generally, for any ring R, we could consider here any preradical of R-modules in lieu of Tor, or even of any finitely complete category with a proper factorization system, in lieu of $\mathcal{A}b\mathcal{G}rp$; see [18].) The map f is $\mathcal{F}$-dense if and only if $B = \text{im} f + \text{Tor}B$.

We end our list of standard examples with an important general procedure:

2.10. Passing to comma categories. Given our standard setting with $\mathcal{X}$, $\mathcal{E}$, $\mathcal{M}$, $\mathcal{F}$ satisfying (F0)-(F5) and a fixed object B in $\mathcal{X}$, let $\mathcal{X}/B$ be the *comma category* (or *sliced category*) of *objects* (X, p) *over* B, simply given by morphisms $p : X \to B$ in $\mathcal{X}$; a morphism $f : (X, p) \to (Y, q)$ in $\mathcal{X}/B$ is a morphism $f : X \to Y$ in $\mathcal{X}$ with $q \cdot f = p$. With the forgetful functor $\Sigma_B : \mathcal{X}/B \to \mathcal{X}$, putting $\mathcal{E}_B := \Sigma_B^{-1}(\mathcal{E})$, $\mathcal{M}_B := \Sigma_B^{-1}(\mathcal{M})$, $\mathcal{F}_B := \Sigma_B^{-1}(\mathcal{F})$, it is easy to check that (F0)-(F5) hold true in $\mathcal{X}/B$.

Proposition. *For every object B of $\mathcal{X}$, $(\mathcal{E}_B, \mathcal{M}_B)$ is a proper factorization system of $\mathcal{X}/B$ and $\mathcal{F}_B$ is an $(\mathcal{E}_B, \mathcal{M}_B)$-closed class.* $\square$

In what follows, whenever we refer to $\mathcal{X}/B$, we think of it as being structured according to the Proposition. This applies particularly to $\mathcal{T}op/B$, $\mathcal{T}op_{\text{open}}/B$, etc. We normally drop the subscript B when referring to $\mathcal{E}_B$, $\mathcal{M}_B$, $\mathcal{F}_B$.

3. Proper maps, compact spaces

3.1. Compact objects. The pullback of a closed map in $\mathcal{T}op$ need not be a closed map. In fact, the only map from the real line $\mathbb{R}$ to a one-point space 1 is closed, but the pullback of this map along itself is not:

$$
\begin{array}{ccc}
\mathbb{R} \times \mathbb{R} & \longrightarrow & \mathbb{R} \\
\downarrow & & \downarrow \\
\mathbb{R} & \longrightarrow & 1
\end{array}
\tag{10}
$$

neither projection of $\mathbb{R} \times \mathbb{R}$ preserves closedness of the subspace $\{(x, y) \mid x \cdot y = 1\}$. Hence, we should pay attention to those closed maps which *are* stable under pullback.

In our finitely complete category $\mathcal{X}$ with its proper $(\mathcal{E}, \mathcal{M})$-factorization system and the distinguished class $\mathcal{F}$ of closed maps (satisfying (F0)-(F5)) we call a morphism f *$\mathcal{F}$-proper* if f belongs *stably* to $\mathcal{F}$, so that in every pullback diagram

$$
\begin{array}{ccc}
W & \xrightarrow{f'} & Z \\
{\scriptstyle g'}\downarrow & & \downarrow{\scriptstyle g} \\
X & \xrightarrow{f} & Y
\end{array}
\tag{11}
$$

$f' \in \mathcal{F}$; laxly we speak of an $\mathcal{F}$-proper map in $\mathcal{X}$ and denote by $\mathcal{F}^*$ the class of all $\mathcal{F}$-proper maps. Note that since g and g' may be chosen as identity morphisms, $\mathcal{F}^* \subseteq \mathcal{F}$.

Checking $\mathcal{F}$-properness of a morphism may be facilitated by the following criterion:

Proposition. *A morphism $f : X \to Y$ is $\mathcal{F}$-proper if and only if every restriction of $f \times 1_Z : X \times Z \to Y \times Z$ (see (8)) is $\mathcal{F}$-closed, for every object Z.*

Proof. Since with $f \times 1_Z$ also every of its restrictions is a pullback of f, the necessity of the condition is clear. That it is sufficient for $\mathcal{F}$-properness follows from a factorization of the pullback diagram (11), as follows:

$$
\begin{array}{ccc}
W & \xrightarrow{f'} & Z \\
{\scriptstyle <g',f'>}\downarrow & & \downarrow{\scriptstyle <g,1_Z>} \\
X \times Z & \xrightarrow{f \times 1_Z} & Y \times Z \\
{\scriptstyle p_1}\downarrow & & \downarrow{\scriptstyle p_1} \\
X & \xrightarrow{f} & Y
\end{array}
\tag{12}
$$

Since both the total diagram and its lower rectangle are pullback diagrams, so is its upper rectangle. Moreover, $n := <g, 1_Z> \in \mathcal{M}$ (see Exercise 3 of 1.3), so that the condition of the Proposition applies here. $\qquad\square$

Often, but not always, the condition of the Proposition can be simplified. Let us say that $\mathcal{F}$ *is stable under restriction* if every restriction of an $\mathcal{F}$-closed morphism is $\mathcal{F}$-closed, i.e., if $\mathcal{F}$ is stable under pullback along morphisms in $\mathcal{M}$. In this case f is $\mathcal{F}$-proper if and only if $f \times 1_Z$ is $\mathcal{F}$-closed for all Z.

Exercises.

1. Show that in $\mathcal{T}op$, $\mathcal{T}op_{\text{open}}$, $\mathcal{T}op_{\text{clopen}}$, $\mathcal{T}op_{\text{Zariski}}$, the class $\mathcal{F}$ is stable under restriction.
2. Using $\mathcal{F} = \mathcal{E}$, show that generally $\mathcal{F}$ fails to be stable under restriction.
3. Let $\mathcal{F} = \mathcal{H}$ with $\mathcal{H}$ as in Exercise 4 of 2.1, and assume that $\mathcal{E}$ is stable under pullback along $\mathcal{M}$, and that every $\mathcal{F}$-closed subobject of M with $m : M \to X$ in $\mathcal{M}$ is the inverse image of an $\mathcal{F}$-closed subobject of X under m. Show that $\mathcal{F}$ is stable under restriction. Revisit Exercise 1.

3.2. Stability properties. We show some important stability properties for $\mathcal{F}$-proper maps:

Proposition.

(1) *The class $\mathcal{F}^*$ contains $\mathcal{F} \cap \mathcal{M}$ and is closed under composition.*
(2) *$\mathcal{F}^*$ is the largest pullback-stable subclass of $\mathcal{F}$.*
(3) *If $g \cdot f \in \mathcal{F}^*$ with g monic, then $f \in \mathcal{F}^*$.*
(4) *If $g \cdot f \in \mathcal{F}^*$ with $f \in \mathcal{E}^*$, then $g \in \mathcal{F}^*$; here $\mathcal{E}^*$ is the class of the morphisms stably in $\mathcal{E}$.*

Proof. (1) follows from (F4) and the fact that a pullback of $g \cdot f$ can be obtained as a composite of a pullback of g preceded by a pullback of f. Likewise, (2) and (4) follow from the composability of adjacent pullback diagrams. (3) is a consequence of Exercise 4 of 1.4. $\qquad \square$

Exercises.

1. Make sure that you understand every detail of the proof of the Proposition.
2. Show that with $f : X \to Y$ also the following morphisms are $\mathcal{F}$-proper:
   (a) $f \cdot m : M \to Y$ for every $\mathcal{F}$-closed embedding $m : M \to X$;
   (b) $f' : f^{-1}[N] \to N$ for every $n \in \text{sub}Y$.
3. Using only closure of $\mathcal{F}^*$ under composition and pullback stability, show that with $f_1 : X_1 \to Y_1$ and $f_2 : X_2 \to Y_2$ also $f_1 \times f_2 : X_1 \times X_2 \to Y_1 \times Y_2$ is $\mathcal{F}$-proper. (We say that $\mathcal{F}^*$ is closed under finite products. Hint: $f_1 \times f_2 = (f_1 \times 1) \cdot (1 \times f_2)$.)
4. Show that in the standard example $\mathcal{T}op$ (see 2.3) one has $\mathcal{E}^* = \mathcal{E}$; likewise for $\mathcal{T}op_{\text{open}}$, $\mathcal{T}op_{\text{clopen}}$, $\mathcal{T}op_{\text{Zariski}}$, and for every topos, but not for $\mathcal{H}aus$ and $\mathcal{L}oc$.

3.3. Compact objects. An object X of $\mathcal{X}$ is called $\mathcal{F}$-*compact* if the unique morphism $!_X : X \to 1$ to the terminal object is $\mathcal{F}$-proper. Since the pullbacks of $!_X$ are the projections $X \times Y \to Y$, $Y \in \mathcal{X}$, to say that X is $\mathcal{F}$-compact means precisely that any such projection must be in $\mathcal{F}$. Before exhibiting this notion in terms of examples, let us draw some immediate conclusions from Prop. 3.2:

Theorem.

    (1) *For any $\mathcal{F}$-proper $f : X \to Y$ with Y $\mathcal{F}$-compact, X is $\mathcal{F}$-compact.*

    (2) *For any $f : X \to Y$ in $\mathcal{E}^*$ with X $\mathcal{F}$-compact, Y is $\mathcal{F}$-compact.*

    (3) *The full subcategory $\mathcal{F}$-Comp of $\mathcal{F}$-compact objects in $\mathcal{X}$ is closed under finite products and under $\mathcal{F}$-closed subobjects.*

Proof. (1) Since
$$!_X = (X \xrightarrow{f} Y \xrightarrow{!_Y} 1),$$
the assertion follows from the compositivity of $\mathcal{F}^*$ (Prop. 3.2(1)).

(2) For the same reason one may apply Prop. 3.2(4) here.

(3) Since $\mathcal{F}^*$ contains all isomorphisms one has $1 \in \mathcal{F}$-Comp, and for $X, Y \in \mathcal{F}$-Comp one can apply (1) to the projection $X \times Y \to Y$. Likewise, one can apply (1) to $f \in \mathcal{F}_0$ to obtain closure under $\mathcal{F}$-closed subobjects.   □

Exercises.

    1. Show that every $\mathcal{F}$-closed subobject of an $\mathcal{F}$-compact object is $\mathcal{F}$-compact.

    2. Find an example in $\mathcal{T}op$ of a morphism $f : X \to Y$ with $X, Y \in \mathcal{F}$-Comp which fails to be $\mathcal{F}$-proper.

    3. Show that $\mathcal{F}$-Comp in $\mathcal{T}op$ fails to be closed under finite limits.

3.4. Categorical compactness in $\mathcal{T}op$. We show that in $\mathcal{T}op$ (with $\mathcal{F} = \{$closed maps$\}$) our notion of $\mathcal{F}$-compactness coincides with the usual notion of compactness given by the Heine-Borel open-cover property:

Theorem (Kuratowski and Mrówka). *For a topological space X, the following are equivalent:*

    (i) *X is $\mathcal{F}$-compact, i.e., the projection $X \times Y \to Y$ is closed for all $Y \in \mathcal{T}op$;*

    (ii) *for every family of open sets $U_i \subseteq X$ ($i \in I$) with $X = \bigcup_{i \in I} U_i$, there is a finite set $F \subseteq I$ with $X = \bigcup_{i \in F} U_i$.*

Proof. (ii) $\Rightarrow$ (i): This part is the well-known Kuratowski Theorem, we only sketch a possible proof here (see also Exercise 1 below). To see that $B = p_2(A)$ is closed for $A \subseteq X \times Y$ closed and $p_2 : X \times Y \to Y$ the second projection, assume $y \in \overline{B} \setminus B$. Then $\{A \cap p_2^{-1}[V] \mid V$ neighborhood of $y\}$ is a base of a filter $\mathbb{F}$ on $X \times Y$. The filter $p_1(\mathbb{F})$ on X must have a cluster point x, hence
$$x \in \bigcap \{\overline{p_1[A \cap p_2^{-1}[V]]} \mid V \text{ neighborhood of } y\},$$
which implies $(x, y) \in \overline{A} \setminus A$: contradiction.

(i) $\Rightarrow$ (ii): With the given open cover $\{U_i \mid i \in I\}$ of X, define a topological space Y with underlying set $X \cup \{\infty\}$ (with $\infty \notin X$) by
$$K \subseteq Y \text{ closed} :\Leftrightarrow \infty \in K \text{ or } K \subseteq \bigcup_{i \in F} U_i \text{ for some finite } F \subseteq I.$$
It now suffices to show that $X \subseteq Y$ is closed in Y, and for that it suffices to prove $p_2(\overline{\Delta_X}) = X$, with the closure of $\Delta_X = \{(x, x) \mid x \in X\}$ formed in $X \times Y$. But the

assumption $\infty \in p_2(\overline{\Delta_X})$ would give an $x \in X$ with $(x, \infty) \in \overline{\Delta_X}$. Since x lies in some U_i we would then have $(x, \infty) \in U_i \times (Y \setminus U_i)$, an open set in $X \times Y$ which does not meet Δ_X, a contradiction! $\qquad \square$

We note that the proof of the Theorem does not require the Axiom of Choice. This applies also to the following exercise.

Exercise. Extending and refining the argumentation in the proof of the Theorem, prove the equivalence of the following statements for a topological space X:

   (i) X is $\mathcal{F}$-compact;
  (ii) $p_2 : X \times Y \to Y$ is closed for every zerodimensional, normal Hausdorff space Y;
 (iii) X has the Heine-Borel open-cover property;
 (iv) every filter $\mathbb{F}$ on X has a cluster point, i.e. $\bigcap\{\overline{F} \mid F \in \mathbb{F}\} \neq \emptyset$.

3.5. Fibres of proper maps. In 3.3 we defined $\mathcal{F}$-compactness via $\mathcal{F}$-properness. We could have proceeded conversely, using the following fact:

Proposition. $f : X \to Y$ *is $\mathcal{F}$-proper if and only if (X, f) is an $\mathcal{F}$-compact object of $\mathcal{X}/Y$ (see 2.10).*

Proof. The terminal object in $\mathcal{X}/Y$ is $(Y, 1_Y)$, and the unique morphism $(X, f) \to (Y, 1_Y)$ is f itself. $\qquad \square$

The question remains whether $\mathcal{F}$-properness can be characterized by $\mathcal{F}$-compactness within the category $\mathcal{X}$. Towards this we first note:

Corollary. *For the following statements on $f : X \to Y$ in $\mathcal{F}$, one has* (i) $\Rightarrow$ (ii) $\Rightarrow$ (iii):

   (i) *f is $\mathcal{F}$-proper;*
  (ii) *for every pullback diagram (11), if Z is $\mathcal{F}$-compact, so is W;*
 (iii) *all fibres of f are $\mathcal{F}$-compact, where a fibre F of f occurs in any pullback diagram*

$$
\begin{array}{ccc}
F & \longrightarrow & 1 \\
\downarrow & & \downarrow \\
X & \xrightarrow{\;f\;} & Y
\end{array}
\tag{13}
$$

Proof. (i) $\Rightarrow$ (ii): Since $\mathcal{F}^*$ is pullback-stable, in (11) with f also f' is $\mathcal{F}$-proper. Hence, the assertion of (ii) follows from Theorem 3.3(1).

(ii) $\Rightarrow$ (iii) follows by putting $Z = 1$, which is $\mathcal{F}$-compact by Theorem 3.3(3). $\quad\square$

We now show (choice free) that in $\mathcal{T}op$ condition (iii) is already sufficient for properness:

Theorem. *In $\mathcal{T}op$ a map is $\mathcal{F}$-proper if and only if it is closed and has compact fibres.*

Proof. Let $f : X \to Y$ be closed with compact fibres, and for any space Z let $A \subseteq X \times Z$ be closed; we must show that $B := (f \times 1_Z)[A] \subseteq Y \times Z$ is closed.

Hence, for any point (y, z) in the complement $(Y \times Z) \setminus B$ we must find open sets $V_0 \subseteq Y$, $W_0 \subseteq Z$ with $(y, z) \in V_0 \times W_0 \subseteq (Y \times Z) \setminus B$. The system of all pairs (U, W) where U is an open neighborhood in X of some $x \in f^{-1}y$ and W is an open neighborhood in Z of z with $U \times W \subseteq (X \times Z) \setminus A$ has the property that its first components cover the compact fibre $f^{-1}y$. Hence, we obtain finitely many open sets $U_1, \cdots, U_n, W_1, \cdots, W_n$ ($n \geq 0$) with $f^{-1}y \subseteq U_0 := \bigcup_{i=1}^{n} U_i$, $z \in W_0 := \bigcap_{i=1}^{n} W_i$ and $U_0 \times W_0 \subseteq (X \times Z) \setminus A$. Since f is closed, the set $V_0 := Y \setminus f(X \setminus U_0)$ is open and has the required properties.   □

However, in general, in $\mathcal{T}op/B$ not even condition (ii) is sufficient for $\mathcal{F}$-properness, as the following example shows:

Example. Let B be any indiscrete topological space with at least 2 points, let X be any non-compact topological space, and let $q : 1 \to B$ be any map. Then, with $p = q \cdot f$, the unique map $f : X \to 1$ becomes a morphism $f : (X, p) \to (1, q)$ in $\mathcal{T}op/B$ which is $\mathcal{F}$-closed but not $\mathcal{F}$-proper. However, condition (ii) of the Corollary is vacuously satisfied since, given any pullback diagram (11) (with $Y = 1$) in $\mathcal{X}$ which then represents a pullback diagram also in $\mathcal{X}/B$, we observe that the object $(Z, q \cdot g)$ is never $\mathcal{F}$-compact in $\mathcal{T}op/B$ unless Z is empty; indeed, for $Z \neq \emptyset$ the constant map $q \cdot g$ is not proper, since its image is not closed in B.

Exercises.

1. In $\mathcal{T}op_{\mathrm{open}}$, prove that every map in $\mathcal{F} = \{f \mid f \text{ open}\}$ is $\mathcal{F}$-proper. Conclude that every object is $\mathcal{F}$-compact.
2. * Prove: a space X in $\mathcal{T}op_{\mathrm{Zariski}}$ is $\mathcal{F}$-compact if and only if if it is compact with respect to its Zariski topology (see 2.7). Conclude that this is the case precisely when (a) every subspace of X is compact (in $\mathcal{T}op$) and (b) every closed subspace of X has the form $\overline{\{x_1\}} \cup \cdots \cup \overline{\{x_n\}}$ for finitely many points $x_1, \cdots, x_n$ in X, $n \geq 0$.

3.6. Proper maps of locales. In $\mathcal{L}oc$, $\mathcal{F}$-proper maps are characterized by:

Theorem (Vermeulen [49, 50]). *For $f : X \to Y$ in $\mathcal{L}oc$, the following are equivalent:*

(i) *f is $\mathcal{F}$-proper;*
(ii) *$f \times 1_Z : X \times Z \to Y \times Z$ is $\mathcal{F}$-closed for all locales Z;*
(iii) *the restriction $f[-] : \mathcal{C}X \to \mathcal{C}Y$ of the direct-image map to closed sublocales is well defined and preserves filtered infima.*

For the proof of this theorem we must refer to Vermeulen's papers [49, 50]. However, we may point out that the equivalence proof for (i), (ii) given in [50] makes use of a refined version of Exercise 3 of 3.1. We leave it as an exercise to show that when exploiting property (iii) in case $Y = 1$, one obtains (see Theorem II.6.5):

Corollary (Pultr-Tozzi [43]). *A locale X is $\mathcal{F}$-compact if and only if every open cover of X contains a finite subcover.*

3.7. Sums of proper maps and compact objects. Next we want to examine the behavior of $\mathcal{F}$-compact objects and $\mathcal{F}$-proper maps under the formation of finite coproducts. Clearly, some compatibility between pullbacks and finite coproducts has to be in place in order to establish any properties, in addition to closedness of $\mathcal{F}$ under finite coproducts, so that with $f_1 : X_1 \to Y_1$, $f_2 : X_2 \to Y_2$ also $f_1 + f_2 : X_1 + X_2 \to Y_1 + Y_2$ is in $\mathcal{F}$. The notion of extensive category (as given in 2.6) fits our requirements:

Proposition. *Assume that the finitely complete category $\mathcal{X}$ has finite coproducts and is extensive, and that $\mathcal{F}$ is closed under finite coproducts. Then:*

(1) *the morphism class $\mathcal{F}^*$ is closed under finite coproducts;*

(2) *the subcategory $\mathcal{F}$-Comp is closed under finite coproducts in $\mathcal{X}$ if and only if the canonical morphisms*

$$0 \longrightarrow X \text{ and } X + X \longrightarrow X$$

are $\mathcal{F}$-closed, for all objects X.

Proof. (1) Diagram (14) shows how one may obtain the pullback f' of $f = f_1 + f_2$ along h: pulling back h along the injections of $Y = Y_1 + Y_2$

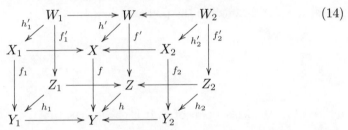 (14)

one obtains h_1 and h_2, along which one pulls back f_1 and f_2 to obtain f_1' and f_2' and the induced arrows $W_1 \to W \leftarrow W_2$ giving the commutative back faces. These are pullback diagrams since all other vertical faces are; by extensivity then, $f' = f_1' + f_2'$. Hence with $f_1, f_2 \in \mathcal{F}^*$ we have $f_1', f_2' \in \mathcal{F}$ and then $f' \in \mathcal{F}$, so that $f \in \mathcal{F}^*$.

(2) Once we have $1 + 1 \in \mathcal{F}$-Comp

$$
\begin{array}{ccccc}
X & \longrightarrow & X + Y & \longleftarrow & Y \\
\downarrow & & \downarrow{\scriptstyle !_X + !_Y} & & \downarrow \\
1 & \longrightarrow & 1 + 1 & \longleftarrow & 1
\end{array}
\qquad (15)
$$

we obtain $X + Y \in \mathcal{F}$-Comp whenever $X, Y \in \mathcal{F}$-Comp since $!_{X+Y} = !_{1+1} \cdot (!_X + !_Y)$, by an application of (1). Now,

$$
\begin{aligned}
1 + 1 \in \mathcal{F}\text{-Comp} \quad &\Leftrightarrow \quad \forall X : ((1+1) \times X \to X) \in \mathcal{F} \\
&\Leftrightarrow \quad \forall X : (X + X \to X) \in \mathcal{F},
\end{aligned}
$$

since $(1 + 1) \times X \cong X + X$ (see Exercise 4 of 2.6). Furthermore,

$$
\begin{aligned}
0 \in \mathcal{F}\text{-Comp} \quad &\Leftrightarrow \quad \forall X : (0 \times X \to X) \in \mathcal{F} \\
&\Leftrightarrow \quad \forall X : (0 \to X) \in \mathcal{F},
\end{aligned}
$$

since $0 \times X \cong 0$ (see Exercise 4 of 2.6). $\qquad\qquad\qquad\qquad\square$

Exercises.

1. Check that the hypotheses of the Theorem are satisfied in $\mathcal{T}op$, $\mathcal{T}op_{\text{open}}$, $\mathcal{T}op_{\text{clopen}}$, $\mathcal{T}op_{\text{Zariski}}$, and in any extensive category with $\mathcal{F}$ as in 2.6. Conclude that in the four listed categories of topological spaces $\mathcal{F}$-Comp is closed under finite coproducts.

2. Show that in $\mathcal{X} = (\mathcal{CR}ng)^{\text{op}}$ with $\mathcal{F}$ as in 2.6, the only $\mathcal{F}$-compact rings (up to isomorphism) are $\mathbb{Z}$ and $\{0\}$, and that $\mathcal{F}$-Comp fails to be closed under finite coproducts in $\mathcal{X}$.

4. Separated maps, Hausdorff spaces

4.1. Separated morphisms. With every morphism f in our category $\mathcal{X}$, structured by $\mathcal{E}$, $\mathcal{M}$, $\mathcal{F}$ as in 2.1, we may associate the morphism

$$\delta_f :< 1_X, 1_X >: X \longrightarrow X \times_Y X$$

where $X \times_Y X$ belongs to the pullback diagram

$$
\begin{array}{ccc}
X \times_Y X & \xrightarrow{\;f_2\;} & X \\
{\scriptstyle f_1}\downarrow & & \downarrow{\scriptstyle f} \\
X & \xrightarrow{\;f\;} & Y
\end{array}
\qquad (16)
$$

representing the kernel pair of f. It is easy to see that

$$X \xrightarrow{\;\delta_f\;} X \times_Y X \overset{f_1}{\underset{f_2}{\rightrightarrows}} X \qquad (17)$$

is an equalizer diagram. Instead of asking whether $f \in \mathcal{F}^*$, in this section we investigate those f with $\delta_f \in \mathcal{F}^*$. Actually, since $\delta_f \in \mathcal{M}$ and $\mathcal{F}_0 = \mathcal{F} \cap \mathcal{M}$ is stable under pullback, it is enough to require $\delta_f \in \mathcal{F}$.

We call a morphism f in $\mathcal{X}$ $\mathcal{F}$-*separated* if $\delta_f \in \mathcal{F}$; laxly we speak of an $\mathcal{F}$-separated map and denote by $\mathcal{F}'$ the class of all $\mathcal{F}$-separated maps.

Example. In $\mathcal{T}op$, to say that $\Delta_X = \{(x, x) \,|\, x \in X\}$ is closed in the subspace $X \times_Y X = \{(x_1, x_2) \,|\, f(x_1) = f(x_2)\}$ of $X \times X$ is to say that $(X \times_Y X) \setminus \Delta_X$ is open, and this means that for all $x_1, x_2 \in X$ with $f(x_1) = f(x_2)$, $x_1 \neq x_2$ there are open neighborhoods $U_1 \ni x_1$, $U_2 \ni x_2$ in X with $U_1 \times_Y U_2 \subseteq (X \times_Y X) \setminus \Delta_X$, i.e., $U_1 \cap U_2 = \emptyset$. Hence, in $\mathcal{T}op$, a map $f : X \to Y$ is $\mathcal{F}$-separated if and only if distinct points in the same fibre of f may be separated by disjoint neighborhoods in X. Such maps are called *separated* in fibred topology.

Exercises. Prove:

1. $f : X \to Y$ is $\mathcal{F}$-separated in $\mathcal{T}op_{\text{open}}$ if and only if f is *locally injective*, so that for every point x in X there is a neighborhood U of x such that $f|_U : U \to Y$ is an injective map.

2. The $\mathcal{F}$-separated maps in $\mathcal{T}op_{\text{clopen}}$ are precisely the separated and locally injective maps. Show that neither of the latter two properties implies the other.
3. $f : X \to Y$ is $\mathcal{F}$-separated in $\mathcal{T}op_{\text{Zariski}}$ if and only if distinct points in the same fibre have distinct neighborhood filters.
4. $f : A \to B$ is $\mathcal{F}$-separated in $\mathcal{A}b\mathcal{G}rp$ if and only if the restriction of f to $\text{Tor}A$ is injective, i.e., $\ker f|_{\text{Tor}A} = 0$.

4.2. Properties of separated maps. When establishing stability properties for $\mathcal{F}'$, it is important to keep in mind that $\mathcal{F}'$ depends only on $\mathcal{F}_0 = \mathcal{F} \cap \mathcal{M}$; hence, only the behavior of $\mathcal{F}_0$ plays a role in what follows.

Proposition.

(1) *The class $\mathcal{F}'$ contains all monomorphisms of $\mathcal{X}$ and is closed under composition.*
(2) *$\mathcal{F}'$ is stable under pullback.*
(3) *Whenever $g \cdot f \in \mathcal{F}'$, then $f \in \mathcal{F}'$.*
(4) *Whenever $g \cdot f \in \mathcal{F}'$ with $f \in \mathcal{E} \cap \mathcal{F}^*$, then $g \in \mathcal{F}'$.*

Proof. (1) Monomorphisms are characterized as those morphisms f with δ_f iso. Let us now consider $f : X \to Y$, $g : Y \to Z$ and their composite $h = g \cdot f$. Then there is a unique morphism $t : X \times_Y X \to X \times_Z X$ with $h_1 \cdot t = f_1$, $h_2 \cdot t = f_2$. It makes the following diagram commutative:

$$\begin{array}{ccccc}
X & \xrightarrow{\delta_f} & X \times_Y X & \xrightarrow{f \cdot f_1} & Y \\
{\scriptstyle 1_X}\downarrow & & {\scriptstyle t}\downarrow & & \downarrow{\scriptstyle \delta_g} \\
X & \xrightarrow{\delta_h} & X \times_Z X & \xrightarrow{f \times f} & Y \times_Z Y
\end{array} \qquad (18)$$

The right square is in fact a pullback diagram since t is the equalizer of $f \cdot h_1, f \cdot h_2$ (with h_1, h_2 the kernelpair of h). Consequently, with δ_g also t is in $\mathcal{F}$, and then with δ_f also $\delta_h = t \cdot \delta_f$ is in $\mathcal{F}$.

(2)

$$\begin{array}{ccccccc}
W & \xrightarrow{\delta_{f'}} & W \times_Z W & \rightrightarrows & W & \xrightarrow{f'} & Z \\
{\scriptstyle k'}\downarrow & \boxed{3} & \downarrow{\scriptstyle k''} & \boxed{2} & \downarrow{\scriptstyle k'} & \boxed{1} & \downarrow{\scriptstyle k} \\
X & \xrightarrow{\delta_f} & X \times_Y X & \rightrightarrows & X & \xrightarrow{f} & Y
\end{array} \qquad (19)$$

Since with $\boxed{1}$ also $\boxed{2}\,\&\,\boxed{1}$ and then $\boxed{3}$ are pullback diagrams, the assertion follows from the pullback stability of $\mathcal{F} \cap \mathcal{M}$.

(3) Going back to (18), if $\delta_h = t \cdot \delta_f$ is in the pullback-stable class $\mathcal{F}_0$, so is δ_f since t is monic (see Exercise 4 of 1.4).

(4) Since $f = f \cdot f_1 \cdot \delta_f$, from (18) one has $\delta_g \cdot f = (f \times f) \cdot \delta_h$. Now, if $h \in \mathcal{F}'$ and $f \in \mathcal{F}^*$, then $(f \times f) \cdot \delta_h = \delta_g \cdot f \in \mathcal{F}$ (see Exercise 1). Hence, if also $f \in \mathcal{E}$, $\delta_f \in \mathcal{F}$ follows with (F5). $\qquad \square$

Exercises.

1. Show that with $f_1, f_2 \in \mathcal{F}'$ also $f_1 \times f_2 \in \mathcal{F}'$ (see Exercise 3 of 3.2).
2. In generalization of 1, prove that $\mathcal{F}'$ is closed under those limits under which $\mathcal{F} \cap \mathcal{M}$ is closed. Hint: In the setting of Exercise 3 of 1.4, prove the formula $\delta_{\lim \mu} = \lim_d \delta_{\mu_d}$.

4.3. Separated objects. An object X of $\mathcal{X}$ is called $\mathcal{F}$-*separated* (or $\mathcal{F}$-*Hausdorff*) if the unique morphism $!_X : X \rightarrow 1$ is $\mathcal{F}$-separated; this simply means that $\delta_X = <1_X, 1_X> : X \rightarrow X \times X$ must be in $\mathcal{F}$.

Theorem. *The following conditions are equivalent for an object X:*

   (i) *X is $\mathcal{F}$-separated;*
  (ii) *every morphism $f : X \rightarrow Y$ is $\mathcal{F}$-separated;*
 (iii) *there is an $\mathcal{F}$-separated morphism $f : X \rightarrow Y$ with Y $\mathcal{F}$-separated;*
 (iv) *for every object Y the projection $X \times Y \rightarrow Y$ is $\mathcal{F}$-separated;*
  (v) *for every $\mathcal{F}$-separated object Y, $X \times Y$ is $\mathcal{F}$-separated;*
 (vi) *for every $\mathcal{F}$-proper morphism $f : X \rightarrow Y$ in $\mathcal{E}$, Y is $\mathcal{F}$-separated;*
(vii) *in every equalizer diagram*

$$E \xrightarrow{\;\;u\;\;} Z \rightrightarrows X,$$

   u is $\mathcal{F}$-closed.

Proof. Since $!_X = !_Y \cdot f$, the equivalence of (i), (ii), (iii) follows from compositivity and left cancellation of $\mathcal{F}'$ (Prop. 4.2(1),(3)). Since the projection $X \times Y \rightarrow Y$ is a pullback of $X \rightarrow 1$, (i) $\Rightarrow$ (iv) follows from pullback stability of $\mathcal{F}'$ (Prop. 4.2(2)), and (iv) $\Rightarrow$ (v) from its compositivity again. For (v) $\Rightarrow$ (i) consider $Y = 1$, which is trivially $\mathcal{F}$-separated. For (i) $\Rightarrow$ (vi) apply Prop. 4.2(4) with $g = !_Y$, and for (vi) $\Rightarrow$ (i) let $f = 1_X$. Finally (i) $\Leftrightarrow$ (vii) follows since such equalizers u are precisely the pullbacks of δ_X. $\qquad\Box$

Corollary. *The full subcategory $\mathcal{F}$-Haus of $\mathcal{F}$-separated objects is closed under finite limits and under subobjects in $\mathcal{X}$. In fact, for every monomorphism $m : X \rightarrow Y$, with Y also X is $\mathcal{F}$-separated.*

Proof. Since every monomorphism is $\mathcal{F}$-separated, consider (iii), (iv) of the Theorem. $\qquad\Box$

Exercises.

1. In $\mathcal{T}op$ and $\mathcal{L}oc$, $\mathcal{F}$-separation yields the usual notion of Hausdorff separation of these categories.
2. Show that in $\mathcal{T}op_{\text{open}}$ and in $\mathcal{T}op_{\text{clopen}}$ the $\mathcal{F}$-separated objects are the discrete spaces. Conclude that, in general, $\mathcal{F}$-Haus fails to be closed under (infinite) products in $\mathcal{X}$.
3. Prove that in $\mathcal{T}op_{\text{Zariski}}$ $\mathcal{F}$-Haus is the category of T0-spaces (i.e., those spaces in which distinct points have distinct neighborhood filters).

4. The quasi-component $q_X(M)$ of a subset M of a topological space X is the intersection of all clopen subsets of X containing M. Let $\mathcal{T}op_{\text{quasicomp}}$ denote the category $\mathcal{T}op$ with its (Epi, $\mathcal{M}$)-factorization structure and take for $\mathcal{F}$ the class all q-preserving maps $f : X \to Y$, i.e., $f[q_X(M)] = q_Y(f[M])$. Show that $\mathcal{F}$-Haus in $\mathcal{T}op_{\text{quasicomp}}$ is the category of totally-disconnected spaces (i.e., spaces in which all components are singletons).

5. Prove that in $\mathcal{A}b\mathcal{G}rp$ $\mathcal{F}$-Haus is the category of torsion-free abelian groups (i.e., groups in which $na = 0$ implies $n = 0$ or $a = 0$).

4.4. Sums of separated maps and objects. Closure of $\mathcal{F}'$ and of $\mathcal{F}$-Haus under finite coproducts requires not only a stability property of $\mathcal{F}_0$ under finite coproducts but, like in 3.7, also extensivity of $\mathcal{X}$:

Proposition. *Assume that the finitely complete category $\mathcal{X}$ has finite coproducts and is extensive, and that the coproduct of two $\mathcal{F}$-closed subobjects is $\mathcal{F}$-closed. Then:*

(1) *the morphism class $\mathcal{F}'$ is closed under finite coproducts;*

(2) *the subcategory $\mathcal{F}$-Haus is closed under finite coproducts in $\mathcal{X}$ if and only if $1 + 1 \in \mathcal{F}$-Haus.*

Proof. (1) For $f_1 : X_1 \to Y_1$, $f_2 : X_2 \to Y_2$, by extensivity, as a coproduct of two pullback squares also the right vertical face of (20) is a pullback square.

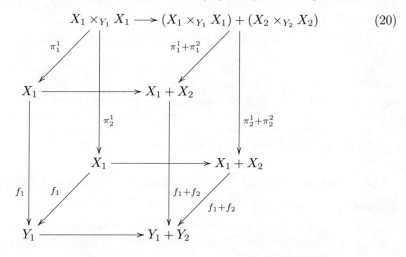

$$(X_1 \times_{Y_1} X_1) + (X_2 \times_{Y_2} X_2) \cong (X_1 + X_2) \times_{Y_1 + Y_2} (X_1 + X_2),$$

and the formula $\delta_{f_1 + f_2} \cong \delta_{f_1} + \delta_{f_2}$ follows immediately. Consequently, $f_1, f_2 \in \mathcal{F}'$ implies $f_1 + f_2 \in \mathcal{F}'$.

(2) We saw in (1) that $(\pi_1^1 + \pi_1^2, \pi_2^1 + \pi_2^2)$ is the kernelpair of $f_1 + f_2$. Exploiting this fact in case $Y_1 = Y_2 = 1$, we see that the canonical morphism u, making diagram

(21) commutative, is actually an equalizer of $(f_1 + f_2) \cdot \pi_1, (f_1 + f_2) \cdot \pi_2$.

$$
\begin{array}{ccc}
X_1 + X_2 & \xrightarrow{\quad 1 \quad} & X_1 + X_2 \\[4pt]
{\scriptstyle \delta_{X_1} + \delta_{X_2}} \downarrow & & \downarrow {\scriptstyle \delta_{X_1 + X_2}} \\[4pt]
(X_1 \times X_1) + (X_2 \times X_2) & \xrightarrow{\quad u \quad} & (X_1 + X_2) \times (X_1 + X_2) \\[4pt]
{\scriptstyle \pi_1^1 + \pi_1^2} \downdownarrows {\scriptstyle \pi_2^1 + \pi_2^2} & & {\scriptstyle \pi_1} \downdownarrows {\scriptstyle \pi_2} \\[4pt]
X_1 + X_2 & \xrightarrow{\quad 1 \quad} & X_1 + X_2
\end{array}
\tag{21}
$$

Hence, if $1 + 1$ is $\mathcal{F}$-separated, then u is $\mathcal{F}$-closed (see (vii) of Theorem 4.3), and with $\delta_{X_1}, \delta_{X_2} \in \mathcal{F}$ we obtain $\delta_{X_1 + X_2} \in \mathcal{F}$. Since $0 \times 0 \cong 0$, trivially $\delta_0 \in \mathcal{F}$. □

Exercises.

1. Show that, in $\mathcal{T}op$, $\mathcal{T}op_{\text{open}}$, $\mathcal{T}op_{\text{clopen}}$ and $\mathcal{T}op_{\text{Zariski}}$, $\mathcal{F}'$ and $\mathcal{F}$-Haus are closed under finite coproducts. Likewise in $\mathcal{A}b\mathcal{G}rp$.
2. Show that, in general, $1+1$ fails to be $\mathcal{F}$-separated. (Consider, for example, $\mathcal{F} = \text{Iso}$.)

4.5. Separated objects in the slices. We have presented $\mathcal{F}$-separation for objects as a special case of the morphism notion. We could have proceeded conversely, using:

Proposition. $f : X \to Y$ *is $\mathcal{F}$-separated if and only if* (X, f) *is an $\mathcal{F}$-separated object of* $\mathcal{X}/Y$ *(see 2.10).*

Proof. $\delta_f : X \to X \times_Y X$ in $\mathcal{X}$ serves as the morphism $\delta_{(X,f)} : (X, f) \to (X, f) \times (X, f)$ in $\mathcal{X}/Y$. □

Exercises.

1. Prove Corollary 4.3 without recourse to the previous Theorem, just using the definition of $\mathcal{F}$-Hausdorff object. Then apply the assertions of the Corollary to $\mathcal{X}/Y$ in lieu of $\mathcal{X}$ and express the assertions in terms of properties of $\mathcal{F}^*$.
2. Prove that with $\mathcal{X}$ also each slice $\mathcal{X}/Y$ is extensive. (Hint: Slices of slices of $\mathcal{X}$ are slices of $\mathcal{X}$.)

We finally mention three important "classical" examples, without proofs.

Examples.

(1) Since commutative unital R-algebras A are completely described by morphisms $R \to A$ in $\mathcal{C}\mathcal{R}ng$, the opposite of the category $\mathcal{C}\mathcal{A}lg_R$ is extensive, by Exercise 2:
$$\mathcal{C}\mathcal{A}lg_R^{\text{op}} \cong \mathcal{C}\mathcal{R}ng^{\text{op}}/R.$$
An $\mathcal{F}$-separated object in this extensive category is precisely a *separable* R-algebra; see [3].

(2) In a topos, considered as an extensive category (see Exercise 5 of 2.6), $\mathcal{F}$-separated means *decidable* (see [2], p.444). To say that every object is decidable is equivalent to the topos being Boolean.

(3) An object in a topos provided with a universal closure operator j (see 2.5) is $\mathcal{F}$-separated precisely when it is *j-separated* in the usual sense (see [36], p.227).

5. Perfect maps, compact Hausdorff spaces

5.1. Maps with compact domain and separated codomain. There is an easy but fundamental property which links compactness and separation, in the general setting provided by 2.1:

Proposition. *If X is $\mathcal{F}$-compact and Y $\mathcal{F}$-separated, then every morphism $f : X \to Y$ is $\mathcal{F}$-proper.*

Proof. One factors f through its "graph":

$$
\begin{array}{ccc}
 & X \times Y & \\
{}^{<1_X,f>}\nearrow & & \searrow{}^{p_2} \\
X & \xrightarrow[\quad f \quad]{} & Y
\end{array}
\tag{22}
$$

Since $< 1_X, f > \cong (f \times 1_Y)^{-1}[\delta_Y]$, this morphism is $\mathcal{F}$-closed, in fact $\mathcal{F}$-proper, when Y is $\mathcal{F}$-separated. Likewise p_2 is $\mathcal{F}$-closed, even $\mathcal{F}$-proper, when X is $\mathcal{F}$-compact. Closure of $\mathcal{F}^*$ under composition shows that f is $\mathcal{F}$-proper. □

Corollary. *If Y is $\mathcal{F}$-separated and $\mathcal{F}$-compact, then $f : X \to Y$ is $\mathcal{F}$-proper if and only if X is $\mathcal{F}$-compact.*

Proof. Combine the Theorem with Theorem 3.3(1). □

Exercises.

1. Show that a subobject of an $\mathcal{F}$-separated object with $\mathcal{F}$-compact domain is $\mathcal{F}$-closed.
2. Show that the assumption of $\mathcal{F}$-separation for Y is essential in the Theorem.

5.2. Perfect maps. Theorem 5.1 leads to a strengthening of Proposition 3.2(3), as follows:

Proposition. *If $g \cdot f \in \mathcal{F}^*$ with $g \in \mathcal{F}'$, then $f \in \mathcal{F}^*$.*

Proof. With $f : X \to Y$, $g : Y \to Z$, if $g \cdot f \in \mathcal{F}^*$ and $g \in \mathcal{F}'$, then $f : (X, g \cdot f) \to (Y, g)$ is a morphism in $\mathcal{X}/Z$ with $\mathcal{F}$-compact domain and $\mathcal{F}$-separated codomain (see Prop. 3.5 and Prop. 4.5). Hence, f is $\mathcal{F}$-proper in $\mathcal{X}/Z$ and also in $\mathcal{X}$. □

We call a morphism $\mathcal{F}$-*perfect* if it is both $\mathcal{F}$-proper and $\mathcal{F}$-separated. With Propositions 3.2, 3.7, 4.2 and 4.4 we then obtain:

Theorem. *The class of $\mathcal{F}$-perfect morphisms contains all $\mathcal{F}$-closed subobjects, is closed under composition and stable under pullback. Furthermore, if a composite morphism $g \cdot f$ is $\mathcal{F}$-perfect, then f is $\mathcal{F}$-perfect whenever g is $\mathcal{F}$-separated, and g is $\mathcal{F}$-perfect whenever f is $\mathcal{F}$-perfect and stably in $\mathcal{E}$. Finally, if the category $\mathcal{X}$*

has finite coproducts, is extensive, and if the coproduct of two $\mathcal{F}$-closed morphisms is $\mathcal{F}$-closed, also the coproduct of two $\mathcal{F}$-perfect morphisms is $\mathcal{F}$-perfect.   □

Example. In $\mathcal{T}op_{\mathrm{open}}$ a map $f : X \to Y$ is $\mathcal{F}$-perfect if and only if f is open and locally injective (see Exercise 1 of 4.1). This means precisely that f is a *local homeomorphism*, so that every point in X has an open neighborhood U such that the restriction $U \to f[U]$ of f is a homeomorphism. In particular, local homeomorphisms enjoy all stability properties described by the Theorem.

5.3. Compact Hausdorff objects. Let $\mathcal{F}$-CompHaus denote the full subcategory of $\mathcal{X}$ containing the objects that are both $\mathcal{F}$-compact and $\mathcal{F}$-Hausdorff.

Theorem. *$\mathcal{F}$-CompHaus is closed under finite limits in $\mathcal{X}$ and under $\mathcal{F}$-closed subobjects. If $\mathcal{E}$ is stable under pullback, then the $(\mathcal{E}, \mathcal{M})$-factorization system of $\mathcal{X}$ restricts to an $(\mathcal{E}, \mathcal{M})$-factorization system of $\mathcal{F}$-CompHaus. If the category $\mathcal{X}$ has finite coproducts, is extensive, and if the coproduct of two $\mathcal{F}$-closed morphisms is $\mathcal{F}$-closed, then $\mathcal{F}$-CompHaus is closed under finite coproducts in $\mathcal{X}$ precisely when $1 + 1 \in \mathcal{F}$-CompHaus.*

Proof. From Theorem 3.3(1) and item (vii) of Theorem 4.3 one sees that $\mathcal{F}$-CompHaus is closed under equalizers in $\mathcal{X}$; likewise for closed subobjects. For closure under finite products, use Theorem 3.3(3) and Corollary 4.3. In the $(\mathcal{E}, \mathcal{M})$-factorization (1) of f, Z is $\mathcal{F}$-compact by Theorem 3.3(2) in case $\mathcal{E} = \mathcal{E}^*$, and $\mathcal{F}$-separated by Corollary 4.3. For the last statement of the Theorem one uses Propositions 3.7 and 4.4, but in order to do so we have to make sure that the morphisms $0 \to X$ and $X + X \to X$ are $\mathcal{F}$-closed if $1 + 1 \in \mathcal{F}$-CompHaus. For $X + X \to X$, this follows directly from $1 + 1 \in \mathcal{F}$-Comp (see the proof of Prop. 3.7), and for $0 \to X$ observe that this is the equalizer of the injections $X \rightrightarrows X + X$ in the extensive category $\mathcal{X}$. Hence, $\mathcal{F}$-closedness follows again with (vii) of Theorem 4.3.   □

Exercises.

1. Using previous exercises, check the correctness of the given characterization of $X \in \mathcal{F}$-CompHaus for each of the following categories: $\mathcal{T}op$: compact Hausdorff, $\mathcal{L}oc$: compact Hausdorff, $\mathcal{T}op_{\mathrm{open}}$: discrete, $\mathcal{T}op_{\mathrm{clopen}}$: finite discrete, $\mathcal{T}op_{\mathrm{Zariski}}$: T0 plus the property of Exercise 2 of 3.5.
2. Find sufficient conditions for $\mathcal{F}$-CompHaus to be closed under finite coproducts.

5.4. Non-extendability of proper maps. An important property of $\mathcal{F}$-proper maps is described by:

Proposition. *An $\mathcal{F}$-proper map $f : M \to Y$ in $\mathcal{X}$ cannot be extended along an $\mathcal{F}$-dense subobject $m : M \to X$ with X $\mathcal{F}$-Hausdorff unless m is an isomorphism.*

Proof. Suppose we had a factorization $f = g \cdot m$ with $g : X \to Y$, X $\mathcal{F}$-Hausdorff and m in $\mathcal{M}$ $\mathcal{F}$-dense. Then $m : (M, f) \to (X, g)$ is an $\mathcal{F}$-dense embedding in $\mathcal{X}/Y$, with $\mathcal{F}$-compact domain and $\mathcal{F}$-separated codomain, by (ii) of Theorem 4.3

and Propositions 3.5, 4.5. Hence, an application of Proposition 5.1 (to $\mathcal{X}/Y$ in lieu of $\mathcal{X}$) gives that m is $\mathcal{F}$-closed, in fact an isomorphism by Corollary 2.2(2).

$\square$

We shall see in 6.6 below under which circumstances the property described by the Proposition turns out to be characteristic for $\mathcal{F}$-properness.

6. Tychonoff spaces, absolutely closed spaces, compactification

6.1. Embeddability and absolute closedness. In our standard setting of 2.1 we consider two important subcategories of $\mathcal{F}$-Haus, both containing $\mathcal{F}$-CompHaus. An object X of $\mathcal{X}$ is called

- *$\mathcal{F}$-Tychonoff* if it is embeddable into an $\mathcal{F}$-compact $\mathcal{F}$-Hausdorff object, so that there is $m : X \to K$ in $\mathcal{M}$ with $K \in \mathcal{F}$-CompHaus;
- *absolutely $\mathcal{F}$-closed* if it is $\mathcal{F}$-separated and $\mathcal{F}$-closed in every $\mathcal{F}$-separated extension object, so that every $m : X \to K$ in $\mathcal{M}$ with $K \in \mathcal{F}$-Haus is $\mathcal{F}$-closed.

Denoting the respective full subcategories of $\mathcal{X}$ by $\mathcal{F}$-Tych and $\mathcal{F}$-AC, with Proposition 5.1, Theorem 3.3(3), Corollary 4.3 we obtain:

Proposition. $\mathcal{F}$-CompHaus $= \mathcal{F}$-Tych $\cap \, \mathcal{F}$-AC.

Examples.

(1) In $\mathcal{T}op$, the $\mathcal{F}$-Tychonoff spaces are precisely the completely regular Hausdorff spaces X, characterized by the property that for every closed set A in X and $x \in X \setminus A$, there is a continuous mapping $g : X \to [0,1]$ into the unit interval with $g[A] \subseteq \{1\}$ and $g(x) = 0$. Using the Stone-Čech compactification $\beta_X : X \to \beta X$ one can prove this assertion quite easily.

(2) In $\mathcal{L}oc$, the $\mathcal{F}$-Tychonoff locales are exactly the completely regular locales (see II.6 and [29], IV.1.7).

(3) In $\mathcal{T}op$, the absolutely $\mathcal{F}$-closed spaces are (by definition) the so-called *H-closed* spaces: see [20], p.223.

(4) In Ω-$\mathcal{S}et$ (see 2.7) the absolutely $\mathcal{F}$-closed objects have been characterized in [15] as the so-called *algebraic* Ω-*sets*. In the case that K is given by the ground field in the category of commutative K-algebras, all such objects are subobjects of K^I for some set I and are called K-*algebraic*. Here the K-algebraic set K^I is equipped with the K-algebra of polynomial functions on K^I, and its K-algebraic subsets are precisely the zero sets of sets of polynomials in $K[X_i]_{i \in I}$.

Exercises.

1. Prove that in $\mathcal{T}op_{\text{open}}$ one has $\mathcal{F}$-Haus $= \mathcal{F}$-CompHaus $= \mathcal{F}$-Tych $= \mathcal{F}$-AC, given by the subcategory of discrete spaces. Prove a similar result for $\mathcal{A}b\mathcal{G}rp$.

2. Prove that in $\mathcal{T}op_{\text{Zariski}}$ $\mathcal{F}$-Haus is the category of $T0$-spaces, whereas $\mathcal{F}$-AC is the category of sober $T0$-spaces (see Chapter II).

3. Find examples in $\mathcal{T}op$ of a Tychonoff space which is not absolutely closed, and of an absolutely closed space which is not Tychonoff. (Following standard praxis we left off the prefix $\mathcal{F}$ here.)

4. Consider the category Chu_2^{ext} of Exercise 3 of 2.7. For an object (X, τ) of this category, let $N : X \to 2^X$, $x \mapsto \{U \in \tau \mid x \in U\}$, be the "neighborhood filter map". Prove that (X, τ) is $\mathcal{F}$-Hausdorff in Chu_2^{ext} if and only if N is injective, and absolutely $\mathcal{F}$-closed if and only if N is bijective. (*Hint:* For (X, τ) absolutely $\mathcal{F}$-closed, assume that N fails to be surjective, witnessed by $\alpha \subseteq \tau$; then consider the structure σ on $Y = X + \{\infty\}$, given by $\sigma = (\tau \setminus \alpha) \cup \{U \cup \{\infty\} \mid U \in \alpha\}$, and verify that (Y, σ) is $\mathcal{F}$-Hausdorff.)

6.2. Stability properties of Tychonoff objects. $\mathcal{F}$-Tych has the expected stability properties:

Proposition. *$\mathcal{F}$-Tych is closed under finite limits and under subobjects in* $\mathcal{X}$.

Proof. With $m_i : X_i \to K_i$ in $\mathcal{M}$ ($i = 1, 2$), also $m_1 \times m_2 : X_1 \times X_2 \to K_1 \times K_2$ is in $\mathcal{M}$ (see 1.4(5)), and we can use 3.3(3) to obtain closure of $\mathcal{F}$-Tych under finite products. Closure under equalizers follows from closure under subobjects, and the latter property is trivial. $\qquad\square$

Remark. $\mathcal{F}$-AC generally fails to be closed under finite products or under equalizers. For example, in $\mathcal{T}op$ $\mathcal{F}$-AC fails to be closed under equalizers, although it is closed under arbitrary products, by a result of Chevalley and Frink [8]. In $\mathcal{T}op_{\text{open}}$ $\mathcal{F}$-AC is still closed under finite products, but not under infinite ones: see Exercise 1 of 6.1.

6.3. Extending the notions to morphisms. Using comma categories, it is natural to extend the notions introduced in 6.1 from objects to morphisms of $\mathcal{X}$. Hence, a morphism $f : X \to Y$ in $\mathcal{X}$ is called

- *$\mathcal{F}$-Tychonoff* if (X, f) is an $\mathcal{F}$-Tychonoff object in $\mathcal{X}/Y$, which means that there is a factorization $f = p \cdot m$ with $m \in \mathcal{M}$ and p $\mathcal{F}$-perfect (see Prop. 3.5, 4.5);
- *absolutely $\mathcal{F}$-closed* if (X, f) is an absolutely $\mathcal{F}$-closed object in $\mathcal{X}/Y$, which means that f is $\mathcal{F}$-separated and, whenever there is a factorization $f = p \cdot m$ with $m \in \mathcal{M}$ and p $\mathcal{F}$-separated, then m is $\mathcal{F}$-closed.

There is some interaction between the object and morphism notions, similarly to what we have seen for compactness and separation.

Proposition. *In each (1) and (2), the assertions (i)-(iii) are equivalent:*

(1) i. $X \in \mathcal{F}$-Tych;

ii. *every morphism $f : X \to Y$ is $\mathcal{F}$-Tychonoff;*

iii. *there is an $\mathcal{F}$-Tychonoff map $f : X \to Y$ with $Y \in \mathcal{F}$-CompHaus;*

(2) i. $X \in \mathcal{F}$-AC;

ii. *every morphism $f : X \to Y$ with $Y \in \mathcal{F}$-Haus is absolutely $\mathcal{F}$-closed;*

   iii. *there is an absolutely $\mathcal{F}$-closed morphism $f : X \to Y$ with $Y \in \mathcal{F}$-CompHaus.*

Proof. For morphisms $f : X \to Y$, $m : M \to K$ in $\mathcal{X}$ consider the diagram

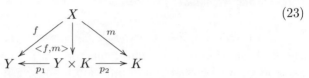

$$\tag{23}$$

If $K \in \mathcal{F}$-CompHaus, then p_1 is $\mathcal{F}$-perfect, and $m = p_2 \cdot < f, m > \in \mathcal{M}$ implies $< f, m > \in \mathcal{M}$; this proves (1) i $\Rightarrow$ ii. Similarly, for (2) iii $\Rightarrow$ i, the hypotheses $m \in \mathcal{M}$ and $K \in \mathcal{F}$-Haus give $< f, m > \in \mathcal{M}$ and $p_1 \in \mathcal{F}'$, so that $< f, m >$ must be $\mathcal{F}$-closed since f is absolutely $\mathcal{F}$-closed; furthermore, p_2 is $\mathcal{F}$-closed since Y is $\mathcal{F}$-compact, so that $m = p_2 \cdot < f, m >$ is $\mathcal{F}$-closed.

Let us now assume $f = p \cdot m$ with $m \in \mathcal{M}$. If $p : K \to Y$ is $\mathcal{F}$-perfect and $Y \in \mathcal{F}$-CompHaus, then also $K \in \mathcal{F}$-CompHaus, by Theorems 3.3(1), 4.3(iii). This proves (1) iii $\Rightarrow$ i. For (2) i $\Rightarrow$ ii, assume both p and Y to be $\mathcal{F}$-separated, so that also K is $\mathcal{F}$-separated; now the hypothesis $X \in \mathcal{F}$-AC gives that m is $\mathcal{F}$-closed, as desired.

The implications ii $\Rightarrow$ iii are trivial in both cases: take $Y = 1$.   □

Of course, the Proposition just proved may be applied to the slices of $\mathcal{X}$ rather than to $\mathcal{X}$ itself and then leads to stability properties for the classes of $\mathcal{F}$-Tychonoff and of absolutely $\mathcal{F}$-closed maps, the proof of which we leave as exercises.

Exercises.

1. The class of $\mathcal{F}$-Tychonoff maps is stable under pullback. If $g \cdot f$ is $\mathcal{F}$-Tychonoff, so is f; conversely, f $\mathcal{F}$-Tychonoff and g $\mathcal{F}$-perfect imply $g \cdot f$ $\mathcal{F}$-Tychonoff.
2. If $g \cdot f$ is absolutely $\mathcal{F}$-closed and g $\mathcal{F}$-separated, then f is absolutely $\mathcal{F}$-closed; conversely, if f is absolutely $\mathcal{F}$-closed and g $\mathcal{F}$-perfect, then $g \cdot f$ is absolutely $\mathcal{F}$-closed.

Remark. Unlike spaces, $\mathcal{F}$-Tychonoff maps in $\mathcal{T}op$ do not seem to allow for an easy characterization in terms of mapping properties into the unit interval. Other authors (see [16, 41, 33]) studied separated maps $f : X \to Y$ with the property that for every closed set $A \subseteq X$ and every $x \in X \setminus A$ there is an open neighborhood U of $f(x)$ in Y and a continuous map $g : f^{-1}[U] \to [0, 1]$ with $g(x) = 0$ and $g[A \cap f^{-1}[U]] \subseteq \{1\}$. In case $Y = 1$ this amounts to saying that X is completely regular Hausdorff, hence Tychonoff. For general Y, maps f with this property are $\mathcal{F}$-Tychonoff, i.e. restrictions of perfect maps, but the converse is generally false: see [51].

6.4. Compactification of objects. An *$\mathcal{F}$-compactification* of an object X is given by an $\mathcal{F}$-dense embedding $X \to K$ with $K \in \mathcal{F}$-CompHaus. Of course, only objects in $\mathcal{F}$-Tych can have $\mathcal{F}$-compactifications. For our purposes it is important to be provided with a *functorial* choice of a compactification for every $X \in \mathcal{F}$-Tych.

Hence, we call an endofunctor $\kappa : \mathcal{F}\text{-Tych} \to \mathcal{F}\text{-Tych}$ which comes with a natural transformation

$$\kappa_X : X \longrightarrow \kappa X \; (X \in \mathcal{F}\text{-Tych})$$

a *functorial $\mathcal{F}$-compactification* if

- each κ_X is an $\mathcal{F}$-dense embedding
- each κX is $\mathcal{F}$-compact and $\mathcal{F}$-separated.

By illegitimally denoting the endofunctor and the natural transformation by the same letter we follow standard praxis in topology. In $\mathcal{T}op$, the prime example for a functorial $\mathcal{F}$-compactification is provided by the Stone-Čech compactification

$$\beta_X : X \longrightarrow \beta X.$$

These morphisms serve as reflexions, showing the reflexivity of $\mathcal{F}$-CompHaus in $\mathcal{F}$-Tych. We shall revisit this theme in Section 11 below; here we just note that the universal property makes the first of the two requirements for a functorial $\mathcal{F}$-compactification redundant, also in our general setting:

Proposition. *If $\mathcal{F}$-CompHaus is reflective in $\mathcal{F}$-Tych, with reflexions $\beta_X : X \to \beta X$, then these provide a functorial $\mathcal{F}$-compactification.*

Proof. We just need to show that each β_X is an $\mathcal{F}$-dense embedding. But for $X \in \mathcal{F}$-Tych we have some $m : X \to K$ in $\mathcal{M}$ with $K \in \mathcal{F}$-CompHaus, which factors as $m = f \cdot \beta_X$, by the universal property of β_X. Hence $\beta_X \in \mathcal{M}$. If $\beta_X = n \cdot g$ with $n : N \to \beta X$ $\mathcal{F}$-closed, then $N \in \mathcal{F}$-CompHaus, and we can apply the universal property again to obtain $h : \beta X \to N$ with $h \cdot \beta_X = g$. Then $n \cdot h \cdot \beta_X = \beta_X$ shows $n \cdot h = 1_{\beta X}$, so that n is an isomorphism. Consequently, β_X is $\mathcal{F}$-dense. $\qquad\square$

6.5. Compactification of morphisms. In order to take full advantage of the presence of a functorial $\mathcal{F}$-compactification κ, we should extend this gadget from objects to morphisms, as follows: for $f : X \to Y$ in $\mathcal{F}$-Tych we form the pullback $P_f = Y \times_{\kappa Y} \kappa X$ and the induced morphism κ_f making the following diagram commutative:

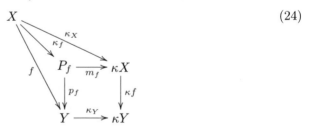

$$(24)$$

Calling a morphism $f : X \to Y$ $\mathcal{F}$-*initial* if every $\mathcal{F}$-closed subobject of X is the inverse image of an $\mathcal{F}$-closed subobject of Y under f, we observe:

Proposition.

(1) $\kappa_f \in \mathcal{M}$, and p_f is $\mathcal{F}$-perfect, with $P_f \in \mathcal{F}$-Tych.
(2) *If every morphism in $\mathcal{M}$ is $\mathcal{F}$-initial, then κ_f is $\mathcal{F}$-dense.*

Proof. (1) Since $\kappa_X = m_f \cdot \kappa_f \in \mathcal{M}$ we have $\kappa_f \in \mathcal{M}$. As a morphism in $\mathcal{F}$-CompHaus, κf is $\mathcal{F}$-perfect (by Prop. 5.1 and Theorem 4.3), and so is its pullback p_f. Since $\mathcal{F}$-Tych is closed under finite limits in $\mathcal{X}$, $P_f \in \mathcal{F}$-Tych.

(2) The given hypothesis implies the following cancellation property for $\mathcal{F}$-dense subobjects: if $n \cdot k$ is $\mathcal{F}$-dense with $n, k \in \mathcal{M}$, then k is $\mathcal{F}$-dense. This may then be applied to $\kappa_X = m_f \cdot \kappa_f$ since $m_f \in \mathcal{M}$, as a pullback of κ_Y. □

Hence, we should think of $\kappa_f : (X, f) \rightarrow (P_f, p_f)$ as of an $\mathcal{F}$-compactification of (X, f) in $\mathcal{F}$-Tych$/Y$, especially under the hypothesis given in (2).

Exercises.

1. Show $P_f \cong \kappa X$ in case $Y = 1$ and $\kappa_1 \cong 1$.
2. Show that by putting $\kappa(X, f) := (P_f, p_f)$ one obtains a functorial $\mathcal{F}$-compactification for $\mathcal{F}$-Tych$/Y$, under the same hypothesis as in (2) of the Proposition. (Here we think of $\mathcal{F}$-Tych$/Y$ as a full subcategory of $\mathcal{X}/Y$ that has inherited the factorization system and the closed system from there.)

6.6. Isbell-Henriksen characterization of perfect maps. In case $\mathcal{X} = \mathcal{T}op$ the following theorem goes back to Isbell and Henriksen (see [23]). Categorical versions of it can be found in [9] and [48].

Theorem. *Let κ be a functorial $\mathcal{F}$-compactification and let every morphism in $\mathcal{M}$ be $\mathcal{F}$-initial. Then the following assertions are equivalent for $f : X \rightarrow Y$ in $\mathcal{F}$-Tych:*

(i) *f is $\mathcal{F}$-perfect;*
(ii) *f cannot be extended along an $\mathcal{F}$-dense embedding $m : X \rightarrow Z$ with Z $\mathcal{F}$-Hausdorff unless m is an isomorphism;*
(iii) *the naturality diagram*

$$\begin{array}{ccc} X & \xrightarrow{\kappa_X} & \kappa X \\ {\scriptstyle f}\downarrow & & \downarrow{\scriptstyle \kappa f} \\ Y & \xrightarrow{\kappa_Y} & \kappa Y \end{array} \qquad (25)$$

is a pullback diagram.

Proof. (i) $\Rightarrow$ (ii) was shown in 5.4.

(ii) $\Rightarrow$ (iii): By Prop. 6.5, κ_f of (24) is an $\mathcal{F}$-dense embedding, hence an isomorphism by hypothesis. Consequently, (25) coincides with the pullback square of (24), up to isomorphism.

(iii) $\Rightarrow$ (i): By hypothesis $f \cong p_f$, hence f is $\mathcal{F}$-perfect by 6.5. □

Example. For $\mathcal{X} = \mathcal{T}op$ and $\kappa = \beta$ the Stone-Čech compactification, by item (iii) of the Theorem perfect maps $f : X \rightarrow Y$ between Tychonoff spaces are characterized by the property that their extension $\beta f : \beta X \rightarrow \beta Y$ maps the remainder $\beta X \setminus X$ into the remainder $\beta Y \setminus Y$.

6.7. Antiperfect-perfect factorization. Suppose that, in our general setting of 2.1, $\kappa = \beta$ is given by reflexivity of $\mathcal{F}$-CompHaus in $\mathcal{F}$-Tych. Then there is another way of thinking of the factorization $f = p_f \cdot \beta_f$ established in diagram (24). With $\mathcal{P}_{\mathcal{F}}$ denoting the class of $\mathcal{F}$-perfect morphisms in $\mathcal{F}$-Tych, and with

$$\mathcal{A}_{\mathcal{F}} := \{m : X \to Y \mid m \in \mathcal{M}, \ X, Y \in \mathcal{F}\text{-Tych}, \ \beta m : \beta X \to \beta Y \ \text{iso}\},$$

whose morphisms are also called $\mathcal{F}$-*antiperfect*, one obtains:

Theorem. $(\mathcal{A}_{\mathcal{F}}, \mathcal{P}_{\mathcal{F}})$ *is a (generally non-proper) factorization system of* $\mathcal{F}$-*Tych, provided that* $\mathcal{F}$-*CompHaus is reflective in* $\mathcal{F}$-*Tych and every morphism in* $\mathcal{M}$ *is* $\mathcal{F}$-*initial.*

Proof. The outer square and the upper triangle of

$$
\begin{array}{ccc}
X & \xrightarrow{\ \beta_X\ } & \beta X \\
{\scriptstyle \beta_f}\big\downarrow & \nearrow{\scriptstyle m_f} & \big\downarrow{\scriptstyle \overline{\beta_f}} \\
P_f & \xrightarrow[\ \beta_{P_f}\ ]{} & \beta P_f
\end{array}
\tag{26}
$$

are commutative, with $\overline{\beta_f} := \beta(\beta_f)$. According to Theorem 2.7 of [27], the assertion of the Theorem follows entirely from the universal property of β once we have established commutativity of the lower triangle of (26). For that, let $u : E \to P_f$ be the equalizer of β_{P_f}, $\overline{\beta_f} \cdot m_f$, which is $\mathcal{F}$-closed since βP_f is $\mathcal{F}$-separated (see Theorem 4.3). Since β_f factors through u by a unique morphism $l : X \to E$, and since β_f is $\mathcal{F}$-dense by Prop. 6.5, u is an isomorphism, and we have $\overline{\beta_f} \cdot m_f = \beta_{P_f}$. $\square$

An important consequence of the Theorem is that *essentially* it allows us to carry the hypothesis of reflectivity of $\mathcal{F}$-CompHaus in $\mathcal{F}$-Tych from $\mathcal{X}$ to its slices, as follows. Consider again a morphism $f : X \to Y$ with $X, Y \in \mathcal{F}$-Tych. Then Prop. 6.5(1) shows that the object (X, f) in $\mathcal{X}/Y$ (in fact: $\mathcal{F}$-Tych/Y) is $\mathcal{F}$-Tychonoff. Now, with the Theorem, one easily shows:

Corollary. *Under the provisions of the Theorem,* $\beta_f : (X, f) \to (P_f, p_f)$ *is a reflexion of the* $\mathcal{F}$-*Tychonoff object* (X, f) *of* $\mathcal{F}$-*Tych/Y into the full subcategory of* $\mathcal{F}$-*compact Hausdorff objects of* $\mathcal{X}/Y$. $\square$

Hence, the construction given by diagram (24) with $\kappa = \beta$ is really the "Stone-Čech $\mathcal{F}$-compactification" of $\mathcal{F}$-Tychonoff objects in $\mathcal{X}/Y$, provided that we restrict ourselves to those objects $(X, f : X \to Y)$ for which X, Y are $\mathcal{F}$-Tychonoff objects in $\mathcal{X}$.

Exercise. Work out the details of the proof of the Theorem and the Corollary.

7. Open maps, open subspaces

7.1. Open maps. In the standard setting of 2.1, a morphism $f : X \to Y$ is said to *reflect* $\mathcal{F}$-*density* if $f^{-1}[-]$ maps $\mathcal{F}$-dense subobjects of Y to $\mathcal{F}$-dense subobjects of X. The *morphism* f is $\mathcal{F}$-*open* if every pullback f' of f (see diagram (11)) reflects $\mathcal{F}$-density. A *subobject* is $\mathcal{F}$-*open* if its representing morphism is $\mathcal{F}$-open. By definition one has

$$\mathcal{F}^+ := \{f \mid f \ \mathcal{F}\text{-open} \} = \{f \mid f \ \text{reflects} \ \mathcal{F}\text{-density}\}^*$$

in the notation of Section 3.

Since reflection of $\mathcal{F}$-density is obviously closed under composition, one obtains immediately some stability properties for $\mathcal{F}$-open maps, just as for $\mathcal{F}$-proper maps.

Proposition.

(1) *The class $\mathcal{F}^+$ contains all isomorphisms, is closed under composition and stable under pullback.*
(2) *If $g \cdot f \in \mathcal{F}^+$ with g monic, then $f \in \mathcal{F}^+$.*
(3) *If $g \cdot f \in \mathcal{F}^+$ with $f \in \mathcal{E}^*$, then $g \in \mathcal{F}^+$.*

Proof. For (1), (2), proceed as in Prop. 3.2. (3): A pullback of $g \cdot f$ with $f \in \mathcal{E}^*$ has the form $g' \cdot f'$ with $f' \in \mathcal{E}$ a pullback of f and g' a pullback of g. So, it suffices to check that g reflects $\mathcal{F}$-density if $g \cdot f$ does and $f \in \mathcal{E}^*$. But for every $\mathcal{F}$-dense subobject d of the codomain of g one obtains from Prop. 1.8 that

$$g^{-1}[d] = f[f^{-1}[g^{-1}[d]]] = f[(g \cdot f)^{-1}[d]]$$

is an image of an $\mathcal{F}$-dense subobject under f. But since $f \in \mathcal{E}$, trivially $f[-]$ preserves $\mathcal{F}$-density (by Cor. 2.2(1),(3)). $\qquad\square$

Analogously to Prop. 2.1 we can state:

Corollary.

(1) *The $\mathcal{F}$-open subobjects of an object X form a (possibly large) subsemilattice of the meet-semilattice $\mathrm{sub}X$.*
(2) *For every morphism f, $f^{-1}[-]$ preserves $\mathcal{F}$-openness of subobjects.*
(3) *For every $\mathcal{F}$-open morphism f, $f[-]$ preserves $\mathcal{F}$-openness of subobjects provided that $\mathcal{E}$ is stable under pullback.*

Proof. (1) follows from the Proposition, and for (2) remember that $\mathcal{F}$-openness of subobjects is, by definition, pullback stable. (3) If f and $m \in \mathcal{M}$ are both open, then $\mathcal{F}$-openness of $f \cdot m = f[m] \cdot e$ with $e \in \mathcal{E}^*$ yields $\mathcal{F}$-openness of $f[m]$, by the Proposition. $\qquad\square$

7.2. Open maps of topological spaces. It is time for a "reality check" in terms of our standard examples.

Proposition. *In $\mathcal{T}op$ (with $\mathcal{F} = \{\text{closed maps}\}$), the following assertions are equivalent for a continuous map $f : X \to Y$:*

(i) *f is $\mathcal{F}$-open;*
(ii) *for all subspaces N of Y, $f^{-1}[\overline{N}] = \overline{f^{-1}[N]}$;*
(iii) *f is open, i.e. $f[-]$ preserves openness of subspaces.*

Proof. (i) $\Rightarrow$ (ii): Given $N \subseteq Y$, by hypothesis, the restriction $f' : f^{-1}[N] \to \overline{N}$ of f reflects the density of N in $\overline{N}$, hence $\overline{f^{-1}[N]} = f^{-1}[\overline{N}]$.

(ii) $\Rightarrow$ (iii): Given $O \subseteq X$ open, $f^{-1}[\overline{Y \setminus f[O]}] = \overline{f^{-1}[Y \setminus f[O]]} \subseteq \overline{X \setminus O} = X \setminus O$, hence $\overline{Y \setminus f[O]} \subseteq Y \setminus f[O]$, that is $f[O]$ is open.

(iii) $\Rightarrow$ (i): Openness of maps (in the sense of preservation of openness of subspaces) is easily shown to be stable under pullback. Hence, it suffices to show that an open map $f : X \to Y$ reflects density of subspaces. But if $D \subseteq Y$ is dense, then $f[X \setminus \overline{f^{-1}[D]}]$ is open, by hypothesis, and this set would have to meet D under the assumption that $X \setminus \overline{f^{-1}[D]}$ is not empty, which is impossible. Hence, $f^{-1}[D]$ is dense in X. □

We emphasize that a map in $\mathcal{T}op$ which reflects $\mathcal{F}$-density need not be open (see Exercise 1 below); hence, the stability requirement in Definition 7.1 of $\mathcal{F}$-openness is essential.

Exercises.

1. Show that in $\mathcal{T}op$ the embedding of the closed unit interval into $\mathbb{R}$ reflects density.
2. In $\mathcal{T}op_{\text{open}}$, $D \subseteq X$ is $\mathcal{F}$-dense if and only if D meets every non-empty closed set in X. Hence, if X is a $T1$-space (so that all singleton sets are closed), only X itself is a dense subobject of X. Show that every proper (=stably-closed) map is $\mathcal{F}$-open, but not conversely.
3. In $\mathcal{T}op_{\text{clopen}}$, $D \subseteq X$ is $\mathcal{F}$-dense if and only if D meets every non-empty clopen set in X. Prove that f is $\mathcal{F}$-open if and only if f is $\mathcal{F}$-proper (i.e., every pullback of f maps clopen subsets onto clopen subsets).
4. Show that in a topos with a universal closure operator (see 2.5), every morphism is $\mathcal{F}$-open.

7.3. Open maps of locales. In $\mathcal{L}oc$, $\mathcal{F}$-open morphisms $f : X \to Y$ are characterized like in $\mathcal{T}op$ as those that have stably the property that $f^{-1}[-]$ commutes with the usual closure, just as in (ii) of Prop. 7.2; this follows formally from the fact that the closure in $\mathcal{L}oc$ is given by an idempotent and hereditary closure operator (see [18]). But more importantly we need to compare this notion with the usual notion of open map of locales (see Chapter II):

Proposition. *If $f[-]$ maps open sublocales to open sublocales (i.e., if f is open), then f is $\mathcal{F}$-open in $\mathcal{L}oc$.*

Proof. Since openness of localic maps is stable under pullback (see Theorem II 5.2), it suffices to show

$$f^{-1}[\overline{n}] = \overline{f^{-1}[n]}$$

for $f : X \to Y$ open and any sublocale $n : N \rightarrowtail Y$. For this it suffices to show

$$f^*[c(n)] = c(f^{-1}[n]),$$

where $c(n) = \bigvee \{b \in OY \mid n^*(b) = 0\}$ (see II, 2.9 and 3.6), which means:

$$\bigvee \{f^*(b) \mid b \in OY, n^*(b) = 0\} = \bigvee \{a \in OX \mid m^*(a) = 0\},$$

with $m = f^{-1}[n] : M \rightarrowtail X$. Now, for this last identity, "$\leq$" is trivial, while "$\geq$" follows when we put $b = f_!(a)$ for every a with $m^*(a) = 0$ and use

$$n^*(f_!(a))) = (f')_!(m^*(a)),$$

where $f_!$ is the left adjoint of f^* and $f' : M \to N$ is the restriction of f. $\square$

After the authors posed the converse statement of the Proposition as an open problem, in April 2002 P.T. Johnstone proved its validity:

Theorem. *The $\mathcal{F}$-open maps in $\mathcal{L}oc$ are precisely the (usual) open maps of locales.*

For its rather intricate proof we must refer the Reader to [31]. The paper also exhibits various subtypes of openness. For example, like in $\mathcal{T}op$, also in $\mathcal{L}oc$ there are examples of morphisms reflecting $\mathcal{F}$-density which fail to be $\mathcal{F}$-open, as also shown in II.5.2.

7.4. Local homeomorphisms. Proposition 7.1 shows that, if $\mathcal{E}$ is stable under pullback in $\mathcal{X}$, then $\mathcal{F}^+$ is a new $(\mathcal{E}, \mathcal{M})$-closed class in $\mathcal{X}$ (see 2.1), and in view of Exercise 1 of 4.3 and Example 5.2, it would make sense to define:

$$\begin{aligned}
X \ \mathcal{F}\text{-}discrete \quad &:\Leftrightarrow \quad X \ \mathcal{F}^+\text{-separated} \\
&\Leftrightarrow \quad \delta_X : X \to X \times X \ \mathcal{F}\text{-open} \\
f : X \to Y \ local \ \mathcal{F}\text{-}homeomorphism \quad &:\Leftrightarrow \quad f \ \mathcal{F}^+\text{-perfect} \\
&\Leftrightarrow \quad f \text{ and } \delta_f : X \to X \times_Y X \ \mathcal{F}\text{-open}
\end{aligned}$$

Here we cannot explore these notions further, but must leave the Reader with:

Exercises.

1. Collect stability properties for $\mathcal{F}$-discrete objects and local $\mathcal{F}$-homeomorphisms, from the properties already shown for $\mathcal{F}$-separated objects and $\mathcal{F}$-proper morphisms.
2. In a topos $\mathcal{S}$ with a universal closure operator (see 2.5) every object is $\mathcal{F}$-discrete and every morphism is a local $\mathcal{F}$-homeomorphism.

7.5. Sums of open maps. The proof of the following Proposition is left as an exercise as well:

Proposition. *Assume that the finitely complete category $\mathcal{X}$ has finite coproducts and is extensive, and that the classes $\mathcal{M}$ and $\mathcal{F}_0$ are closed under finite coproducts. Then:*

$$\begin{aligned}
f_1 + f_2 \ \mathcal{F}\text{-}dense \quad &\Leftrightarrow \quad f_1, f_2 \ \mathcal{F}\text{-}dense, \\
f_1 + f_2 \ \mathcal{F}\text{-}open \quad &\Leftrightarrow \quad f_1, f_2 \ \mathcal{F}\text{-}open.
\end{aligned}$$

$\square$

8. Locally perfect maps, locally compact Hausdorff spaces

8.1. Locally perfect maps. In the setting of 2.1, a morphism $f : X \to Y$ is *locally $\mathcal{F}$-perfect* if it is a restriction of an $\mathcal{F}$-perfect morphism $p : K \to Y$ to an $\mathcal{F}$-open subobject $u : X \to K$:

$$\begin{array}{ccc} & K & \\ {}^{u}\nearrow & & \searrow {}^{p} \\ X & \xrightarrow{\quad f \quad} & Y \end{array} \qquad (27)$$

Such morphisms are in particular $\mathcal{F}$-Tychonoff. An object X is *locally $\mathcal{F}$-compact Hausdorff* if $X \to 1$ is locally $\mathcal{F}$-perfect, that is: if there is an $\mathcal{F}$-open embedding $u : X \to K$ with $K \in \mathcal{F}$-CompHaus. If it is clear that X is $\mathcal{F}$-Hausdorff, we may simply call X *locally $\mathcal{F}$-compact*.

Examples.

(1) In $\mathcal{T}op$, X is locally $\mathcal{F}$-compact Hausdorff if and only if X is Hausdorff and locally compact, in the sense that every point in X has a base of compact neighborhoods. By constructing the Alexandroff one-point compactification for such spaces, one sees that they are locally $\mathcal{F}$-compact. Conversely, compact Hausdorff spaces are locally compact, and local compactness is open-hereditary.

(2) Every locally compact Hausdorff locale (in the sense of Chapter II, 7) is an open sublocale of its Stone-Čech compactification (see II.6.7), hence the embedding is $\mathcal{F}$-open by Prop. 7.3 and X is a locally $\mathcal{F}$-compact Hausdorff object in $\mathcal{L}oc$. We conjecture that also the converse proposition is true.

8.2. First stability properties. The following properties are easy to prove:

Proposition.

(1) *Every morphism representing an $\mathcal{F}$-closed or an $\mathcal{F}$-open subobject is a locally $\mathcal{F}$-perfect morphism, and so is every $\mathcal{F}$-perfect morphism.*

(2) *The class of locally $\mathcal{F}$-perfect morphisms is stable under pullback; moreover, if in the pullback diagram (11) both f and g are locally $\mathcal{F}$-perfect, so is $g \cdot f' = f \cdot g'$.*

Proof. (1) $\mathcal{F}$-closed subobjects give $\mathcal{F}$-perfect morphisms. Hence, the assertion follows by choosing u or p in (27) to be an identity morphism.

(2) Both $\mathcal{F}$-open subobjects and $\mathcal{F}$-perfect morphisms are stable under pullback. Furthermore, if $f = p \cdot u$, $g = q \cdot v$ are both locally $\mathcal{F}$-perfect, the pullback diagram

(11) is decomposed into four pullback diagrams, as follows:

$$
\begin{array}{ccccc}
\cdot & \xrightarrow{\;u'\;} & \cdot & \xrightarrow{\;p'\;} & \cdot \\
{\scriptstyle v'}\downarrow & & {\scriptstyle v''}\downarrow & & \downarrow{\scriptstyle v} \\
\cdot & \xrightarrow{\;u''\;} & \cdot & \xrightarrow{\;p''\;} & \cdot \\
{\scriptstyle q'}\downarrow & & {\scriptstyle q''}\downarrow & & \downarrow{\scriptstyle q} \\
\cdot & \xrightarrow{\;u\;} & \cdot & \xrightarrow{\;p\;} & \cdot
\end{array}
\tag{28}
$$

Hence $g \cdot f' = (q \cdot v) \cdot (p' \cdot u') = (q \cdot p'') \cdot (v'' \cdot u')$, with $q \cdot p''$ $\mathcal{F}$-perfect and $v'' \cdot u'$ an $\mathcal{F}$-open subobject. $\qquad\square$

Corollary. *The full subcategory $\mathcal{F}$-LCompHaus of locally $\mathcal{F}$-compact Hausdorff objects in $\mathcal{X}$ is closed under finite products and under $\mathcal{F}$-open subobjects in $\mathcal{X}$.*

Proof. For the first statement, choose $Y = 1$ in diagram (11) and apply assertion (2) of the Proposition. The second statement is trivial. $\qquad\square$

Exercise. Prove: if $U \to X$ in subX is $\mathcal{F}$-open and $A \to X$ in subX $\mathcal{F}$-closed, then $U \wedge A \to X$ is locally $\mathcal{F}$-perfect.

8.3. Local compactness via Stone-Čech. For the remainder of this Section, *we assume that $\mathcal{F}$-CompHaus is reflective in $\mathcal{F}$-Tych*, so that in particular we have a functorial $\mathcal{F}$-compactification

$$
\beta_X : X \longrightarrow \beta X \quad (X \in \mathcal{F}\text{-Tych})
$$

at our disposal. We shall also use its extension to morphisms, as described in diagram (24), with $\kappa = \beta$.

Theorem (Clementino-Tholen [13]). *An object X is locally $\mathcal{F}$-compact Hausdorff if and only if X is $\mathcal{F}$-Tychonoff and β_X is $\mathcal{F}$-open.*

Proof. Sufficiency of the condition is trivial. For its necessity, let $u : X \to K$ be an $\mathcal{F}$-open subobject with $K \in \mathcal{F}$-CompHaus. With the unique morphism $f : \beta X \to K$ with $f \cdot \beta_X = u$ we can then form the pullback diagram

$$
\begin{array}{ccc}
P & \xrightarrow{\;u'\;} & \beta X \\
{\scriptstyle f'}\downarrow & & \downarrow{\scriptstyle f} \\
X & \xrightarrow{\;u\;} & K
\end{array}
\tag{29}
$$

and have an induced morphism $d : X \to P$ with $f' \cdot d = 1_X$ and $u' \cdot d = \beta_X$. Since u' is $\mathcal{F}$-open and β_X $\mathcal{F}$-dense (by Cor. 7.1(2) and Prop. 6.4), the pullback diagram

$$
\begin{array}{ccc}
X & \xrightarrow{\ 1_X\ } & X \\
{\scriptstyle d}\big\downarrow & & \big\downarrow{\scriptstyle \beta_X} \\
P & \xrightarrow{\ u'\ } & \beta X
\end{array}
\qquad (30)
$$

shows that d is $\mathcal{F}$-dense. On the other hand, the equalizer diagram

$$
X \xrightarrow{\ d\ } P \underset{d \cdot f'}{\overset{1_P}{\rightrightarrows}} P
\qquad (31)
$$

shows that d is $\mathcal{F}$-closed (since $P \in \mathcal{F}$-Haus, by Cor. 4.3). Therefore, d is an isomorphism (see Cor. 2.2(2)), and $\beta_X \cong u'$ is $\mathcal{F}$-open. $\qquad\square$

With Corollary 6.7 we obtain immediately from the Theorem:

Corollary. *If every morphism in $\mathcal{M}$ is $\mathcal{F}$-initial, then a morphism $f : X \to Y$ with $X, Y \in \mathcal{F}$-Tych is locally $\mathcal{F}$-perfect if and only if the morphism $\beta_f : X \to P_f$ making diagram (24) (with $\beta = \kappa$) commutative is $\mathcal{F}$-open.* $\qquad\square$

8.4. Composites of locally perfect maps. We can now embark on improving some of the properties given in 8.2.

Proposition. *Let every morphism in $\mathcal{M}$ be $\mathcal{F}$-initial. Then the class of locally $\mathcal{F}$-perfect morphisms with $\mathcal{F}$-Tychonoff domain and codomain is closed under composition.*

Proof. It obviously suffices to show that any composite morphism $h = v \cdot p$ with $p : K \to Y$ $\mathcal{F}$-perfect and $v : Y \to L$ an $\mathcal{F}$-open subobject is locally $\mathcal{F}$-perfect. For that we must show that its $\mathcal{F}$-antiperfect factor $\beta_h : K \to P_h$ is $\mathcal{F}$-open. But the β-naturality diagram for h decomposes as

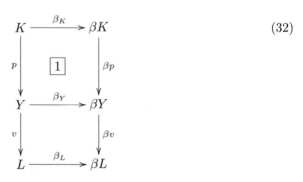

$$
\begin{array}{ccc}
K & \xrightarrow{\ \beta_K\ } & \beta K \\
{\scriptstyle p}\big\downarrow & \boxed{1} & \big\downarrow{\scriptstyle \beta p} \\
Y & \xrightarrow{\ \beta_Y\ } & \beta Y \\
{\scriptstyle v}\big\downarrow & & \big\downarrow{\scriptstyle \beta v} \\
L & \xrightarrow{\ \beta_L\ } & \beta L
\end{array}
\qquad (32)
$$

with the upper square a pullback, since p is $\mathcal{F}$-perfect (see Theorem 6.6). Now (32) decomposes further, as

$$
\begin{array}{ccccc}
K & \xrightarrow{\;\beta_h\;} & P_h & \xrightarrow{\;m_h\;} & \beta K \\[2pt]
\scriptstyle p\Big\downarrow & \boxed{2} & \Big\downarrow & \boxed{3} & \Big\downarrow\scriptstyle \beta p \\[2pt]
Y & \xrightarrow{\;\beta_v\;} & P_v & \xrightarrow{\;m_v\;} & \beta Y \\[2pt]
\scriptstyle v\Big\downarrow & \scriptstyle p_v\Big\downarrow & & \boxed{4} & \Big\downarrow\scriptstyle \beta v \\[2pt]
L & \xrightarrow{\;1_L\;} & L & \xrightarrow{\;\beta_L\;} & \beta L
\end{array}
\tag{33}
$$

Here $\boxed{3}$ is a pullback diagram, since $\boxed{4}$ and the concatenation $\boxed{3}\,\&\,\boxed{4}$ are pullback diagrams. Now, with $\boxed{1} = \boxed{2}\,\&\,\boxed{3}$ also $\boxed{2}$ is a pullback diagram. Since the $\mathcal{F}$-antiperfect factor β_v of the $\mathcal{F}$-open (and therefore locally $\mathcal{F}$-perfect) morphism v is $\mathcal{F}$-open, also its pullback β_h must be $\mathcal{F}$-open. $\qquad\square$

Corollary. *If every morphism in $\mathcal{M}$ is $\mathcal{F}$-initial, then the following assertions are equivalent for $X \in \mathcal{F}$-Tych:*

   (i) *X is locally $\mathcal{F}$-compact;*
  (ii) *every morphism $f : X \to Y$ with $Y \in \mathcal{F}$-Tych is locally $\mathcal{F}$-perfect;*
 (iii) *there is a locally $\mathcal{F}$-perfect map $f : X \to Y$ with $Y \in \mathcal{F}$-LCompHaus.*

Proof. (i) $\Rightarrow$ (ii): In diagram (24) with $\kappa = \beta$, β_f is open since β_X is, and since m_f is monic, by Prop. 7.1(2). (ii) $\Rightarrow$ (iii) is trivial. (iii) $\Rightarrow$ (i): $!_X = !_Y \cdot f$ is locally $\mathcal{F}$-perfect, by the Proposition. $\qquad\square$

8.5. Further stability properties. We can now improve the assertions made in 8.2:

Theorem. *If every morphism in $\mathcal{M}$ is $\mathcal{F}$-initial, then the full subcategory $\mathcal{F}$-LCompHaus of locally $\mathcal{F}$-compact Hausdorff objects is closed under finite limits and under $\mathcal{F}$-closed or $\mathcal{F}$-open subobjects in $\mathcal{X}$.*

Proof. We already stated closure under finite products and $\mathcal{F}$-open subobjects in Cor. 8.2. Closure under $\mathcal{F}$-closed subobjects follows from Prop. 8.4, and this fact then implies closure under equalizers, by Thm. 4.3(viii). $\qquad\square$

8.6. Invariance theorem for local compactness. We finally turn to invariance of local $\mathcal{F}$-compactness under $\mathcal{F}$-perfect maps. For that purpose let us call a morphism $f : X \to Y$ $\mathcal{F}$-*final* if a subobject of Y is $\mathcal{F}$-closed whenever its inverse image under f is $\mathcal{F}$-closed. Clearly, with Prop. 1.8(2) one obtains:

Lemma. *If every pullback of $f \in \mathcal{F}$ along a morphism in $\mathcal{M}$ lies in $\mathcal{E}$, then f is $\mathcal{F}$-final.*

Translated into standard topological terms, the Lemma asserts that in $\mathcal{T}op$ closed surjective maps are quotient maps. Now, in $\mathcal{T}op$, to say that the map

$f : X \to Y$ has the property that $N \subseteq Y$ is closed as soon as $f^{-1}[N] \subseteq X$ is closed is trivially equivalent to saying that N is open whenever $f^{-1}[N]$ is open. In general, however, we may not assume such equivalence.

Example. In $\mathcal{T}op_{\mathrm{open}}$, $\mathcal{F}$-finality takes on the usual meaning: $N \subseteq Y$ is open ($=$ $\mathcal{F}$-closed) as soon as $f^{-1}[N]$ is open in X. However, every map $f : X \to Y$ in $\mathcal{T}op_{\mathrm{open}}$ has the property that $N \subseteq Y$ is $\mathcal{F}$-open as soon as $f^{-1}[N]$ is $\mathcal{F}$-open, provided that X and Y are $T1$-spaces. Indeed, in this case all subobjects are $\mathcal{F}$-open, because there are only isomorphic $\mathcal{F}$-dense subobjects: see Exercise 2 of 7.2.

Theorem. *Let every morphism in $\mathcal{M}$ be $\mathcal{F}$-initial, let $\mathcal{E}$ be stable under pullback along $\mathcal{M}$-morphisms and let every $\mathcal{F}$-final morphism in $\mathcal{E}$ have the property that a subobject of its codomain is $\mathcal{F}$-open whenever its inverse image is $\mathcal{F}$-open. Then, for every $\mathcal{F}$-perfect morphism $f : X \to Y$ with $X, Y \in \mathcal{F}$-Tych, one has:*

(1) *if Y is locally $\mathcal{F}$-compact, so is X;*
(2) *if $f \in \mathcal{E}$ and X is locally $\mathcal{F}$-compact, so is Y.*

Proof. (1) is a special case of Cor. 8.4 (iii) $\Rightarrow$ (i). For (2), first observe that $\mathcal{F}$-density of f and β_Y gives the same first for $\beta f \cdot \beta_X = \beta_Y \cdot f$ and then for βf. But $\beta f : \beta X \to \beta Y$ is also $\mathcal{F}$-closed (see Prop. 5.1), hence $\beta f \in \mathcal{E}$ (see Exercise 1 of 2.2). Furthermore, by the Lemma, βf is $\mathcal{F}$-final. Since the $\mathcal{F}$-perfect morphism f makes

$$X \xrightarrow{\;\beta_X\;} \beta X \qquad\qquad (34)$$
$$\downarrow{\scriptstyle f} \qquad\qquad \downarrow{\scriptstyle \beta f}$$
$$Y \xrightarrow{\;\beta_Y\;} \beta Y$$

a pullback diagram with

$$\beta_X = (\beta f)^{-1}[\beta_Y]$$

$\mathcal{F}$-open, by hypothesis also β_Y must be $\mathcal{F}$-open. $\square$

Remark. Analyzing the condition of pullback stability of $\mathcal{E}$ along embeddings one observes that this stability property is needed only along $\mathcal{F}$-closed embeddings.

Exercise. Using Prop. 7.5, give sufficient conditions for $\mathcal{F}$-LCompHaus to be closed under finite coproducts in $\mathcal{X}$.

9. Pullback stability of quotient maps, Whitehead's Theorem

9.1. Quotient maps. The formation of quotient spaces is an important tool in topology when constructing new spaces from old: given a space X and a surjective mapping $f : X \to Y$, one provides the set Y with a topology such that any $B \subseteq Y$ is open (closed) if $f^{-1}[B] \subseteq X$ is open (closed). In our general setting of 2.1, we have called a morphism $f : X \to Y$ in $\mathcal{X}$ *$\mathcal{F}$-final* (see 8.6) if any $b \in$ sub$Y$ is $\mathcal{F}$-closed whenever $f^{-1}[b] \in$ sub$X$ is $\mathcal{F}$-closed, and we refer to an $\mathcal{F}$-final morphism in $\mathcal{E}$ as an *$\mathcal{F}$-quotient map*. Certain pullbacks of $\mathcal{F}$-final morphisms are $\mathcal{F}$-final:

Proposition. *The restriction $f' : f^{-1}[B] \to B$ of an $\mathcal{F}$-final morphism $f : X \to Y$ to an $\mathcal{F}$-closed subobject $b : B \to Y$ is $\mathcal{F}$-final.*

Proof. For a subobject $c : C \to B$ one has
$$f^{-1}[b \cdot c] \cong f^{-1}[b] \cdot (f')^{-1}[c].$$
Hence, if $(f')^{-1}[c]$ is $\mathcal{F}$-closed, so is $f^{-1}[b \cdot c]$ and also $b \cdot c$, by hypothesis. But $\mathcal{F}$-closedness of $b \cdot c$ implies the same for c, by Prop. 3.2(3). □

In general, $\mathcal{F}$-finality fails badly to be stable under pullback, already for $\mathcal{X} = \mathcal{T}op$ and in very elementary situations.

Example. We consider the quotient map $f : \mathbb{R} \to \mathbb{R}/\mathbb{Z}$, so that $f(x) = f(y)$ if and only if $x - y \in \mathbb{Z}$, and the subspace $S := \mathbb{R} \setminus \{\frac{1}{n} \mid n \in \mathbb{N}, \ n \geq 2\}$ of $\mathbb{R}$. Then the map
$$g := \mathrm{id}_S \times f : S \times \mathbb{R} \to S \times \mathbb{R}/\mathbb{Z}$$
is the pullback of f along the projection $S \times \mathbb{R}/\mathbb{Z} \to \mathbb{R}/\mathbb{Z}$, but fails to be a quotient map. Indeed, the set
$$B := \{\frac{1}{i} + \frac{\pi}{j}, \frac{i+1}{j}) \mid i, j \geq 2\}$$
is closed in $S \times \mathbb{R}$ and $g^{-1}[g[B]] = B$, but $g[B]$ is not closed in $S \times \mathbb{R}/\mathbb{Z}$ (the point $(0, f(1))$ lies in $\overline{g[B]} \setminus g[B]$).

Let us also note that the map f is closed; hence: the product of two closed quotient maps needs neither to be closed nor a quotient map.

Exercise. Verify the claims made in the Example and show that the space S fails to be locally compact.

9.2. Beck-Chevalley Property.

In what follows we would like to extend Proposition 9.1 greatly by showing that $\mathcal{F}$-quotient maps are stable under pullback along $\mathcal{F}$-perfect maps. But this needs some preparations and extra conditions. The condition given in the following proposition is known as (an instance of) the Beck-Chevalley Property.

Proposition. *The class $\mathcal{E}$ of the factorization system $(\mathcal{E}, \mathcal{M})$ of $\mathcal{X}$ is stable under pullback if and only if, for every pullback diagram*

$$
\begin{array}{ccc}
U & \xrightarrow{g} & V \\
{\scriptstyle q}\downarrow & & \downarrow{\scriptstyle p} \\
X & \xrightarrow{f} & Y
\end{array}
\tag{35}
$$

and every $a \in$ sub$X$, $g[q^{-1}[a]] \cong p^{-1}[f[a]]$.

Proof. Considering $a \cong 1_X$ one sees that the condition is sufficient for pullback stability: with Exercise 2 of 1.6, $f \in \mathcal{E}$ implies $g \in \mathcal{E}$. For its necessity we consider the commutative diagram

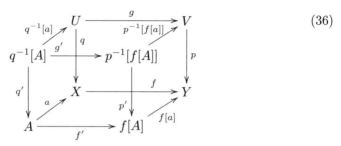

$$(36)$$

It suffices to show that the morphism g' (induced by the pullback property of $p^{-1}[f[A]]$) is in $\mathcal{E}$. But since all other vertical faces are pullback diagrams, also the front face of (36) is a pullback diagram; consequently, with $f' \in \mathcal{E}$ we obtain $q' \in \mathcal{E}$, by hypothesis on $\mathcal{E}$. □

9.3. Fibre-determined categories.

We showed in 3.5 that in $\mathcal{T}op$ proper maps are characterized as closed maps with compact fibres, but that this characterization fails in general. Since in what follows we would like to make crucial use of it, we say that our category $\mathcal{X}$ is *fibre-determined* (with respect to its $(\mathcal{E}, \mathcal{M})$-closed system $\mathcal{F}$) if the following two conditions are satisfied:

1. a morphism f lies in $\mathcal{E}$ if and only if all of its fibres have *points*, that is: if for every fibre F of f there is a morphism $1 \to F$;
2. a morphism f is in $\mathcal{F}^*$ if and only if $f[-]$ preserves $\mathcal{F}$-closedness of subobjects and f has $\mathcal{F}$-compact fibres.

Of course, in condition 2 "only if" comes for free (see Prop. 2.1(3) and Cor. 3.5). Condition 1 means equivalently that $\mathcal{E}$ contains exactly those morphisms with respect to which the terminal object 1 is projective, as the Reader will readily check:

Exercises.

1. Prove that for a morphism $f : X \to Y$ all fibres have points if and only if the hom-map $\mathcal{X}(1, f) : \mathcal{X}(1, X) \to \mathcal{X}(1, Y)$ is surjective.
2. Prove that the class of all morphisms f for which $\mathcal{X}(1, f)$ is surjective is stable under pullback in $\mathcal{X}$.

As a consequence we see that condition 1 forces $\mathcal{E}$ to be stable under pullback. We note that, granted pullback stability of $\mathcal{E}$, in the remainder of this section we shall use only the "only if" part of condition 1.

Observe that $\mathcal{T}op$ is fibre-determined (by Theorem 3.5), and so is $\mathcal{T}op_{\mathrm{open}}$ (trivially, see Exercise 1 of 3.5), but that $\mathcal{T}op/B$ is not (by Example 3.5).

9.4. Pullbacks of quotient maps.

Next we establish a result which, even in the case $\mathcal{X} = \mathcal{T}op$, was established only recently (see [10]):

Theorem (Richter-Tholen [46]). *If $\mathcal{X}$ is fibre-determined, then every pullback of an $\mathcal{F}$-quotient map along an $\mathcal{F}$-perfect map is an $\mathcal{F}$-quotient map.*

Proof. We consider the pullback diagram (35) with an $\mathcal{F}$-perfect map f and an $\mathcal{F}$-quotient map p. In order to show that q is an $\mathcal{F}$-quotient map as well, for $a \in \mathrm{sub}X$ we assume $q^{-1}[a]$ $\mathcal{F}$-closed and must show that a is $\mathcal{F}$-closed.

First, with Prop. 9.2 and the Exercises 9.3 we obtain

$$g[q^{-1}[a]] \cong p^{-1}[f[a]],$$

and this subobject is $\mathcal{F}$-closed since g (as a pullback of $f \in \mathcal{F}^*$) is $\mathcal{F}$-closed. Whence $f[a]$ is $\mathcal{F}$-closed, by hypothesis on p. Therefore it suffices to show that the morphism f' of diagram (36) is $\mathcal{F}$-proper, since then $\mathcal{F}$-properness of

$$f[a] \cdot f' = f \cdot a$$

and $\mathcal{F}$-separatedness of f give $\mathcal{F}$-closedness of a, with Prop. 5.2.

Now, in order to show $\mathcal{F}$-properness of f' we invoke condition 2 of 9.3 and first show that $f'[-]$ preserves $\mathcal{F}$-closedness of subobjects. Hence we consider $b \in \mathrm{sub}A$ and note

$$g'[(q')^{-1}[b]] \cong (p')^{-1}[f'[b]]$$

with the Beck-Chevalley Property again, applied to the front face of (36). This subobject is $\mathcal{F}$-closed since the morphism g' is $\mathcal{F}$-closed; in fact, since

$$g \cdot q^{-1}[a] = p^{-1}[f[a]] \cdot g'$$

is $\mathcal{F}$-proper, so is g', by Prop. 3.2(3). Furthermore, p' is an $\mathcal{F}$-quotient map, by Prop. 9.1, and we can conclude that $f'[b]$ is $\mathcal{F}$-closed.

Lastly, it remains to be shown that f' has $\mathcal{F}$-compact fibres. But by condition 1 and Exercises 9.3, any point $z : 1 \to f[A]$ factors as $z = p' \cdot w$, with $w : 1 \to p^{-1}[f[A]]$. With F, G denoting the fibres of f', g' belonging to z, w, respectively, we can form the commutative diagram

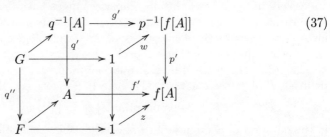

$$(37)$$

Its front face is a pullback diagram since the back-, top-, and bottom-faces are pullback diagrams. Hence, q'' is an isomorphism, and $F \cong G$ is $\mathcal{F}$-compact since g' is $\mathcal{F}$-proper, as shown earlier.   □

Corollary. *For $\mathcal{X}$ fibre-determined and K an $\mathcal{F}$-compact $\mathcal{F}$-Hausdorff object, with $p : V \to Y$ also $1_K \times p : K \times V \to K \times Y$ is an $\mathcal{F}$-quotient map.*

Proof. Apply the Theorem to the $\mathcal{F}$-perfect projection $K \times Y \to Y$.   □

Exercises.

1. By applying the Theorem to $\mathcal{T}op_{\mathrm{open}}$, conclude that pullbacks of quotient maps along local homeomorphisms are quotient maps.
2. Verify that when $\mathcal{E}$ is stable under pullback, the class $\mathcal{F}^+$ of $\mathcal{F}$-open maps in $\mathcal{X}$ is $(\mathcal{E}, \mathcal{M})$-closed. What does the Theorem say when we exchange $\mathcal{F}$ for $\mathcal{F}^+$?

9.5. Whitehead's Theorem. An $\mathcal{F}^+$-final morphism $f : X \to Y$ has, by definition, the property that $b \in \mathrm{sub}Y$ is $\mathcal{F}$-open whenever $f^{-1}[b] \in \mathrm{sub}X$ is $\mathcal{F}$-open. From Prop. 9.1 applied to $\mathcal{F}^+$ in lieu of $\mathcal{F}$ one obtains immediately:

Lemma. *The restriction $f' : f^{-1}[B] \to B$ of an $\mathcal{F}^+$-final morphism $f : X \to Y$ to an $\mathcal{F}$-open subobject $b : B \to Y$ is $\mathcal{F}^+$-final.* $\square$

In $\mathcal{T}op$, $\mathcal{F}^+$-quotient maps ($=\mathcal{F}^+$-final maps in $\mathcal{E}$) coincide with $\mathcal{F}$-quotient maps. If this happens in $\mathcal{X}$, we can use the lemma in order to upgrade Theorem 9.4 to:

Theorem. *If $\mathcal{X}$ is fibre-determined and if $\mathcal{F}^+$-quotient maps coincide with $\mathcal{F}$-quotient maps, then every pullback of an $\mathcal{F}$-quotient map along a locally $\mathcal{F}$-perfect map is an $\mathcal{F}$-quotient map.* $\square$

Corollary. *If $\mathcal{X}$ is fibre-determined and if $\mathcal{F}^+$-quotient maps coincide with $\mathcal{F}$-quotient maps, then for every locally $\mathcal{F}$-compact $\mathcal{F}$-Hausdorff object K and every $\mathcal{F}$-quotient map p, also $1_K \times p$ is an $\mathcal{F}$-quotient map.* $\square$

In $\mathcal{T}op$, the assertion of the Corollary is known as *Whitehead's Theorem*.

10. Exponentiable maps, exponentiable spaces

10.1. Reflection of quotient maps. In the previous section we have seen that, under certain hypotheses, locally $\mathcal{F}$-perfect maps have (stably) the property that $\mathcal{F}$-quotients pull back along them. This is a fundamental property which is worth investigating separately. Analogously to the terminology employed in the definition of $\mathcal{F}$-open maps, we say that a morphism *reflects $\mathcal{F}$-quotients* if in every pullback diagram (35) with p also q is an $\mathcal{F}$-quotient map, and f is *$\mathcal{F}$-exponentiable* if every pullback f' of f (see diagram (11)) reflects $\mathcal{F}$-quotients. Hence,

$$\mathcal{F}^{\mathrm{exp}} := \{f \mid f \ \mathcal{F}\text{-exponentiable}\} = \{f \mid f \text{ reflects } \mathcal{F}\text{-quotients}\}^*$$

in the notation of Section 3. The reason for this terminology will become clearer at the end of 10.2 below, and fully transparent in 10.9.

In this terminology, since (locally) $\mathcal{F}$-perfect maps are pullback stable, we proved in Theorems 9.4 and 9.5:

Corollary. *Let $\mathcal{X}$ be fibre-determined. Then every $\mathcal{F}$-perfect map is $\mathcal{F}$-exponentiable. Even every locally $\mathcal{F}$-perfect map has this property, if the notions of $\mathcal{F}^+$-quotient map and $\mathcal{F}$-quotient map are equivalent in $\mathcal{X}$.* $\square$

Exercise. Conclude that in $\mathcal{T}op$ both perfect maps and local homeomorphisms are $\mathcal{F}$-exponentiable.

Example. For a space X, the map $X \to 1$ in $\mathcal{T}op$ always reflects $\mathcal{F}$-quotients (since the pullback of $V \to 1$ along it is the projection $X \times V \to X$), but generally not every pullback of $X \to 1$ reflects $\mathcal{F}$-quotients: see Example 9.1. Hence, in $\mathcal{T}op$, reflection of $\mathcal{F}$-quotients is a property properly weaker than $\mathcal{F}$-exponentiability.

Before developing a theory of $\mathcal{F}$-exponentiability, we should provide a first justification for the terminology.

10.2. Exponentiable maps. For every morphism $f : X \to Y$ we have the pullback functor
$$f^* : \mathcal{X}/Y \to \mathcal{X}/X$$
which sends an object (V,p) in $\mathcal{X}/Y$ to $(U,q) = (X \times_Y V, \mathrm{proj}_1)$ in $\mathcal{X}/X$ (see diagram (35)), and for a morphism $h : (V,p) \to (V',p')$, $f^*(h) : (U,q) \to (U',q')$ has underlying $\mathcal{X}$-morphism $1_X \times h : X \times_Y V \to X \times_Y V'$. Of course, the functor $f^{-1}[-]$ of 1.7 is just a restriction of f^*.

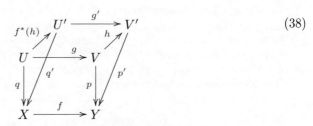

$$(38)$$

Proposition. *f is $\mathcal{F}$-exponentiable if and only if the functor f^* maps an $\mathcal{F}$-quotient morphism of $\mathcal{X}/Y$ to an $\mathcal{F}$-quotient morphism of $\mathcal{X}/X$.*

Proof. If f is $\mathcal{F}$-exponentiable and $h : (V,p) \to (V',p')$ an $\mathcal{F}$-quotient map, then g' (as a pullback of f) reflects the $\mathcal{F}$-quotient map h in $\mathcal{X}$, hence its pullback $f^*(h)$ along g' is an $\mathcal{F}$-quotient map in $\mathcal{X}$ and also in $\mathcal{X}/X$. Conversely, considering any pullback g' of f and an $\mathcal{F}$-quotient map h in $\mathcal{X}$, we can consider this as a morphism in $\mathcal{X}/Y$ and argue in the same way as before. $\qquad\square$

Exercises.
1. Prove that if $\mathcal{X}$ has a certain type of colimits, so does $\mathcal{X}/B$, and the forgetful functor $\Sigma_B : \mathcal{X}/B \to \mathcal{X}$ preserves them.
2. For $\mathcal{X} = \mathcal{T}op$, prove that f^* preserves coproducts for every f.
3. For $\mathcal{X} = \mathcal{T}op$, prove that f^* preserves coequalizers if and only if f^* preserves $\mathcal{F}$-quotient morphisms.
4. Conclude that in $\mathcal{T}op$ (as well as in $\mathcal{T}op_{\mathrm{open}}$) the $\mathcal{F}$-exponentiable morphisms f are exactly those for which f^* preserves all small colimits.
5. Apply *Freyd's Special Adjoint Functor Theorem* (see Mac Lane [35]) to prove that for an $\mathcal{F}$-exponentiable morphism f in $\mathcal{T}op$ the functor f^* actually has a right adjoint.

Exercise 5 identifies the $\mathcal{F}$-exponentiable morphisms f in $\mathcal{T}op$ (and in $\mathcal{T}op_{\mathrm{open}}$) as those for which f^* has a right adjoint functor, a property that may be considered in any category $\mathcal{X}$ with pullbacks and that is equivalent to the absolute ($=$"$\mathcal{F}$-free") categorical notion of *exponentiability*; see 10.8 below.

10.3. Stability properties. We shall collect some properties of the class $\mathcal{F}^{\mathrm{exp}}$ of $\mathcal{F}$-exponentiable morphisms. First we note:

Lemma. *If $g \cdot f$ is $\mathcal{F}$-final, so is g.*

Proof. If for a subobject c of the codomain of g $g^{-1}[c]$ is $\mathcal{F}$-closed, so is $f^{-1}[g^{-1}[c]]$ $= (g \cdot f)^{-1}[c]$. Hence $\mathcal{F}$-closedness of c follows with the hypothesis. $\qquad\square$

Following the terminology used in topology we call a morphism f in $\mathcal{X}$ an $\mathcal{F}$-*biquotient map* if every pullback of f is an $\mathcal{F}$-quotient map. In $\mathcal{T}op$ other names in use are *universal quotient map* and *descent map*. In order not to divert too much, we leave it to the Reader to establish their characterization in terms of points and open sets.

Exercises.

1. Show that every $\mathcal{F}$-proper map in $\mathcal{E}^*$ is an $\mathcal{F}$-biquotient map.
2. Conclude that if the notions of $\mathcal{F}^+$-quotient map and $\mathcal{F}$-quotient map are equivalent in $\mathcal{X}$, then every $\mathcal{F}$-open map in $\mathcal{E}^*$ is an $\mathcal{F}$-biquotient map.
3. * Show that in $\mathcal{T}op$ $f : X \to Y$ is an $(\mathcal{F}\text{-})$biquotient map if and only if, for every point $y \in Y$ and every open cover $\{U_i \mid i \in I\}$ of the fibre $f^{-1}y$, the system $\{f[U_i] \mid i \in F\}$ covers some neighborhood of y in Y, for some finite $F \subseteq I$. (See Day-Kelly [14].)

Proposition.

(1) *The class $\mathcal{F}^{\mathrm{exp}}$ contains all isomorphisms, is closed under composition and stable under pullback.*
(2) *If $g \cdot f \in \mathcal{F}^{\mathrm{exp}}$ with g monic, then $f \in \mathcal{F}^{\mathrm{exp}}$.*
(3) *If $g \cdot f \in \mathcal{F}^{\mathrm{exp}}$ with f an $\mathcal{F}$-biquotient map, then $g \in \mathcal{F}^{\mathrm{exp}}$.*

Proof. Statements (1) and (2) are trivial, see Prop. 3.2. A pullback of $g \cdot f$ with f an $\mathcal{F}$-biquotient map has the form $g' \cdot f'$ with f' and $\mathcal{F}$-(bi)quotient map and $g' \cdot f' \in \mathcal{F}^{\mathrm{exp}}$. Hence it suffices to show that g reflects $\mathcal{F}$-quotients if $g \cdot f$ does, with f an $\mathcal{F}$-quotient map. But if $r : W \to Z$ is an $\mathcal{F}$-quotient map, we can form the consecutive pullback diagram

$$
\begin{array}{ccccc}
U & \xrightarrow{\ f'\ } & V & \xrightarrow{\ g'\ } & W \\
{\scriptstyle q}\downarrow & & {\scriptstyle p}\downarrow & & \downarrow{\scriptstyle r} \\
X & \xrightarrow{\ f\ } & Y & \xrightarrow{\ g\ } & Z
\end{array}
\qquad (39)
$$

and have that q is an $\mathcal{F}$-quotient map, by hypothesis on $g \cdot f$. But then also $f \cdot q = p \cdot f'$ is an $\mathcal{F}$-quotient map, whence p is one too, by the Lemma and Exercise 2 of 1.4. $\qquad\square$

10.4. Exponentiable objects. An object X of $\mathcal{X}$ is $\mathcal{F}$-*exponentiable* if the unique morphism $!_X : X \to 1$ is $\mathcal{F}$-exponentiable. This means precisely that for every $\mathcal{F}$-quotient map $p : W \to V$ also

$$1_X \times p : X \times W \to X \times V$$

is an $\mathcal{F}$-quotient map. From Corollaries 9.4 and 9.5 (or from 9.1) we have:

Corollary. *Let $\mathcal{X}$ be fibre-determined. Then every $\mathcal{F}$-compact Hausdorff object is $\mathcal{F}$-exponentiable. Even every locally $\mathcal{F}$-compact Hausdorff object has this property, if the notions of $\mathcal{F}^+$-quotient map and $\mathcal{F}$-quotient map are equivalent.* □

Just as Theorem 3.3 follows from Prop. 3.2, one can conclude the following Theorem from Prop. 10.3:

Theorem.

(1) *For any $\mathcal{F}$-exponentiable $f : X \to Y$ with Y $\mathcal{F}$-exponentiable, also X is $\mathcal{F}$-exponentiable.*
(2) *For any $\mathcal{F}$-biquotient map $f : X \to Y$ with X $\mathcal{F}$-exponentiable, also Y is $\mathcal{F}$-exponentiable.*
(3) *With X and Y also $X \times Y$ is $\mathcal{F}$-exponentiable.*

□

10.5. Exponentiability in $\mathcal{T}op$. It is time to shed more light on the notion of exponentiability in case $\mathcal{X} = \mathcal{T}op$. A characterization in traditional topological terms is, however, not quite obvious, and to a large extent we must refer the interested Reader to the literature. Reasonably manageable is the case of a subspace embedding.

Proposition (Niefield [38]). *For a subspace A of a topological space X, the following are equivalent:*

(i) *the inclusion map $A \hookrightarrow X$ is exponentiable;*
(ii) *A is open in its closure $\overline{A}$ in X;*
(iii) *A is locally closed in X, that is: $A = O \cap C$, for an open set O and a closed set C in X.*

Proof (Richter [44]). The equivalence (iii) $\Leftrightarrow$ (ii) is straightforward, while (ii) $\Rightarrow$ (i) follows easily from the characterization of quotients in $\mathcal{T}op$ via open subsets. (i) $\Rightarrow$ (ii): Let the inclusion map $A \hookrightarrow X$ be exponentiable. Then the inclusion $f : A \hookrightarrow \overline{A} := Y$ is exponentiable by Prop. 10.3(2). Consider the inclusion of the complement $g : Y \setminus A \hookrightarrow Y$ and the pushout $(j_1, j_2 : Y \to Y +_{Y \setminus A} Y)$ of (g, g). We

therefore have the following diagram

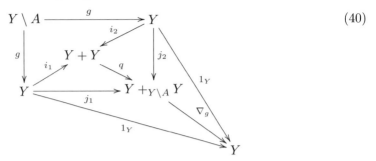

$$(40)$$

where q and ∇_g are quotient maps, since q is the coequalizer of $(i_1 \cdot g, i_2 \cdot g)$ and ∇_g is by construction a split epimorphism. Form now the pullback diagrams

$$
\begin{array}{ccccc}
A + A & \xrightarrow{\ r\ } & B & \xrightarrow{\ p\ } & A \\
\downarrow & & \downarrow{\scriptstyle s} & & \downarrow{\scriptstyle f} \\
Y + Y & \xrightarrow{\ q\ } & Y +_{Y \backslash A} Y & \xrightarrow{\ \nabla_g\ } & Y
\end{array}
\qquad (41)
$$

where $s : B \hookrightarrow Y +_{Y \backslash A} Y$ is an inclusion and the map $r : A + A \to B$ is an identity. Since with f also s is exponentiable, then $r : A + A \to B$ is a quotient, hence an homeomorphism. The inclusion $k_1 : A \hookrightarrow A + A \cong B$ is open, and therefore there is an open subset O of $Y +_{Y \backslash A} Y$ such that $O \cap (A + A) = k_1(A)$, hence $j_1^{-1}(O) \supseteq A$. Moreover, $j_2^{-1}(O) = \emptyset = j_1^{-1}(O) \cap (Y \backslash A)$ since otherwise $j_2^{-1}(O) \cap A \neq \emptyset$ because A is dense in Y. Therefore $A = j_1^{-1}(O)$ is open in Y as claimed. $\square$

For the characterization of exponentiable spaces we refer to [14, 26, 21]:

Theorem (Day-Kelly). *A topological space is exponentiable if and only if for every neighborhood V of a point there is a smaller neighborhood U such that every open cover of V contains a finite subcover of U.* $\square$

Spaces satisfying the condition of the Theorem are called *core-compact*. These are precisely the spaces whose system of open sets forms a *continuous lattice* (see [47, 26]). We already know that locally compact Hausdorff spaces are of this type. But in fact, without any separation condition, already Brown [4] proved that local compactness (in the sense of Example 1 of 8.1) implies exponentiability. The converse proposition in case of a Hausdorff space goes back to Michael [37]; actually, the separation condition may be eased from Hausdorff to sober, see [24]. But in general, exponentiable spaces fail to be locally compact, although no constructive example is known (see Isbell [26]).

Niefield [38] established a characterization of exponentiable maps in $\mathcal{T}op$ in the Day-Kelly style and thereby greatly generalized the Theorem above. For recent accounts of this characterization we refer the Reader to Niefield [40] and Richter [45], with the latter paper giving a smooth extension of the Day-Kelly result in terms of *fibrewise core-compactness*.

10.6. Local separatedness. We briefly look at morphisms in the class $(\mathcal{F}^{\exp})'$. Hence, we call $f : X \to Y$ *locally $\mathcal{F}$-separated* if $\delta_f : X \to X \times_Y X$ is $\mathcal{F}$-exponentiable. With Prop. 4.2 applied to $\mathcal{F}^{\exp}$ in lieu of $\mathcal{F}$ and with Prop. 10.3(3) we obtain:

Proposition.

(1) *The class of locally $\mathcal{F}$-separated maps contains all monomorphisms, is closed under composition and stable under pullback.*

(2) *If $g \cdot f$ is locally $\mathcal{F}$-separated, so is f.*

(3) *If $g \cdot f$ is locally $\mathcal{F}$-separated with an $\mathcal{F}$-exponentiable $\mathcal{F}$-biquotient map f, also g is locally $\mathcal{F}$-separated.*

$\square$

Example. The terminology justifies itself in case $\mathcal{X} = \mathcal{T}op$. With Prop. 10.5 one easily verifies that $f : X \to Y$ is locally $(\mathcal{F}$-$)$separated if and only if every point in X has a neighborhood U such that $f|_U : U \to Y$ is separated. Both separated and locally injective maps (see Exercise 1 of 4.1) are locally separated.

An object X of $\mathcal{X}$ is *locally $\mathcal{F}$-separated* (or *locally $\mathcal{F}$-Hausdorff*) if $X \to 1$ is locally $\mathcal{F}$-separated, i.e., if $\delta_X : X \to X \times X$ is $\mathcal{F}$-exponentiable. For $\mathcal{X} = \mathcal{T}op$, this means that every point in X has a Hausdorff neighborhood.

Exercises.

1. Apply Theorem 4.3 and Corollary 4.3 with $\mathcal{F}^{\exp}$ in lieu of $\mathcal{F}$ in order to establish stability properties for locally $\mathcal{F}$-separated spaces. (Attention: special care needs to be given when "translating" condition (vi) of Theorem 4.3: see (3) of the Proposition above.)

2. Show that in $\mathcal{T}op$ locally Hausdorff spaces are sober T1-spaces, but not conversely.

10.7. Maps with exponentiable domain and locally separated codomain. A powerful sufficient criterion for $\mathcal{F}$-exponentiability of morphisms (a weaker version of which was established in [39]) is obtained "for free" when we apply Prop. 5.1 to $\mathcal{F}^{\exp}$ in lieu of $\mathcal{F}$:

Theorem. *Every morphism $f : X \to Y$ in $\mathcal{X}$ with X $\mathcal{F}$-exponentiable and Y locally $\mathcal{F}$-Hausdorff is $\mathcal{F}$-exponentiable.* $\square$

Just as Prop. 5.2 follows from Prop. 5.1, we conclude from the Theorem the following strengthening of the assertion of Prop. 10.3(2):

Corollary. *If $g \cdot f$ is $\mathcal{F}$-exponentiable with g locally $\mathcal{F}$-separated, then f is $\mathcal{F}$-exponentiable.*

Exercise. Prove that the full subcategory of $\mathcal{F}$-exponentiable and locally $\mathcal{F}$-separated objects is closed in $\mathcal{X}$ under finite limits and under $\mathcal{F}$-exponentiable subobjects.

10.8. Adjoints describing exponentiability. We saw in 10.2 that $f : X \to Y$ is $\mathcal{F}$-exponentiable in $\mathcal{T}op$ if and only if $f^* : \mathcal{T}op/Y \to \mathcal{T}op/X$ has a right adjoint functor. It is a slightly tricky exercise on adjoint functors to show that the latter property may equivalently be expressed by the adjointness of two other functors (see Niefield [38]):

Proposition. *In a finitely complete category $\mathcal{X}$, the following assertions are equivalent for a morphism $f : X \to Y$:*

(i) $f^* : \mathcal{X}/Y \to \mathcal{X}/X$ *has a right adjoint;*

(ii) *the functor $X \times_Y (-) : \mathcal{X}/Y \to \mathcal{X}$, $(V, p) \mapsto X \times_Y V$, has a right adjoint;*

(iii) *the functor $(X, f) \times (-) : \mathcal{X}/Y \to \mathcal{X}/Y$, $(V, p) \mapsto (X \times_Y V, f \cdot f^*(p))$, has a right adjoint.*

Proof. Note that the functor described by (ii) is simply $\Sigma_X f^*$, with the forgetful $\Sigma_X : \mathcal{X}/X \to \mathcal{X}$, and the functor described by (iii) may be written as $f_! f^*$, where

$$f_! : \mathcal{X}/X \to \mathcal{X}/Y, \quad (U, q) \mapsto (U, f \cdot q),$$

is left adjoint to f^*. □

Property (iii) identifies (X, f) as an exponentiable object of the category $\mathcal{X}/Y$ since, in general, one calls an object X of a category $\mathcal{X}$ with finite products *exponentiable* if $X \times (-) : \mathcal{X} \to \mathcal{X}$ has a right adjoint.

10.9. Function space topology. As important as establishing the existence of the right adjoints in question is their actual description. In case $\mathcal{X} = \mathcal{T}op$, the right adjoint to $X \times (-)$ for an exponentiable (=core-compact, see 10.5) space is described by the *function-space functor*

$$(-)^X : \mathcal{T}op \to \mathcal{T}op,$$

where the space Y^X has underlying set $C(X, Y) = \mathcal{T}op(X, Y)$ (the set of continuous functions from X to Y); its open sets are generated by the sets

$$N(U, V) = \{f \in C(X, Y) \mid U \ll f^{-1}(V)\},$$

where $U \ll f^{-1}(V)$ for open sets $U \subseteq X$, $V \subseteq Y$ means that every open cover of $f^{-1}(V)$ has a finite subcover of U (see [21]). In case X is a locally compact Hausdorff space (hence exponentiable, see Cor. 10.4) the topology of Y^X has generating open sets

$$N(C, V) = \{f \in C(X, Y) \mid f(C) \subseteq V\},$$

with $C \subseteq X$ compact and $V \subseteq Y$ open, which is known by the name *compact-open topology* (see [20]).

One of the great advantages of Scott's *way-below relation* $\ll$ (which we already encountered in the formulation 10.5 of the Day-Kelly Theorem) is that it translates smoothly from topological spaces to locales and actually leads to corresponding results. For example, a locale L is exponentiable in $\mathcal{L}oc$ if and only if it is a continuous lattice; see [25].

10.10. Partial products. How does the function-space functor look like in $\mathcal{T}op/Y$, rather than in $\mathcal{T}op$? Here we just indicate how the right adjoint of $X \times_Y (-) :$ $\mathcal{T}op/Y \to \mathcal{T}op$ (see Prop. 10.8(ii)) looks like for $f : X \to Y$ exponentiable in $\mathcal{T}op$.

Such right adjoint functor must produce, for every object Z in $\mathcal{T}op$, an object $(P, p : P \to Y)$ in $\mathcal{T}op/Y$ and a morphism $\varepsilon : X \times_Y P \to Z$ which makes

$$Z \xleftarrow{\ \varepsilon\ } X \times_Y P \longrightarrow P \qquad (42)$$
$$\downarrow \qquad\qquad \downarrow p$$
$$X \xrightarrow{\ f\ } Y$$

terminal amongst all diagrams

$$Z \xleftarrow{\ d\ } X \times_Y T \longrightarrow T \qquad (43)$$
$$\downarrow \qquad\qquad \downarrow t$$
$$X \xrightarrow{\ f\ } Y$$

such that there is a unique morphism $h : T \to P$ with $p \cdot h = t$ and $\varepsilon \cdot (1_X \times h) = d$ (see [19]). Exploiting the bijective correspondence

$$\frac{h : T \to P}{t : T \to Y,\ d : X \times_Y T \to Z}$$

in case $T = 1$, one sees that $P = P(f, Z)$ should have underlying set

$$P = \{(t, d) \mid t \in Y,\ d : f^{-1}t \to Z \text{ continuous}\},$$

with projection $p : P \to Y$, and that $\varepsilon = \varepsilon_{f,Z}$ is the evaluation map $(x, d) \mapsto d(x)$ when we realize the underlying set of the pullback $X \times_Y P$ as

$$X \times_Y P = \{(x, d) \mid x \in X,\ d : f^{-1}(f(x)) \to Z \text{ continuous}\}.$$

Now, for f exponentiable, there is a coarsest topology on P which makes both p and ε continuous. This topology has been described quite succinctly in terms of ultrafilter convergence in [10].

The case that $f : U \hookrightarrow Y$ is the embedding of an open subspace of Y (hence exponentiable, by Prop. 10.5) deserves special mentioning. In this case, the underlying sets of P and $U \times_Y P$ may be realized as $(U \times Z) + (Y \setminus U)$ and $U \times Z$, respectively, with $p|_{U \times Z}$ and ε projection maps and $p|_{Y \setminus U}$ the inclusion map. Here the coarsest topology on P making p and ε continuous provides the pullback $U \times_Y P = U \times Z$ with the product topology, and it was first described by Pasynkov [41] who called P the *partial product of Y over U with fibre Z*. He gave various interesting examples for such spaces; for instance, the n-dimensional sphere S^n can be obtained recursively via partial products from the n-dimensional cube I^n and the discrete doubleton D_2, as

$$S^n \cong P(I^n \setminus S^{n-1} \hookrightarrow I^n, D_2).$$

Exercises.

1. Identify the functors described by Prop. 10.8 in the commutative diagram

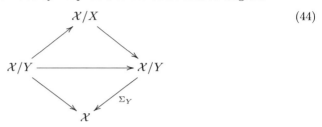

$$(44)$$

2. Show that $Z \mapsto (Y \times Z, Y \times Z \to Y)$ is right adjoint to Σ_Y.
3. Conclude that when we denote by $(-)^{(X,f)}$ the right adjoint of $(X, f) \times (-)$ in $\mathcal{T}op/Y$, then $P(f, Z) \cong (Y \times Z, Y \times Z \to Y)^{(X,f)}$.

11. Remarks on the Tychonoff Theorem and the Stone-Čech compactification

11.1. Axioms giving closures. If $\mathcal{X}$ is $\mathcal{T}op$ or $\mathcal{L}oc$, in addition to (F3)-(F5) the class $\mathcal{F}$ satisfies also the following condition:

(F6) arbitrary intersections of morphisms in $\mathcal{F} \cap \mathcal{M}$ exist and belong to $\mathcal{F} \cap \mathcal{M}$.

Hence, $\mathcal{F} \cap \mathcal{M}$ is *stable under intersection* (see Prop. 1.4(4); we allow the indexing system to be arbitrarily large). Equipped with this additional condition in our category $\mathcal{X}$ satisfying the conditions of 2.1, one introduces the $\mathcal{F}$-*closure*

$$c_X(m) : c_X(M) \to X$$

of a subobject $m : M \to X$ in X by

$$c_X(m) := \bigwedge \{k \in \mathrm{sub}X \mid k \geq m,\ k\ \mathcal{F}\text{-closed}\}.$$

The following properties are easily checked:

1. $m \leq c_X(m)$,
2. $m \leq n \Rightarrow c_X(m) \leq c_X(n)$,
3. $c_X(m) \cong c_X(c_X(m))$,
4. $f[c_X(m)] \leq c_Y(f[m])$,

for all $m, n \in \mathrm{sub}X$ and $f : X \to Y$ in $\mathcal{X}$. In other words, $c = (c_X)_{X \in \mathcal{X}}$ is an *idempotent closure operator* in the sense of [17]. In what follows we therefore assume that $\mathcal{F}$ is given as in 2.5, that is:

(F7) a morphism f lies in $\mathcal{F}$ if and only if $f[-]$ preserves $\mathcal{F}$-closed subobjects.

The condition for f means equivalently that image commutes with closure, i.e., in assertion 4 above we may write $\cong$ instead of $\leq$ when f is closed. Of course, (F7) implies (F5) when $\mathcal{E}$ is stable under pullback along morphisms in $\mathcal{F}_0 \cap \mathcal{M}$ (see Exercise 4 of 2.1). In addition, we assume the closure operator c to be *hereditary*, that is:

(F8) $c_Y(m) \cong y^{-1}[c_X(y \cdot m)]$ for all $m : M \to Y$, $y : Y \to X$ in $\mathcal{M}$.

Exercises.

1. Verify that $\mathcal{T}op$ and $\mathcal{L}oc$ satisfy conditions (F6)-(F8).
2. Show that (F8) implies that every morphism in $\mathcal{M}$ is $\mathcal{F}$-initial (see 6.5).
3. Prove that when $\mathcal{X}$ satisfies (F6)-(F8), so does $\mathcal{X}/B$.

11.2. Products as inverse limits of finite products. Every product of objects in $\mathcal{X}$ is an inverse limit of its "finite subproducts". This means, given a product

$$X_I = \prod_{i \in I} X_i$$

of objects in $\mathcal{X}$, one has

$$X_I \cong \lim_{F \subseteq I \text{finite}} X_F,$$

where $X_F = \prod_{i \in F} X_i$, with canonical bonding morphisms $X_G \to X_F$ for $F \subseteq G \subseteq I$.
It is then natural to expect that the closure operator defined in 11.1 commutes with this limit presentation:

(F9) $c_{X_I}(M) \cong \lim_{F \subseteq I \text{finite}} c_{X_F}(\pi_F[M])$ for all $m : M \to X_I$ in $\mathcal{M}$;

here $\pi_F : X_I \to X_F$ is the canonical morphism. Equivalently, this formula may be written as:

$$c_{X_I}(m) \cong \bigwedge_{F \subseteq I \text{finite}} \pi_F^{-1}[c_{X_F}(\pi_F[m])]$$

(see [9]). Hence, by (F9) the "topological structure" of X_I is completely determined by the structure of the "finite subproducts".

Exercises.

1. Verify that $\mathcal{T}op$ and $\mathcal{L}oc$ satisfy condition (F9).
2. Prove that when $\mathcal{X}$ satisfies (F6)-(F9), so does $\mathcal{X}/B$.

11.3. Towards proving the Tychonoff Theorem. Let us now assume that all X_i $(i \in I)$ are $\mathcal{F}$-compact. In order to show that the projection

$$p : X_I \times Y \to Y$$

is $\mathcal{F}$-closed for every object Y in $\mathcal{X}$, by (F7) we would have to establish

$$p[c(m)] \cong c(p[m]) \tag{$*$}$$

for all $m : M \to X_I \times Y$ in $\mathcal{M}$. For that consider the commutative diagram

$$(45)$$

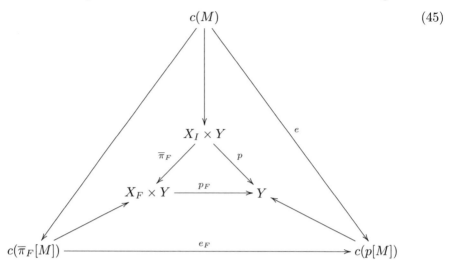

Here e, e_F are determined by the projections p, p_F, respectively, and $\overline{\pi}_F = \pi_F \times 1_Y$ represents $X_I \times Y$ as an inverse limit of $X_F \times Y$ ($F \subseteq I$ finite). Formula $(*)$ says precisely that e should be in $\mathcal{E}$. By Theorem 3.3(3), p_F is $\mathcal{F}$-closed for every finite $F \subseteq I$, hence $e_F \in \mathcal{E}$. Now, by (F9),

$$c(M) \cong \lim_F c(\overline{\pi}_F[M])$$

in $\mathcal{X}$, which implies

$$(c(M), e) \cong \lim_F (c(\overline{\pi}_F[M]), e_F)$$

in $\mathcal{X}/c(p[M])$.

11.4. The Tychonoff Theorem. We call $\mathcal{E}$ a *surjectivity class* if for some class $\mathcal{P}$ of objects in $\mathcal{X}$ one has $f : X \to Y$ in $\mathcal{E}$ exactly when all maps $\mathcal{X}(P, f) : \mathcal{X}(P, X) \to \mathcal{X}(P, Y)$ ($P \in \mathcal{P}$) are surjective. Note that if $\mathcal{E}$ in $\mathcal{X}$ is a surjectivity class, so is $\mathcal{E}_B$ in $\mathcal{X}/B$. Furthermore, when $\mathcal{X}$ is fibre-determined (see 9.3), then $\mathcal{E}$ is a surjectivity class. In $\mathcal{T}op$, $\mathcal{E}$ is a surjectivity class (take $\mathcal{P} = \{1\}$), but in $\mathcal{L}oc$ it is not (since every surjectivity class is pullback stable).

Theorem (Clementino-Tholen [11]). *In addition to* (F1)-(F5), *let $\mathcal{X}$ satisfy conditions* (F6)-(F9) *of 11.1/2. Then each of the following two assumptions makes $\mathcal{F}$-Comp closed under direct products in $\mathcal{X}$:*

(L) *if $(X, \pi_\nu : X \to X_\nu)_\nu$ is an inverse limit in $\mathcal{X}$ and $(e_\nu : X_\nu \to Y)_\nu$ a compatible family of morphisms in $\mathcal{E}$, then the induced morphism $e : X \to Y$ lies in $\mathcal{E}$;*

(T) *$\mathcal{E}$ is a surjectivity class, and the Axiom of Choice holds true.*

Proof (sketch). By 11.3, assumption (L) makes $\mathcal{F}$-Comp closed under products in $\mathcal{X}$. Regarding assumption (T), it is sufficient to deduce that for the *particular* inverse system of diagram (45), $e_F \in \mathcal{E}$ for all finite $F \subseteq I$ implies e in $\mathcal{E}$. This is shown (using an ordinal induction) in [11] and [9]. $\qquad\square$

Examples.

(1) Via (T) one obtains the classical *Tychonoff Theorem* in $\mathcal{T}op$. Note that the validity of the Tychonoff Theorem in $\mathcal{T}op$ is logically equivalent to the Axiom of Choice (see [32]).

(2) While (L) fails in $\mathcal{T}op$, the condition is satisfied in $\mathcal{L}oc$ (see [49], 2.3), and the Theorem gives a choice-free proof for the Tychonoff Theorem in $\mathcal{L}oc$. (See also Chapter II.)

Exercises.

1. Assuming (F1)-(F6) show that for every monic family $(p_i : X \to X_i)_{i \in I}$ (so that for $x, y : P \to X$, $p_i \cdot x = p_i \cdot y$ for all $i \in I$ always implies $x = y$) $X_i \in \mathcal{F}$-Haus implies $X \in \mathcal{F}$-Haus.

2. Show that under the assumptions of the Theorem, $\mathcal{F}$-CompHaus is closed under all small limits in $\mathcal{X}$, if $\mathcal{X}$ has them.

11.5. Products of proper maps. Since all assumptions of Theorem 11.4 are invariant under "slicing", we may apply it to the slices $\mathcal{X}/B$ rather than to $\mathcal{X}$ and obtain:

Corollary. *If $\mathcal{X}$ satisfies* (F1)-(F9) *and condition* (L) *or* (T) *of Theorem 11.4, then the direct product of $\mathcal{F}$-proper ($\mathcal{F}$-perfect) morphisms in $\mathcal{X}$ is again $\mathcal{F}$-proper ($\mathcal{F}$-perfect, respectively).*

Proof. Given $\mathcal{F}$-proper ($\mathcal{F}$-perfect) maps $f_i : X_i \to Y_i$, $i \in I$, the morphism
$$\prod_{i \in I} f_i : \prod_{i \in I} X_i \to \prod_{i \in I} Y_i = Y$$
may be regarded as an object of $\mathcal{X}/Y$ and is then a product of the objects (P_i, f_i'), where $f_i' : P_i \to Y$ is the pullback of f_i along the projection $Y \to Y_i$, $i \in I$. $\qquad\square$

In case $\mathcal{X} = \mathcal{T}op$ the Corollary (which formally generalizes Theorem 11.4) is known as *Frolik's Theorem*.

11.6. Existence of the Stone-Čech compactification. A full subcategory $\mathcal{A}$ of $\mathcal{X}$ is said to be $\mathcal{F}$-*cowellpowered* if every object $X \in \mathcal{A}$ admits only a small set of non-isomorphic $\mathcal{F}$-dense morphisms with domain X and codomain in $\mathcal{A}$; that is: if there is a set-indexed family $d_i : X \to A_i$ ($i \in I$) of $\mathcal{F}$-dense morphisms in $\mathcal{A}$ such that every $\mathcal{F}$-dense morphism $d : X \to A$ in $\mathcal{A}$ factors as $d = j \cdot d_i$ for some isomorphism j and $i \in I$.

An easy application of *Freyd's General Adjoint Functor Theorem* gives (see [35]):

Theorem. *Let $\mathcal{X}$ be small-complete, and let $\mathcal{F}$-Haus be $\mathcal{F}$-cowellpowered. Then $\mathcal{F}$-CompHaus is reflective in $\mathcal{F}$-Haus with $\mathcal{F}$-dense reflexions if and only if $\mathcal{F}$-CompHaus is closed under products in $\mathcal{X}$.*

Proof (sketch). The necessity of the condition is clear (see also Exercise 1 of 11.4). For its sufficiency, let $X \in \mathcal{F}$-Haus and consider a representative system $d_i : X \to A_i$ $(i \in I)$ of $\mathcal{F}$-dense morphisms as in the definition of $\mathcal{F}$-cowellpoweredness, with the additional condition that $A_i \in \mathcal{F}$-CompHaus. By (F6), the induced morphism $f : X \to \prod_{i \in I} A_i$ factors through the $\mathcal{F}$-closed subobject

$$\beta X := c(f[X]) \longrightarrow \prod_{i \in I} A_i$$

by a unique morphism $\beta_X : X \to \beta X$, which is $\mathcal{F}$-dense by (F8). Note that βX is $\mathcal{F}$-compact Hausdorff, by Theorem 3.3(3), since $\prod_{i \in I} A_i$ is.

An arbitrary morphism $g : X \to A$ with $A \in \mathcal{F}$-CompHaus factors through the $\mathcal{F}$-dense morphism $g' : X \to c(g[X])$, which must be isomorphic to some d_i and must therefore factor through β_X. The resulting factorization is unique since β_X is $\mathcal{F}$-dense and A is $\mathcal{F}$-Hausdorff.   □

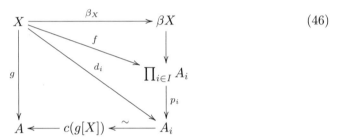

(46)

Of course, reflectivity of $\mathcal{F}$-CompHaus in $\mathcal{F}$-Haus gives in particular reflectivity of $\mathcal{F}$-CompHaus in $\mathcal{F}$-Tych and therefore the existence of the *Stone-Čech compactification*, as needed in Section 6.

Exercise. Show that the cardinal number of the underlying set of a Hausdorff space X with dense subset A cannot exceed $2^{2^{\operatorname{card} A}}$. Conclude that the full subcategory $\mathcal{F}$-Haus in $\mathcal{T}op$ is $\mathcal{F}$-cowellpowered, so that Theorem 11.6 in conjunction with the Tychonoff Theorem gives the existence of the classical Stone-Čech compactification of Tychonoff spaces.

References

[1] N. Bourbaki, *Topologie Générale*, Ch. I et II, Third ed. (Paris 1961).
[2] F. Borceux, *Handbook of Categorical Algebra 3. Categories of Sheaves*, Encyclopedia of Mathematics and its Applications **52** (Cambridge University Press, Cambridge 1994).
[3] F. Borceux and G. Janelidze, *Galois Theories*, Cambridge Studies in Advanced Mathematics **72** (Cambridge University Press, Cambridge 2001).

[4] R. Brown, Function spaces and product topologies, *Quart. J. Math.* (Oxford) **15** (1964) 238-250.

[5] A. Carboni and G. Janelidze, Decidable (= separable) objects and morphisms in lextensive categories, *J. Pure Appl. Algebra* **110** (1996) 219-240.

[6] A. Carboni, S. Lack and R.F.C. Walters, Introduction to extensive and distributive categories, *J. Pure Appl. Algebra* **84** (1993) 145-158.

[7] A. Carboni, M.C. Pedicchio and J. Rosický, Syntatic characterizations of various classes of locally presentable categories, *J. Pure Appl. Algebra* **161** (2001) 65-90.

[8] C. Chevalley and O. Frink Jr., Bicompactness of cartesian products, *Bull. Amer. Math. Soc.* **47** (1941) 612-614.

[9] M.M. Clementino, E. Giuli and W. Tholen, Topology in a category: compactness, *Port. Math.* **53** (1996) 397-433.

[10] M.M. Clementino, D. Hofmann and W. Tholen, The convergence approach to exponentiable maps, *Port. Math.* **60** (2003) 139-160.

[11] M.M. Clementino and W. Tholen, Tychonoff's Theorem in a category, *Proc. Amer. Math. Soc.* **124** (1996) 3311-3314.

[12] M.M. Clementino and W. Tholen, Separation versus connectedness, *Topology Appl.* **75** (1997) 143-181.

[13] M.M. Clementino and W. Tholen, A note on local compactness, *Rend. Istit. Mat. Univ. Trieste*, Suppl. 2, 32 (2001) 17-31.

[14] B.J. Day and G.M. Kelly, On topological quotient maps preserved by pullbacks or products, *Proc. Cambridge Phil. Soc.* **67** (1970) 553-558.

[15] Y. Diers, Categories of algebraic sets, *Appl. Categ. Structures* **4** (1996) 329-341.

[16] R. Dyckhoff, Categorical cuts, *Topology Appl.* **6** (1976) 291-295.

[17] D. Dikranjan and E. Giuli, Closure operators I, *Topology Appl.* **27** (1987) 129-143.

[18] D. Dikranjan and W. Tholen, *Categorical Structure of Closure Operators* (Kluwer Academic Publishers 1995).

[19] R. Dyckhoff and W. Tholen, Exponentiable morphisms, partial products and pullback complements, *J. Pure Appl. Algebra* **49** (1987) 103-116.

[20] R. Engelking, *General Topology*, revised and completed edition (Heldermann Verlag, Berlin 1989).

[21] M. Escardó and R. Heckmann, On spaces of continuous functions in topology (preprint, London 1999).

[22] E. Giuli, Zariski closure, completeness and compactness (preprint, L'Aquila 2001).

[23] H. Henriksen and J.R. Isbell, Some properties of compactification, *Duke Math. J.* **25** (1958) 83-106.

[24] K.H. Hofmann and J.D. Lawson, The spectral theory of distributive continuous lattices, *Trans. Amer. Math. Soc.* **246** (1978) 285-310.

[25] M. Hyland, Function spaces in the category of locales, in: *Continuous Lattices*, Lecture Notes in Mathematics **871** (Springer, Berlin 1981), pp. 264-281.

[26] J.R. Isbell, General function spaces, products and continuous lattices, *Math. Proc. Camb. Phil. Soc.* **100** (1986) 193-205.

[27] G. Janelidze and W. Tholen, Functorial factorization, well-pointedness and separability, *J. Pure Appl. Algebra* **142** (1999) 99-130.

[28] P.T. Johnstone, *Topos Theory*, London Mathematical Society Monographs, Vol. **10** (Academic Press, London 1977).

[29] P.T. Johnstone, *Stone Spaces*, Cambridge Studies in Advanced Mathematics, **3** (Cambridge University Press, Cambridge 1982).

[30] P.T. Johnstone, Factorization theorems for geometric morphisms, II, *Lecture Notes in Mathematics* **915** (Springer, Berlin 1982), pp. 216-233.

[31] P.T. Johnstone, Complemented sublocales and open maps, preprint (Cambridge University, 2002).

[32] J.L. Kelley, The Tychonoff product theorem implies the axiom of choice, *Fund. Math.* **37** (1950) 75-76.

[33] H.P. Künzi and B.A. Pasynkov, Tychonoff compactifications and *R*-completions of mappings and rings of continuous functions, in: *Categorical Topology* (Kluwer Acad. Publ., Dordrecht 1996) pp. 175-201.

[34] E. Lowen-Colebunders and G. Richter, An elementary approach to exponentiable spaces, *Appl. Categ. Structures* **9** (2001) 303-310.

[35] S. Mac Lane, *Categories for the Working Mathematician*, 2nd. ed. (Springer, New York 1998).

[36] S. Mac Lane and I. Moerdijk, *Sheaves in Geometry and Logic* (Springer, Berlin 1992).

[37] E. Michael, Bi-quotient maps and cartesian products of quotient maps, *Ann. Inst. Fourier* (Grenoble) **18** (1969) 287-302.

[38] S.B. Niefield, Cartesianness: topological spaces, uniform spaces, and affine schemes, *J. Pure Appl. Algebra* **23** (1982) 147-167.

[39] S.B. Niefield, A note on the local Hausdorff property, *Cahiers Topologie Géom. Différentielle Catégoriques* **24** (1983) 87-95.

[40] S.B. Niefield, Exponentiable morphisms: posets, spaces, locales, and Grothendieck toposes, *Theory Appl. Categories* **8** (2001) 16-32.

[41] B.A. Pasynkov, On extension to mappings of certain notions and assertions concerning spaces, in: *Mapping and Functors* (Eds. V.V. Fedorčuk et al.), Izdat. MGU (Moscow 1984), pp. 72-102 (in Russian).

[42] V. Pratt, Chu spaces, in: *School on category theory and applications*, Textos Mat., Sér. B. **21** (University of Coimbra 1999) pp. 39-100.

[43] A. Pultr and A. Tozzi, Notes on Kuratowski-Mrówka theorems in pointfree context, *Cahiers Topologie Géom. Différentielle Catég.* **33** (1992) 3-14.

[44] G. Richter, Exponentiable maps and triquotients in *Top*, *J. Pure Appl. Algebra* **168** (2002) 99-105.

[45] G. Richter, Exponentiability for maps means fibrewise core-compactness (preprint, Bielefeld 2001).

[46] G. Richter and W. Tholen, Perfect maps are exponentiable – categorically, *Theory Appl. Categories* **8** (2001) 457-464.

[47] D. Scott, Continuous lattices, in: *Toposes, Algebraic Theories and Logic*, Lecture Notes in Mathematics **274** (Springer, Berlin 1972), pp. 97-136.

[48] W. Tholen, A categorical guide to separation, compactness and perfectness, *Homology Homotopy Appl.* **1** (1999) 147-161.

[49] J.J.C. Vermeulen, Proper maps of locales, *J. Pure Appl. Algebra* **92** (1994) 79-107.

[50] J.J.C. Vermeulen, A note on stably closed maps of locales, *J. Pure Appl. Algebra* **157** (2001) 335-339.

[51] A. Zouboff, Herrlich-type embeddings, partial products and compact maps, preprint (1998).

Departamento de Matemática
Universidade de Coimbra
Apartado 3008
3001-454 Coimbra, Portugal
E-mail: mmc@mat.uc.pt

Dip. di Matematica Pura ed Applicata
Università degli Studi di L'Aquila
L'Aquila, Italy
E-mail: giuli@univaq.it

Department of Mathematics and Statistics
York University
Toronto, Canada M3J 1P3
E-mail: tholen@mathstat.yorku.ca

IV

Regular, Protomodular, and Abelian Categories

Dominique Bourn and Marino Gran

The aim of this chapter is mainly to introduce some basic categorical concepts dealing with General Algebra, and to illustrate in this context many important properties of one of the most remarkable categories in this area, namely the category *Grp* of groups.

These concepts are basic in the sense that they deal with notions as elementary as, for instance, those of epimorphism, monomorphism, kernel, cokernel and equivalence relation.

The first section is devoted to setting basic facts concerning the internal equivalence relations in a given finitely complete category.

In the second section different notions of epimorphisms are considered and classified, and their stability properties are studied. We prove a new result concerning strong epimorphisms, which turns out to be a powerful tool in many situations. We call this result the Barr-Kock Theorem, in reference to an important and weaker version of it given in [1]. The notions of *regular* and *Barr-exact categories* are then recalled, and their main basic properties are proved. In this context the previous distinctions of different kinds of epimorphisms are pretty well simplified.

In Algebra, among the several notions of monomorphism (or, equivalently, of subobject), the class of normal subgroups in the category *Grp* of groups, or the class of ideals in the category *Rng* of rings naturally deserve great attention. The third section deals with the general notion of normal monomorphism. It introduces the concept of *protomodular category*, where this notion of normal monomorphism becomes intrinsic, precisely as this happens in *Grp* or *Rng*. As an illustration of the efficiency of the concept of protomodular category we shall see that in such categories a morphism $f : X \to Y$ is a monomorphism if and only if the kernel of f is trivial. Moreover, we characterize the abelian objects in a pointed protomodular category as those objects X which have the property that the diagonal $X \to X \times X$ is normal.

The fourth section shows that when the two axioms of regular and protomodular category are put together, then we are in a position to define what an exact sequence is, in full generality, thus extrapolating from the classical model of the category of groups. In particular, we show that in any pointed regular protomodular category the Short Five Lemma holds. Again, as in the category *Grp*, when the basic category is not only regular, but also Barr-exact, then the notion of normal monomorphism and the notion of kernel coincide.

In the early development of category theory, the category Ab of abelian groups, and the category R-Mod of modules on a ring R became also the object of great attention because of their similar and specific features. One of the most remarkable of these features is certainly the fact that the hom-sets $Hom(A, B)$ are themselves endowed with an abelian group structure. It is to this notion of additive category that the fifth section is devoted, with special attention to the notion of biproduct or direct sum. Among the many examples, the additive subcategory of abelian objects in a protomodular category is studied: in this context, our characterization of the abelian objects allows us to prove the following conceptual "equation":

$$Additive = Protomodular + pointed + every\, mono\, is\, normal.$$

Historically the main fact about the categories Ab and R-Mod is that they are the most natural environment to make homological and cohomological calculations. The development of sheaf theory and the study of the category of sheaves of abelian groups subsequently showed that the context of these calculations could be widely enlarged. This allowed to set up categorical (co)homological tools, mainly dealing with kernel/cokernel techniques and led to the notion of *abelian category*, where these tools were put together under their simplest formulation.

In the sixth section the notion of abelian category is recalled. Thanks to the concept of protomodular category a new proof of the classical Tierney conceptual equation

$$Abelian = Exact + Additive$$

can be given, which tightly links the various notions introduced in this chapter. In our terms, this equation can be also rewritten as

$$Abelian = Exact + Protomodular + pointed + every\, mono\, is\, normal.$$

The reader will find in [16] many useful links of the concepts presented here to the literature of the fifties and sixties.

1. Internal equivalence relations

1.A Basic definitions and properties

In this section we briefly introduce some basic facts on internal relations in categories. In this introductory section we shall always assume that the category $\mathcal{C}$ has finite limits. The unique arrow from any object $X \in \mathcal{C}$ to the terminal object 1 is denoted by $\tau_X : X \to 1$.

A *relation* R on an object X in a category $\mathcal{C}$ is a pair of arrows represented by the diagram

$$R \underset{d_1}{\overset{d_0}{\rightrightarrows}} X,$$

which is jointly monic, in the sense that the factorization $(d_0, d_1): R \to X \times X$ is a monomorphism. The arrow d_0 is called the domain arrow, while d_1 is called the codomain arrow. In the category *Sets* of sets, this is equivalent to saying that the arrow $(d_0, d_1): R \to X \times X$ determines a relation in the usual sense on the set X.

A relation R in a category $\mathcal{C}$ is *reflexive* if the diagonal $(1_X, 1_X) = s_0: X \to X \times X$ factorizes through R.

A relation R in $\mathcal{C}$ is *symmetric* if there is an arrow $\sigma: R \to R$ with $d_0 \circ \sigma = d_1$ and $d_1 \circ \sigma = d_0$.

Let $(R \times_X R, p_0, p_2)$ denote the pullback of d_0 along d_1:

$$
\begin{array}{ccc}
R \times_X R & \xrightarrow{\;p_2\;} & R \\
{\scriptstyle p_0}\downarrow & & \downarrow{\scriptstyle d_0} \\
R & \xrightarrow[\;d_1\;]{} & X
\end{array}
$$

A relation R in $\mathcal{C}$ is *transitive* if there is an arrow $p_1: R \times_X R \to R$ satisfying $d_0 \circ p_0 = d_0 \circ p_1$ and $d_1 \circ p_1 = d_1 \circ p_2$.

A relation is an *equivalence relation* if it is reflexive, symmetric and transitive. An equivalence relation $(d_0, d_1): R \to X \times X$ will be also briefly denoted by (R, X), when there is no ambiguity.

1.1. Exercise. Check that an internal equivalence relation (as just defined) in the category $\mathcal{C} = Sets$ of sets is an equivalence relation in the usual sense.

1.2. Exercise. Let R be a relation on the set X. Prove that R is an equivalence relation if and only if (1) R is reflexive and (2) $\forall x, y, z \in X \quad xRy, xRz \Rightarrow yRz$.

1.3. Remark. Any equivalence relation (R, X) has the property that

$R \times_X R \underset{p_1}{\overset{p_0}{\rightrightarrows}} R$ is the kernel pair of $d_0: R \to X$, and $R \times_X R \underset{p_2}{\overset{p_1}{\rightrightarrows}} R$ is the kernel pair of $d_1: R \to X$.

Let us denote by $Eq(\mathcal{C})$ the category whose objects are the equivalence relations in $\mathcal{C}$ and whose maps from (R, X) to (R', X') are the arrows $f: X \to X'$ in $\mathcal{C}$ such that there is a (necessarily unique) factorization $\tilde{f}$ making the following diagram commutative:

$$
\begin{array}{ccc}
R & \xrightarrow{\;\tilde{f}\;} & R' \\
{\scriptstyle (d_0, d_1)}\downarrow & {\scriptstyle (1)} & \downarrow{\scriptstyle (d_0, d_1)} \\
X \times X & \xrightarrow[\;f \times f\;]{} & X' \times X'
\end{array}
$$

In the category *Set*, this is a map $f: X \to X'$ such that $\alpha R \beta$ implies $f(\alpha) R f(\beta)$. There is a forgetful functor $F: Eq(\mathcal{C}) \to \mathcal{C}$ associating the object X with any equivalence relation (R, X) in $Eq(\mathcal{C})$ (with the obvious definition on arrows).

1.4. Definition. If $f\colon (R, X) \to (R', X')$ is an arrow in $Eq(\mathcal{C})$, we then say that R is the inverse image of R' along f if the commutative square (1) above is a pullback. In this case we write $R = f^{-1}(R')$ and we call this arrow *cartesian* with respect to the functor F.

Let us denote by $Eq_X(\mathcal{C})$ the subcategory of $Eq(\mathcal{C})$ with objects the equivalence relation on a fixed object X, and arrows of the form

$$
\begin{array}{ccc}
R & \longrightarrow & R' \\
{\scriptstyle (d_0,d_1)}\big\downarrow & & \big\downarrow{\scriptstyle (d_0,d_1)} \\
X \times X & = & X \times X
\end{array}
$$

Any category $Eq_X(\mathcal{C})$ has an initial object Δ_X and a terminal object ∇_X:

1.5. Definition. The *discrete* relation on a given object X is the smallest equivalence relation Δ_X on X:

$$
X \underset{1}{\overset{1}{\underset{\longrightarrow}{\overset{\longrightarrow}{\longleftarrow}}}} X \ .
$$

The *coarse* relation on X is the largest equivalence relation ∇_X on X:

$$
X \times X \underset{p_1}{\overset{p_0}{\underset{\longrightarrow}{\overset{\longrightarrow}{\underset{s_0}{\longleftarrow}}}}} X \ .
$$

1.6. Exercise. Prove that any arrow $f\colon (\nabla_X, X) \to (R, X')$ in $Eq(\mathcal{C})$ is cartesian.

Another important class of distinguished arrows in $Eq(\mathcal{C})$ will be considered:

1.7. Definition. An arrow $f\colon (R, X) \to (R', X')$ is *fibrant* if and only if the square

$$
\begin{array}{ccc}
R & \overset{\tilde{f}}{\longrightarrow} & R' \\
{\scriptstyle d_1}\big\downarrow & & \big\downarrow{\scriptstyle d_1} \\
X & \underset{f}{\longrightarrow} & X'
\end{array}
$$

is a pullback.

1.8. Exercise. Prove that an arrow $f\colon (R, X) \to (R', X')$ is fibrant if and only if the commutative square $f \circ d_0 = d_0 \circ \tilde{f}$ (with the same notations as above) is a pullback.

It is not difficult to check that the class of fibrant arrows and cartesian arrows are stable under composition, under pullbacks, and under products in $Eq(\mathcal{C})$.

1.B Kernel relations

In a finitely complete category the kernel pair of any arrow $f\colon X \to Y$ is an equivalence relation (we leave the verification to the reader).

1.9. Definition. The equivalence relation arising as the kernel pair of an arrow $f\colon X \to Y$ is called the *kernel relation of f*, and is denoted by $(R[f], X)$.

When an equivalence relation R on X is a kernel relation, it is said to be *effective*. The discrete and the coarse equivalence relations on a given object X are always effective, because $\Delta_X = R[1_X]$ and $\nabla_X = R[\tau_X]$.

1.10. Exercise. Let $f\colon X \to Y$ be an arrow in $\mathcal{C}$.

(1) Prove that $R[f] = f^{-1}(\Delta_Y)$.

(2) Prove that $f\colon X \to Y$ is a monomorphism if and only if its kernel relation

$$R[f] \underset{\underset{\pi_1}{\longrightarrow}}{\overset{\overset{\pi_0}{\longrightarrow}}{\xleftarrow{\sigma_0}}} X$$

is isomorphic to Δ_X, or, equivalently, if and only if any of the arrows π_0, π_1 or σ_0 is an isomorphism, and if and only if $\pi_0 = \pi_1$.

Let us then make a useful observation concerning kernel relations. Given an arrow $f\colon X \to Y$ in $\mathcal{C}$, let ∇_f be the following arrow from ∇_X to ∇_Y:

$$
\begin{array}{ccc}
X \times X & \xrightarrow{\ f \times f\ } & Y \times Y \\
\Big\downarrow{\scriptstyle p_0}\ \Big\downarrow{\scriptstyle p_1} & & \Big\downarrow{\scriptstyle p_0}\ \Big\downarrow{\scriptstyle p_1} \\
X & \xrightarrow{\ f\ } & Y
\end{array}
$$

The kernel relation $R[f] \underset{\underset{\pi_1}{\longrightarrow}}{\overset{\overset{\pi_0}{\longrightarrow}}{\xleftarrow{\sigma_0}}} X$ of f is obtained by the following pullback in $Eq(\mathcal{C})$:

$$
\begin{array}{ccc}
R[f] & \xrightarrow{\ \phi\ } & \Delta_Y \\
\big\downarrow & & \big\downarrow \\
\nabla_X & \xrightarrow{\ \nabla_f\ } & \nabla_Y
\end{array}
$$

1.11. Exercise. Prove that the arrow f is a mono if and only if $\phi\colon R[f] \to \Delta_Y$ is fibrant (hint: use Exercise 1.10).

2. Epimorphisms and regular categories

We are now going to study the subtle and important distinctions between different kinds of epimorphism. This is a first and elementary analysis, which will be deepened in Chapter VIII.

2.A Variations on epimorphisms

2.1. Definition. An arrow $f\colon A \to B$ is a *strong epimorphism* if, given any commutative square as below, with m a monomorphism,

$$
\begin{array}{ccc}
A & \xrightarrow{\ f\ } & B \\
{\scriptstyle g}\big\downarrow & {\scriptstyle t}\nearrow & \big\downarrow{\scriptstyle h} \\
C & \xrightarrow[m]{} & D
\end{array}
$$

there exists a unique $t\colon B \to C$ with $t \circ f = g$ and $m \circ t = h$.

2.2. Lemma. *In a category with binary products, any strong epimorphism is an epimorphism.*

Proof. Let $f\colon A \to B$ be a strong epimorphism, and let $u, v\colon B \to C$ be two arrows with $u \circ f = v \circ f$. The existence of a factorization $t\colon B \to C$ along the diagonal $s_0\colon C \to C \times C$

$$
\begin{array}{ccc}
A & \xrightarrow{\ f\ } & B \\
{\scriptstyle u\circ f=v\circ f}\big\downarrow & {\scriptstyle t}\nearrow & \big\downarrow{\scriptstyle (u,v)} \\
C & \xrightarrow[s_0]{} & C \times C
\end{array}
$$

forces $u = t = v$, and then f is an epimorphism. □

2.3. Lemma. *An arrow $f\colon X \to Y$ is an isomorphism if and only if f is both a strong epi and a monomorphism.*

Proof. If f is a mono and a strong epi, one just needs to apply the definition of a strong epi to the commutative square

$$
\begin{array}{ccc}
X & \xrightarrow{\ f\ } & Y \\
\big\| & {\scriptstyle t}\nearrow & \big\| \\
X & \xrightarrow[f]{} & Y
\end{array}
$$

□

Remark. In the category $U\!Rng$ of unitary rings the embedding $\mathbb{Z} \to \mathbb{Q}$ from the ring of integers to the ring of rationals is an epi which is not strong. Accordingly, there are arrows which are monomorphic and epimorphic without being isomorphic.

The strong epimorphisms have some important stability properties (see for instance [2]). The following ones are easy to prove:

2.4. Lemma.

(1) *If $f\colon X \to Y$ and $g\colon Y \to Z$ are strong epis, then $g \circ f$ is a strong epi.*

(2) *If $g \circ f$ is a strong epi, then g is a strong epi.*

2.5. Corollary. *If $q \colon X \to Y$ is a strong epi such that $q = i \circ g$*

$$
\begin{array}{ccc}
X & \xrightarrow{\;g\;} & I \\
\Big\| & & \Big\downarrow{\scriptstyle i} \\
X & \xrightarrow[\;q\;]{} & Y
\end{array}
$$

with i a mono, then i is an isomorphism.

Proof. It follows by Lemma 2.4 (2) and Lemma 2.3. □

An important class of strong epimorphisms is represented (see the Lemma 2.10 here below) by the regular epimorphisms. Let us recall the definition:

2.6. Definition. An arrow $f \colon A \to B$ is a *regular epimorphism* when it is the coequalizer of a pair of arrows.

2.7. Definition. A *split epimorphism* is an arrow $f \colon X \to Y$ such that there is an arrow $i \colon Y \to X$ with $f \circ i = 1_Y$.

2.8. Exercise. 1. Prove that if the kernel relation $(R[f], \pi_0, \pi_1)$ of a regular epi f exists, then f is the coequalizer of its kernel relation.

$$
R[f] \;\underset{\pi_1}{\overset{\pi_0}{\rightrightarrows}}\; A \xrightarrow{\;f\;} B.
$$

2. Prove that a regular epimorphism is an epimorphism.
3. Prove that any split epimorphism is a regular epimorphism.
4. Prove that an arrow $f \colon A \to B$ is a strong epimorphism if and only if the following property holds: in any pullback of a monomorphism i along f

$$
\begin{array}{ccc}
A \times_B C & \xrightarrow{\;\pi_C\;} & C \\
{\scriptstyle \pi_A}\Big\downarrow & & \Big\downarrow{\scriptstyle i} \\
A & \xrightarrow[\;f\;]{} & B
\end{array}
$$

if π_A is an iso, then i is an iso.

2.9. Exercise. Prove that regular epimorphisms are stable under pushouts.

2.10. Lemma. *Any regular epimorphism is a strong epimorphism.*

Proof. Let $f \colon A \to B$ be a regular epimorphism, i.e. the coequalizer of some pair of arrows (π_0, π_1)

$$
\begin{array}{ccccc}
X & \underset{\pi_1}{\overset{\pi_0}{\rightrightarrows}} & A & \xrightarrow{\;f\;} & B \\
& & {\scriptstyle q}\Big\downarrow & & \Big\downarrow{\scriptstyle p} \\
& & C & \xrightarrow[\;m\;]{} & D
\end{array}
$$

and let the right hand square be commutative, with m a mono. Since $m \circ q \circ \pi_0 = m \circ q \circ \pi_1$ and m is a monomorphism, it follows that $q \circ \pi_0 = q \circ \pi_1$. There exists then a unique $\alpha \colon B \to C$ with $\alpha \circ f = q$. Moreover the arrow f being an epimorphism,

$$m \circ \alpha \circ f = m \circ q = p \circ f$$

implies $m \circ \alpha = p$, as desired. □

Remark. In summary: when $\mathcal{C}$ has binary products, we have the following string of implications:

$$\textit{split epi} \Rightarrow \textit{regular epi} \Rightarrow \textit{strong epi} \Rightarrow \textit{epi}.$$

If a category $\mathcal{C}$ has an initial object 0, we shall denote by $\alpha_X \colon 0 \to X$ the unique arrow from 0 to X.

2.11. Definition. $\mathcal{C}$ is a *pointed category* when $\mathcal{C}$ has a zero object (i.e. the map $\alpha_1 = \tau_0 \colon 0 \to 1$ is an isomorphism).

In a pointed category there is a unique arrow from any object X to any Y that factorizes through the zero object 0, namely $\alpha_Y \circ \tau_X$. Such an arrow will be denoted by $0_{X,Y}$, or by 0 when there is no ambiguity.

In a pointed category we also have the notions of kernel and of cokernel, which we recall below:

2.12. Definition. The *kernel* of a map $f \colon A \to B$ is given by the pullback of f along the initial arrow $\alpha_B \colon 0 \to B$:

$$
\begin{array}{ccc}
K[f] & \xrightarrow{\ Ker(f)\ } & A \\
\downarrow & & \downarrow{\scriptstyle f} \\
0 & \xrightarrow[\ \alpha_B\ ]{} & B
\end{array}
$$

2.13. Definition. The *cokernel* of a map $f \colon A \to B$ is given by the pushout of f along the terminal arrow τ_A

$$
\begin{array}{ccc}
A & \xrightarrow{\ f\ } & B \\
{\scriptstyle \tau_A}\downarrow & & \downarrow{\scriptstyle Coker(f)} \\
0 & \longrightarrow & Cok[f]
\end{array}
$$

2.14. Exercise. Given any pullback diagram

$$
\begin{array}{ccc}
E \times_B A & \longrightarrow & A \\
{\scriptstyle f'}\downarrow & & \downarrow{\scriptstyle f} \\
E & \longrightarrow & B
\end{array}
$$

prove that the kernel of f is isomorphic to the kernel of f', i.e. $K[f] \simeq K[f']$. What is the dual result?

When the cokernel of an arrow $f: A \to B$ exists, this cokernel is a regular epimorphism: indeed, in any finitely complete category, any map $A \to 0$ is a regular epimorphism (being a split epi), and the regular epimorphisms are stable under pushout (by Exercise 2.9).

2.15. Remark. It is not true in general that a regular epi or a split epi in a pointed category is a cokernel. For instance, in the category of pointed sets, the split epis fail to be the cokernels.

2.B The Barr-Kock Theorem

The following result is needed in order to prove the main result of this section:

2.16. Lemma. *In a category with pullbacks, let us consider a commutative triangle*

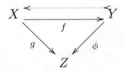

where f is a split epi. Then $R[g] \simeq R[f]$ if and only if ϕ is a mono.

Proof. Clearly, if ϕ is a monomorphism one has $R[g] \simeq R[\phi \circ f] \simeq R[f]$. Conversely, if $s: Y \to X$ is a splitting of f, one has $g \circ s = \phi$, since $g \circ s \circ f = \phi \circ f \circ s \circ f = \phi \circ f$ and f is an epi. Then, in the following diagram, where $s_1 = (s \circ f, 1)$ and $s_{-1} = (1, s \circ f)$,

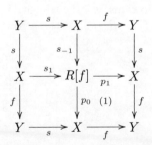

all the squares are pullbacks. By replacing f by g in the square (1) one still obtains a pullback diagram by the assumption $R[g] \simeq R[f]$, so that the kernel relation of $g \circ s = \phi$ is discrete, since $f \circ s = 1_Y$. Thus, the arrow ϕ is a mono (Exercise 1.10). $\qquad \square$

2.17. Theorem. *Let C be a category with pullbacks. Consider the following commutative diagram:*

$$R[f] \underset{\pi_1}{\overset{\pi_0}{\rightrightarrows}} X \xrightarrow{f} Y$$

$$\gamma \downarrow \qquad \qquad g \downarrow \quad (1) \quad \downarrow h$$

$$R[f'] \underset{p_1}{\overset{p_0}{\rightrightarrows}} X' \xrightarrow{f'} Y'$$

The following properties hold:

(1) *Let $(g, \gamma)\colon R[f] \to R[f']$ be a fibrant arrow, i.e. any of the commutative left hand squares is a pullback. If f is a strong epimorphism stable under pullback, then the square (1) is a pullback.*

(2) *Moreover, if g is a monomorphism, then h is a monomorphism.*

Proof. (1) Let us denote by

$$
\begin{array}{ccc}
\overline{X} & \xrightarrow{\overline{f}} & Y \\
\overline{h} \downarrow & & \downarrow h \\
X' & \xrightarrow{f'} & Y'
\end{array}
$$

the pullback of f' along h; let ϕ denote the unique arrow $\phi\colon X \to \overline{X}$ with $\overline{h} \circ \phi = g$ and $\overline{f} \circ \phi = f$. We denote by π_0^1, π_2^1 the projections and by π_1^1 the arrow giving the transitivity of the kernel relation $R[f]$:

$$R^2[f] \underset{\pi_2^1}{\overset{\pi_0^1}{\underset{\pi_1^1}{\rightrightarrows}}} R[f] \underset{\pi_1}{\overset{\pi_0}{\rightrightarrows}} X \xrightarrow{f} Y,$$

where we write $(R^2[f], \pi_0^1, \pi_2^1)$ for the pullback of π_0 along π_1. We use similar notations p_0^1, p_1^1, p_2^1 for the projections and the transitivity of the equivalence relation $R[f']$, and we write $\gamma^2\colon R^2[f] \to R^2[f']$ for the arrow naturally induced by γ and g. Consider the diagram:

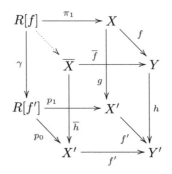

Clearly the arrow $\phi \circ \pi_0$ has the property that $\overline{f} \circ (\phi \circ \pi_0) = f \circ \pi_0 = f \circ \pi_1$ and $\overline{h} \circ (\phi \circ \pi_0) = g \circ \pi_0 = p_0 \circ \gamma$, so that the arrow $\phi \circ \pi_0$ is precisely the dotted arrow induced by the universal property of the front pullback in the cube above. The upper square is a pullback by a classical result (see Exercise 8 p. 72 in [17]) because the back, the bottom and the front squares are pullbacks. By the assumption that f is a strong epi stable under pullback we then know that the dotted arrow $\phi \circ \pi_0$ is a strong epimorphism; this, of course, implies that ϕ is a strong epi as well. In order to show that ϕ is an iso, and then that the square (1) a pullback, it is sufficient to prove that ϕ is also a mono.

For this, let us take the kernel pairs of p_0, f' and f, respectively:

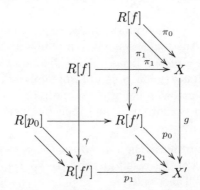

certainly $R[p_0] \simeq R^2[f']$, the relation $R[f']$ being an equivalence relation (see Remark 1.3). By taking the pullback of $p_2^1 \colon R^2[f'] \to R[f']$ along γ on the back square one gets the following diagram:

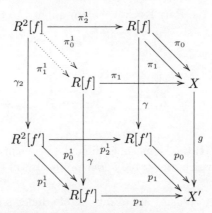

By a commutation property of finite limits, it follows that the dotted pair of arrows $(R^2[f], \pi_0^1, \pi_1^1)$ is the kernel relation of the dotted arrow $\phi \circ \pi_0$ in our first cube. Moreover, the arrow π_0 has the same equivalence relation $(R^2[f], \pi_0^1, \pi_1^1)$ as kernel relation. Now, the arrow π_0 is a split epi, and by the Lemma 2.16 it follows that ϕ is a mono.

(2) In order to prove the second part of the statement, consider the following diagram

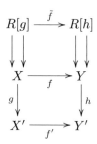

The arrow f is a strong epimorphism, the lower commutative square a pullback by (1) and then any of the upper squares is a pullback: accordingly the induced arrow $\tilde{f}$ is a strong epimorphism. Let $s_0 \colon X \to R[g]$ and $\sigma_0 \colon Y \to R[h]$ denote the subdiagonals; the fact that g is a monomorphism implies that s_0 is an isomorphism (by Exercise 1.10); the arrow $\tilde{f}$ being an epimorphism, it follows that the split monomorphism σ_0 is an isomorphism, and then h a monomorphism, as desired. □

We shall refer to this result as the *Barr-Kock Theorem*, in reference to a preliminary and weaker version of this result given in [1]. We think that it is a significant basic result in Category Theory.

2.18. Corollary. *Let C be a category with pullbacks. If $f \colon A \to B$ is an arrow in C that factors as $f = i \circ q$, with q a strong epi stable under pullback, then $R[q] \simeq R[f]$ if and only if i is a mono.*

Proof. It follows from the Barr-Kock Theorem applied to the diagram

$$
\begin{array}{ccc}
R[q] \rightrightarrows A \xrightarrow{\ q\ } I \\
\| \qquad \| \qquad \downarrow i \\
R[f] \rightrightarrows A \xrightarrow{\ f\ } B
\end{array}
$$
□

Remark. In Chapter VIII the (sometimes difficult) problem of characterizing those strong epimorphisms which are stable under pullback will be considered.

Since the split epimorphisms are always stable under pullback, the Barr-Kock Theorem obviously holds for any split epi f in any category with pullbacks C. Let us assert this explicitly:

2.19. Corollary. *Let C be a category with pullbacks and let us consider the same commutative diagram as in Theorem 2.17. Then, if f is a split epi, the following properties hold:*

(1) *If* $(g, \gamma) \colon R[f] \to R[f']$ *is a fibrant arrow, then the square* (1) *is a pullback.*
(2) *Moreover, if g is a monomorphism, then h is a monomorphism.*

2.C Regular categories

2.20. Definition. A finitely complete category $\mathcal{C}$ is *regular* if
   (1) the strong epis are stable under pullback in $\mathcal{C}$
   (2) $\mathcal{C}$ has coequalizers of effective equivalence relations

The following classical result proves the existence of a "good" factorization of any arrow in a regular category (see also [13]).

2.21. Theorem. *Let $\mathcal{C}$ be a regular category. Any arrow $f \colon X \to Y$ factorizes in a regular epimorphism q followed by a monomorphism i.*

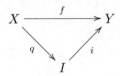

This factorization is unique (up to isomorphism) and is called the "regular image factorization of f".

Proof. Let $(R[f], \pi_0, \pi_1)$ be the kernel relation of f, and let q be the coequalizer of π_0 and π_1 which, as such, is a strong epi. It is simple to check that $R[f] = R[q]$, so that by Lemma 2.10 and by Corollary 2.18 it follows that the induced arrow i such that $f = i \circ q$ is a monomorphism. The factorization regular epimorphism-monomorphism is easily seen to be unique (up to isomorphism). $\qquad\Box$

It is a remarkable fact that in the context of regular categories the notions of regular and strong epis coincide:

2.22. Corollary. *Let $\mathcal{C}$ be a regular category. Then an arrow $f \colon X \to Y$ is a regular epi if and only if f is a strong epi.*
Proof. If $f \colon X \to Y$ is a strong epi in a regular category, we can consider its regular image factorization $i \circ q$ as in the previous theorem. The arrow i is a mono, and Lemma 2.3 implies that i is an iso. $\qquad\Box$

Remark. 1. According to the previous result, in any regular category the regular epis fullfill the assumptions of the Barr-Kock Theorem.
2. The Definition 2.20 is to be credited to Joyal. The alternative definition of regular category in which regular epimorphisms are pullback stable was given by Barr and Grillet [1] [14]. In the presence of finite limits these two definitions are equivalent.

Moreover, regular epis inherit the good stability properties of strong epis:

2.23. Corollary. *Let $\mathcal{C}$ be a regular category.*
   (1) *Regular epis are pullback stable.*

(2) If $f\colon X \to Y$ and $g\colon Y \to Z$ are regular epis, then $g \circ f$ is a regular epi.
(3) If $g \circ f$ is a regular epi, then g is a regular epi.

Proof. By Lemma 2.4 and by Corollary 2.22.   □

2.24. Lemma. *If $\mathcal{C}$ is a regular category and $f\colon A \to B$ and $g\colon C \to D$ are regular epimorphisms, then $f \times g\colon A \times C \to B \times D$ is a regular epimorphism.*

Proof. Consider the pullbacks

$$
\begin{array}{ccc}
A \times C & \xrightarrow{\;f \times 1_C\;} & B \times C \\
{\scriptstyle p_A}\big\downarrow & & \big\downarrow{\scriptstyle p_B} \\
A & \xrightarrow{\quad f \quad} & B
\end{array}
$$

and

$$
\begin{array}{ccc}
B \times C & \xrightarrow{\;1_B \times g\;} & B \times D \\
{\scriptstyle p_C}\big\downarrow & & \big\downarrow{\scriptstyle p_D} \\
C & \xrightarrow{\quad g \quad} & B
\end{array}
$$

Since regular epis are stable under pullbacks in regular categories, both $(f \times 1_C)$ and $(1_B \times g)$ are regular epimorphisms; then by Corollary 2.23, $f \times g = (1_B \times g) \circ (f \times 1_C)$ is a regular epimorphism.   □

In a category, we shall call a *fork* a diagram

$$
P \; \underset{\pi_1}{\overset{\pi_0}{\rightrightarrows}} \; A \xrightarrow{\;f\;} B. \qquad (*)
$$

with $f \circ \pi_0 = f \circ \pi_1$. A fork is reflexive when there is an arrow $s_0\colon A \to P$ with $\pi_0 \circ s_0 = 1_A = \pi_1 \circ s_0$. In a regular category we say that a fork is exact when $f\colon A \to B$ is a regular epi and P is its kernel relation: $P = R[f]$.

The *exact forks* have some nice stability properties in any regular category:

2.25. Corollary. *In a regular category $\mathcal{C}$ the exact forks are stable under products.*

Proof. By Lemma 2.24.   □

2.26. Proposition. *Let $\mathcal{C}$ be a regular category and let us consider the commutative diagram*

$$
\begin{array}{ccccc}
P & \underset{\pi_1}{\overset{\pi_0}{\rightrightarrows}} & A' & \xrightarrow{\;f'\;} & B' \\
{\scriptstyle k}\big\downarrow & & {\scriptstyle h}\big\downarrow & & \big\downarrow{\scriptstyle g} \\
R[f] & \underset{\pi_1}{\overset{\pi_0}{\rightrightarrows}} & A & \xrightarrow{\;f\;} & B
\end{array}
$$

where all the commutative squares are pullbacks and the lower fork is exact. Then

$$P \overset{\pi_0}{\underset{\pi_1}{\rightrightarrows}} A' \overset{f'}{\longrightarrow} B'$$

is an exact fork.

Proof. (1) By Corollary 2.23, we know that f' is a regular epimorphism.
(2) We still have to prove that (P, π_0, π_1) is the kernel relation of f'. This relies
on a classical commutation property of finite limits: for this, let us consider two
arrows $x, y \colon E \to A'$ with $f' \circ x = f' \circ y$. Then

$$f \circ h \circ x = g \circ f' \circ x = g \circ f' \circ y = f \circ h \circ y$$

gives a unique $w \colon E \to R[f]$ with $\pi_0 \circ w = h \circ x$ and $\pi_1 \circ w = h \circ y$. This arrow w
induces $z_1, z_2 \colon E \to P$ with $k \circ z_1 = w, \pi_0 \circ z_1 = x$, $k \circ z_2 = w$ and $\pi_1 \circ z_2 = y$.
This easily implies that $z_1 = z_2 = z$, with $\pi_0 \circ z = x$ and $\pi_1 \circ z = y$. One can then
check that such a z is unique, and this completes the proof.   □

2.27. Definition. A regular category is *Barr-exact* [1] if any equivalence relation is
effective.

2.28. Examples.

(1) *Set*

The category *Set* of sets is complete and cocomplete. Regular epimorphisms are surjective maps. They are pullback stable: indeed, if p is a
surjection, in the pullback of p along f

$$
\begin{array}{ccc}
E \times_B A & \overset{\pi_A}{\longrightarrow} & A \\
{\scriptstyle \pi_E} \downarrow & & \downarrow {\scriptstyle f} \\
E & \underset{p}{\longrightarrow} & B
\end{array}
$$

π_A is necessarily a surjection: for a given $a \in A$, there exists $e \in E$ such
that $p(e) = f(a)$, so that $(e, a) \in E \times_B A$ and $\pi_A(e, a) = a$. Equivalence
relations in *Set* are effective, so that the category of sets is exact.

(2) *Presheaves*

If $\mathcal{C}$ is a small category, the category $Set^{\mathcal{C}^{op}}$ of functors from $\mathcal{C}^{op}$ to *Set*,
with natural transformations between such functors as arrows, is called a
category of presheaves. This kind of categories are exact, because limits
and colimits are constructed pointwise as in *Set*. In particular, a natural
transformation $\gamma \colon F \to F'$ is a regular epi if and only if, for each object
X in $\mathcal{C}$, the map $\gamma_X \colon FX \to F'X$ is surjective. The exactness of $Set^{\mathcal{C}^{op}}$
then follows from the exactness of the category *Set* of sets.

(3) *Algebraic varieties*

Any variety in the sense of universal algebra is an exact category, with
regular epimorphisms given by surjective homomorphisms. Hence, the

category Grp of groups, Ab of abelian groups, Rng of rings, Lie algebras on a given ring, lattices, and Heyting algebras are all examples of exact categories [2].

(4) *Torsion-free abelian groups*

The category of torsion-free abelian groups Ab_{tf} is regular, but not exact. An abelian group G is torsion-free when for any element $g \in G$ and any non-zero natural number $n \in \mathbb{N}^*$

$$n \cdot g = g + g + \cdots + g = 0 \Rightarrow g = 0.$$

Let us first remark that the category of torsion-free abelian groups is closed under finite limits and subobjects in the category Ab. The regularity of the category Ab of abelian groups then implies the regularity of the category Ab_{tf}. In the category Ab_{tf} equivalence relations are not always effective: for instance, consider the equivalence relation R_2 on $\mathbb{Z}$ defined by

$$m R_2 n \text{ if and only if there is a } k \in \mathbb{Z} \text{ with } m - n = 2k.$$

R_2 is an abelian group, it belongs to Ab_{tf}, being a subobject of the torsion free $\mathbb{Z} \times \mathbb{Z}$. Now, the inclusion functor $R \colon Ab_{tf} \to Ab$ has a left adjoint $L \colon Ab \to Ab_{tf}$ defined by $L(A) = \frac{A}{A_t}$, where A_t is the subgroup of the torsion elements. Accordingly, the quotients in Ab_{tf} exist, they are images by L of quotients in Ab. Consider then the quotient $q \colon \mathbb{Z} \to Q$ in Ab_{tf}: it is the trivial group, since $q(1) + q(1) = q(2) = 0$ implies $q(1) = 0$ and then $q = \tau_{\mathbb{Z}}$ in Ab_{tf}. It follows that the kernel relation $R[q]$ of q is $\mathbb{Z} \times \mathbb{Z}$ and not R_2 (see Exercise 2.8), so that the latter relation is not effective.

(5) *Topological spaces*

The category Top is not regular. Here there is a simple example of a regular epimorphism which is not pullback stable [2]. Let $A = \{a, b, c, d\}, B = \{l, m, n\}$ and $C = \{x, y, z\}$, where $\{a, b\}$ is open in A, $\{l, m\}$ open in B and let C have the indiscrete topology. Define $f \colon A \to C$ with $f(a) = x, f(b) = y = f(c), f(d) = z$, $g \colon B \to C$ with $g(l) = x, g(m) = z = g(n)$. The arrow f is a quotient map, and its pullback along g is not a quotient map.

(6) *Topological groups*

The category $Grp(Top)$ of topological groups is regular with regular epimorphisms given by open homomorphic surjections, which are stable under pullback. But, as it is explained in [11] the equivalence relations are not necessarily effective, so that $Grp(Top)$ is not exact.

(7) *The category of small categories*

Cat is not regular, since strong and regular epimorphisms do not coincide (see Chapter VIII for such an example).

3. Normal monomorphisms and protomodular categories

The importance of the category Grp of groups in Algebra is universally recognized. There, a very important role is played by the notion of normal subgroup. In this section we shall introduce the notion of protomodular category, where there is an intrinsic notion of normal monomorphism. We shall then prove some simple properties of protomodular categories extending some familiar ones of the category Grp, and we shall then characterize the internal abelian objects.

3.A Normal subobjects

Let $\mathcal{C}$ be a finitely complete category.

3.1. Definition. An arrow $f \colon X \to Y$ is said to be normal to the equivalence relation S when

    (1) $f^{-1}(S) = \nabla_X$
    (2) the arrow $f \colon (\nabla_X, X) \to (S, Y)$ in $Eq(\mathcal{C})$ is fibrant

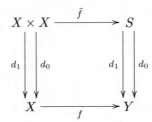

3.2. Lemma. *If $f \colon X \to Y$ is normal to (S, Y), then f is a monomorphism.*

Proof. Let us consider the following diagram in $Eq(\mathcal{C})$:

$$
\begin{array}{ccc}
R[f] & \xrightarrow{\ \phi\ } & \Delta_Y \\
{\scriptstyle \alpha}\Big\downarrow & {\scriptstyle (1)} & \Big\downarrow \\
\nabla_X & \xrightarrow[\ f\ ]{} & S \\
\Big\| & {\scriptstyle (2)} & \Big\downarrow \\
\nabla_X & \xrightarrow[\ \nabla_f\ ]{} & \nabla_Y
\end{array}
$$

The squares (2) and (1) + (2) are pullbacks, then (1) is a pullback. The arrow ϕ is a fibrant because f is fibrant. Then, by Exercise 1.10, f is a monomorphism. $\qquad\square$

3.3. Remark. A normal arrow is then always a *normal monomorphism*. Accordingly, when f is normal, we can think of X as a part of Y

$$X \rightarrowtail Y.$$

In the category *Set* of sets, the two conditions in the definition of a normal monomorphism mean, respectively,

(1) $\forall x_1, x_2 \in X \qquad x_1 S x_2$
"X is contained in an equivalence class"
(2) $\forall x, y \in Y, \qquad (x \in X) \wedge (x S y) \qquad \Rightarrow \qquad y \in X$
"there is nothing but the elements of X in that class".

So, the notion of normal monomorphism provides an intrisic way to express what an equivalence class of an equivalence relation is. If $f \colon X \to Y$ is normal to an equivalence relation (S, Y) we shall write $f \dashv S$.

Given two subobjects $f_1 \colon X_1 \to X$ and $f_2 \colon X_2 \to X$ of an object X we denote by $f_1 \cap f_2$ the unique subobject given by the dotted composite in the following pullback:

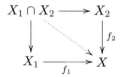

3.4. Lemma.

(1) *Normal monomorphisms are stable under pullbacks:*

$$
\begin{array}{ccc}
X & \xrightarrow{\ h\ } & Y \\
\downarrow{\scriptstyle g} & & \downarrow{\scriptstyle f} \\
X' & \xrightarrow[\ h'\ ]{} & Y'
\end{array}
$$

if f is normal to R, then g is normal to $h'^{-1}(R)$.
(2) *Normal monomorphisms are stable under products: if $f_1 \dashv R$ and $f_2 \dashv R'$, then $f_1 \times f_2 \dashv R \times R'$.*
(3) *Normal monomorphisms are stable under intersections: if $f_1 \dashv R$ and $f_2 \dashv S$, then $f_1 \cap f_2 \dashv R \cap S$.*

Proof. The proof is left to the reader. $\qquad\qquad\qquad\qquad\qquad\qquad\qquad\qquad$ □

3.5. Proposition. *Let $\mathcal{C}$ be a finitely complete pointed category. The arrow f is a kernel if and only if f is normal to an effective equivalence relation.*

Proof. If $f \colon X \to Y$ is the kernel of an arrow $g \colon Y \to Z$, then $f \dashv R[g]$. Conversely, if $f \dashv R[g]$, then by applying the Barr-Kock Theorem (Corollary 2.19) to the diagram

$$\begin{array}{ccc} X \times X & \longrightarrow & R[g] \\ \downarrow\downarrow & & \downarrow\downarrow \\ X & \xrightarrow{\ f\ } & Y \\ {\scriptstyle \tau_X}\downarrow & (1) & \downarrow {\scriptstyle g} \\ 0 & \longrightarrow & Z \end{array}$$

where τ_X is a split epi, it follows that (1) is a pullback and f is the kernel of g. $\qquad\square$

In any pointed category we then have the following string of implications:

$$kernel \Rightarrow normal\,mono \Rightarrow mono.$$

3.6. Proposition. *If $\mathcal{C}$ is a finitely complete pointed category, then any equivalence relation determines a canonical normal monomorphism.*

Proof. Let

$$R \times_X R \begin{array}{c} \xrightarrow{p_0} \\ \xrightarrow{p_1} \\ \xrightarrow[p_2]{} \end{array} R \begin{array}{c} \xrightarrow{d_0} \\ \xleftarrow{s_0} \\ \xrightarrow[d_1]{} \end{array} X$$

be an equivalence relation on X. The arrows

$$R \times_X R \begin{array}{c} \xrightarrow{p_0} \\ \xrightarrow[p_1]{} \end{array} R$$

determine the kernel relation of d_0 and the arrow $(d_1, p_2) \colon R[d_0] \to R$ determines a fibrant arrow in $Eq(\mathcal{C})$ (by Remark 1.3):

$$\begin{array}{ccc} R \times_X R \simeq R[d_0] & \xrightarrow{\ p_2\ } & R \\ {\scriptstyle p_0}\downarrow\downarrow {\scriptstyle p_1} & & {\scriptstyle d_0}\downarrow\downarrow {\scriptstyle d_1} \\ R & \xrightarrow{\ d_1\ } & X \end{array}$$

By taking the kernel $K[d_0]$ of d_0, one obtains a fibrant arrow $(d_1 \circ K[d_0], p_2 \circ k)$ from $\nabla_{K[d_0]}$ to R

$$\begin{array}{ccccc} K[d_0] \times K[d_0] & \xrightarrow{\ k\ } & R[d_0] & \xrightarrow{\ p_2\ } & R \\ {\scriptstyle p_0}\downarrow\downarrow {\scriptstyle p_1} & & {\scriptstyle p_0}\downarrow\downarrow {\scriptstyle p_1} & & {\scriptstyle d_0}\downarrow\downarrow {\scriptstyle d_1} \\ K[d_0] & \xrightarrow{K[d_0]} & R & \xrightarrow{\ d_1\ } & X \\ \downarrow & & \downarrow {\scriptstyle d_0} & & \\ 0 & \longrightarrow & X & & \end{array}$$

showing that $d_1 \circ K[d_0]$ is normal to R.   □

3.7. Exercise. Prove that the canonical normal mono $d_1 \circ K[d_0]$ can be obtained directly by the following pullback

$$
\begin{array}{ccc}
K[d_0] & \longrightarrow & R \\
{\scriptstyle d_1 \circ K[d_0]} \downarrow & & \downarrow {\scriptstyle (d_0, d_1)} \\
X & \underset{r_X}{\longrightarrow} & X \times X
\end{array}
$$

where $r_X = (0_{X,X}, 1_X)$.

3.B Protomodular categories

Let us now introduce the main notion of this section [3]:

3.8. Definition. A finitely complete category $\mathcal{C}$ is *protomodular* if the following property holds: in any commutative diagram

$$
\begin{array}{ccccc}
A & \xrightarrow{\ f\ } & B & \xrightarrow{\ h\ } & C \\
{\scriptstyle \alpha}\downarrow & {\scriptstyle (1)} & {\scriptstyle \beta}\big\downarrow\big\uparrow{\scriptstyle j} & {\scriptstyle (2)} & \downarrow{\scriptstyle \gamma} \\
D & \underset{g}{\longrightarrow} & E & \underset{l}{\longrightarrow} & F
\end{array}
\qquad (*)
$$

where β is a split epimorphism, if (1) and (1) + (2) are pullbacks, then (2) is a pullback.

Let us denote by $Pt(\mathcal{C})$ the category with objects the split epis with a given splitting in $\mathcal{C}$, and arrows the commutative squares between such data

$$
\begin{array}{ccc}
A & \xrightarrow{\ g\ } & D \\
{\scriptstyle p}\big\downarrow\big\uparrow{\scriptstyle i} & & {\scriptstyle q}\big\downarrow\big\uparrow{\scriptstyle j} \\
E & \underset{f}{\longrightarrow} & B
\end{array}
$$

i.e. $q \circ g = f \circ p$, $g \circ i = j \circ f$.

There is a functor $\pi \colon Pt(\mathcal{C}) \to \mathcal{C}$ associating its codomain with any split epi. We call it the *fibration of pointed objects*. The previous square is called cartesian in $Pt(\mathcal{C})$ when the previous downward square is a pullback. We denote by $Pt_B(\mathcal{C})$ the subcategory of $Pt(\mathcal{C})$ whose objects are split epimorphisms with codomain B and whose arrows from (A, B, p, i) to (C, B, q, j) are commutative diagrams of the following kind:

$$
\begin{array}{ccc}
A & \xrightarrow{\ g\ } & C \\
{\scriptstyle p}\big\downarrow\big\uparrow{\scriptstyle i} & & {\scriptstyle q}\big\downarrow\big\uparrow{\scriptstyle j} \\
B & = & B
\end{array}
$$

We call the category $Pt_B(\mathcal{C})$ the fiber over B. If $f\colon E \to B$ is an arrow in $\mathcal{C}$, we denote by $f^\star\colon Pt_B(\mathcal{C}) \to Pt_E(\mathcal{C})$ the functor defined by pulling back along f, and we call it the change-of-base functor with respect to π.

Before we give a characterization of protomodular categories, let us first recall that a pair of arrows $(\overline{f}, i)$ with the same codomain A is *jointly strongly epic* if the following property holds: whenever the pullbacks of a monomorphism g with codomain A along $\overline{f}$ and i are isos, then g is an iso (see Exercice 2.8, part 4).

3.9. Proposition. *The following conditions are equivalent:*

(1) *$\mathcal{C}$ is protomodular*

(2) *any change-of-base functor $f^\star\colon Pt_B(\mathcal{C}) \to Pt_E(\mathcal{C})$ is conservative (i.e. it reflects isomorphisms)*

(3) *in any pullback with $g\colon A \to B$ a split epi*

$$
\begin{array}{ccc}
E \times_B A & \xrightarrow{\ \overline{f}\ } & A \\
{\scriptstyle \overline{g}}\downarrow & {\scriptstyle (3)}\quad {\scriptstyle g}\downarrow\ \uparrow {\scriptstyle i} & \\
E & \xrightarrow[\ f\ ]{} & B
\end{array}
$$

the pair $(\overline{f}, i)$ is jointly strongly epimorphic.

Proof. $(1) \Rightarrow (2)$
Consider the diagram

$$
\begin{array}{ccccc}
A & \xrightarrow{\ f\ } & B & \xrightarrow{\ h\ } & C \\
{\scriptstyle \alpha}\downarrow\ \uparrow {\scriptstyle i} & {\scriptstyle (4)} & {\scriptstyle \beta}\downarrow\ \uparrow {\scriptstyle j} & {\scriptstyle (5)} & {\scriptstyle \gamma}\downarrow\ \uparrow {\scriptstyle k} \\
D & \xrightarrow[\ g\ ]{} & E & = & E
\end{array}
$$

To say that $g^\star(h)$ is an isomorphism means that (4) and $(4) + (5)$ are pullbacks, while the fact that (5) is a pullback means that h is then an isomorphism, and thus the functor $g^\star\colon Pt_E(\mathcal{C}) \to Pt_D(\mathcal{C})$ is conservative.

$(2) \Rightarrow (3)$
Let us consider the diagram (3) and let us assume that $h\colon H \to A$ is a monomorphism with the property that the pullbacks h_1 and h_2 along $\overline{f}$ and i are isomorphisms. Then the fact that h_2 is an iso means that h is an iso in $Pt_B(\mathcal{C})$. The fact that h_1 is an iso precisely means that $f^\star(h)$ is an iso. Accordingly, the arrow h itself is an iso by condition (2).

$(3) \Rightarrow (1)$
Our condition (3) implies that any change-of-base functor is conservative on monomorphisms (i.e. any mono which is mapped onto an iso is an iso). It follows that $f^\star$ is conservative, since $f^\star$ preserves finite limits. Indeed, let h be an arrow with the property that $f^\star(h)$ is an iso. Then, since $f^\star$ preserves kernel relations, the subdiagonal $s_0\colon X \to R[h]$ is a mono which is sent by $f^\star$ to an iso (by Exercise

1.10). Accordingly, $f^\star(s_0)$ is an iso and then s_0 is an iso, proving that h is a mono. Consider then any diagram as $(*)$ in Definition 3.8 in which (1) and (1) + (2) are pullbacks. There is a unique factorization $\alpha\colon B \to E \times_F C$ with $\pi_E \circ \alpha = \beta$ and $h = \pi_C \circ \alpha$. This arrow α is such that $g^\star(\alpha)$ is an iso, since (1) + (2) is a pullback. Then α is itself invertible and the square (2) is a pullback.   □

In the pointed case, the protomodularity property becomes equivalent to the classical *Split Short Five Lemma*. We recall that the Split Short Five Lemma asserts that in any commutative diagram

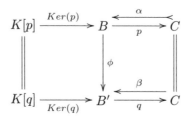

where $p \circ \alpha = 1_C$ and $q \circ \beta = 1_C$, the arrow ϕ is an iso.

3.10. Lemma. *Let C be a finitely complete pointed category. Then the following conditions are equivalent:*

    (1) *C is protomodular*
    (2) *the Split Short Five Lemma holds in C*

Proof. (1) $\Rightarrow$ (2) In order to prove the Split Short Five Lemma we need to prove that the arrow ϕ in the diagram above is an isomorphism. Now, the same diagram precisely asserts that $\alpha_C^\star(\phi)$ is an iso, and then ϕ is an iso by protomodularity. (2) $\Rightarrow$ (1) Let us assume that $\alpha_C^\star\colon Pt_C(C) \to Pt_0(C) \simeq C$ is conservative (= Split Short Five Lemma). If $p\colon E \to C$ is any arrow in C with $p^\star(f)$ an isomorphism, then $\alpha_E^\star(p^\star(f)) = \alpha_C^\star(f)$ is an iso, and then f is an iso, as desired.   □

3.11. Remark. Let C and D be two finitely complete categories and let $F\colon C \to D$ be a pullback preserving functor that reflects isomorphisms. It is easy to check that, under these assumptions, if D is protomodular, then also C is protomodular. This is often the quickest method to prove that a given category is protomodular.

3.12. Examples.

    (1) The categories *Grp* of groups, *Ab* of abelian groups, *Rng* of (not necessarily unitary) rings are protomodular. More generally, any variety V of Ω-groups [15] is protomodular, being equipped with an obvious forgetful functor $U\colon V \to Grp$ which is pullback preserving and conservative.

(2) The category of presheaves Grp^C and Rng^C on the category of groups or rings are protomodular. Indeed, consider the following diagram in Grp^C:

$$K[p] \xrightarrow{Ker(p)} B \underset{p}{\overset{\alpha}{\rightleftarrows}} C$$

with ϕ the vertical map from B to B' and

$$K[q] \xrightarrow[Ker(q)]{} B' \underset{q}{\overset{\beta}{\rightleftarrows}} C$$

It induces, for each object X in C the following diagram in Grp:

$$K[p](X) \xrightarrow{Ker(p(X))} B(X) \underset{p(X)}{\overset{\alpha(X)}{\rightleftarrows}} C(X)$$

with $\phi(X)$ and

$$K[q](X) \xrightarrow[Ker(q(X))]{} B'(X) \underset{q(X)}{\overset{\beta(X)}{\rightleftarrows}} C(X)$$

Now, $\phi(X)$ is an iso for all X, so that ϕ itself is an iso in Grp^C, and Grp^C protomodular.

(3) The category $Grp(C)$ of internal groups, $Ab(C)$ of internal abelian groups in any finitely complete category C (see Definition 3.25) are protomodular. In particular the category $Grp(Top)$ of topological groups is protomodular. We shall prove this fact in Proposition 3.24.

(4) It is interesting to mention that also the variety of Heyting algebras is protomodular [4]. By using this fact one can show that the dual category of any elementary topos is protomodular.

3.13. Lemma. *In a protomodular category, pullbacks reflect monomorphisms.*

Proof. Consider a pullback

$$
\begin{array}{ccc}
\overline{X} & \xrightarrow{k} & X \\
\overline{f} \downarrow & & \downarrow f \\
\overline{Y} & \xrightarrow{h} & Y
\end{array}
$$

where $\overline{f}$ is a monomorphism. We have to prove that f is a mono as well. The diagram $(1) + (2)$ here below

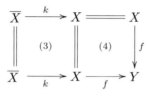

is a pullback ($\overline{f}$ is a mono), and then the diagram $(3)+(4)$ here below is a pullback:

The square (3) is a pullback and the vertical central arrow is a split epi; so, by protomodularity, (4) is a pullback, and f is a monomorphism (Exercise 1.10). □

The following characterization of monos by the triviality of their kernel, as in the category Grp, is a special case of the general reflection of monos by pullbacks (Lemma 3.13).

3.14. Corollary. *Let C be a pointed protomodular category.*
Then an arrow $f \colon A \to B$ is a monomorphism if and only if $K[f] \simeq 0$.

Proof. We only need to prove that if $K[f] \simeq 0$ then f is a mono, since the other implication holds in any pointed category. Consider the pullback

$$
\begin{array}{ccc}
K[f] & \xrightarrow{\ Ker f\ } & A \\
\downarrow & & \downarrow f \\
0 & \longrightarrow & B
\end{array}
$$

If $K[f] \simeq 0$, then Lemma 3.13 implies that the arrow f is a monomorphism. □

It is well-known that in the category Grp of groups there is a perfect correspondence between the normal subobjects of a group G and the congruence relations (= internal equivalence relations) on G. One can find a similar situation in the category Rng of rings: there, the correspondence is between ideals of a ring R and congruences on R. These phenomena are all special cases of a general categorical property, which is a direct consequence of protomodularity:

3.15. Theorem. *Let C be a protomodular category. If $f \colon X \to Y$ is normal to an equivalence relation (S, Y), then S is unique (up to isomorphism).*

Proof. Let (S, Y) and (S', Y) be two equivalence relations to which the arrow f is normal. Let $S \cap S'$ denote the equivalence relation which is the intersection of

S and S'. If $j \colon S \cap S' \to S'$ is the natural inclusion, one clearly has the following commutative diagram:

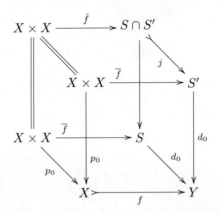

It indicates that $f^{-1}(S \cap S') = f^{-1}(S) \cap f^{-1}(S') = \nabla_X \cap \nabla_X = \nabla_X = f^{-1}(S')$, which means that the inclusion $j \colon S \cap S' \to S'$, considered as a map in the fiber $Pt_Y(\mathcal{C})$, has the property that $f^\star(j)$ is an isomorphism. By protomodularity j is an iso and $S' \simeq S \cap S' \simeq S$. $\square$

3.16. Remark. Thanks to the previous theorem, in the presence of the protomodularity assumption, being normal, for a monomorphism, becomes a property.

3.17. Exercise. Check that in the category *Grp* our definition of normal subobject coincides with the usual one.

In the category *Grp* of groups any internal reflexive relation is actually an equivalence relation. This important property, which provides the definition of a *Maltsev category* [10] (see also Chapter VI), holds more generally in any protomodular category:

3.18. Theorem. *Any reflexive relation in a protomodular category $\mathcal{C}$ is an equivalence relation.*

Proof. Let

$$R \underset{d_1}{\overset{d_0}{\underset{\longleftarrow}{\rightrightarrows}}}\, X$$

be a reflexive relation in $\mathcal{C}$, and let us prove that it is an equivalence relation. Let us call *simplicial kernel* of this relation the limit $K[d_0, d_1]$ of the following diagram

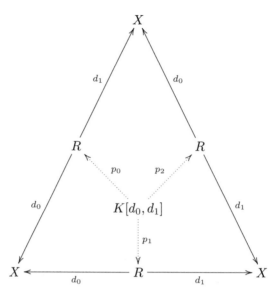

In the category *Set* of sets $K[d_0, d_1]$ is just the set of triples (x, y, z) such that xRy, yRz and xRz, and, in the following diagram, $R[d_0]$ is the set of triples (x, y, z) such that xRy and xRz. As in the category *Set*, there is a factorization $j \colon K[d_0, d_1] \to R[d_0]$ in any category $\mathcal{C}$:

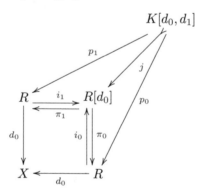

This induced arrow $j \colon K[d_0, d_1] \to R[d_0]$ is a monomorphism, since R is a relation. The arrow $s_0 \colon X \to R$ giving the reflectivity of R induces two arrows $s_0, s_1 \colon R \to K[d_0, d_1]$ with the properties that $j \circ s_0 = i_0$ and $j \circ s_1 = i_1$. This means that the pullbacks of j along i_1 and i_0 are isomorphisms, and then j itself is an iso by protomodularity (see Proposition 3.9(3)). The simplicial kernel $K[d_0, d_1]$ is then isomorphic to the kernel pair $R[d_0]$: in the set theoretical context this means that each triple (x, y, z) such that xRy and xRz is such that yRz. By this property and by the assumption that R is reflexive, it follows that R is an equivalence relation (see Exercise 1.2). □

3.C Internal groups and internal abelian groups

It is clear that inside the category *Grp* of groups, the fact of being an abelian group is a property, and not an additional structure. We are now going to show that it is also the case in any pointed protomodular category. In order to explain this important fact, we first need to recall the notions of internal monoid and of internal group in a category $\mathcal{C}$ with finite products.

3.19. Definition. An *internal monoid* in a category $\mathcal{C}$ with finite products is a triple (X, m, e), where X is an object in $\mathcal{C}$, $m \colon X \times X \to X$ and $e \colon 1 \to X$ are two arrows in $\mathcal{C}$ such that the following identities are satisfied:

$$(1) \quad m \circ (e \times 1) = a, \quad m \circ (1 \times e) = b$$

$$(2) \quad m \circ (m \times 1) = m \circ (1 \times m)$$

where $a \colon 1 \times X \to X$ and $b \colon X \times 1 \to X$ are the canonical isomorphisms.

An internal monoid (X, m) is abelian when m also satisfies the identity

$$(3) \quad m \circ \chi = m$$

where $\chi \colon X \times X \to X \times X$ is the "twisting" isomorphism interchanging the first and the second components in the product (i.e. $\chi(a, b) = (b, a)$).

3.20. Definition. An internal (abelian) monoid (X, m) is an *internal (abelian) group* when, moreover, there is an arrow $\sigma \colon X \to X$ with the property

$$(4) \quad m \circ (1 \times \sigma) \circ s_0 = e \circ \alpha_X.$$

3.21. Exercise. (1) Prove that, in the category *Set* of sets, an internal monoid (X, m) is an internal group exactly when the commutative square

$$
\begin{array}{ccc}
X \times X & \xrightarrow{\;m\;} & X \\
{\scriptstyle p_0}\downarrow & & \downarrow \\
X & \longrightarrow & 1
\end{array}
$$

is a pullback.

(2) Extend this result to any category $\mathcal{C}$ with products.

Later on, the following tool will be needed: with any internal group structure $m \colon X \times X \to X$ a ternary operation $p \colon X \times X \times X \to X$ is associated, which is defined by $p = m \circ (m \times 1) \circ (1 \times \sigma \times 1)$. In the set theoretical context it is the operation defined by $p(x, y, z) = x \cdot y^{-1} \cdot z$, where $\cdot$ is the group operation.

3.22. Exercise. Let (X, m) be a group (in *Set*) and let us define the following relation on the set $X \times X$, which is called the *Chasles relation of* m : $(x, t)Ch[m](y, z)$ if and only if $t = x \cdot y^{-1} \cdot z$. Prove that $Ch[m]$ is an equivalence relation on the set $X \times X$.

3.23. Definition. If $\mathcal{C}$ is a category with finite products, the category $Grp(\mathcal{C})$ of *internal groups* in $\mathcal{C}$ has as objects the internal groups in $\mathcal{C}$ and as arrows $f \colon (X, m_X, e_X) \to (Y, m_Y, e_Y)$ the arrows $f \colon X \to Y$ in $\mathcal{C}$ such that $f \circ m_X = m_Y \circ (f \times f)$ and $f \circ e_X = e_Y$.

Similarly, $Ab(\mathcal{C})$ denotes the category of *internal abelian groups* in $\mathcal{C}$.

As promised, we now sketch the proof of the protomodularity of $Grp(\mathcal{C})$:

3.24. Proposition. *Let $\mathcal{C}$ be a category with finite limits. Then $Grp(\mathcal{C})$ is protomodular.*

Proof. First remark that any internal group X induces for all $Z \in \mathcal{C}$, a group structure on the hom-set $\mathcal{C}(Z, X)$: indeed, if m denotes the internal group structure on X, one defines a binary operation $*$ on $\mathcal{C}(Z, X)$ by setting, for any $f, g \in \mathcal{C}(Z, X)$, $f * g = m \circ (f, g)$:

$$Z \xrightarrow{(f,g)} X \times X \xrightarrow{m} X.$$

This operation defines a group structure on $\mathcal{C}(Z, X)$. Now, the functor

$$\mathcal{C}(-, X) \colon \mathcal{C}^{op} \to Sets$$

factorizes through the category Grp:

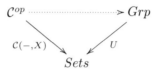

In the diagram above U denotes the forgetful functor, while the dotted functor from $\mathcal{C}^{op}$ to Grp will be denoted by $\overline{\mathcal{C}}(-, X)$, so that for any Z in $\mathcal{C}$, $\overline{\mathcal{C}}(Z, X)$ is the set $\mathcal{C}(Z, X)$ equipped with this group structure $*$. Since the the functor $\overline{Y} \colon Grp(\mathcal{C}) \to Grp^{\mathcal{C}^{op}}$ associating with any internal group X the functor $\overline{\mathcal{C}}(-, X)$ preserves finite limits and is conservative, the protomodularity of $Grp^{\mathcal{C}^{op}}$ (see Example 3.12(2)) can be lifted to $Grp(\mathcal{C})$. $\qquad \square$

3.D   Abelian objects in pointed protomodular categories

In a pointed category we shall denote by $l_X = (1_X, 0_{X,Y}) \colon X \to X \times Y$ and $r_Y = (0_{Y,X}, 1_Y) \colon Y \to X \times Y$ the canonical inclusions in the product $X \times Y$. A strong consequence of the protomodularity assumption is given by the following fact:

3.25. Proposition. *In any pointed protomodular category the pair $l_X = (1_X, 0_{X,Y})$ and $r_Y = (0_{Y,X}, 1_Y)$ is jointly strongly epimorphic:*

$$X \xrightarrow{l_X} X \times Y \xleftarrow{r_Y} Y \ .$$

Proof. The pair (l_X, r_Y) in the following pullback is clearly jointly strongly epimorphic, because the following square is a pullback (see Proposition 3.9(3)):

$$
\begin{array}{ccc}
X & \xrightarrow{\;l_X\;} & X \times Y \\
\downarrow & & {\scriptstyle r_Y} \big\Vert \big\Vert {\scriptstyle p_Y} \\
0 & \longrightarrow & Y
\end{array}
$$

$\square$

3.26. Exercise. If $\mathcal{C}$ is finitely complete and protomodular, then the pair l_X and r_Y is jointly epimorphic: given two maps $h, k \colon X \times Y \to Z$, $h = k$ if and only if $h \circ l_X = k \circ l_X$ and $h \circ r_Y = k \circ r_Y$.

It is clear that in a pointed category $\mathcal{C}$ an internal monoid is simply an object X equipped with an arrow $m \colon X \times X \to X$ satisfying the usual identities

$$(1) \quad m \circ l_X = 1_X = m \circ r_X \quad \textit{(unitary magma)}$$

$$(2) \quad m \circ (m \times 1) = m \circ (1 \times m) \quad \textit{(associative)}$$

An internal monoid X is an internal group if, moreover, there is an arrow $\sigma \colon X \to X$ such that

$$(3) \quad m \circ (\sigma, 1_X) = 1_X = m \circ (1_X, \sigma).$$

Then, according to Exercise 3.26, given an object X in a pointed protomodular category, there is at most one arrow $m \colon X \times X \to X$ such that $m \circ l_X = 1_X = m \circ r_X$. Therefore, the existence of such an arrow m for an object X becomes a property. It is precisely this last observation that explains why, for an object X in a pointed protomodular category, the fact of being an abelian object is a property, and not an additional structure. Furthermore, it turns out that any internal unitary magma in a pointed protomodular category is always an internal abelian group:

3.27. Proposition. *Any internal unitary magma in a pointed protomodular category is associative and commutative.*

Proof. In order to check the associativity, thanks to Exercise 3.26, it is sufficient to show that

(a) $\quad m \circ (1 \times m) \circ l_{X \times X} = m \circ (m \times 1) \circ l_{X \times X},$

(b) $\quad m \circ (1 \times m) \circ r_{X \times X} = m \circ (m \times 1) \circ r_{X \times X}.$

In a pointed protomodular category, to prove (a) it is enough to check (a_1) and (a_2) here below:

$(a_1) \quad [m \circ (1 \times m) \circ l_{X \times X} \circ l_X] = 1_X = [m \circ (m \times 1) \circ l_{X \times X} \circ l_X]$

and

$(a_2) \quad [m \circ (1 \times m) \circ l_{X \times X} \circ r_X] = 1_X = [m \circ (m \times 1) \circ l_{X \times X} \circ r_X].$

These equalities are easy to check, and, similarly, one can prove (b). In order to show that any internal magma in a protomodular category is abelian, one has to prove that

$$m \circ l_X = m \circ \sigma \circ l_X$$

and

$$m \circ r_X = m \circ \sigma \circ r_X,$$

where $\sigma \colon X \times X \to X \times X$ is the twisting isomorphism. This is straightforward, since $\sigma \circ l_X = r_X$ and $\sigma \circ r_X = l_X$. $\square$

3.28. Theorem. *Any internal unitary magma in a pointed protomodular category* $\mathcal{C}$ *is actually an internal abelian group.*

Proof. By Proposition 3.27, we only need to prove that any element in the abelian monoid X has an inverse or, equivalently, we can prove that the following square is a pullback (thanks to Exercise 3.21):

$$
\begin{array}{ccc}
X \times X & \xrightarrow{\ m\ } & X \\
{\scriptstyle p_0}\downarrow & & \downarrow \\
X & \longrightarrow & 0
\end{array}
$$

This follows by protomodularity and by the fact that (1) and (1) + (2) here below are pullbacks with p_0 a split epi:

$$
\begin{array}{ccccc}
X & \xrightarrow{\ r_X\ } & X \times X & \xrightarrow{\ m\ } & X \\
\downarrow & {\scriptstyle (1)} & {\scriptstyle p_0}\downarrow & {\scriptstyle (2)} & \downarrow \\
0 & \longrightarrow & X & \longrightarrow & 0
\end{array}
\qquad \square
$$

The previous result then justifies the following

3.29. Definition. An object X in a pointed protomodular category $\mathcal{C}$ is *abelian* when there is a (necessarily unique) arrow $m\colon X \times X \to X$ such that $m \circ l_X = 1_X = m \circ r_X$.

In any pointed category we can associate with any internal abelian group structure (X, m) a ternary operation $p\colon X \times X \times X \to X$ by setting

$$
p = m \circ (m \times 1) \circ (1 \times \sigma \times 1).
$$

It determines a relation on $X \times X$, called the *Chasles relation of* m (see also Exercise 3.22) defined by the following diagram:

$$
X \times X \times X \underset{\pi_0^2}{\overset{(\pi_0,p)}{\rightrightarrows}} X \times X,
$$

where, internally speaking, $(\pi_0, p)(x, y, z) = (x, p(x, y, z))$ and $\pi_0^2(x, y, z) = (y, z)$. The fact that it is a relation in any category is easily checked by composing the pair $((\pi_0, p), \pi_0^2)$ with the projections from X^4 to X (i.e. the pair $((\pi_0, p), \pi_0^2)$ is jointly monic). In the category *Set* the previous checking consists in proving the implication $(x, p(x, y, z), y, z)) = (x', p(x', y', z'), y', z')) \Rightarrow (x, y, z) = (x', y', z')$.

3.30. Lemma. *If* (X, m) *is an internal abelian group in a pointed protomodular category, then*

    (1) *the Chasles relation* $Ch[m]$ *of* m *is an equivalence relation*
    (2) *the diagonal* $s_0\colon X \to X \times X$ *is normal to* $Ch[m]$.

Proof. (1) $Ch[m]$ is reflexive because $p \circ (s_0, 1_X) = p_1$ (i.e. $p(x, x, y) = y$), and then it is an equivalence relation because $\mathcal{C}$ is protomodular and then Maltsev.

(2) s_0 is normal to $Ch[m]$, because the diagram

$$
\begin{array}{ccc}
X \times X & \xrightarrow{\ 1 \times s_0\ } & X \times X \times X \\
{\scriptstyle d_1}\big\downarrow\big\downarrow{\scriptstyle d_0} & & {\scriptstyle \pi_0^2}\big\downarrow\big\downarrow{\scriptstyle (\pi_0, p)} \\
X & \xrightarrow[\ s_0\ ]{} & X \times X
\end{array}
$$

commutes, and determines a fibrant map in $Eq(\mathcal{C})$. $\qquad\square$

We shall now extend to any pointed protomodular category the classical result in the category Grp asserting that a group G is abelian if and only if the diagonal $s_0 \colon G \to G \times G$ is normal in Grp:

3.31. Theorem. *Let $\mathcal{C}$ be a pointed protomodular category. An object X is abelian if and only if $s_0 \colon X \to X \times X$ is normal.*

Proof. If $m \colon X \times X \to X$ is an abelian group operation, then s_0 is normal to $Ch[m]$ by the previous lemma. Conversely, let $(d_0, d_1) \colon R \to X \times X$ be the equivalence relation to which s_0 is normal, and let $\phi \colon R \to X \times X \times X$ be the arrow $(\pi_0 \circ d_0, \pi_0 \circ d_1, \pi_1 \circ d_1)$, which, internally speaking, associates the triple (x, y, z) with $((x, t), (y, z))$.

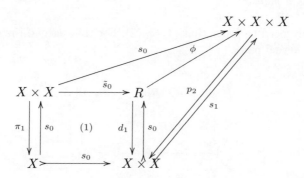

We have

$$p_2 \circ \phi = (\pi_0 \circ d_1, \pi_1 \circ d_1) = d_1$$

$$\phi \circ s_0 = (\pi_0 \circ d_0 \circ s_0, \pi_0 \circ d_1 \circ s_0, \pi_1 \circ d_1 \circ s_0) = (\pi_0, \pi_0, \pi_1) = s_1$$

$$\phi \circ \tilde{s}_0 = (\pi_0 \circ d_0 \circ \tilde{s}_0, \pi_0 \circ d_1 \circ \tilde{s}_0, \pi_1 \circ d_1 \circ \tilde{s}_0) = (\pi_0, \pi_1, \pi_1) = s_0$$

Accordingly, ϕ is a map in the fibre $Pt_{X \times X}(\mathcal{C})$. The square (1) is a pullback by assumption, the outer downward diagram is always a pullback. Thus, $s_0^{\star}(\phi) = 1_{X \times Y}$ is an iso in $Pt_X(\mathcal{C})$, and then ϕ is an iso. By defining

$$m = \pi_1 \circ d_0 \circ \phi^{-1} \circ (\pi_0, 0_{X \times X, X}, \pi_1)$$

one gets an internal unitary magma structure on X in a pointed protomodular category. Accordingly, X is then abelian. $\qquad\qquad\qquad\qquad\qquad\square$

4. Regular protomodular categories

If C is regular, then the pullback cancellation property defining a protomodular category can be extended from split epimorphisms to regular epimorphisms:

4.A The normalized Barr-Kock Theorem

4.1. Lemma. *If C is regular, then the following conditions are equivalent*

(1) *C is protomodular*
(2) *In any commutative diagram*

$$
\begin{array}{ccccc}
A & \xrightarrow{\ f\ } & B & \xrightarrow{\ h\ } & C \\
\Big\downarrow{\alpha} & (1) & \Big\downarrow{\beta} & (2) & \Big\downarrow{\gamma} \\
D & \xrightarrow[\ g\ ]{} & E & \xrightarrow[\ l\ ]{} & F
\end{array}
$$

where β is a regular epimorphism, if (1) and (1) + (2) are pullbacks, then (2) is a pullback.

Proof. We only need to prove that (1) $\Rightarrow$ (2). For this, extend the diagram above by taking the kernel pairs:

$$
\begin{array}{ccccc}
R[\alpha] & \xrightarrow{\ \tilde{f}\ } & R[\beta] & \xrightarrow{\ \tilde{h}\ } & R[\gamma] \\
{\scriptstyle p_0}\Big\downarrow\Big\downarrow{\scriptstyle p_1} & (3) \ {\scriptstyle p_0} & \Big\downarrow\Big\downarrow{\scriptstyle p_1} & (4) \ {\scriptstyle p_0} & \Big\downarrow\Big\downarrow{\scriptstyle p_1} \\
A & \xrightarrow[\ f\ ]{} & B & \xrightarrow[\ h\ ]{} & C
\end{array}
$$

Clearly both (3) and (3) + (4) are pullbacks, so that, by protomodularity, (4) is a pullback and determines a fibrant arrow $(h, \tilde{h}) : (R[\beta], B) \to (R[\gamma], C)$. By the Barr-Kock Theorem it follows that (2) is a pullback. $\qquad\square$

In the pointed situation the protomodularity condition becomes now equivalent to the *Short Five Lemma*.

4.2. Theorem. *Let C be a regular pointed category. Then the following conditions are equivalent*

(1) *C is protomodular*

(2) *the "normalized Barr-Kock" property holds: in any commutative diagram*

$$
\begin{array}{ccccc}
K[f'] & \xrightarrow{\ k'\ } & B & \xrightarrow{\ f'\ } & C \\
\downarrow{\scriptstyle u} & & \downarrow{\scriptstyle v}\ \ (1) & & \downarrow{\scriptstyle w} \\
K[f] & \xrightarrow[\ k\ ]{} & \overline{B} & \xrightarrow[\ f\ ]{} & \overline{C}
\end{array}
$$

with f' *a regular epi, u is an isomorphism if and only if* (1) *is a pullback*

(3) *The Short Five Lemma holds: this means that in any diagram as above (with f' a regular epi), if u and w are isomorphisms, then v is an isomorphism.*

Proof. (1) $\Rightarrow$ (2) The non-trivial part is to show that if u is an isomorphism, then the square (1) is a pullback. For this, let us consider the diagram

$$
\begin{array}{ccccc}
K[f'] & \xrightarrow{\ k'\ } & B & \xrightarrow{\ v\ } & \overline{B} \\
\downarrow & \ (3)\ \ f'\downarrow & & \ (1)\ \ \downarrow f & \\
0 & \xrightarrow{\hspace{2em}} & C & \xrightarrow[\ w\ ]{} & \overline{C}
\end{array}
$$

The square (3) is a pullback by definition, the map $v \circ k' = k \circ u$ is the kernel of f, so that (3) + (1) is a pullback. The result then follows by Lemma 4.1.

(2) $\Rightarrow$ (3) Obviously, when w is an isomorphism and the square (1) is a pullback (since u is an iso), then v is an isomorphism.

(3) $\Rightarrow$ (1) In order to prove that any change-of-base functor is conservative, it is sufficient to check that the change-of-base functors $\alpha_X^\star \colon Pt_X(\mathcal{C}) \to Pt_0(\mathcal{C}) \simeq \mathcal{C}$ are conservative for any X. This latter condition is precisely guaranteed by the validity of the Split Short Five Lemma, which is clearly a special case of the Short Five Lemma (any split epi being a regular epi). $\qquad\square$

4.3. Proposition. *Let $\mathcal{C}$ be a regular pointed protomodular category. Then the following conditions are equivalent:*

(1) *Any equivalence relation is effective (i.e. $\mathcal{C}$ is Barr-exact)*

(2) *Any normal monomorphism is a kernel*

Proof. (1) $\Rightarrow$ (2) follows by Proposition 3.5.

(2) $\Rightarrow$ (1) Let (R, Y) be an equivalence relation in $\mathcal{C}$, and let $x \colon X \to Y$ be the (unique) normal monomorphism associated with (R, Y) (Proposition 3.6). By assumption x is the kernel of some arrow $y \colon Y \to Z$, so that $x \dashv R[y]$ and $x \dashv R$. By Theorem 3.15 it follows that $R \simeq R[y]$, so that (R, Y) is effective. $\qquad\square$

Remark. Accordingly, in any Barr-exact pointed protomodular category we have that the normal monomorphisms are precisely the kernel maps, as in *Grp* (see also Chapter VIII).

4.B Exact sequences in regular protomodular categories

The main purpose of this section is to establish some basic facts on exact sequences in regular pointed protomodular categories. Once again, our leading model will be the category Grp of groups, where our general results admit a "classical" interpretation. We shall here restrict ourselves to the basic aspects of the theory, and we refer to [6] and [7] for further developments.

In a pointed category we shall write

$$0 \longrightarrow A \overset{k}{\longrightarrow} B \overset{f}{\longrightarrow} C$$

when $k \colon A \to B$ is the kernel of f. Dually,

$$A \overset{k}{\longrightarrow} B \overset{f}{\longrightarrow} C \longrightarrow 0$$

will indicate that $f \colon B \to C$ is the cokernel of k. If both conditions are satisfied, we shall call

$$0 \longrightarrow A \overset{k}{\longrightarrow} B \overset{f}{\longrightarrow} C \longrightarrow 0$$

an *exact sequence*.

In the following theorem we shall prove that the notion of exact sequence in a protomodular category behaves as in the category Grp of groups, since it corresponds to a regular epi equipped with its kernel. This important fact, which allows one to identify cokernels and regular epis, is one of the main reasons why regular pointed protomodular categories are a suitable categorical setting for (non abelian) homological algebra.

4.4. Theorem. *Let C be a pointed protomodular category. An arrow $f \colon A \to B$ is a regular epi if and only if f is the cokernel of its kernel.*

Proof. Consider the commutative diagram

$$
\begin{array}{ccccc}
K[f] \times K[f] & \underset{p_1}{\overset{p_0}{\rightrightarrows}} & K[f] & \longrightarrow & 0 \\
\overline{k} \downarrow & & \downarrow k & & \downarrow \\
R[f] & \underset{p_1}{\overset{p_0}{\rightrightarrows}} & A & \underset{f}{\longrightarrow} & B
\end{array}
$$

and let us prove that the right hand commutative square is a pushout if and only if f is a regular epi. Indeed, an arrow $\alpha \colon A \to C$ is such that it factorizes through 0 if and only if $\alpha \circ k \circ p_0 = \alpha \circ k \circ p_1$ if and only if $\alpha \circ p_0 \circ \overline{k} = \alpha \circ p_1 \circ \overline{k}$. Now, the left hand squares are pullbacks and, by protomodularity (Proposition 3.9(3)), the pair $\overline{k} \colon K[f] \times K[f] \to R[f]$ and $s_0 \colon A \to R[f]$ is jointly strongly epic. Then $\alpha \circ p_0 = \alpha \circ p_1$ if and only if $\alpha \circ p_0 \circ \overline{k} = \alpha \circ p_1 \circ \overline{k}$ and $\alpha \circ p_0 \circ s_0 = \alpha \circ p_1 \circ s_0$. This latter equality being trivially satisfied, we have proved that α factorizes through 0 precisely when α coequalizes p_0 and p_1. □

Remark. Thus, in any regular pointed protomodular category, one has the following equalities:

$$strong\ epi\ =\ regular\ epi\ =\ cokernel$$

where the first equality is due to regularity, and the second one is due to Theorem 4.4.

4.5. Corollary. *In a pointed protomodular category*

$$A \xrightarrow{\ k\ } B \xrightarrow{\ f\ } C$$

is an exact sequence if and only if

    (1) f *is a regular epi*
    (2) k *is the kernel of* f.

When the pointed category is protomodular, the notion of exact sequence corresponds precisely to the notion of exact fork given in the second section, as we can see in the following

4.6. Corollary. *If* C *is a regular pointed protomodular category, then a reflexive fork*

$$G \underset{\underset{d_1}{\longrightarrow}}{\overset{\overset{d_0}{\longrightarrow}}{\xleftarrow{s_0}}} A \xrightarrow{\ f\ } B$$

with $f \circ d_0 = f \circ d_1$ *and* f *a regular epi is exact if and only if*

$$0 \longrightarrow K[d_0] \xrightarrow{d_1 \circ Ker(d_0)} A \xrightarrow{\ f\ } B \longrightarrow 0$$

is an exact sequence.

Proof. We first remark that in both conditions f is assumed to be a regular epi. Then consider the following diagram, where $d_0 \colon G \to A$ is split by $s_0 \colon A \to G$:

$$
\begin{array}{ccccc}
K[d_0] & \xrightarrow{Ker(d_0)} & G & \xrightarrow{d_1} & A \\
\downarrow & & \;\;(1)\quad d_0\downarrow & & \;\;(2)\quad f\downarrow \\
0 & \xrightarrow{\quad\quad} & A & \xrightarrow{\;f\;} & B
\end{array}
$$

The result follows from the fact that under our assumptions, the square (2) is a pullback if and only if (1) + (2) is a pullback. $\square$

4.7. Corollary. *Exact sequences are stable under products.*

Proof. By Corollary 2.25 and by Corollary 4.6. $\square$

4.8. Corollary. *Exact sequences are stable under pullback: this means that in any commutative diagram*

$$
\begin{array}{ccccc}
K[f] & \xrightarrow{\ k'\ } & A' & \xrightarrow{\ f'\ } & B' \\
\| & & \downarrow{\scriptstyle g} \quad (1) & & \downarrow{\scriptstyle h} \\
0 \longrightarrow K[f] & \xrightarrow[\ Ker(f)\ ]{} & A & \xrightarrow{\ f\ } & B \longrightarrow 0
\end{array}
$$

when (1) is a pullback and k' is the unique arrow such that $f' \circ k' = 0$ and $g \circ k' = Ker(f)$, then the upper sequence is exact.

Proof. This follows from Theorem 4.4 and from the fact that k' is the kernel of f' in any pointed category (Exercise 2.14). $\qquad\qquad\qquad\qquad\qquad\qquad\qquad$ □

5. Additive categories

5.A Additive and linear categories

5.1. Definition. A pointed category $\mathcal{C}$ with finite limits is *additive* if any hom-set $\mathcal{C}(A, B)$ is an additive abelian group, and the composition of arrows is a group homomorphism by composing both on the left and on the right.

5.2. Examples. The categories Ab of abelian groups and the category $R\text{-}Mod$ of modules over a ring R are the major examples of additive categories. The category $Ab(\mathcal{C})$ of internal abelian groups in any category $\mathcal{C}$ with finite limits (see Definition 3.23) is additive. Indeed, if we consider any object A in $Ab(\mathcal{C})$, we know already that each hom-set $\mathcal{C}(Z, A)$ is a group (Proposition 3.24). This group $\mathcal{C}(Z, A)$ is abelian because A is itself abelian, as we can see by looking at the following commutative diagram, where $\chi\colon A \times A \to A \times A$ is the twisting isomorphism:

$$
\begin{array}{ccccc}
Z & \xrightarrow{\ (f,g)\ } & A \times A & \xrightarrow{\ +\ } & A \\
\| & & \downarrow{\scriptstyle \chi} & & \| \\
Z & \xrightarrow[\ (g,f)\ ]{} & A \times A & \xrightarrow[\ +\ ]{} & A
\end{array}
$$

We leave it to the reader to check that the composition of arrows is a group homomorphism both on the left and on the right.

As a particular case, the category of topological abelian groups $Ab(Top)$ is additive.

5.3. Definition. A pointed category $\mathcal{C}$ with finite limits is *linear* if for all objects $X, Y \in \mathcal{C}$ the diagram

$$
X \xrightarrow{\ l_X\ } X \times Y \xleftarrow{\ r_Y\ } Y
$$

is a coproduct.

5.4. Lemma. *Any additive category is linear.*

Proof. Consider the product of two objects X and Y:

$$X \xrightarrow[l_X]{\pi_X} X \times Y \xleftarrow[r_X]{\pi_Y} Y.$$

Given any object Z and any pair of arrows $f \colon X \to Z$ and $g \colon Y \to Z$, we define $h = (f \circ \pi_X + g \circ \pi_Y) \colon X \times Y \to Z$. This arrow h has the property that $h \circ l_X = (f \circ \pi_X + g \circ \pi_Y) \circ l_X = f + 0 = f$ and $h \circ r_Y = (f \circ \pi_X + g \circ \pi_Y) \circ r_Y = 0 + g = g$. In order to prove the unicity of this factorization, let us first remark that the equalities

$$\pi_X \circ (l_X \circ \pi_X + r_Y \circ \pi_Y) = \pi_X \circ l_X \circ \pi_X + \pi_X \circ r_Y \circ \pi_Y = \pi_X + 0 = \pi_X$$

and

$$\pi_Y \circ (l_X \circ \pi_X + r_Y \circ \pi_Y) = 0 + \pi_Y = \pi_Y$$

imply that $l_X \circ \pi_X + r_Y \circ \pi_Y = 1_{X \times Y}$. Then, if $h' \colon X \times Y \to Z$ is any arrow such that $h' \circ l_X = f$ and $h' \circ r_Y = g$, one has $h' = h' \circ 1_{X \times Y} = h' \circ (l_X \circ \pi_X + r_Y \circ \pi_Y) = h' \circ l_X \circ \pi_X + h' \circ r_Y \circ \pi_Y = f \circ \pi_X + g \circ \pi_Y = h$, so that the required factorization is unique, and

$$X \xrightarrow{l_X} X \times Y \xleftarrow{r_Y} Y$$

is a coproduct. □

By the previous result any product in an additive category is a coproduct; we shall call a diagram

$$X \xrightarrow[l_X]{\pi_X} X \times Y \xleftarrow[r_X]{\pi_Y} Y$$

a *biproduct* of X and Y. The object $X \times Y$ is then usually denoted by $X \oplus Y$; of course, in the category Ab of abelian groups or $R\text{-}Mod$ of modules over a ring R, it gives the classical notion of *direct sum* of X and Y.

5.B Additive categories and protomodularity

5.5. Proposition. *Any split epimorphism* $X \xrightarrow{f} Y$ *in an additive category produces an isomorphism* $X \simeq K[f] \oplus Y$.

Proof. Let us consider the diagram

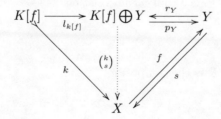

where $s\colon Y \to X$ is a section of f and

$$\binom{k}{s}\colon K[f] \oplus Y \to X$$

is the unique factorization induced by k and s through the direct sum $K[f] \oplus Y$. Since $f \circ s = 1_Y$, we have $f \circ (1_X - s \circ f) = 0_{X,Y}$. There is then a unique $g\colon X \to K[f]$ such that $k \circ g = 1_X - s \circ f$. Therefore, by the universal property of the product there is a unique arrow

$$(g, f)\colon X \to K[f] \oplus Y.$$

One can then check that (g, f) is the inverse of $\binom{k}{s}$. $\square$

Remark. The property proved in Proposition 5.5 implies the validity of the Split Short Five Lemma. Accordingly:

5.6. Corollary. *Any additive category is protomodular.*

In the following, we shall write $\nabla\colon X \oplus X \to X$ for the codiagonal.

5.7. Remark. The arrow $\nabla\colon X \oplus X \to X$ provides X with the structure of a unitary magma. Since $\mathcal{C}$ is protomodular, this operation endows any object of the additive category $\mathcal{C}$ with an internal abelian group structure in $\mathcal{C}$.

5.8. Corollary. *In an additive category, the diagram*

$$
\begin{array}{ccc}
X \oplus X & \xrightarrow{\ \nabla\ } & X \\
{\scriptstyle p_0}\downarrow & & \downarrow \\
X & \longrightarrow & 0
\end{array}
$$

is a pullback.

Proof. By Exercise 3.21 and Theorem 3.28. $\square$

5.9. Lemma. *In an additive category any monomorphism is normal.*

Proof. Consider any monomorphism $j\colon X' \to X$ in an additive category $\mathcal{C}$. Let us prove that j is normal to the equivalence relation

$$
X \oplus X' \underset{\pi_1}{\overset{p_0}{\underset{\longrightarrow}{\overset{\longrightarrow}{\xleftarrow{\ 1_X\ }}}}} X \qquad (1)
$$

where $\pi_1\colon X \oplus X' \to X$ is the composite

$$X \oplus X' \xrightarrow{\ 1_X \times j\ } X \oplus X \xrightarrow{\ \nabla\ } X.$$

Indeed, the reflexive graph (1) is a relation by the Corollary 5.8. Since any additive category is protomodular and therefore Maltsev (see Theorem 3.18), this reflexive relation is an equivalence relation. Consider then the diagram

$$
\begin{array}{ccccc}
X' \oplus X' & \xrightarrow{(p_0,p_1-p_0)} & X' \oplus X' & \xrightarrow{\;j\times 1\;} & X \oplus X' \\
p_0 \Big\downarrow \Big\downarrow p_1 & & p_0 \Big\downarrow & & p_0 \Big\downarrow \Big\downarrow \pi_1 \\
X' & =\!=\!=\!=\!= & X' & \xrightarrow{\qquad j \qquad} & X
\end{array}
$$

where $p_1 - p_0$ is calculated in the abelian group $\mathcal{C}(X', X')$. The exterior diagram determines a map in $Eq(\mathcal{C})$. It is fibrant, because the two commutative squares with the p_0 are clearly pullbacks (the arrow $(p_0, p_1 - p_0)$ being invertible with inverse $(p_0, p_1 + p_0)$). Accordingly, $j\colon X' \to X$ is normal to the equivalence relation $(X \oplus X', X)$. $\qquad\square$

5.10. Theorem. *A finitely complete category $\mathcal{C}$ is additive if and only if*

(1) *$\mathcal{C}$ is pointed and protomodular*
(2) *any monomorphism is normal.*

Proof. By Corollary 5.6 and Lemma 5.9, we only need to prove that if any mono is normal in a pointed protomodular category $\mathcal{C}$, then $\mathcal{C}$ is additive. For this, remark that if any diagonal $s_0\colon X \to X \times X$ is normal, then any object X in $\mathcal{C}$ is abelian (Theorem 3.31). It follows that $\mathcal{C}$ is equivalent to $Ab(\mathcal{C})$, and consequently $\mathcal{C}$ is additive. $\qquad\square$

6. The Short Five Lemma and the Tierney equation

6.A Abelian categories

6.1. Definition. [9] [12] A category $\mathcal{C}$ is *abelian* if

(1) $\mathcal{C}$ has a zero object
(2) $\mathcal{C}$ has binary products and binary coproducts
(3) every arrow has a kernel and a cokernel
(4) every monomorphism is a kernel, every epimorphism is a cokernel.

Clearly, this definition is selfdual, i.e. every axiom in $\mathcal{C}$ also holds in its dual category $\mathcal{C}^{op}$. On the other hand, (1), (2) and (3) are completeness and cocompleteness conditions, while (4) is of a different nature: it identifies the two ends of the two strings of implications

$$cokernel \;\Rightarrow\; regular\ epi \;\Rightarrow\; strong\ epi \;\Rightarrow\; epi$$

and

$$kernel \;\Rightarrow\; normal\ mono \;\Rightarrow\; mono.$$

6.2. Examples.

(1) The categories Ab of abelian groups and the category $R\text{-}Mod$ of right (or left) modules over a ring R are abelian.

(2) If R is a ring and $\mathcal{C}$ a small category, then the category $R\text{-}Mod^{\mathcal{C}^{op}}$ of presheaves of R-modules is abelian.

(3) If R is a ring and $(\mathcal{C}, \tau)$ a site, the category of sheaves of R-Mod on $(\mathcal{C}, \tau)$ is a localization of the corresponding category of presheaves of R-modules on $\mathcal{C}$, thus the category of sheaves of R-Mod is abelian [2].

(4) The category $Ab(\mathcal{C})$ of internal abelian groups in an exact category is abelian; as a special case, the category of internal abelian groups in the category of compact Haussdorff spaces is an abelian category, see Theorem 6.5 below.

(5) The category $Ab(Top)$ is additive, regular, but not abelian.

6.3. Remark. As a cokernel, any epimorphism in an abelian category is necessarily a regular epi; as a kernel, any monomorphism is normal.

6.4. Corollary. *Let $f\colon X \to Y$ be an arrow in an abelian category $\mathcal{C}$. Then f is an isomorphism if and only if f is both a monomorphism and an epimorphism.*

Proof. By Lemma 2.3 and by the Remark 6.3 above.   □

We have an immediate class of examples:

6.5. Theorem. *If $\mathcal{C}$ is Barr-exact and additive, then $\mathcal{C}$ is abelian.*

Proof. By the previous results we only need to prove that any arrow has a cokernel and that every epimorphism is a cokernel. Let $f\colon X \to Y$ be any arrow in $\mathcal{C}$ and let $f = m \circ q$ be the regular epi-mono factorization of f. Since m is normal to an effective equivalence relation (R, Y) (by Corollary 5.6 and by Lemma 5.9), the quotient of this equivalence relation R is the cokernel of m and then of f.

If $g\colon X \to Y$ is an epimorphism, let $g = m \circ p$ be the regular epi-mono factorization of g. If $q\colon Y \to Cok[g]$ is the cokernel of g, then from the fact that g is an epimorphism and $q \circ g = 0 \circ g$ we deduce that $q = 0$; now, $m = Ker(q) = Ker(0)$ is an isomorphism and g a regular epimorphism.   □

6.B Abelian categories and protomodularity

Actually, it turns out that the previous condition is a characterization. This is the meaning of the classical Tierney equation

$$Abelian = Exact + Additive,$$

which will allow us to measure the strength of the identifications:

$$kernel = mono \quad \text{and} \quad cokernel = epi.$$

Our aim now is to establish step by step the converse result of Theorem 6.5 (see also [12]). From now on, we shall always assume that the category is abelian. Let us begin with the following:

6.6. Lemma. *Every pair of subobjects has an intersection.*

Proof. Let $i: A_2 \to A$ and $j: A_1 \to A$ be two subobjects of A, let $p: A \to Cok[j]$ be the cokernel of j. Consider the following diagram:

$$
\begin{array}{ccc}
K[p \circ i] & \xrightarrow{\;k\;} & A_1 \\
\downarrow{\scriptstyle l} & (1) & \downarrow{\scriptstyle j} \\
A_2 & \xrightarrow[\;i\;]{} & A \xrightarrow[\;p\;]{} Cok[j]
\end{array}
$$

From $p \circ i \circ l = 0$ it follows that there exists a unique $k: K[p \circ i] \to A_1$ with $i \circ l = j \circ k$. Let $\alpha: X \to A_1$ and $\beta: X \to A_2$ be two arrows such that $j \circ \alpha = i \circ \beta$; then $p \circ i \circ \beta = p \circ j \circ \alpha = 0$ implies that there is a unique $\gamma: X \to K[p \circ i]$ with $l \circ \gamma = \beta$. Moreover, $j \circ k \circ \gamma = i \circ l \circ \gamma = i \circ \beta = j \circ \alpha$ gives $k \circ \gamma = \alpha$, and the square (1) is a pullback. □

6.7. Corollary. C *is finitely complete.*

Proof. C has finite products and intersections of subobjects. Given any pair of parallel arrows f and g from A to B their equalizer is given by the intersection of the subobjects $(1, f): A \to A \times B$ and $(g, 1): A \to A \times B$. The existence of finite products and equalizers implies that all finite limits exist. □

6.8. Lemma. *Any monomorphism is the kernel of its cokernel.*

Proof. Let $f: X \to Y$ be a monomorphism; by assumption f is the kernel of some arrow, say, $g: Y \to Z$. Let $p: Y \to Cok[f]$ be the cokernel of f and let $j: K[p] \to Y$ be the kernel of p. Now, since $g \circ f = 0$, there is a unique $z: Cok[f] \to Z$ with $z \circ p = g$. Similarly, $p \circ f = 0$ gives a unique $s: X \to K[p]$ with $j \circ s = f$; the equality $g \circ j = z \circ p \circ j = 0$ gives a unique $t: K[p] \to X$ with $f \circ t = j$. One then concludes that $X \simeq K[p]$. □

By duality, it follows:

6.9. Corollary.

(1) C *is finitely cocomplete.*
(2) *Every epimorphism is the cokernel of its kernel.*

6.10. Lemma. *If* $f: X \to Y$ *is any arrow, then* $q = Coker(Ker(f))$ *coequalizes the pair of parallel arrows* p_0, p_1 *from* $R[f]$ *to* X. *Consequently,* $R[f] \subseteq R[q]$.

Proof. If $Ker(f)$ is the kernel of f, then, by Lemma 6.8 it follows that $Ker(f)$ is also the kernel of q. The factorization $\bar{k}: K[f] \times K[f] \to R[f]$ making the following diagram

$$K[f] \times K[f] \xrightarrow{\ \overline{k}\ } R[f]$$

$$
\begin{array}{ccc}
K[f] \times K[f] & \xrightarrow{\ \overline{k}\ } & R[f] \\
{\scriptstyle p_0}\big\downarrow\big\downarrow{\scriptstyle p_1} & & {\scriptstyle p_0}\big\downarrow\big\downarrow{\scriptstyle p_1} \\
K[f] & \xrightarrow[\ Ker(f)\ ]{} & X
\end{array}
$$

commutative, also makes pullbacks the two commutative squares. Thus, $\overline{k}$ is at the same time the kernel of $q \circ p_0$ and of $q \circ p_1$, so that $q \circ p_0$ and $q \circ p_1$ are both the cokernels of $\overline{k}$. There is then a unique

$$\epsilon \colon Cok[Ker(f)] \to Cok[Ker(f)]$$

with $\epsilon \circ q \circ p_0 = q \circ p_1$. Then $\epsilon \circ q \circ p_0 \circ s_0 = q \circ p_1 \circ s_0$ gives $\epsilon \circ q = q$, and then $\epsilon = 1_{Cok[Ker(f)]}$. Consequently, $q \circ p_0 = q \circ p_1$. $\qquad\square$

6.11. Corollary. *If $f \colon X \to Y$ is any arrow and $q = Coker(Ker(f))$, then $R[f] \simeq R[q]$. Consequently, q as a regular epi, is the quotient of $R[f]$.*

Proof. There is a factorization i such that $f = i \circ q$, so that $R[q] \subseteq R[f]$. $\qquad\square$

Our first aim will be to show step by step that any abelian category is protomodular.

6.12. Corollary. *$f \colon X \to Y$ is a mono if and only if $K[f] \simeq 0$.*

Proof. With the same notations as in the Lemma 6.10 above, if $K[f] \simeq 0$, then $q = Coker(Ker(f))$ is an isomorphism. Whence $R[f] \simeq R[q] \simeq \Delta_X$, and f is a monomorphism. $\qquad\square$

6.13. Corollary. *Pullbacks reflect monomorphisms.*

Proof. Consider any pullback

$$
\begin{array}{ccc}
A \times_C B & \xrightarrow{\ f'\ } & B \\
{\scriptstyle g'}\big\downarrow & & \big\downarrow{\scriptstyle g} \\
A & \xrightarrow[\ f\ ]{} & C
\end{array}
$$

in which f' is a monomorphism, and let us prove that f is a monomorphism as well. Indeed, this follows from $0 \simeq K[f'] \simeq K[f]$ (Exercise 2.14 and Corollary 6.12). $\qquad\square$

By duality:

6.14. Corollary.

(1) *$f \colon X \to Y$ is an epimorphism if and only if $Cok[f] \simeq 0$.*
(2) *Pushouts reflect epimorphisms.*

6.15. Theorem. *The Short Five Lemma holds in $\mathcal{C}$.*

Proof. Consider a commutative diagram

$$
\begin{array}{ccccccccc}
0 & \longrightarrow & A & \overset{k}{\rightarrowtail} & B & \overset{f}{\longrightarrow} & C & \longrightarrow & 0 \\
 & & \| & (1) & \phi\downarrow & (2) & \| & & \\
0 & \longrightarrow & A & \underset{k'}{\rightarrowtail} & B' & \underset{f'}{\longrightarrow} & C & \longrightarrow & 0
\end{array}
$$

In the diagram

$$
\begin{array}{ccccc}
A & =\!=\!= & A & \longrightarrow & 0 \\
k\downarrow & (1)\ \ k'\downarrow & & (3) & \downarrow \\
B & \underset{\phi}{\longrightarrow} & B' & \underset{f'}{\longrightarrow} & C
\end{array}
$$

the outer rectangle and the square (3) are pullbacks, so that (1) is a pullback and ϕ is a monomorphism by Corollary 6.13. By duality, (2) is a pushout and the arrow ϕ is an epimorphism. By Corollary 6.4, ϕ is an isomorphism. $\qquad\square$

6.16. Corollary. *Any abelian category is protomodular.*

Proof. $\mathcal{C}$ is finitely complete, pointed and the (Split) Short Five Lemma holds. $\qquad\square$

6.17. Theorem. *Any abelian category is additive.*

Proof. We can apply Theorem 5.10: indeed, $\mathcal{C}$ is pointed protomodular and any monomorphism is a kernel and, consequently, it is normal. $\qquad\square$

6.C Abelian categories and Barr-exactness

6.18. Theorem. *Any abelian category satisfies the "normalized Barr-Kock property": in any diagram*

$$
\begin{array}{ccccccccc}
0 & \longrightarrow & K[f] & \overset{k}{\longrightarrow} & X & \overset{f}{\longrightarrow} & Y & \longrightarrow & 0 \\
 & & \tau\downarrow & & g\downarrow & (1) & g'\downarrow & & \\
0 & \longrightarrow & K[f'] & \underset{k'}{\longrightarrow} & X' & \underset{f'}{\longrightarrow} & Y' & &
\end{array}
$$

(1) *is a pullback if and only if τ is an isomorphism.*

Proof. When (1) is a pullback, then $\tau\colon K[f] \to K[f']$ is an isomorphism in any category with kernels (Exercise 2.14). Conversely, let $(X' \times_{Y'} Y, \pi_{X'}, \pi_Y)$ denote the pullback of g' along f', and let $\phi\colon X \to X' \times_{Y'} Y$ be the induced factorization with $\pi_Y \circ \phi = f$ and $\pi_{X'} \circ \phi = g$. The arrow $\pi_Y\colon X' \times_{Y'} Y \to Y$ is an epi because $f\colon X \to Y$ is an epi.

The assumption that τ is an iso asserts that the epimorphisms f and π_Y have isomorphic kernels. The Short Five Lemma then implies that ϕ is an isomorphism, and the square (1) a pullback.   □

6.19. Theorem. *Any arrow* $f \colon X \to Y$ *has a factorization* $f = i \circ q$:

$$K[f] \xrightarrow{\ k\ } X \xrightarrow{\ q\ } Cok[Ker(f)]$$

with the arrows f from X to Y and i from $Cok[Ker(f)]$ to Y.

with $i \colon Cok[Ker(f)] \to Y$ *a monomorphism (and* q *an epimorphism).*

Proof. Since k is a mono, it is the kernel of its cokernel q:

$$
\begin{array}{ccccccccc}
0 & \longrightarrow & K[q] & \longrightarrow & X & \xrightarrow{\ q\ } & Cok[Ker(f)] & \longrightarrow & 0 \\
& & \| & & \| & (1) & \downarrow i & & \\
0 & \longrightarrow & K[f] & \xrightarrow{\ k\ } & X & \xrightarrow{\ f\ } & Y & &
\end{array}
$$

The square (1) is then a pullback by the normalized Barr-Kock property, and i is a mono because in any protomodular category pullbacks reflect monomorphisms.   □

The dual result of the normalized Barr-Kock condition will also be needed:

6.20. Corollary. *In any diagram*

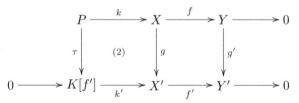

$$
\begin{array}{ccccccccc}
& & P & \xrightarrow{\ k\ } & X & \xrightarrow{\ f\ } & Y & \longrightarrow & 0 \\
& & \tau\downarrow & (2) & g\downarrow & & g'\downarrow & & \\
0 & \longrightarrow & K[f'] & \xrightarrow{\ k'\ } & X' & \xrightarrow{\ f'\ } & Y' & \longrightarrow & 0
\end{array}
$$

the square (2) is a pushout if and only if g' *is an isomorphism.*

6.21. Proposition.

(1) *If* $f \colon X \to Y$ *is an epimorphism, then*

$$X \times Z \xrightarrow{\ f \times 1_Z\ } Y \times Z$$

is an epimorphism.

(2) *If* $f \colon X \to Y$ *and* $f' \colon X' \to Y'$ *are epimorphisms, then*

$$f \times f' \colon X \times X' \to Y \times Y'$$

is an epimorphism.

Proof.

(1) In the diagram

$$
\begin{array}{ccccccccc}
0 & \longrightarrow & X & \xrightarrow{\ l_X\ } & X \times Z & \xrightarrow{\ \pi_Z\ } & Z & \longrightarrow & 0 \\
 & & \downarrow{\scriptstyle f} & & \downarrow{\scriptstyle f \times 1_Z} & & \parallel & & \\
0 & \longrightarrow & Y & \xrightarrow[\ l_Y\ ]{} & Y \times Z & \xrightarrow[\ \pi_Z\ ]{} & Z & \longrightarrow & 0
\end{array}
$$

with (1) in the middle square.

$\pi_Z \colon X \times Z \to Z$ is epimorphic, then $\pi_Z = Coker(l_X)$ and (1) is a pushout. Accordingly, the stability of regular epis by pushouts implies that $f \times 1_Z$ is an epi.

(2) The arrow $f \times f'$ is the composite of the epimorphisms $(1_Y \times f') \circ (f \times 1_{X'})$. □

6.22. Proposition.

(1) *Any pullback of an epi $f \colon A \to B$ along a monomorphism $m \colon B' \to B$*

$$
\begin{array}{ccc}
A \times_B B' & \xrightarrow{\ f'\ } & B' \\
{\scriptstyle m'}\downarrow & & \downarrow{\scriptstyle m} \\
A & \xrightarrow[\ f\ ]{} & B
\end{array}
$$

is a pushout.

(2) *Epimorphisms are stable under pullbacks along monomorphisms.*

Proof.

(1) Let $q \colon B \to Cok[m]$ be the cokernel of m. The fact that m is a monomorphism implies that m is the kernel of q, and then m' is $Ker(q \circ f)$, with $q \circ f$ an epimorphism. It follows that $q \circ f$ is the cokernel of m':

$$
\begin{array}{ccccccccc}
0 & \longrightarrow & A \times_B B' & \xrightarrow{\ m'\ } & A & \xrightarrow{\ q \circ f\ } & Cok[m] & \longrightarrow & 0 \\
 & & \downarrow{\scriptstyle f'} & & \downarrow{\scriptstyle f} & & \parallel & & \\
0 & \longrightarrow & B' & \xrightarrow[\ m\ ]{} & B & \xrightarrow[\ q\ ]{} & Cok[m] & \longrightarrow & 0
\end{array}
$$

with (1) in the middle square.

By Corollary 6.20, the square (1) is a pushout.

(2) It follows by (1) and by Corollary 6.14. □

Thanks to Theorems 6.19 and 6.22, in order to prove that regular epimorphisms are pullback stable, it will be sufficient to show that the pullback of an epimorphism along an epimorphism is an epimorphism:

6.23. Proposition. *Epimorphisms are stable under pullbacks along epimorphisms.*

Proof. Consider a pullback

$$
\begin{array}{ccc}
P & \xrightarrow{g'} & A \\
{\scriptstyle f'}\downarrow & & \downarrow{\scriptstyle f} \\
B & \xrightarrow{g} & C
\end{array}
$$

where both g and f are regular epis. We shall prove that g' is an epi. The commutative square above is a pullback precisely when the commutative square

$$
\begin{array}{ccc}
P & \xrightarrow{(g',f')} & A \times B \\
{\scriptstyle h}\downarrow & & \downarrow{\scriptstyle f\times g} \\
B & \xrightarrow{s_0} & C \times C
\end{array}
$$

where $h = g \circ f' = f \circ g'$, is itself a pullback. Now, by Theorems 6.21 and 6.22, h is a regular epimorphism, being the pullback of the epimorphism $f \times g$ along the monomorphism s_0. Moreover, the kernel of h is $K[f] \times K[g]$, since $K[f] \times K[g]$ is the kernel of $f \times g$. By applying Corollary 6.20 to the diagram

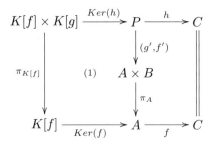

it follows that (1) is a pushout and then $\pi_A \circ (g', f') = g'$ an epimorphism, since $\pi_{K[f]}$ is a (split) epimorphism.   □

6.24. Theorem. *Any abelian category is Barr-exact and additive.*

Proof. By 6.7, 6.9, 6.19, 6.22 and 6.23 any abelian category is regular. By 6.17, $\mathcal{C}$ is additive, and then by 4.3 $\mathcal{C}$ is Barr-exact.   □

References

[1] M. Barr, *Exact categories*, Springer Lecture Notes in Mathematics 236 (1971) 1–120.

[2] F. Borceux, *Handbook of Categorical Algebra*, Cambridge University Press (1994).

[3] D. Bourn, *Normalization Equivalence, Kernel Equivalence and Affine Categories*, Springer Lecture Notes in Mathematics 1488 (1991) 43–62.

[4] D. Bourn, *Mal'cev categories and fibration of pointed objects,* Appl. Categ. Structures 4 (1996) 307–327.

[5] D. Bourn, *Normal subobjects and abelian objects in protomodular categories,* J. Algebra 228 (2000) 143–164.

[6] D. Bourn, 3×3 *Lemma and Protomodularity,* J. Algebra 236 (2001) 778–795.

[7] D. Bourn, *The denormalized* 3×3 *lemma,* J. Pure Appl. Algebra 177 (2003) 113–129.

[8] D. Bourn, *Intrinsic centrality and associated classifying properties,* J. Algebra 256 (2002) 126–145.

[9] D. Buchsbaum, *Exact categories and duality,* Trans. Amer. Math. Soc. 80 (1955) 1–34.

[10] A. Carboni, J. Lambek, M.C. Pedicchio, *Diagram chasing in Mal'cev categories,* J. Pure Appl. Algebra 69 (1991) 271–284.

[11] A. Carboni, G.M. Kelly, M.C. Pedicchio, *Some remarks on Maltsev and Goursat categories,* Appl. Categ. Structures 1 (1993) 385–421.

[12] P. Freyd, *Abelian categories,* Harper and Row, (1964).

[13] P. Gabriel, F. Ulmer, *Lokal präsentierbare Kategorien,* Springer Lecture Notes in Mathematics 221 (1971).

[14] P. Grillet, *Regular categories,* Springer Lecture Notes in Mathematics 236 (1971).

[15] P.J. Higgins, *Groups with multiple operators,* Proc. London Math. Soc. (3) 6 (1956) 366–416.

[16] G. Janelidze, L. Marki, W. Tholen, *Semi-abelian categories,* J. Pure Appl. Algebra 168 (2002) 367–386.

[17] S. Mac Lane, *Categories for the Working Mathematician,* Second Ed., Springer (1998).

Université du Littoral,
Lab. Math. Pures Appl. J. Liouville,
50 av. Ferdinand Buisson BP 699
62228 Calais, France
E-mail: bourn@lmpa.univ-littoral.fr
        gran@lmpa.univ-littoral.fr

V

Aspects of Monads

John MacDonald and Manuela Sobral

The objective of this chapter is to give the reader an understanding of monads and their associated algebraic structures, as well as some of their applications.

Monads first arose explicitly in homological algebra in the late 1950s and early 1960s. Comonads, i.e. monads in the dual of a category, first appear in Godement [15] who used the flabby sheaf comonad for resolving sheaves to compute sheaf cohomology. These structures were formalized and then applied to homotopy theory and homology theory by Huber, Barr, and Beck in [17, 4, 8]. In fact the history of the subject has been closely connected with applications. The applications to algebraic structures were initiated by the description of the algebras for a monad by Eilenberg and Moore [14] and closely related work of Kleisli [19].

Those mathematical structures which can be defined using only operations and equations have always attracted the mathematicians by their beauty, simplicity and wide range of applications. The first published proof that monads capture equational classes of algebras is the isomorphism theorem of Linton [23], extending Lawvere's description of algebraic theories [21]. Beck realized that monads describe universal algebra in the category of sets and proved his famous monadicity theorem [8].

More recently important applications of monads to topos theory have been discovered (see e.g. [5]). They have also been used in applications to computer science by Manes [30, 31] and later by Barr and Wells [6] and Moggi [33].

Monads were first known as standard constructions and later as triples.

The approach we take is that monads can be best understood, initially, as monoids generalized to the setting of endofunctors, as we indicate in the following description of the chapter contents.

We begin in section 1 with monoids, which are simply sets with an associative binary operation with identity. This definition, when expressed by a commutative diagram, can be used to define monoids in other categories thus leading to the definition of generalized monoid. A monad on a category can then be described as a generalized monoid of endofunctors.

The correspondence between monads and adjoint functors is given, including the universal properties of the Eilenberg-Moore and Kleisli categories. Limits and colimits in Eilenberg-Moore categories are described and a number of examples of monads are given.

For the subsequent sections we just mention some of the highlights. Section 2 considers the question of when a category **A** is monadic over a category **X**, that is, when it is equivalent to the category of Eilenberg-Moore algebras for some monad on **X**. Beck's criterion for monadicity is given in detail and applied to show, among other examples, that varieties of general algebras and the category of frames are monadic over **Set**.

Other examples are considered which arise from adjunctions quite distinct from those of the algebraic "free-forgetful" type. These lead to important and unexpected applications of monads, as do the cases where operations and equations are defined over a base category different from **Set**, as in the category of categories with equalizers over the category of graphs. Duskin's criterion for monadicity and a characterization of the monadic categories over **Set** are given. A monadicity criterion due to Janelidze and an application proving a theorem of Joyal and Tierney are then presented.

In section 3 we examine conditions for the monicity of the unit morphism for monads defined on a specific category **C** defined by operations and equations. We begin by presenting the answer when **C** = **Set**, a result going back to the thesis of Manes. A general set of conditions for monicity is then developed and applied to show how the Birkhoff-Witt Theorem for Lie-algebras and the Schreier Theorem for groups follow. A connection with coherence is indicated.

Section 4 describes Kleisli triples in a formulation due to Manes. This is an alternative description of monad used by many computer scientists. We show exactly how these two notions correspond and give an alternative description of the Kleisli category used as well. Some examples and a theorem illustrating this point of view are also given.

In section 5 we describe Eilenberg-Moore and Kleisli objects for a monad defined on an object of a 2-category, using results of Kelly and Street [18] as a point of departure. In the first part we consider algebras for an endomorphism (i.e. a 1-cell) in a 2-category. This is closely related to some ideas appearing in the chapter on algebras for an endofunctor in Barr-Wells [6]. Later in the section we illustrate some of the ideas by showing how the co-Kleisli object of a slice 2-category is related to the co-Kleisli object of a 2-category when a monad is defined on an object of the slice.

In section 6 we show further how the notions of monoid and monad are connected by showing how the category of monads, the category of commutative monads and the category of idempotent monads may each be regarded as the category of algebras for a suitable strict 2-monad (different in each case). Some of these ideas are derived from the thesis of Peter Stone [37] and others go back earlier. They also involve a notion of canonical form for morphisms in the simplicial category Δ in one case (Mac Lane [29]) and in the other two cases involve a similar canonical form for morphisms of categories closely related to Δ.

1. Monoids and monads

1.1. Monoids. Defining a monoid $\mathbf{M} = (M, e, m)$, one usually writes $e = 1$ for the identity element and $m(x, y) = xy$ for the multiplication with the familiar axioms

$$x(yz) = (xy)z \quad \text{and} \quad 1x = x = x1, \tag{1}$$

which otherwise would be written as

$$m(x, m(y, z)) = m(m(x, y), z) \quad \text{and} \quad m(e, x) = x = m(x, e). \tag{2}$$

A third way to express these axioms is to display them as the commutative diagram

$$
\begin{array}{ccccc}
M \times M \times M & \xrightarrow{m \times 1_M} & M \times M & \xleftarrow{e \times 1_M} & M \\
{\scriptstyle 1_M \times m} \downarrow & & \downarrow {\scriptstyle m} & {\scriptstyle 1_M} & \downarrow {\scriptstyle 1_M \times e} \\
M \times M & \xrightarrow{m} & M & \xleftarrow{m} & M \times M
\end{array}
\tag{3}
$$

where the maps are defined by $(m \times 1_M)(x, y, z) = (m(x, y), z)$, $(1_M \times m)(x, y, z) = (x, m(y, z))$, $(e \times 1_M)(x) = (e, x)$ and $(1_M \times e)(x) = (x, e)$, for $x, y, z \in M$.

This diagram shows that the definition of monoid in fact uses a "higher order monoid structure" given by the cartesian product of sets. It satisfies the monoid axioms up to "good" isomorphisms: the associativity isomorphism of the cartesian product $a : (M \times M) \times M \cong M \times (M \times M)$ and the isomorphisms $l : 1 \times M \cong M$ and $r : M \times 1 \cong M$ defined by the projections of the cartesian product with the one point set 1.

Any monoid determines a functor $T = M \times - : \mathbf{Set} \to \mathbf{Set}$ and natural transformations $\eta : 1_{\mathbf{Set}} \to T$ defined by $\eta_X(x) = (e, x)$ and $\mu : T^2 \to T$ defined by $\mu_X(n, n', x) = (nn', x)$ making the diagram

$$
\begin{array}{ccccc}
T \circ T \circ T & \xrightarrow{\mu T} & T \circ T & \xleftarrow{\eta T} & T \\
{\scriptstyle T\mu} \downarrow & & \downarrow {\scriptstyle \mu} & {\scriptstyle 1_T} & \downarrow {\scriptstyle T\eta} \\
T \circ T & \xrightarrow{\mu} & T & \xleftarrow{\mu} & T \circ T
\end{array}
\tag{4}
$$

commute.

This follows from the commutativity of (3). Furthermore, (4) is the "same" as (3) with the cartesian product replaced by the functor composition. Indeed, both systems $M = (M, e, m)$ and $T = (T, \eta, \mu)$ are examples of the more general concept of *a monoid in a monoidal category* . These are categories equipped with an additional structure providing the proper context for the definition of monoid and they are studied in detail in Section 6.

The monoidal category $(\mathbf{Set}, \times, 1, a, l, r)$ provides the proper setting or context for definition (3).

Replacing M by an endofunctor, the cartesian product by the composition of functors and the one element set by the identity functor, thus making the transition from (3) to (4), we find that (T, η, μ) is a monoid in the *strict* monoidal category $(\mathbf{Set}^{\mathbf{Set}}, \circ, \mathrm{Id}, 1, 1, 1)$, in the sense that a, l, and r are identity transformations.

No intrinsic structure of sets is used here. We only need to know that sets and functions form a category. Then we use the properties of functors to define the category of endofunctors of $\mathbf{Set}$. In the same way, for each category $\mathbf{X}$ we have the strict monoidal category $(\mathbf{X}^{\mathbf{X}}, \circ, \mathrm{Id})$ of its endofunctors.

Monoids in the category of endofunctors of $\mathbf{X}$ are called *monads*. That is, monads are defined by analogy with monoids, or, more precisely, as a special case of a generalized monoid.

1.2. Monads. A monad on a category $\mathbf{X}$ is a system $T = (T, \eta, \mu)$ consisting of a functor $T : \mathbf{X} \to \mathbf{X}$ and natural transformations $\eta : 1_{\mathbf{X}} \to T$ and $\mu : T^2 \to T$ making the diagram

$$
\begin{array}{ccccc}
T^3 & \xrightarrow{\mu T} & T^2 & \xleftarrow{\eta T} & T \\
{\scriptstyle T\mu}\downarrow & & {\scriptstyle \mu}\downarrow \;\; \swarrow {\scriptstyle 1_T} & & \downarrow {\scriptstyle T\eta} \\
T^2 & \xrightarrow{\mu} & T & \xleftarrow{\mu} & T^2
\end{array}
\tag{5}
$$

commute.

Examples. (1) *The power set monad.* This monad $\mathcal{P} = (\mathcal{P}, \eta, \mu)$ is determined by the following data.

– The power set functor $\mathcal{P} : \mathbf{Set} \to \mathbf{Set}$ which assigns to each set X the set $\mathcal{P}X$ of all subsets of X, i.e. $\mathcal{P}(X) \cong 2^X$, and to each function $f : X \to Y$ the *direct image* map $\mathcal{P}f : \mathcal{P}X \to \mathcal{P}Y$ defined by $\mathcal{P}(f)(S) = f(S)$ for each subset S of X.

– Insertion and union. There are two natural transformations η and μ associated with $\mathcal{P}$ called the *insertion* and the *union*. The insertion $\eta_X : X \to \mathcal{P}X$ takes value $\{x\}$ on $x \in X$. The union $\mu : \mathcal{PP}X \to \mathcal{P}X$ takes each set $\{A_i\}_{i \in I}$ of subsets of X to its union $\cup A_i$ in $\mathcal{P}X$.

(2) *The list monad.* This monad $\mathcal{L} = (\mathcal{L}, \eta, \mu)$ is defined by

– The list functor $\mathcal{L} : \mathbf{Set} \to \mathbf{Set}$ which sends each set X to the set $\mathcal{L}X$ of all finite strings of elements of X, including the empty string. If $f : X \to Y$, then $\mathcal{L}f : \mathcal{L}X \to \mathcal{L}Y$ is defined by $\mathcal{L}f(x_1 x_2 \cdots x_n) = (f x_1 f x_2 \cdots f x_n)$.

– Embedding of sets into strings of length one and the merging of strings of strings to strings. There is a natural map $\eta_X : X \to \mathcal{L}X$ taking each element to itself considered as a string of length one. As soon as we apply $\mathcal{L}$ twice, then we have strings of strings and by removing inner brackets we get a natural map μ_X : $\mathcal{LLX} \to \mathcal{LX}$.

(3) *Adding an element.* Let $A : \mathbf{Set} \to \mathbf{Set}$ be defined on objects by $A(X) = X + 1$, the disjoint union (=coproduct) of X with the one element set. Then there are natural maps $\eta_X : X \to X + 1$ and $\mu_X : (X + 1) + 1 \to X + 1$. The morphisms η_X are coproduct injections and the morphisms μ_X collapse two copies of 1 to one copy.

(4) *Closure operations.* On a poset X a monotone function $C : X \to X$ with the properties $x \le C(x)$ and $C^2(x) \le C(x)$ for all $x \in X$ is called a *closure operation*. Indeed, it is an idempotent closure operation since C monotone implics $C^2 = C$. Considering X as a category, then C is a functor and the two properties define natural transformations $\eta : \mathrm{Id} \to C$ and $\mu : C^2 \to C$.

The following example shows how wide is the generalization of monoid just described.

Proposition. *Any adjunction* $(F, G, \eta, \varepsilon) : \mathbf{X} \rightharpoonup \mathbf{A}$ *determines a monad* $T = (T, \eta, \mu)$ *on the category* $\mathbf{X}$, *where*

(1) $T = GF : \mathbf{X} \to \mathbf{X}$;
(2) $\eta : 1_{\mathbf{X}} \to T$ *is the unit of the adjunction;*
(3) $\mu = G\varepsilon F : T^2 \to T$.

Proof. The commutativity of (5) easily follows from the naturality of ε and the triangular identities of the adjunction. □

1.3. The Eilenberg-Moore construction. To an arbitrary monoid $M = (M, e, m)$ we can associate the category M-**Set** of M-actions (on the left) of M on sets and of equivariant maps. This is the category of pairs $(X, \xi : M \times X \to X)$ such that $ex = x$ and $m(nx) = (mn)x$, where $\xi(m, x) = mx$, with morphisms $f : (X, \xi) \to (Y, \theta)$ all maps $f : X \to Y$ such that $f \cdot \xi = \theta \cdot (1 \times f)$.

Considering an arbitrary monad on any category as a generalized monoid, do we still have anything like the T-actions for the monoid T? Yes, we do – they are exactly the T-*structure maps* and the corresponding M-sets are the Eilenberg-Moore algebras. They are the pairs (X, ξ), where X is an object and $\xi : TX \to X$ is a morphism of $\mathbf{X}$, such that

$$\xi \cdot \eta_X = 1_X \text{ and } \xi \cdot \mu_X = \xi \cdot T\xi.$$

Morphisms of T-algebras $f : (X, \xi) \to (Y, \theta)$ are $\mathbf{X}$-morphisms which are compatible with the T-action in an appropriate sense: they are the morphisms f such that $f \cdot \xi = \theta \cdot Tf$. In this context the category M-**Set** is replaced by the Eilenberg-Moore category of T-algebras which will be denoted by $\mathbf{X}^T$.

Examples. We may identify the Eilenberg-Moore categories for the monads presented in 1.2 as follows:

(1) The $\mathcal{P}$-algebras are the complete lattices – $(X, \xi : \mathcal{P}(X) \to X)$ where $\xi(S)$ is the supremum of $S \subseteq X$ – and the morphisms are the maps that preserve arbitrary suprema, the so called category of sup-complete semilattices.

(2) The $\mathcal{L}$-algebras are the monoids: the sets equipped with a structure map $\xi : \mathcal{L}(X) \to X$, where $\mathcal{L}(X) = X^*$ is the free monoid generated by X, are monoids and their morphisms are monoid homomorphisms.

(3) The category of pointed sets and functions preserving the base point, for the monad obtained by adding an element.

(4) The category of the closed elements for a closure operation.

Again, if $M = (M, e, m)$ is an arbitrary monoid the forgetful functor $G : M\text{-}\mathbf{Set} \to \mathbf{Set}$ has a left adjoint F defined by

$$F(X) = (M \times X, \ \xi : M \times M \times X \to M \times X), \ \text{with } \xi(m, n, x) = (mn, x),$$

and $F(f) = 1_M \times f$; and the free-forgetful adjunction between $M\text{-}\mathbf{Set}$ and $\mathbf{Set}$

$$M\text{-}\mathbf{Set} \underset{F}{\overset{G}{\rightleftarrows}} \mathbf{Set}$$

induces in $\mathbf{Set}$ the monad considered in 1.1.

For generalized monoids of endofunctors we have a similar result:

Proposition. *For a monad $T = (T, \eta, \mu)$ on $\mathbf{X}$ there is a free-forgetful adjunction*

$$\mathbf{X}^T \underset{F^T}{\overset{G^T}{\rightleftarrows}} \mathbf{X}$$

which induces the monad T in $\mathbf{X}$.

Proof. The functor defined by $F^T(X) = (TX, \mu_X : T^2X \to TX)$ and $F^T(f) = T(f)$ is the left adjoint of G^T with unit $\eta_X : X \to G^T F^T(X) = TX$. Indeed, (TX, μ_X) is a T-algebra by the defining axioms of the monad and Tf is a T-morphism by the naturality of μ.

Let us prove now that η_X is universal from X to G^T. For each morphism $f : X \to G^T(Y, \theta)$, there is a T-morphism $\overline{f} = \theta \cdot Tf$ such that $\overline{f} \cdot \eta_X = f$. Furthermore it is unique because, if $g \cdot \eta_X = f$ for a T-morphism g, then

$$g = g \cdot \mu_X \cdot T\eta_X = \theta \cdot Tg \cdot T\eta_X = \theta \cdot Tf.$$

The counit of the adjunction is $\varepsilon^T = \xi : (TX, \mu_X) \to (X, \xi)$ for each T-algebra (X, ξ). Thus, $T = (G^T F^T, \eta, G^T \varepsilon^T F^T = \mu)$. $\square$

1.4. Limits in Eilenberg-Moore categories. Let $T = (T, \eta, \mu)$ be a monad on a category $\mathbf{X}$. If $\mathbf{X}$ has a certain type of limit, so does its Eilenberg-Moore category of algebras. In particular $\mathbf{X}^T$ is at least as complete as $\mathbf{X}$ is.

Theorem. *For each monad $T = (T, \eta, \mu)$ on $\mathbf{X}$, the forgetful functor $G^T : \mathbf{X}^T \to \mathbf{X}$ creates all limits.*

Proof. Given a diagram $D : \mathbf{I} \to \mathbf{X}^T$ let $\alpha : L \to G^T \circ D$ be the limit of $G^T \circ D$ in $\mathbf{X}$. Let us denote $D(i)$ by (X_i, ξ_i), for each object i of $\mathbf{I}$. Since the morphisms $\xi_i \cdot T\alpha_i$ form a cone in $\mathbf{X}$ from TL to $G^T \circ D$, there exists a unique $\mathbf{X}$-morphism ξ such that $\alpha_i \cdot \xi = \xi_i \cdot T\alpha_i$.

We are going to show that (L, ξ) is a T-algebra and that $\alpha_i : (L, \xi) \to D(i)$ are the elements of the limit cone of D in $\mathbf{X}^T$.

$-$ We have that $\xi \cdot \eta_L = 1_L$ because

$$\alpha_i \cdot (\xi_i \cdot \eta_L) = \xi_i \cdot T\alpha_i \cdot \eta_L = \xi \cdot \eta_{X_i} \cdot \alpha_i = \alpha_i$$

and, by the universal property of the limit, the morphisms α_i are jointly monomorphic.

$-$ The commutativity of the diagram

$$
\begin{array}{ccc}
T^2 L & \xrightarrow{\;T^2 \alpha_i\;} & T^2 X_i \\[2pt]
{\scriptstyle T\xi}\big\Vert\;{\scriptstyle \mu_L} & & {\scriptstyle T\xi_i}\big\Vert\;{\scriptstyle \mu_{X_i}} \\[2pt]
TL & \xrightarrow{\;T\alpha_i\;} & TX_i \\[2pt]
{\scriptstyle \xi}\big\downarrow & & \big\downarrow\,{\scriptstyle \xi_i} \\[2pt]
L & \xrightarrow[\;\alpha_i\;]{} & X_i
\end{array}
\tag{6}
$$

and the universal property of the limit tell us that

$$\xi \cdot \mu_L = \xi \cdot T\xi.$$

Therefore, α_i is a T-morphism from $(L, \xi) \to (X_i, \xi_i)$.

$-$ If $\beta_i : (Y, \theta) \to (X_i, \xi_i)$ are the elements of a cone in $\mathbf{X}^T$ and $f : Y \to L$ there is a unique $\mathbf{X}$-morphism such that $\alpha_i \cdot f = \beta_i$, for each i in $\mathbf{I}$. Since, for each i,

$$\alpha_i \cdot f \cdot \theta = \alpha_i \cdot \xi \cdot Tf,$$

using again the universal property of the product, we conclude that $f \cdot \theta = \xi \cdot Tf$. Thus $\alpha_i \cdot f = \beta_i$ in $\mathbf{X}^T$ and f is unique such by the faithfulness of G^T. $\qquad\square$

1.5. Colimits in Eilenberg-Moore categories. The behavior of $G^T : \mathbf{X}^T \to \mathbf{X}$ with respect to colimits is not as good as it is for limits: the existence of a certain type of colimit on $\mathbf{X}$ does not imply, in general, its existence in $\mathbf{X}^T$. This was first proved by J. Adámek in [1], exhibiting an example of an Eilenberg-Moore category of algebras over a cocomplete category which is not cocomplete.

However, various sufficient conditons for cocompleteness of $\mathbf{X}^T$ when $\mathbf{X}$ is cocomplete have been found which cover essentially all the important cases.

In the sequel, we are mainly interested in the existence in $\mathbf{X}^T$ of (certain) colimits which are *absolute colimits* in $\mathbf{X}$ (i.e. those which are preserved by any functor). The following result guarantees their existence:

Theorem. *The forgetful functor $G^T : \mathbf{X}^T \to \mathbf{X}$ creates all those colimits which exist in $\mathbf{X}$ and are preserved by T and T^2.*

Proof. Given a diagram $D : \mathbf{I} \to \mathbf{X}^T$, let $\alpha : G^T \circ D \to L$ be the colimit of $G^T \circ D$ in $\mathbf{X}$. Like in 1.4, we take $D(i) = (X_i, \xi_i)$. Since $\alpha_i \cdot \xi_i$ are the components of a cone from $T \circ G^T \circ D$ to L in $\mathbf{X}$ and $T\alpha_i$ are the elements of a colimit cone, there exists a unique $\mathbf{X}$-morphism ξ such that $\xi \cdot T\alpha_i = \alpha_i \cdot \xi_i$.

Using the fact that each family $T(\alpha_i)$ and $T^2(\alpha_i)$, for $i \in \mathbf{I}$, is jointly epimorphic since it is a colimit, one easily concludes that (L, ξ) is a T-algebra and that $\alpha_i : D(i) \to (X, \xi)$ are the elements of the colimit cone of D in $\mathbf{X}^T$. $\qquad\square$

Corollary. *The functor $G^T : \mathbf{X}^T \to \mathbf{X}$ creates absolute colimits.*

1.6. The Kleisli construction. Let T be a monad on a category $\mathbf{X}$. Then the *Kleisli category* $\mathbf{X}_T$ is given as follows. It has the same objects as $\mathbf{X}$ and for morphisms $f : X \to Y$ the morphisms $X \to TY$ in $\mathbf{X}$. Given $g : Y \to Z$ in $\mathbf{X}_T$, the composite $gf : X \to Z$ in $\mathbf{X}_T$ is the following $\mathbf{X}$ morphism

$$X \xrightarrow{\ f\ } TY \xrightarrow{\ Tg\ } TTZ \xrightarrow{\ \mu_Z\ } TZ$$

Then $\eta_X : X \to TX$ turns out to be the identity morphism for X in $\mathbf{X}_T$.

Examples. (1) *The category* **Rel** *of relations.* In traditional approaches a relation from A to B is a subset of the product $A \times B$ and has been denoted in various ways by $i : X \hookrightarrow A \times B$, $i : A \rightharpoonup B$, or

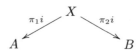

A function is then just a special kind of relation with a unique pair (a, b) in X for each a in A.

Composition of relations $i : A \rightharpoonup B$ and $j : B \rightharpoonup C$ can then be described directly in terms of elements of $A \times C$ as follows: $(a, c) \in ji : A \rightharpoonup C$ if there exists $b \in B$ such that $(a, b) \in i$ and $(b, c) \in j$. Alternatively, the composition can be described as the relation given by constructing the pullback diagram on $\pi_2 i$ and $\pi_1 j$ in the following diagram

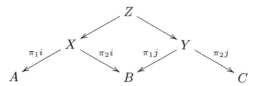

and take the image of the resulting morphism in $A \times C$.

Lemma. **Rel** *is the Kleisli category for the power set monad* $\mathcal{P} = (\mathcal{P}, \eta, \mu)$.

Proof. The Kleisli category $\mathbf{Set}_{\mathcal{P}}$ has sets $A, B, C, \cdots$ for objects. A morphism $A \to B$ in $\mathbf{Set}_{\mathcal{P}}$ is a function $A \to \mathcal{P}B$. This corresponds to the relation $i : A \rightharpoonup B$ which, as a part $i : X \hookrightarrow A \times B$, consists of all pairs (a, b) such that $b \in fa$. Conversely, given such an i, we define $f : A \to \mathcal{P}B$ by letting fa be the part of B consisting of all b with $(a, b) \in X$. The composite $gf : A \to B \to C$ in $\mathbf{Set}_{\mathcal{P}}$ is the composite $\mu_C \cdot \mathcal{P}g \cdot f : A \to \mathcal{P}B \to \mathcal{P}^2C \to \mathcal{P}C$. This determines the same part of $A \times C$ as the composition in $\mathbf{Rel}$ given previously using pullbacks. $\qquad \square$

(2) *The Kleisli category for the list monad.* Morphisms $f : A \to \mathcal{L}B$ and $g : B \to \mathcal{L}C$ are A-tuples $\{f_a | a \in A\}$ of strings in B and B-tuples $\{g_b | b \in B\}$ of strings in C whose composition gf in $\mathbf{Set}_{\mathcal{L}}$ takes each a to a string in C in the following steps:

- For a in A let $b_1 b_2 \cdots b_n$ be the string f_a in B.
- $\mathcal{L}(g)(f(a)) = g(b_1)g(b_2) \cdots g(b_n)$, an element of $\mathcal{L}^2 C$, that is, a list of lists in C.
- Finally, applying μ_C we obtain the string $(g \circ f)_a$ in C by removing inner brackets, as in 1.2(2).

For a given monad T there exists the *Kleisli adjunction* $(F_T, G_T, \eta_T, \varepsilon_T) : \mathbf{X} \rightharpoonup \mathbf{X}_T$ which, like the Eilenberg-Moore adjunction described earlier, induces the monad T on $\mathbf{X}$. On objects let $F_T X = X$ and on morphisms $F_T(f : X \to Y) : X \to Y$ in $\mathbf{X}_T$ is the $\mathbf{X}$-morphism

$$(X \xrightarrow{F_T f} TY) = (X \xrightarrow{f} Y \xrightarrow{\eta_Y} TY).$$

Similarly $G_T X = TX$ and given $g : X \to Y$ in $\mathbf{X}_T$, which is $g : X \to TY$ in $\mathbf{X}$, we let

$$G_T g = (TX \xrightarrow{Tg} T^2 Y \xrightarrow{\mu_Y} TY)$$

in $\mathbf{X}$. The unit $\eta_T : 1 \to U_T F_T = T$ is η and the counit $\varepsilon_T : F_T U_T \to 1_{\mathbf{X}_T}$ is given by $\varepsilon_T X : F_T U_T X = F_T TX = TX \to X$ in $\mathbf{X}_T$ which is just $1_{TX} : TX \to TX$ as an $\mathbf{X}$-morphism.

1.7. The category of adjunctions for a given monad. Proposition 1.2 tells us that every adjunction $(F, G, \eta, \varepsilon) : \mathbf{X} \rightharpoonup \mathbf{A}$ induces a monad $T = (GF, \eta, G\varepsilon F)$ on $\mathbf{X}$. Conversely each monad on $\mathbf{X}$ is induced by the free-forgetful adjunction from the Eilenberg-Moore category $\mathbf{X}^T$ as well as from the Kleisli adjunction.

For a monad $T = (T, \eta, \mu)$ on $\mathbf{X}$ let T-**Adj** be the category with objects all adjunctions $(F, G, \eta, \varepsilon) : \mathbf{X} \rightharpoonup \mathbf{A}$ such that $GF = T$ and $G\varepsilon F = \mu$. A morphism in T-**Adj** from $(F, G, \eta, \varepsilon) : \mathbf{X} \rightharpoonup \mathbf{A}$ to $(F_1, G_1, \eta, \varepsilon_1) : \mathbf{X} \rightharpoonup \mathbf{A}_1$ is a functor $L : \mathbf{A} \to \mathbf{A}_1$ such that $G_1 L = G$ and $LF = F_1$.

The Kleisli and Eilenberg-Moore adjunctions are the "minimal" and the "maximal" solution to the problem of finding an adjunction which induces a given monad in the following sense:

Theorem. *The category T-$\mathbf{Adj}$ has an initial object and a terminal object. The initial object is the Kleisli adjunction and the terminal object is the Eilenberg-Moore adjunction.*

Proof. Given an adjunction $(F, G, \eta, \varepsilon) : \mathbf{X} \rightharpoonup \mathbf{A}$, we have $L : \mathbf{X}_T \to \mathbf{A}$ defined by $LX = FX$ on objects and by $L(f : X \to Y) = \varepsilon_{FY} \cdot Ff : FX \to FGFY \to FY$ on morphisms. Then L is the unique morphism in T-$\mathbf{Adj}$ from $(F_T, G_T, \eta_T, \varepsilon_T)$: $\mathbf{X} \rightharpoonup \mathbf{X}_T$ to $(F, G, \eta, \varepsilon) : \mathbf{X} \rightharpoonup \mathbf{A}$.

Similarly $K : (F, G, \eta, \varepsilon) \to (F^T, G^T, \eta^T, \varepsilon^T)$ is determined by $K : \mathbf{Y} \to \mathbf{X}^T$ with $KY = (GY, G\varepsilon_Y)$ and $Kf = Gf$. The functor K is called the *Eilenberg-Moore comparison functor*.   □

We leave it as an exercise to prove that the functor L is full and faithful and that its image is the full subcategory of $\mathbf{A}$ generated by the image of F. Note that this implies, in particular, that the comparison functor $K : \mathbf{X}_T \to \mathbf{X}^T$ is full and faithful.

2. Conditions for monadicity

2.1. Monadic functors. Many familiar categories including those of groups, rings and all other varieties of universal algebras are of the form $\mathbf{Set}^T$ for the monad T induced by the corresponding free-forgetful adjunction. But there are adjunctions $(F, G; \eta, \varepsilon) : \mathbf{X} \rightharpoonup \mathbf{A}$ whose corresponding monads do not carry any information about the category $\mathbf{A}$. Among them are not only the trivial examples where $\mathbf{X} = \mathbf{1}$ but also several free-forgetful adjunctions such as the one where $\mathbf{A}$ is the category **Top** of topological spaces and continuous maps and $\mathbf{X} = \mathbf{Set}$.

However, as explained in Theorem 1.7, for every adjunction $(F, G; \eta, \varepsilon) : \mathbf{X} \rightharpoonup \mathbf{A}$ there exists a unique functor $K : \mathbf{A} \to \mathbf{X}^T$ with $G^T \cdot K = G$ and $K \cdot F = F^T$, the so called *comparison functor*. In the cases where K is an equivalence, like in the algebraic free-forgetful adjunctions mentioned above, the functor G is said to be *monadic*.

A category $\mathbf{A}$ is said to be *monadic over* $\mathbf{X}$ whenever it is equivalent to $\mathbf{X}^T$ for some monad T in $\mathbf{X}$. In this case, $\mathbf{A}$ comes equipped with a monadic functor $G : \mathbf{A} \to \mathbf{X}$ which, in particular, is a faithful right adjoint functor.

Historically, monads appeared before their algebras were known. However, here we are going to use the forgetful functor of algebraic varieties, whose monadicity will be proved later, as the guiding example for characterizing monadic functors.

The forgetful functor of every variety of universal algebras $G : \mathbf{Alg}(\Omega, E) \to \mathbf{Set}$ reflects isomorphisms. This is exactly what we mean when we say that isomorphisms, for example, of groups are the homomorphisms whose underlying functions are bijective. Forgetful functors of the form $G^T : \mathbf{X}^T \to \mathbf{X}$, for an arbitrary category **X**, also satisfy this condition. Indeed, for $f : (X, \xi) \to (Y, \theta)$, if $G^T(f)$ is an isomorphism in **X** then $g = f^{-1}$ is also a morphism of $\mathbf{X}^T$:

$$\xi \cdot Tg = g \cdot f \cdot \xi \cdot Tg = g \cdot \theta \cdot T(f \cdot g) = g \cdot \theta.$$

Since equivalences of categories reflect isomorphisms this proves that:

Proposition. *Monadic functors reflect isomorphisms.*

This already enables us to prove that some functors are *not* monadic. This is the case of the underlying functor from the category **Pos** of partially ordered sets and monotone functions to **Set** which does not reflects isomorphisms.

However, this is clearly not enough to characterize monadic functors. The crucial fact for such characterization is that *algebra structures are coequalizers*. Like in varieties of general algebras, each algebra is a quotient of a free algebra. Indeed, the T-structure map of each algebra is the coequalizer of a pair of morphisms between free algebras. Furthermore, the image of this coequalizer diagram in the base category has important features that we describe next.

For each T-algebra (X, ξ), we have the equality $\xi \cdot \mu_X = \xi \cdot T\xi$ in $\mathbf{X}^T$ and the image of this diagram in **X**

$$T^2 X \underset{T\xi}{\overset{\mu_X}{\rightrightarrows}} TX \xrightarrow{\xi} X \tag{7}$$

comes equipped with two morphisms $\eta_X : X \to TX$ and $\eta_{TX} : TX \to T^2 X$ satisfying the following conditions:

- $\xi \cdot \eta_X = 1$;
- $\mu_X \cdot \eta_{TX} = 1$;
- $T\xi \cdot \eta_{TX} = \eta_X \cdot \xi$.

Considering the obvious abstract situation we will prove that ξ is the coequalizer of the pair $(\mu_X, T\xi)$ in $\mathbf{X}^T$.

2.2. Split forks and contractible coequalizers. In a category a *fork* is a diagram

$$X \underset{g}{\overset{f}{\rightrightarrows}} Y \xrightarrow{h} Z \tag{8}$$

such that $h \cdot f = h \cdot g$. Such a fork is called a *split fork* if there exist two morphisms $i : Z \to Y$ and $j : Y \to X$ such that $h \cdot i = 1$, $f \cdot j = 1$ and $g \cdot j = i \cdot h$; (i, j) is also called a splitting of the fork.

Examples. (1) Given an adjoint situation $(F, G; \eta, \varepsilon) : \mathbf{X} \rightharpoonup \mathbf{A}$, for each object A in **A**, the fork $(GFG\varepsilon_A, G\varepsilon_{FGA}, G\varepsilon_A)$ has the splitting (η_{GA}, η_{GFGA}). Also, for each $X \in \mathbf{X}$, $(\varepsilon_{FGFX}, FG\varepsilon_{FX}, \varepsilon_{FX})$ has the splitting $(F\eta_X, FGF\eta_X)$.

(2) In a category with kernel pairs a morphism h is a split epimorphism if and only if the following fork

$$Y \times_Z Y \begin{array}{c} \pi_1 \\ \rightrightarrows \\ \pi_2 \end{array} Y \xrightarrow{\ h\ } Z \tag{9}$$

splits, where (π_1, π_2) is the kernel pair of h.

(3) Let N be a normal subgroup of G. We denote by $G \rtimes N$ its semidirect product, i.e. the cartesian product of the underlying sets with the operation $(x, n)(y, m) = (xy, y^{-1}nym)$. We define a fork

$$G \rtimes N \begin{array}{c} f \\ \rightrightarrows \\ g \end{array} G \xrightarrow{\ p\ } G/N \tag{10}$$

taking $f(x, n) = x$, $g(x, n) = xn$ and p the canonical projection. Then $p = \operatorname{coeq}(f, g)$ but this fork, in general, does not split in the category **Grp** of groups. However, its image by the forgetful functor $G : \textbf{Grp} \to \textbf{Set}$ is a split fork with $i(gN) = g$ and $j(x) = (x, x^{-1}(ih(x)))$.

(4) For a small category C, let C_0 and C_1 denote its sets of objects and of morphisms and $d, c : C_1 \to C_0$ be the functions domain and codomain, respectively. In the category **Cat** of small categories and functors the fork (of functors between discrete categories)

$$C_1 \begin{array}{c} d \\ \rightrightarrows \\ c \end{array} C_0 \xrightarrow{\ h\ } 1 \tag{11}$$

is a split fork provided C has a terminal object t. In this case $i(1) = t$ and $j(x) = x \to t$.

Proposition. *For every split fork $(f, g, h; i, j)$, (f, g, h) is an absolute coequalizer diagram (i.e. a coequalizer diagram preserved by every functor).*

Proof. If $t \cdot f = t \cdot g$ then $t' = t \cdot i$ is the unique morphism such that $t' \cdot h = t$ and so h is the coequalizer of the parallel pair (f, g). Furthermore, since functors preserve equations the image by any functor of a split fork is a split fork and so a coequalizer diagram. □

Let us characterize now the parallel pairs that are part of a split fork.

A parallel pair (f, g) is said to be *contractible* if there exists a morphism j such that $f \cdot j = 1$ and $g \cdot j \cdot f = g \cdot j \cdot g$. Those are exactly the parallel pairs (f, g) whose coequalizer splits, provided it exists. Indeed, for every split fork $(f, g, h; i, j)$ we have that

$$f \cdot j = 1 \quad \text{and} \quad g \cdot j \cdot f = i \cdot h \cdot f = i \cdot h \cdot g = g \cdot j \cdot g,$$

and so (f, g) is a contractible pair.

Conversely, if a contractible pair $(f, g; j)$ has a coequalizer h, since $g \cdot j \cdot g = g \cdot j \cdot f$, there is a unique morphism i such that $i \cdot h = g \cdot j$. Then

$$h \cdot i \cdot h = h \cdot g \cdot j = h \cdot f \cdot j = h$$

and so $h \cdot i = 1$. Therefore $(f, g, h; i, j)$ is a split fork.

Example. Let $(f, g : X \to Y)$ be a relation R in **Set**, i.e. assume that (f, g) is a monopair and R is the relation $\{(f(x), g(x)) | x \in X\}$ in Y. Its coequalizer is the canonical projection $h : Y \to Y/E = Z$, where E is the equivalence relation generated by R. Since h is a split epimorphism the fork

$$Y \times_Z Y \underset{\pi_2}{\overset{\pi_1}{\rightrightarrows}} Y \overset{h}{\longrightarrow} Z \tag{12}$$

has a splitting (i, j).

For each $y \in Y$ let $y^* = i \cdot h(y)$. Then (f, g) is a contractible pair if and only if the equivalence relation E satisfies the following condition

$$yEy' \Leftrightarrow yRy^* \text{ and } y'Ry^*,$$

Consequently, if R is a reflexive relation then $E = R \circ R^{\mathrm{op}}$.

Coequalizers of contractible pairs will be called *contractible coequalizers*. A parallel pair of morphisms in **A** whose image by G is a contractible pair which has a coequalizer will be called a *G-contractible coequalizer pair*.

By Corollary 1.5 and Proposition 2.2 we conclude that forgetful functors of the form G^T create coequalizers of G^T-contractible coequalizer pairs. In particular, this proves what we claimed at the end of 2.1.

Theorem. *For each T-algebra (X, ξ), the structure map ξ is the coequalizer of the pair $(\mu_X, T\xi)$ of T-morphisms in the category of T-algebras.*

Proof. For each T-algebra (X, ξ), the pair

$$\mu_X, T\xi : (T^2 X, \mu_{TX}) \to (TX, \mu_X)$$

is a G^T- contractible coequalizer pair. Hence $\xi : (TX, \mu_X) \to (X, \sigma)$ is its coequalizer in $\mathbf{X}^T$. It remains to prove that the T-structure map σ is exactly ξ and this follows from

$$\sigma = \xi \cdot \mu_X \cdot T\eta_X = \xi.$$

$\square$

Summing up, underlying functors $G^T : \mathbf{X}^T \to \mathbf{X}$ are conservative (i.e. they reflect isomorphisms) faithful right adjoints, $\mathbf{X}^T$ has coequalizers of G^T-contractible pairs and G^T preserves them.

2.3. The comparison functor. Given an arbitrary adjoint situation $(F, G, \eta, \varepsilon)$: $\mathbf{X} \rightharpoonup \mathbf{A}$ we prove some important and well-known results about the comparison functor $K : \mathbf{A} \to \mathbf{X}^T$.

Lemma. *The counit ε_A is a regular epimorphism for each $A \in \mathbf{A}$ if and only if ε_A is the coequalizer of $(\varepsilon_{FGA}, FG\varepsilon_A)$.*

Proof. The nontrivial part follows from the fact that if $\varepsilon_A = \mathrm{coeq}(f, g)$ then

$$t \cdot f = t \cdot g \Leftrightarrow t \cdot FG\varepsilon_A = t \cdot \varepsilon_{FGA}.$$

This clearly implies that $\varepsilon_A = \mathrm{coeq}(\varepsilon_{FGA}, FG\varepsilon_A)$.   □

Proposition (Beck). *The comparison functor K is full and faithful if and only if, for each $A \in \mathbf{A}$, the counit of the adjunction ε_A is a regular epimorphism.*

Proof. If ε_A is a regular epimorphism for all objects $A \in \mathbf{A}$, then G is faithful. Since $G^T K = G$ so is K.

For $h : K(A) \to K(B)$ in $\mathbf{X}^T$, we have that

$$\varepsilon_B \cdot Fh \cdot \varepsilon_{FGA} = \varepsilon_B \cdot Fh \cdot FG\varepsilon_A.$$

Then, by the previous lemma, this implies the existence of a unique morphism h' in $\mathbf{A}$ such that $h' \cdot \varepsilon_A = \varepsilon_B \cdot Fh$. Then $Gh' \cdot G\varepsilon_A = h \cdot G\varepsilon_A$ and so $Gh' = G^T Kh' = Gh$, because $G\varepsilon_A$ is an epimorphism.

Conversely, if K is full and faithful, since G^T creates coequalizers of G^T-contractible pairs, we have that

$$K(\varepsilon_A) = \mathrm{coeq}(K\varepsilon_{FGA}, KFG\varepsilon_A)$$

and so that $\varepsilon_A = \mathrm{coeq}(\varepsilon_{FGA}, FG\varepsilon_A)$.   □

Theorem (Dubuc). *The functor K has a left adjoint if and only if for each T-algebra (X, ξ) the pair $(\varepsilon_{FX}, F\xi)$ has a coequalizer $c : FX \to Q$ in $\mathbf{A}$. In this case $L(X, \xi) = Q$ is the function of objects of the left adjoint L. Furthermore, in $(L, K; \alpha, \beta) : \mathbf{X}^T \rightharpoonup \mathbf{A}$,*

- *$\alpha_{(X,\xi)}$ is the unique morphism of $\mathbf{X}^T$ such that $\alpha_{(X,\xi)} \cdot \xi = Kc$ and*
- *β_A is unique such that $\beta_A \cdot \varepsilon_A = c'$, where $c' = \mathrm{coeq}(\varepsilon_{FGA}, FG\varepsilon_A)$.*

Proof. For $(X, \xi) \in \mathbf{X}^T$ let $c : FX \to Q$ be the coequalizer of $(\varepsilon_{FX}, F\xi)$ in $\mathbf{A}$. Since ξ is the coequalizer of $(K\varepsilon_{FX}, KF\xi)$ there exists a unique T-morphism $\alpha_{(X,\xi)}$ such that $\alpha_{(X,\xi)} \cdot \xi = Kc$.

If $g : (X, \xi) \to KA$ is a T-morphism, in the diagram

$$
\begin{array}{ccccc}
FGFX & \underset{F\xi}{\overset{\varepsilon_{FX}}{\rightrightarrows}} & FX & \overset{c}{\longrightarrow} & Q \\
{\scriptstyle FGFg}\downarrow & & {\scriptstyle Fg}\downarrow & & \downarrow {\scriptstyle h} \\
FGFGA & \underset{FG\varepsilon_A}{\overset{\varepsilon_{FGA}}{\rightrightarrows}} & FGA & \underset{\varepsilon_A}{\longrightarrow} & A
\end{array}
\tag{13}
$$

the existence of a unique morphism h for which the square commutes follows from the fact that

$$\varepsilon_A \cdot Fg \cdot \varepsilon_{FX} = \varepsilon_A \cdot Fg \cdot F\xi.$$

Hence

$$Kh \cdot \alpha_{(X,\xi)} \cdot \xi = Kh \cdot Kc = K\varepsilon_A \cdot KFg = g \cdot \xi$$

and so $Kh \cdot \alpha_{(X,\xi)} = g$, because ξ is an epimorphism. Furthermore, if $h' : Q \to A$ is such that $Kh' \cdot \alpha_{(X,\xi)} = g$ then $Kh \cdot Kc = Kh' \cdot Kc$. Since $G(h \cdot c) \cdot \eta_X = G(h' \cdot c) \cdot \eta_X$ we have that $h \cdot c = h' \cdot c$ and so that $h = h'$. Therefore $\alpha_{(X,\xi)}$ is universal from (X,ξ) to K.

Conversely, let K be part of an adjunction $(L, K; \alpha, \beta) : \mathbf{X}^T \rightharpoonup \mathbf{A}$.

Since $G^T K = G$ there is a natural isomorphism $\theta : LF^T \to F$ such that

$$G\theta_X \cdot G^T \alpha_{F^T X} \cdot \eta_X^T = \eta_X$$

for each $X \in \mathbf{X}$. Thus

$$K\theta_X \cdot \alpha_{KFX} = 1_{KFX}$$

and so the counit of the comparison adjunction $\beta_{FX} = \theta_X$ is an isomorphism for each $X \in \mathbf{X}$.

For each T-algebra (X,ξ), $L\xi = \mathrm{coeq}(LK\varepsilon_{FX}, LKF\xi)$ in $\mathbf{A}$ because left adjoints preserve colimits, in particular coequalizers. Therefore, $L\xi \cdot \beta_{FX}^{-1}$ is the coequalizer of the pair $(\varepsilon_{FX}, F\xi)$. □

Example. If $P : \mathbf{E} \to \mathbf{B}$ is a functor between small categories then the induced functor $P^* : [\mathbf{B}, \mathbf{Set}] \to [\mathbf{E}, \mathbf{Set}]$ is monadic if and only it is faithful.

The condition is necessary because monadic functors are faithful.

Conversely, observe that the functor P^* has a left adjoint L which is defined on objects by assigning to each functor its left Kan extension along P. Hence, the functor P^* is faithful if and only if the counit of the adjunction is an epimorphism and, since epimorphisms in the functor category $[\mathbf{B}, \mathbf{Set}]$ are regular epimorphisms, by the proposition above, the comparison functor is full and faithful. Thus the counit β of the adjunction defined in Theorem 2.3 is an isomorphism.

Furthermore, the right Kan extensions along P define a right adjoint to P^*. Consequently, P^* preserves coequalizers and so $G^T(\alpha)$ is pointwise an isomorphism. Since monadic functors reflect isomorphisms, the unit α of the comparison adjunction is an isomorphism.

Thus K is an equivalence of categories.

It is now easy to prove that, for functors between small categories, the following conditions are equivalent:

   (i) P^* is faithful;
  (ii) P^* is conservative;
 (iii) P^* is monadic.

2.4. Beck's criterion. In the previous example, monadicity was proved by direct inspection of the comparison functor. Now we are going to give conditions for a functor G to be monadic, an important criterion due to Beck. We point out an important feature of this criterion which is the fact that it does not involve the description of the left adjoint of G. It just assumes its existence.

Theorem. *A functor $G : \mathbf{A} \to \mathbf{X}$ is monadic if and only if*

(i) *G has a left adjoint;*
(ii) *G reflects isomorphisms;*
(iii) *$\mathbf{A}$ has and G preserves coequalizers of all G-contractible coequalizer pairs.*

Proof. Monadic functors are right adjoint conservative functors by Propositions 1.3 and 2.1, respectively. Furthermore, G monadic implies that it creates coequalizers of G-contractible pairs, by Corollary 1.5. Then $\mathbf{A}$ has and G preserves coequalizers of such pairs.

The sufficiency of the conditions follow from the fact that

$$(\varepsilon_{FX}, F\xi) \quad \text{and} \quad (\varepsilon_{FGA}, FG\varepsilon_A) \tag{14}$$

are G-contractible coequalizers. This fact together with the second and the third condition imply that the unit α and the counit β of the comparison adjunction defined in Theorem 2.3 are natural isomorphisms and so that G is monadic. $\square$

Corollary. *Let $G : \mathbf{A} \to \mathbf{X}$ be a monadic functor and $\mathbf{B}$ a full reflective subcategory of $\mathbf{A}$ with embedding E. Then*

(i) *$E : \mathbf{B} \to \mathbf{A}$ is monadic.*
(ii) *GE is monadic if and only if $\mathbf{B}$ is closed under the formation of coequalizers in $\mathbf{A}$.*

Proof. (i) Consider a left adjoint F to the embedding E so chosen that the counit is an identity $FE = \mathrm{Id}$. If

$$E(X) \underset{Eg}{\overset{Ef}{\rightrightarrows}} E(Y) \overset{h}{\longrightarrow} Z \tag{15}$$

is a split fork in $\mathbf{A}$ then (f, g, Fh) is a split fork in $\mathbf{B}$ and so Fh is the coequalizer of (f, g). Furthermore, being an absolute coequalizer, it is preserved by E.

(ii) If $\mathbf{B}$ is closed under the formation of quotients then it is clear that GE satisfies the conditions of the monadicity theorem.

Conversely, if GE is monadic then it is enough to notice that the coequalizer of a parallel pair (f, g) in $\mathbf{B}$ is the composite $r \cdot h$, where h is its coequalizer in $\mathbf{A}$ (whenever it exists) and r is the reflexion of the codomain of h in $\mathbf{B}$. Thus, since G is monadic, r is an isomorphism if Gr is an isomorphism. $\square$

Beck's theorem has been extended to pseudo-monads by Le Creurer, Marmolejo, and Vitale in [22].

2.5. Examples of monadic categories. Equationally definable categories of algebras are often but not always monadic over **Set**, with respect to the forgetful functor.

Examples. (1) *Varieties of general algebras are monadic over* **Set**. The forgetful functor $G : \mathbf{Alg}(\Omega) \to \mathbf{Set}$, where the domain is the category of algebras defined by a set of operations indexed by Ω and their homomorphisms, has a left adjoint which assigns to each set X the set of all Ω-terms in X defined by finitely many applications of

    (a) every $x \in X$ is an Ω-term;
    (b) given Ω-terms $t_1, t_2, \cdots, t_n$ and an n-ary operation symbol ω, then $\omega t_1, t_2, \cdots, t_n$ is a term.

If $f : (A, \alpha) \to (B, \beta)$ is bijective then its inverse g commutes with the Ω-operations: $g \cdot \beta_\omega = \alpha_\omega \cdot g^n$ for each $\omega \in \Omega$.

Finally, if

$$(A, \alpha) \underset{g}{\overset{f}{\rightrightarrows}} (B, \beta) \tag{16}$$

is a G-contractible pair with coequalizer $h : B \to C$ then $h^n : A^n \to B^n$ is the coequalizer of (f^n, g^n) in **Set**, for all natural number n. This implies the existence of a function $\gamma_\omega : C^n \to C$ for each n-ary operation symbol $\omega \in \Omega_n$ such that $h \cdot \beta_\omega = \gamma_\omega \cdot h^n$. It follows that $h : (B, \beta) \to (C, \gamma)$ is the coequalizer of the pair (f, g). This finishes the proof of the monadicity of G.

Now the conclusion follows from Corollary 2.4 and from Birkhoff's famous theorem, since subcategories of (Ω, E)-algebras for any set of equations E are exactly the full reflective subcategories of $\mathbf{Alg}(\Omega)$ which are closed under the formation of quotients.

Each full reflective subcategory of a finitary variety which is not closed under quotients provides an example of a composite of two monadic functors which is not monadic. This is the case of the composite **TFAb** $\to$ **Ab** $\to$ **Set**, where **TFAb** is the full subcategory of torsion-free abelian groups.

(2) *For a monoid M the forgetful functor $U : M$-**Set** $\to$ **Set** is monadic.* The first example shows that the forgetful functor of each category of sets with a structure defined by finitary operations and equations is monadic. In this case, it is enough to notice that the objects of M-**Set** can equivalently be defined as sets with one unary operation for each element of $m \in M$ – the function $X \to X$ defined by $\xi_m(x) = m \cdot x$ for each $x \in X$ – satisfying the obvious equations.

What can we say whenever infinitary operations are involved? The answer is that the corresponding forgetful functor into **Set** is monadic if and only if it has a left adjoint. This is explained in Chapter 1 of [30] where the following two examples are presented in detail. There it is also proved that, for every monad T over **Set**, the category of T-algebras can be described by operations and equations provide we allow infinitary operations.

(3) *The category of complete lattices and morphisms preserving arbitrary joins and meets is not monadic over* **Set**. Complete lattices are defined by operations α^k, β^k for each cardinal k and equations telling us that, for each set L, $\alpha^k{}_L : L^k \to L$ and $\beta^k{}_L : L^k \to L$ assign to each element the join and the meet, respectively, and that these operations define the same order relation on L.

Though this category is presentable by operations and equations, there is no left adjoint to the forgetful functor – the free complete lattice on a three element set does not exist – and so it is not monadic. In this case the infinitary operations cannot be reduced to a set: the class of natural transformations from U^k in U, where U is the forgetful functor into **Set**, is not a set for all cardinal k.

(4) *The category of sup-complete semilattices is monadic over* **Set**. The category of complete lattices and sup-preserving maps is monadic over **Set**: it is exactly the Eilenberg-Moore category of algebras for the power set monad.

(5) *The category of frames is monadic over* **Set**. The category of frames has as objects the complete lattices with finite meets distributing over arbitrary joins and as morphisms the functions preserving finite meets and arbitrary joins (see Chapter II, 2.5). The forgetful functor $U : \mathbf{Frm} \to \mathbf{Set}$ has a left adjoint which assigns to each set X the set of all subsets $\mathcal{A}$ of the set of all finite subsets $\mathcal{P}_f(X)$ of X, such that

$$H \supseteq K \in \mathcal{A} \Rightarrow H \in \mathcal{A}$$

for each $H \in \mathcal{P}_f(X)$. The unit $\eta_X : X \to UFX$ is defined by

$$\eta_X(x) = \{H | x \in H \subseteq X \text{ and } H \text{ is finite}\}.$$

Therefore U is monadic. The construction of the left adjoint F is due to Bénabou [7].

2.6. Other examples. Powerful applications of the theory of monads in several areas of algebra, geometry, and logic have been discovered when adjunctions whose construction was very far from something like the algebraic "free-forgetful" turned out to be monadic. For those adjunctions nobody could imagine that the objects of the domain category **A** were as simple as "algebraic structures in **X**". This also gives an answer to the question of: *What can we say when we have operations and equations and the base category is not* **Set**?

(1) Let $(M, \in)$ be a model of the ZF set theory and $\mathbf{Set}_M$ be the corresponding theory of internal sets. Then $\mathbf{Set}_M$ is an elementary topos and the internal power set functor $\mathcal{P} : (\mathbf{Set}_M)^{\mathrm{op}} \to \mathbf{Set}_M$ is monadic by a well known theorem of Paré. This tells us that the colimits in $\mathbf{Set}_M$, which are the limits in the dual category, can be described using limits in $\mathbf{Set}_M$. Translating from the categorical to the

set-theoretic language, this theorem tells us that the "right axioms" for the theory of sets should not involve unions of sets.

(2) Other than the functor $\mathcal{P} : (\mathbf{Set})^{\mathrm{op}} \to \mathbf{Set}$, there are many other functors from $\mathbf{Set}^{\mathrm{op}}$ to a category $\mathbf{A}$ which are monadic. Indeed, a functor $G : \mathbf{Set}^{\mathrm{op}} \to \mathbf{A}$ has a left adjoint if and only if its function of objects assigns to each set X the product of X copies of a fixed object $A \in \mathbf{A}$. Consequently, any category with arbitrary powers of an object admits such a right adjoint functor into it. Then $G = A^-$ is faithful if and only if A is not preterminal, i.e. there exist at least two distint $\mathbf{A}$-morphisms from some object to A. If, furthermore, $\mathbf{A}(1, A)$ is non-empty then G is monadic (see [36]). Therefore, there exist a "bunch" of monadic functors from $\mathbf{Set}^{\mathrm{op}}$ to many categories. For example, every functor $A^- : \mathbf{Set}^{\mathrm{op}} \to \mathbf{Top}$ is monadic provided that A has at least two points. Thus, the category $\mathbf{CABool}$ of complete atomic Boolean algebras, being equivalent to $\mathbf{Set}^{\mathrm{op}}$, is monadic over several categories.

Note that the existence of at least one morphism from the terminal object to A is not a necessary condition. O. Wyler [39] proved that a right adjoint functor $A^- : \mathbf{Set}^{\mathrm{op}} \to \mathbf{A}$ is monadic if and only if A is not preterminal and all preterminal objects with a T-structure, for the monad induced by the adjunction, are terminal.

(3) Another remarkable example is provided by A. Grothendieck's descent theory which is developed in Chapter VIII of this book. In the language of monads descent theory studies morphisms $p : E \to B$ in an abstract category $\mathbf{C}$ for which the pullback functor $p^* : C \downarrow B \to C \downarrow E$ and/or some similar functor is monadic.

(4) Let $\mathbf{Grph}$ be the category of one-sorted universal algebras for the two unary operations Sce and Tgt, with four axioms:

$$\mathrm{Sce}(\mathrm{Sce}x) = \mathrm{Sce}x \quad \text{and} \quad \mathrm{Sce}(\mathrm{Tgt}x) = \mathrm{Tgt}x.$$

$$\mathrm{Tgt}(\mathrm{Sce}x) = \mathrm{Sce}x \quad \text{and} \quad \mathrm{Tgt}(\mathrm{Tgt}x) = \mathrm{Tgt}x.$$

Here we speak of a morphism x as an *object* if it satisfies the equation $\mathrm{Sce}x = x$. A two-sorted exposition, with disjoint classes of objects and morphisms, as presented in [29], is more standard but would involve a longer presentation. In the following examples over the category of graphs, a key idea is describing the free structures in each example in terms of operations and axioms given in $\mathbf{Grph}$. In particular we show that a category is monadic over $\mathbf{Grph}$ by presenting it in terms of operations and equations. The operations are defined only when certain equations in the operations of $\mathbf{Grph}$ hold. In order to be specific about how this works we give the following operations and axioms for the category of categories with equalizers.

Proposition ([26]). *The category of categories with equalizers is monadic over the category* $\mathbf{Grph}$ *of graphs.*

Proof. In this result and its extensions (a), (b), and (c) below specific operations and equations are given in terms of graphs. Furthermore in equations involving partial operations, as in our axioms, no assumption is implied regarding the existence of the element denoted by either side.

The operations are as follows:

O1. Composition $c(x, y) = y \cdot x$ is a partially defined binary operation defined when $\text{Tgt}x = \text{Sce}y$.

O2. Equalizer $e(x, y)$ is defined when

$$\text{Sce}x = \text{Sce}y \quad \text{and} \quad \text{Tgt}x = \text{Tgt}y.$$

This is the *equalizer morphism*. The *equalizer object* is its source.

O3. The universal morphism for equalizers $h(w, x, y)$ is defined when

$$\text{Tgt}w = \text{Sce}x = \text{Sce}y \quad \text{and} \quad \text{Tgt}x = \text{Tgt}y.$$

This is slightly different from the usual universal morphism which is defined when $x \cdot w = y \cdot w$ (using language that is not available to us at this point).

O4. The equalizer inverse $i(x)$ is defined for all x. This will be an inverse for $e(x, x)$.

The following axioms are needed:

A1,2. $\text{Sce}(y \cdot x) = \text{Sce}x$ and $\text{Tgt}(y \cdot x) = \text{Tgt}y$.

A3. Associativity: $(z \cdot y) \cdot x = z \cdot (y \cdot x)$.

A4,5. Unit axioms: $x \cdot \text{Sce}x = x$ and $(\text{Tgt}x) \cdot x = x$.

A6. $\text{Tgt}(e(x, y)) = \text{Sce}x$.

A7. $x \cdot e(x, y) = y \cdot e(x, y)$.

A8. $\text{Tgt}(h(w, x, y)) = \text{Sce}(e(x, y))$.

A9. $e(x, y) \cdot h(w, x, y) = w \cdot e(x \cdot w, y \cdot w)$ (so $\text{Sce}(h(w, x, y)) = \text{Sce}(e(x \cdot w, y \cdot w))$).

A10,11. $\text{Sce}(i(x)) = \text{Sce}x$ and $\text{Tgt}(i(x)) = \text{Sce}(e(x, x))$.

A12,13. $e(x, x) \cdot i(x) = \text{Sce}x$ and $i(x) \cdot e(x, x) = \text{Sce}(e(x, x))$.

A14. $h(w \cdot v, x, y) = h(w, x, y) \cdot h(v, x \cdot w, y \cdot w)$.

A15. $e(x \cdot e(x, y), y \cdot e(x, y)) = h(e(x, y), x, y)$.

□

The proposition still holds when we replace the category of categories with equalizers by a category **A** where **A** is one of the following:

(a) finitely complete categories or

(b) cartesian closed categories or

(c) toposes.

Early work on algebras over **Grph** was done by Burroni [11], with a number of other contributors including Dubuc-Kelly [12], Lambek [20] and MacDonald-Stone [26]. The preceding presentation follows that in [26].

In another approach we let $\mathcal{M}$ be a set of graphs, i.e. a diagram in **Set** of the form $s, t : A \to N$. A category is said to be $\mathcal{M}$-complete if it has limits of all diagrams $D : M \to \mathbf{C}$ for $M \in \mathcal{M}$. Let $\mathbf{Cat}_\mathcal{M}$ be the category of all $\mathcal{M}$-complete small categories, and the morphisms are the $\mathcal{M}$-continuous functors.

In the following we will say that $\mathbf{Cat}_\mathcal{M}$ is monadic over **Grph** if the forgetful functor $\mathbf{Cat}_\mathcal{M} \to \mathbf{Cat} \to \mathbf{Grph}$ is monadic.

Theorem ([2]). *If $\mathcal{M}$ consists of finite graphs and the empty graph, then $\mathbf{Cat}_\mathcal{M}$ is monadic over **Grph** if and only if $\mathcal{M}$ specifies either*

(i) *all finite limits,*
(ii) *equalizers and the terminal objects, or*
(iii) *just terminal objects.*

2.7. Duskin's criterion. Condition (iii) in Theorem 2.4 is, in general, hard to check. In the following we are going to state a monadicity criterion which, under some completeness assumptions, enables us to replace condition (iii) by a more convenient one.

If $G : \mathbf{A} \to \mathbf{X}$ is monadic then $\mathbf{A}$ has and G preserves coequalizers of G-contractible equivalence relations. The question is when is it possible, instead of arbitrary G-contractible coequalizers, to consider just those parallel pairs (f, g) whose image (Gf, Gg) is a contractible equivalence relation which has a coequalizer.

Duskin's theorem assumes the existence in the domain category of the intersection of the kernel pairs of f and g whenever (f, g) is a parallel pair. This is called the *separator* of (f, g).

Theorem. *Let $\mathbf{A}$ have separators and $\mathbf{X}$ have kernel pairs of split epimorphisms. Then $G : \mathbf{A} \to \mathbf{X}$ is monadic if and only if*

(i) *G has a left adjoint;*
(ii) *G reflects isomorphisms;*
(iii)' *$\mathbf{A}$ has and G preserves coequalizers of G-contractible equivalence relations.*

Proof. See [13] or [5]. □

Over **Set** things simplify considerably by observing that

   – monadic categories over **Set** are complete, and
   – equivalence relations in **Set** are always contractible equivalence pairs because they are effective and their coequalizers, being split epimorphims, are part of split forks.

Corollary. *The functor $G : \mathbf{A} \to \mathbf{Set}$ is monadic if and only if*

(i) *G has a left adjoint;*
(ii) *G reflects isomorphisms;*

(iii)' **A** *has and G preserves coequalizers of G-equivalence relations.*

2.8. Characterization of monadic categories over Set. Over **Set** it is possible to "internalize" the monadicity conditions in the sense that one can give necessary and sufficient conditions for a category **A** to be monadic over **Set**. We state and sketch the proof of this fact using Corollary 2.7. (See also Theorem 2.2.4 of Chapter VI of this book.)

We say that an object P is a *regular generator* if the induced map

$$\varepsilon_A : \mathbf{A}(P, A) \cdot P \to A,$$

from the coproduct of $\mathbf{A}(P, A)$ copies of P to A, is a regular epimorphism for all $A \in \mathbf{A}$.

The object P is called *regular projective* whenever $\mathbf{A}(P, -) : \mathbf{A} \to \mathbf{Set}$ preserves regular epimorphisms.

If P is simultaneously a regular generator and a regular projective is called a *regular projective generator*.

A *category* is *exact* if it has finite limits, coequalizers of kernel pairs, pullback stable regular epimorphisms and all equivalence relations are effective (see Chapter VI, 2.1).

Theorem. *The category* **A** *is monadic over* **Set** *if and only if* **A** *is an exact category, has a regular projective generator P and arbitrary copowers of P.*

Proof. (See also Chapter VI, 2.2.) The functor $G : \mathbf{A} \to \mathbf{Set}$ has a left adjoint F if and only if, for some object P, $G \cong \mathbf{A}(P, -)$ and **A** has coproducts $X \cdot P$ of X copies of P for every set X. Then, $P = F1$ and $F(X) = X \cdot P$, F being the left adjoint to G.

If G is monadic, conditions (ii) and (iii)' of Corollary 2.7 imply that

- G reflects regular epimorphisms and so, in particular, ε_A is a regular epimorphism for all A, i.e. P is a regular generator;
- G preserves regular epimorphisms which means that P is a regular projective;
- **A** has finite limits (indeed it is complete), and
  - **A** has coequalizers of kernel pairs because kernel pairs are G-contractible equivalence relations;
  - regular epimorphisms in **A** are stable under pullback because in **Set** the stability condition holds and G preserves pullbacks and reflects regular epimorphisms;
  - and equivalence relations are effective since they are G-contractible equivalence relations.

  Therefore, **A** is an exact category.

Conversely, if **A** satisfies the prescribed conditions, we have that $G = \mathbf{A}(P, -)$ is a faithful right adjoint functor with left adjoint defined by $FX = X \cdot P$ for every set X.

Let Gf be an isomorphism for $f : A \to B$ in **A**. Then $f \cdot \varepsilon_A = \varepsilon_B \cdot FGf$ is a regular epimorphism and so f is also a regular epimorphism. Furthermore, f is a monomorphism because G, being faithful, reflects monomorphims. Hence f is an isomorphism.

It remains to prove that if (f, g) is a G-contractible equivalence relation then it has a coequalizer which is preserved by G.

We observe that a parallel pair (f, g) is an equivalence relation in **A** whenever

$$(Gf = \mathbf{A}(P, f), Gg = \mathbf{A}(P, g))$$

is an equivalence relation. Indeed, if (Gf, Gg) is an equivalence relation then $(\mathbf{Set}(X, Gf), \mathbf{Set}(X, Gg))$ is an equivalence relation for all set X and, since there is a natural isomorphism $\mathbf{Set}(FX, A) \cong \mathbf{A}(X, GA)$, $(\mathbf{A}(FX, f), \mathbf{A}(FX, g))$ is an equivalence relation. So, in particular, this holds for $X = G(C)$ and $X = GFG(C)$, for each $C \in \mathbf{A}$. Now, for an arbitrary object C, one can prove that

$$(\mathbf{A}(C, f), \mathbf{A}(C, g))$$

is an equivalence relation using the fact that $\mathbf{A}(\varepsilon_C, D)$ is the equalizer of $(\mathbf{A}(\varepsilon_{FG(C)}, D), \mathbf{A}(FG(\varepsilon_C), D))$ for every object D in **A**.

If (f, g) is a G-contractible equivalence relation, it has a coequalizer h in **A** because it is an equivalence relation and equivalence relations are effective. Then this kernel pair/coequalizer sequence is preserved by G, as required. □

2.9. Janelidze's criterion and a theorem by Joyal and Tierney. For full reflective subcategories, since the embedding "reflects contractibility", condition (iii) of Theorem 2.4 always holds. Next we describe a situation where, by the same reason, condition (iii) becomes a triviality and give an example of application.

First we remark that every retract of a contractible pair is again a contractible pair. Indeed, let us consider the diagram

$$
\begin{array}{ccc}
X & \underset{g}{\overset{f}{\rightrightarrows}} & Y \\
s \big\uparrow\big\downarrow t & & u \big\uparrow\big\downarrow v \\
X' & \underset{g'}{\overset{f'}{\rightrightarrows}} & Y'
\end{array}
\tag{17}
$$

where

- $v \cdot f = f' \cdot t, v \cdot g = g' \cdot t$;
- $f \cdot s = u \cdot f', g \cdot s = u \cdot g'$;
- $t \cdot s = 1$ and $v \cdot u = 1$.

If there exists a morphism $j : Y \to X$ such that

$$f \cdot j = 1 \quad \text{and} \quad g \cdot j \cdot f = g \cdot j \cdot g,$$

then $j' = t \cdot j \cdot u$ makes (f', g') a contractible pair.

Lemma. *Let us consider the above diagram where (f, g) is a contractible pair in a category with pullbacks. Then (f', g') has a coequalizer provided (f, g) has a coequalizer. Moreover (f', g') is part of a split fork.*

Proof. We consider the diagram

$$
\begin{array}{ccccc}
X & \underset{g}{\overset{f}{\rightrightarrows}} & Y & \overset{i}{\longleftarrow} & Z \\
s \big\| \big\downarrow t & & u \big\| \big\downarrow v & h & \big\uparrow u' \\
X' & \underset{g'}{\overset{f'}{\rightrightarrows}} & Y' & \overset{i'}{\longleftarrow} & Z'
\end{array}
\tag{18}
$$

where $(f, g, h; i, j))$ is a split fork, (u', i') is the pullback of (u, i), and h' is the induced map $< g'j', hu >$, which is well-defined because $ug'j' = ihu$.

Furthermore, since

$$h'f' = h'g', h'i' = 1_{Z'} \text{ and } i'h' = g'j'$$

$(f', g', h'; i', j')$ is a split fork and so h' is the coequalizer of (f', g'), as required.

$\square$

Proposition. *If $\mathbf{X}$ has pullbacks, a functor $G : \mathbf{A} \to \mathbf{X}$ is monadic whenever the following conditions hold:*

(i) *G has a left adjoint;*
(ii) *G reflects isomorphisms;*
(iii) *the counit $\varepsilon : FG \to \mathrm{Id}$ is a split epimorphism.*

Proof. If (f, g) is a G-contractible coequalizer pair then $(FG(f), FG(g))$ is part of a split fork. Since ε is a split epimorphism then, by Lemma 2.9, (f, g) has a coequalizer which is preserved by every functor. Thus $\mathbf{A}$ has and G preserves coequalizers of those pairs (f, g) for which (Gf, Gg) has a contractible coequalizer.

$\square$

We consider now a more general situation where the image of the counit by some functor $H\varepsilon : HFG \to G$ is a split epimorphism.

Theorem. *If $\mathbf{X}$ has pullbacks a functor $G : \mathbf{A} \to \mathbf{X}$ is monadic if and only if the following conditions hold:*

(i) *G has a left adjoint F;*
(ii) *G reflects isomorphisms;*

(iii) *there exist a (possibly up to isomorphism) commutative diagram*

$$
\begin{array}{ccc}
\mathbf{A} & \xrightarrow{G} & \mathbf{X} \\
H \downarrow & & \downarrow H' \\
\mathbf{B} & \xrightarrow[G']{} & \mathbf{Y}
\end{array}
\tag{19}
$$

such that

- $H\varepsilon : HFG \to H$ *is a split epimorphism;*
- **A** *has and* H *preserves coequalizers of all* G-*contractible coequalizer pairs;*
- H' *reflects isomorphisms.*

Proof. The "only if" part is immediate: take $H = G$ and $G' = H' = \mathrm{Id}$.

To prove that the conditions are sufficient, we consider a G-contractible coequalizer pair (f, g). Its image by H is a contractible pair (Hf, Hg) because it is a retract of $(HFG(f), HFG(g))$, and has a coequalizer $H(c)$ which, being part of a split fork, is absolute.

Let us assume that $c : A \to A'$ and that $q : G(A) \to X$ is the coequalizer of $(G(f), G(g))$ in **X**. Then the induced morphism from X to GA' becomes an isomorphism in **Y**. Since H' reflects isomorphisms we comclude that $G(c)$ is the coequalizer of $(G(f), G(g))$. $\qquad\square$

Example. Let $p : R \to S$ be a homomorphism of commutative rings with identity such that $p(1) = 1$. Then S has an R-module structure with the same abelian group and taking $r \cdot s = p(r) \cdot s$, for all $r \in R$ and $s \in S$:

The morphism p, considered as a morphism of R-modules, is an said to be

- *an effective descent morphism* if $G = S \otimes_R - : (R\text{-}\mathbf{Mod})^{\mathrm{op}} \to (S\text{-}\mathbf{Mod})^{\mathrm{op}}$ is monadic and
- *a pure monomorphism* when, for every R-module A, the canonical map

$$
p \otimes 1 : R \otimes_R A \to S \otimes_R A
$$

is a monomorphism.

A theorem by Joyal and Tierney (unpublished) says that for commutative rings with identity, the effective descent morphisms (see Chapter VIII) in this category are exactly the pure monomorphisms. The first published proof of this result is due to B. Mesablishvili in [32]. There the author uses the definition of monadic functors, i.e. he proves that the corresponding comparison functor is an equivalence if and only if p is pure.

Using the monadicity criterion of the second theorem above, a simpler proof of that theorem was given by G. Janelidze, choosing the data involved there as follows:

- $\mathbf{A} = (R\text{-}\mathbf{Mod})^{\mathrm{op}}$;

- $\mathbf{X} = (S\text{-}\mathbf{Mod})^{\mathrm{op}}$
- $G = S \otimes_R - : \mathbf{A} \to \mathbf{X}$ the extension-of-scalars-functor;
- F is thus the restriction-of-scalars-functor, sending each S-module to the R-module with the same abelian group and the R-module structure defined by $r \cdot x = p(r) \cdot x$;
- $\varepsilon_A : FG(A) \to A$ is the R-module homomorphism $A \to S \otimes_R A$ which sends each a to $1 \otimes a$;
- $\mathbf{B} = R\text{-Mod}$;
- $\mathbf{Y} = \mathbf{Ab}$;
- $H = \mathrm{Hom}_{\mathbb{Z}}(-, \mathbb{Q}/\mathbb{Z})$;
- $G' = \mathrm{Hom}_R(S, -)$;
- $H' = \mathrm{Hom}_{\mathbb{Z}}(-, \mathbb{Q}/\mathbb{Z})$;

Then we delineate Janelidze's proof:

(a) For each R-module A there is a canonical isomorphism

$$\mathrm{Hom}_{\mathbb{Z}}(S \otimes_R A, \mathbb{Q}/\mathbb{Z}) \cong \mathrm{Hom}_R(S, \mathrm{Hom}_{\mathbb{Z}}(A, \mathbb{Q}/\mathbb{Z}))$$

which tell us that, in this case, the diagram (16) commutes up to an isomorphism.

(b) Since the forgetful functor from a category of modules to $\mathbf{Ab}$ is exact and reflects isomorphisms, and since $\mathbb{Q}/\mathbb{Z}$ is an injective cogenerator in $\mathbf{Ab}$, the functors H and H' preserve all coequalizers (in fact they preserve all limits and all finite colimits) and reflect isomorphisms.

(c) If

$$\mathrm{Hom}_{\mathbb{Z}}(p, \mathbb{Q}/\mathbb{Z}) : \mathrm{Hom}_{\mathbb{Z}}(S, \mathbb{Q}/\mathbb{Z}) \to \mathrm{Hom}_{\mathbb{Z}}(R, \mathbb{Q}/\mathbb{Z}) \tag{20}$$

is a split epimorphism of R-modules then so is each

$$\mathrm{Hom}_{\mathbb{Z}}(\varepsilon_A, \mathbb{Q}/\mathbb{Z}) : \mathrm{Hom}_{\mathbb{Z}}(S \otimes_R A, \mathbb{Q}/\mathbb{Z}) \to \mathrm{Hom}_{\mathbb{Z}}(A, \mathbb{Q}/\mathbb{Z}). \tag{21}$$

Moreover, the splitting is natural in A since, up to a natural isomorphism, (21) can be rewritten as

$$\mathrm{Hom}_{\mathbb{Z}}(p \otimes_R A, \mathbb{Q}/\mathbb{Z}) : \mathrm{Hom}_{\mathbb{Z}}(S \otimes_R A, \mathbb{Q}/\mathbb{Z}) \to \mathrm{Hom}_{\mathbb{Z}}(R \otimes_R A, \mathbb{Q}/\mathbb{Z}) \tag{22}$$

and then as

$$\mathrm{Hom}_R(A, \mathrm{Hom}_{\mathbb{Z}}(S, \mathbb{Q}/\mathbb{Z})) \xrightarrow{\mathrm{Hom}_R(A, \mathrm{Hom}_{\mathbb{Z}}(p, \mathbb{Q}/\mathbb{Z}))} \mathrm{Hom}_R(A, \mathrm{Hom}_{\mathbb{Z}}(R, \mathbb{Q}/\mathbb{Z})). \tag{23}$$

Together with (a)-(c), Theorem 2.8 tell us that the functor G is monadic whenever it reflects isomorphisms and the morphism (17) is a split epimorphism of R-modules. And this holds if and only if p is pure.

Remark. If the morphism $p : R \to S$ considered in the example is itself a split monomorphism of R-modules, i.e. if there exists an homomorphism $q : S \to R$ of the underlying abelian groups such that $q \cdot p = 1$ and $q(p(r)s) = r(q(s))$, then all the requirements above are obviously satisfied and so the extension-of-scalars functor $S \otimes_R - : (R\text{-}\mathbf{Mod})^{\mathrm{op}} \to (S\text{-}\mathbf{Mod})^{\mathrm{op}}$ is monadic.

However, in this case, the monadicity of $S \otimes_R - : (R\text{-}\mathbf{Mod})^{\mathrm{op}} \to (S\text{-}\mathbf{Mod})^{\mathrm{op}}$ can be obtained in a much easier way directly from Proposition 2.8 since its assumptions are fulfilled.

3. Conditions for the unit of a monad to be a monomorphism

We give necessary and sufficient conditions for the monicity of any given unit morphism of a monad on $\mathbf{C}$, where $\mathbf{C}$ is a specific category of algebras defined by operations and equations. The degenerate case when $\mathbf{C}$ is $\mathbf{Set}$ has a straightforward description. The direct verification of these conditions in specific instances when $\mathbf{C}$ is not $\mathbf{Set}$ can be subtle. A stronger set of sufficient conditions may be given which is easier to check directly. These conditions are still sufficiently general to provide a categorical form for the proof of the Birkhoff-Witt Theorem for Lie-algebras, closely related to Birkhoff's original proof, as well as one for the Schreier Theorem on free products of groups with amalgamated subgroups.

3.1. Monads over Set. Let $T = (T, \eta, \mu)$ be a monad defined on $\mathbf{Set}$. The category $\mathbf{Set}^T$ of Eilenberg-Moore algebras is said to be *nontrivial* if there is at least one algebra (A, α) whose underlying set A has more than one element.

Lemma. *If the category of Eilenberg-Moore algebras for a monad* (T, η, μ) *on* $\mathbf{Set}$ *is nontrivial, then* η_X *is monic for all sets* X.

Proof. Let x and y be distinct elements of a set X with more than one element and suppose that (A, α) is an algebra with more than one element. Then there is a function $f : X \to A$ with $f(x) \neq f(y)$. By the universal mapping property there is a unique algebra morphism $g : (TX, \mu_X) \to (A, \alpha)$ such that the following diagram commutes

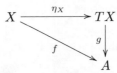

and so $\eta_X(x) \neq \eta_X(y)$. $\qquad\qquad\qquad\square$

It is clearly rare for the category of Eilenberg-Moore algebras to be trivial since, if trivial, each TX is either empty or consisting of one element being the underlying object of the algebra (TX, μ_X).

Now, we remark that the unique function $T1 \to 1$ is a T-algebra, where 1 is the one point set, and the empty set has a T-structure if and only if $T(\emptyset) = \emptyset$. Accordingly, up to isomorphism, there are only two monads (T_1, η_1, μ_1) and (T_2, η_2, μ_2) yielding just trivial algebras. They are determined by $T_1 X = 1$ for each set X and $T_2(\emptyset) = \emptyset$ otherwise $T_2(X) = 1$.

3.2. Diamonds, strong embeddings, and connectedness. We make use of the following principles in later sections.

Let **G** be a subcategory of a category **C** and let $\mathcal{P}(\mathbf{G})$ be the *power category* of **G**. The objects of $\mathcal{P}(\mathbf{G})$ are the subclasses of objects of **G** and the morphisms are the inclusions.

Diamond Principle for G. Let **G** be a subcategory of **C**, then the objects $\alpha : X \to A$ of $X/\mathbf{C}$ with A an object of **G** are terminal in $X/\mathbf{C}$ for each object X of **C**.

The *reduction functor* $\mathcal{R}_{\mathbf{G}} : \mathbf{C}^{\mathrm{op}} \to \mathcal{P}(\mathbf{G})$ is defined by

$$\mathcal{R}_{\mathbf{G}}X = \{A | A \text{ is an object of } \mathbf{G} \text{ and there exists a } \mathbf{C}\text{-morphism } X \to A\}$$

with the obvious definition on morphisms.

An object A is *reduced* in **C** if the only **C**-morphism with domain A is the identity. The subcategory **G** is *reduced in* **C** if each if its objects is reduced in **C**. An object X of **C** is **G**-*reducible* if $\mathcal{R}_{\mathbf{G}}X$ is nonempty.

Lemma A. *Let* **G** *be a reduced subcategory of* **C**. *Then the following statements are equivalent:*

(a) *The Diamond Principle for* **G**.
(b) *Each pair $Y \leftarrow X \to Z$ of* **C**-*morphisms with X a* **G**-*reducible object can be completed to a commutative diamond in* **C**.

Proof. Suppose that (b) holds and let $\alpha : X \to A$ be an object of $X/\mathbf{C}$ with A in **G** and $\beta : X \to Y$ be any other object. Then, by hypothesis there exists a commutative diagram

in **C**. But A in **G** implies that $\delta = 1$. Thus $\gamma : \beta \to \alpha$. If $\gamma' : \beta \to \alpha$ then, by hypothesis, $A \xleftarrow{\gamma} Y \xrightarrow{\gamma'} A$ can be completed to a commutative square since Y is **G**-reducible. Thus $\gamma = \gamma'$ since A is reduced and so α is terminal.

It is trivial to show that (a) implies (b).   □

The *component class* $[X]$ of an object X of a category **C** (or a graph **C**) is the class of all objects Y which can be connected to X by a finite sequence of morphisms (e.g. $X \to X_1 \leftarrow X_2 \to Y$). We let Comp**C** denote the collection of component classes.

Strong Embedding Principle for G. If $[X] = [Y]$ in Comp**C**, then $\mathcal{R}_{\mathbf{G}}X = \mathcal{R}_{\mathbf{G}}Y$. Furthermore there is at most one morphism $X \to A$ for each pair (X, A) consisting of an object X of **C** and an object A of **G**.

Lemma B. *The Strong Embedding Principle for* **G** *implies the Diamond Principle for* **G**.

Proof. Let $A \xleftarrow{\alpha} X \xrightarrow{\beta} Y$ be a diagram in **C** with A in **G**. Thus $[A] = [Y]$ in Comp**C** and $\mathcal{R}_\mathbf{G}X = \mathcal{R}_\mathbf{G}Y$, by hypothesis. Hence A is in $\mathcal{R}_\mathbf{G}Y$ and there exists $\gamma : Y \to A$. But then $\gamma\beta$ and α are morphisms $X \to A$ and $\alpha = \gamma\beta$ by hypothesis. Thus $\gamma : \beta \to \alpha$ in $X/\mathbf{C}$ and γ is unique since as a **C** morphism it is the only morphism $Y \to A$ by hypothesis. □

The implication also goes in the other direction. For a proof see [24]. ⌐

Given an object X of **C** let $(X/\mathbf{C})_\mathcal{P}$ be the full subcategory of the slice category $X/\mathbf{C}$ obtained by omitting the object $1_X : X \to X$.

Principle of Connectedness for T. The categories $(X/\mathbf{C})_\mathcal{P}$ are connected for each object X of **T**, where **T** is a subcategory of **C**.

Lemma C. *If $\alpha : X \to A$ is a terminal object of $X/\mathbf{C}$ with A reduced, then $(X/\mathbf{C})_\mathcal{P}$ is connected.*

Proof. If X is reduced, then $X/\mathbf{C}$ contains only one object, namely 1_X, and $(X/\mathbf{C})_\mathcal{P}$ is empty, hence trivially connected. If X is not reduced and $\alpha : X \to A$ is terminal in $X/\mathbf{C}$ with A reduced, then $\alpha \neq 1_X$ and α is terminal in $(X/\mathbf{C})_\mathcal{P}$. Thus $(X/\mathbf{C})_\mathcal{P}$ is connected. □

Lemma D. *If the Diamond Principle holds for a reduced subcategory* **G** *of* **C**, *then the Principle of Connectedness holds for the full subcategory* **T_G** *of* **C** *consisting of all* **G**-*reducible objects of* **C**.

Proof. Let X be in **T_G**. Then there is a morphism $X \to A$ in **C** with A in **G**. By the Diamond Principle for **G** the morphism $X \to A$ is terminal in $X/\mathbf{C}$. By Lemma C then $(X/\mathbf{C})_\mathcal{P}$ is connected. □

Definition. Let $\mathbb{N}$ be the preorder of nonnegative integers with $n \to m$ iff $n \geq m$. A *rank functor* for a category **C** is a functor $R : \mathbf{C} \to \mathbb{N}$ with $R\alpha \neq 1$ whenever $\alpha \neq 1$.

Theorem. *Let* **C** *be a category with rank functor given and* **G** *a subcategory which is reduced in* **C**. *Then the following are equivalent.*

(a) *The Principle of Connectedness for the full subcategory* **T_G** *of* **C** *determined by all* **G**-*reducible objects of* **C**.

(b) *The Diamond Principle for* **G**.

(c) *The Strong Embedding Principle for* **G**.

3.3. Classical coherence. In considering categories with operations and natural equivalences (replacing the equations of algebras) we immediately discover a relationship between connectedness and the commutativity of diagrams arising from the isomorphisms. This is illustrated by the following example.

Let $\mathbf{V}$ be a category and $\otimes : \mathbf{V} \times \mathbf{V} \to \mathbf{V}$ a functor associative up to a natural isomorphism

$$a : A \otimes (B \otimes C) \to (A \otimes B) \otimes C.$$

Since we have isomorphisms and not (in general) equalities it is then natural to ask whether all diagrams built up from the natural isomorphism a commute. This is an example of a *coherence* question. It turns out that the answer is affirmative in this case if a certain type of *pentagon* diagram commutes in a subcategory $\mathbf{C}$ of the category of shapes $\mathcal{N}(\mathbf{V})$ that we are going to define next.

Shapes are defined inductively by

    S1. 1 is a shape
    S2. If T and S are shapes, so is $T \oslash S$.

For each shape there is a variable set $\nu(T)$ defined inductively by

    V1. $\nu(1)$ is a chosen one element set.
    V2. $\nu(T \oslash S)$ is the disjoint union of $\nu(T)$ and $\nu(S)$.

Given $\mathbf{V}$, then for each shape T there is a functor $|T|$ given inductively by

    F1. $|1|: \mathbf{V} \to \mathbf{V}$ is the identity functor.
    F2. $|T \oslash S| = \otimes \cdot (|T| \times |S|) : \mathbf{V}_T \times \mathbf{V}_S \to \mathbf{V} \times \mathbf{V} \to \mathbf{V}$ for $\mathbf{V}_T$ and $\mathbf{V}_S$ products of $\nu(T)$ and $\nu(S)$ copies of $\mathbf{V}$, respectively.

Let $\mathcal{N}(\mathbf{V})$ be the category whose objects are all shapes and whose morphisms $F : T \to S$ are the natural transformations $F : |T| \to |S|$.

Given T, S and R we obtain $\alpha_{TSR} : T \oslash (S \oslash R) \to (T \oslash S) \oslash R$ in $\mathcal{N}(\mathbf{V})$ by letting $\alpha_{TSR}(X, Y, Z)$ be the component

$$|T|(X) \otimes (|S|(Y) \otimes |R|(Z)) \to (|T|(X) \otimes |S|(Y)) \otimes |R|(Z)$$

of the natural transformation a on $\mathbf{V}$ for X, Y, Z objects of the domain categories for $|T|$, $|S|$ and $|R|$, respectively.

Let $\mathbf{C}$ $(=\mathbf{C}(\mathbf{V}))$ be the subcategory of $\mathcal{N}(\mathbf{V})$ whose objects are all shapes and whose morphisms, called the *allowable* morphisms of $\mathcal{N}(\mathbf{V})$, are given by

    AM1. $1 : T \to T$ and $\alpha : T \oslash (S \oslash R) \to (T \oslash S) \oslash R$ are in $\mathbf{C}$ for any shapes T, S, R.
    AM2. If $f : T \to T'$ and $g : S \to S'$ are in $\mathbf{C}$, then so is $f \oslash g : T \oslash S \to T' \oslash S'$.
    AM3. If $f : T \to S$ and $g : S \to R$ are in $\mathbf{C}$, then so is $gf : T \to R$.

Lemma. *Let **D** be a category with a rank functor and assume that $(X/\mathbf{D})_{\mathcal{P}}$ is connected for each object X of **D**. Then **D** is a preorder if and only if the morphisms of **D** are all monomorphisms.*

Proof. Let **G** be the subcategory of all reduced objects. Then, since **D** has a rank functor, every object of **D** is **G**-reducible. Thus, if $f, g : X \to Y$ in **D**, then there is $h : Y \to G$ with G in **G**. By Theorem 3.2 the Diamond Principle for **G** holds. Thus objects of $X/\mathbf{D}$ with codomain in **G** are terminal. Thus $h \cdot f = h \cdot g$ since G is reduced and $f = g$ since h is monic.   □

A rank functor ρ for the subcategory **C** of $\mathcal{N}(\mathbf{V})$ described above is defined recursively by $\rho(1) = 0$ and $\rho(T \oslash S) = \rho(T) + \rho(S) + |\nu(S)| - 1$, where $|\nu(S)| = \operatorname{card} \nu(S)$.

Theorem. *The category $(X/\mathbf{C})_{\mathcal{P}}$ is connected for every shape $X \in |\mathbf{C}|$ provided that the pentagon diagram*

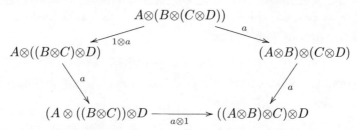

*commutes for all A, B, C, D in **V**.*

Corollary. *The category **C** is a preorder provided all pentagon diagrams commute.*

A functor $T : \mathbf{C} \to \mathbf{D}$ whose domain is a preorder is called a *coherence functor* for **D**. Intuitively, T describes a class of commutative diagrams in **D**.

The subcategory of $\mathcal{N}(\mathbf{V})$ generated by **C** and the inverses to the associativity isomorphisms is also a preorder. For further details see [24] or [28].

3.4. Necessary and sufficient conditions. Let $\mathbf{Alg}(\Omega, E)$ denote the category of algebras defined by a set of operators Ω and identities E as described in Mac Lane [29], already considered in Example 2.5(1). In this section we describe a set of necessary and sufficient conditions for a monad defined on a category of algebras $\mathbf{Alg}(\Omega, E)$ to have its unit a monomorphism. In particular we see that this condition involves the Strong Embedding Principle, which is actually the same principle used in case of classical coherence (as shown in Theorem 3.2).

Let

$$\mathbf{A} \xrightarrow{\;U\;} \mathbf{B} \xrightarrow{\;V\;} \mathbf{D} \tag{24}$$

be a diagram such that

(a) **A**, **B**, and **D** are the categories of (Ω, E), (Ω', E') and (Ω'', E'') algebras, respectively, with $\Omega'' \subseteq \Omega'$ and $E'' \subseteq E'$, and

(b) V is the forgetful functor on operators $\Omega' - \Omega''$ and identities $E' - E''$ and U is a functor commuting with the underlying set functors on **A** and **B**. Note that U is *not* necessarily a functor forgetting part of Ω and E.

We next describe a functor $C_V : \mathbf{B} \to \mathbf{Grph}$ associated to each pair consisting of a diagram (24) of algebras and an adjunction $(L, VU, \eta', \varepsilon') : \mathbf{D} \rightharpoonup \mathbf{A}$, where **Grph** is the category of directed graphs in the sense of [29].

Given an object G of **B** let the objects of the graph $C_V(G)$ be the elements of the underlying set $|LVG|$ of LVG.

Recursive definition of the arrows of $C_V(G)$:

$$\omega_{ULVG}(|\eta'_{VG}|x_1, \cdots, |\eta'_{VG}|x_n) \to |\eta'_{VG}|\omega_G(x_1, \cdots, x_n)$$

is an arrow if ω is in the set $\Omega' - \Omega''$ of operators forgotten by V and $(x_1, \cdots, x_n)$ is an n-tuple of elements of $|G|$ for which $\omega_G(x_1, \cdots, x_n)$ is defined.

If $d \to e$ is an arrow of $C_V(G)$, then so is

$$\rho_{LVG}(d_1, \cdots, d, \cdots, d_q) \to \rho_{LVG}(d_1, \cdots, e, \cdots, d_q)$$

for ρ an operator of arity q in Ω and $d_1, \cdots, d_{i-1}, d_{i+1}, \cdots, d_q$ arbitrary elements of $|LVG|$.

If $\beta : G \to G' \in \mathbf{B}$, then $C_V(\beta) : C_V(G) \to C_V(G')$ is the graph morphism which is just the function $|LV\beta| : |LVG| \to |LVG'|$ on objects and defined recursively on arrows in the obvious way.

In the following proposition note that if, in the diagram (24), **D** is the category of sets, then an adjunction $(L, VU, \eta', \varepsilon')$ is given by letting LX be the free (Ω, E)-algebra on the set X.

Proposition. *Suppose*

$$\mathbf{A} \underset{L}{\overset{U}{\rightleftarrows}} \mathbf{B} \overset{V}{\longrightarrow} \mathbf{D}$$

is a diagram of categories of algebras as in (24), with given adjunction $(L, VU, \eta', \varepsilon') : \mathbf{D} \rightharpoonup \mathbf{A}$. Then there is an adjunction $(F, U, \eta, \varepsilon) : \mathbf{B} \rightharpoonup \mathbf{A}$ with the following specific properties:

(a) *The underlying set of FG is $\mathrm{Comp}\, C_V(G)$.*

(b) *If ρ is an operator of arity n in Ω, then ρ_{FG} is defined by*

$$\rho_{FG}([c_1], \cdots, [c_n]) = [\rho_{LVG}(c_1, \cdots, c_n)]$$

where $c_1, \cdots, c_n$ are members of the set $|LVG|$ of objects of the graph $C_V(G)$.

(c) *The unit morphism $\eta_G : G \to UFG$ of $(F, U, \eta, \varepsilon)$ has an underlying set map which is the composition $[\ ] \cdot |\eta'_{VG}|$, where $|\eta'_{VG}| : |G| \to |LVG| = \mathrm{Obj}C_V(G)$ is the set map underlying the unit $\eta'_{VG} : VG \to VULVG$ of the adjunction $(L, VU, \eta', \varepsilon')$ and $[\ ] : \mathrm{Obj}C_V(G) \to \mathrm{Comp}C_V(G)$ is the function which sends each object to its component.*

Suppose the hypotheses of the proposition hold. Let S_V be a subgraph of $C_V(G)$ having the same objects $|LVG|$ and the same components as $C_V(G)$. Then the proposition remains valid under substitution of S_V for $C_V(G)$ throughout. This allows us to "picture" the adjoint using a possibly smaller set of arrows than those present in $C_V(G)$. Accordingly, we define a *V picture of the adjoint F to U at* $G \in |\mathbf{B}|$ to be any quotient category $\mathbf{C}(= \mathbf{C}(S_V))$ of the free category generated by such a subgraph S_V of $C_V(G)$. This proposition is then valid upon substitution of the underlying graph of a V picture $\mathbf{C}$ for $C_V(G)$ throughout.

Theorem. *Let*

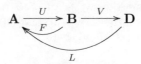

be given with adjunctions $(L, VU, \eta', \varepsilon') : \mathbf{D} \rightharpoonup \mathbf{A}$ and $(F, U, \eta, \varepsilon) : \mathbf{B} \rightharpoonup \mathbf{A}$ as described in Proposition 3.4.

Given $G \in |\mathbf{B}|$ let $C(S_V)$ be any V picture of the adjoint F to U at G.

Then the unit morphism $\eta_G : G \to UFG$ of the adjunction $(F, U, \eta, \varepsilon)$ is monic if and only if the following hold

(a) *The discrete subcategory $\mathbf{G} = \eta'_{VG}(|G|)$ is reduced in $\mathbf{C}(S_V)$ for η'_{VG} the unit of $(L, VU, \eta', \varepsilon')$.*

(b) *If $[A] = [B]$ in $\mathrm{Comp}C(S_V)$ with $A, B \in |\mathbf{G}|$, then $\mathcal{R}_{\mathbf{G}}A = \mathcal{R}_{\mathbf{G}}B$, where $\mathcal{R}_{\mathbf{G}} : C(S_V)^{\mathrm{op}} \to \mathcal{P}(\mathbf{G})$ is the reduction functor.*

(c) *The unit morphism η'_{VG} is monic.*

3.5. Sufficient conditions. In the presence of a rank functor we have seen in Theorem 3.2 that the three principles are equivalent. We apply Theorem 3.2 to the hypotheses of Theorem 3.4 to obtain the following result:

Theorem. *Let*

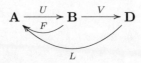

be given with hypotheses as in Theorem 3.4.

Let G be an object of $\mathbf{B}$. Then the unit morphism $\eta_G : G \to UFG$ of $(F, U, \eta, \varepsilon)$ is monic provided there exists a V picture $\mathbf{C}$ of the adjoint F to U at G for which the following conditions hold:

   (a) **C** *has a rank functor.*
   (b) *The discrete subcategory* $\mathbf{G} = \eta'_{VG}(|G|)$ *is reduced in* **C** *for* η'_{VG} *the unit of* $(L, VU, \eta', \varepsilon')$.
   (c) *The categories* $(X/\mathbf{C})_{\mathcal{P}}$ *are connected for each* **G**-*reducible object* $X \in |\mathbf{C}|$.
   (d) *The unit morphism* η'_{VG} *of* $(L, VU, \eta', \varepsilon')$ *is monic.*

3.6. Associative Embedding of Lie Algebras. Let $U : \mathbf{A} \to \mathbf{L}$ be the usual algebraic functor from associative algebras over K to Lie algebras over K, for K a commutative ring. That is, for $A \in |\mathbf{A}|$, UA is the same as A except that a new multiplication $[a, b] = ab - ba$ replaces the associative multiplication ab of $A \in |\mathbf{A}|$.

It is well known that an adjunction $(F, U, \eta, \varepsilon) : \mathbf{L} \rightharpoonup \mathbf{A}$ exists. The question of embeddability of a Lie algebra in its universal associative algebra (i.e. the question as to whether the unit morphisms η_G of $(F, U, \eta, \varepsilon)$ are monomorphisms) has been investigated by various authors (cf. Birkhoff [10] and Serre [35]). Not all Lie algebras can be so embedded (cf. Higgins [16]).

We demonstrate how such a question can be put in the context of the previous sections. Let $V : \mathbf{L} \to K$-**Mod** be the functor forgetting the Lie multiplication. We then have the diagram

$$\mathbf{A} \xrightarrow{U} \mathbf{L} \xrightarrow{V} K\text{-}\mathbf{Mod}$$
$$L$$

where the conditions (a) and (b) of diagram (24) hold. The existence of the adjunction $(L, VU, \eta', \varepsilon') : K$-**Mod** $\rightharpoonup \mathbf{A}$ is assured on theoretical grounds, but can also be described explicitly as follows.

Given $G \in |\mathbf{L}|$ it is known that LVG is the tensor algebra of VG. Thus

$$LVG = \oplus_{n \geq 0}(\otimes^n_{i=1}(VG)).$$

Furthermore $\eta'_{VG} : VG \to K \oplus VG \oplus (VG \otimes VG) \oplus \cdots$ is monic. In this section let **G** be the discrete subgraph $\eta'_{VG}(|G|)$ of $C_V(G)$. Thus **G** is a discrete subcategory of any V picture of F at G. Applying 3.4 the following Lie-algebra embedding result holds.

Theorem A. *Let G be a Lie-algebra and* **C** *any V picture of F at G. Then a Lie-algebra G can be embedded in its universal associative algebra FG if and only if the following hold:*

   (a) $[A] = [B]$ *in* $\mathrm{Comp}\,\mathbf{C}$ *implies that* $\mathcal{R}_\mathbf{G}A = \mathcal{R}_\mathbf{G}B$ *for all* $A, B \in |\mathbf{G}|$ *where* $\mathcal{R}_\mathbf{G} : \mathbf{C}^{\mathrm{op}} \to \mathcal{P}(\mathbf{G})$ *is the reduction functor.*
   (b) **G** *is a reduced subcategory of* **C**.

Similarly, by applying Theorem 3.5 and noting that η'_{VG} is monic we have the following sufficient conditions:

Theorem B. *A Lie-algebra G can be embedded in its universal associative algebra FG if there exists any V picture $\mathbf{C}$ of the adjoint F to U at G with the following properties.*

(a) *The categories $(X/\mathbf{C})_{\mathcal{P}}$ are connected for each $X \in |\mathbf{C}|$ which is $\mathbf{G}$-reducible.*

(b) *$\mathbf{C}$ has a rank functor.*

(c) *The discrete subcategory $\mathbf{G} = \eta'_{VG}(|G|)$ is reduced in $\mathbf{C}$ for η'_{VG} the unit of $(L, VU, \eta', \varepsilon')$.*

Theorem (Birkhoff-Witt). *A Lie-algebra G whose underlying module VG is free can be embedded in its universal associative algebra FG.*

Proof. The conditions of the previous theorem are to be verified for the following V picture of the adjoint F to U at G. Let $\mathbf{C}$ be the preorder which is a quotient of the free category on the following subgraph S_V of $C_V(G)$. The objects of S_V are the elements of the free K module LVG on all finite strings $x_{i_1} \cdots x_{i_n}$ of elements from a basis $(x_i)_{i \in I}$ of the free K module VG. Given a well ordering of I we let the arrows of S_V be those of the form

$$k_i x_{i_1} \cdots x_{i_n} + \alpha \to k_i x_{i_1} \cdots x_{i_{j+1}} x_{i_j} \cdots x_{i_n} + k_i x_{i_1} \cdots [x_{i_j}, x_{i_{j+1}}] \cdots x_{i_n} + \alpha$$

for $i_{j+1} < i_j$, $k_i \in K$, and α any element of LVG (not involving $x_{i_1} \cdots x_{i_n}$).

To show that the categories $(X/\mathbf{C})_{\mathcal{P}}$ are connected for each $\mathbf{G}$-reducible object in $\mathbf{C}$, it turns out that the key idea is to show that for $c < b < a$ in I the objects

$$\beta : x_a x_b x_c \to x_b x_a x_c + [x_a, x_b] x_c$$

and

$$\gamma : x_a x_b x_c \to x_a x_c x_b + x_a [x_b, x_c]$$

can be connected in $((x_a x_b x_c)/\mathbf{C})_{\mathcal{P}}$. This is done by further reduction of the ranges of β and γ and use of the Jacobi identity and the identity $[x, y] = -[y, x]$.

The rank functor for $\mathbf{C}$ is given as follows. Given $X = k x_{a_1} \cdots x_{a_n}$ let $R(X) = (R_n(X))$ be a sequence of nonnegative integers defined by $R_n(X) = \sum_{i=1}^{n} p_{a_i}$ where p_{a_i} is the number of x_{a_j} to the right of x_{a_i} with $a_j < a_i$ and $R_s(X) = 0$ for $s \neq n$. We extend by linearity to all elements of $LVG = |\mathbf{C}|$. If $X \to Y$ is an arrow, then $R(Y) < R(X)$ where the latter inequality means that $R_n(Y) < R_n(X)$ for n the largest integer with $R_n(Y) \neq R_n(X)$. Thus R extends to a rank functor.

Finally, we verify condition (c) by observing that any element of $|G|$ may be expressed in the form $\sum_{i \in I} k_i x_i$ in terms of the basis $(x_i)_{i \in I}$ of G, which is regarded as a subset $\mathbf{G}$ of LVG via the embedding η'_{VG}. From the preceding description of arrows of S_V there is no arrow with domain an element of $\mathbf{G} = \eta'_{VG}(|G|)$. Thus $\mathbf{G}$ is reduced in $\mathbf{C}$. $\qquad\square$

3.7. Sets with a partially defined binary operation and a result of Schreier. We show how the following classical theorem follows from 3.4 and 3.5.

Theorem (Schreier). *If S is a common subgroup of the groups X and Y and if*

$$
\begin{array}{ccc}
S & \xrightarrow{\ \subseteq\ } & X \\
{\scriptstyle\subseteq}\downarrow & & \downarrow{\scriptstyle\alpha} \\
Y & \xrightarrow[\ \beta\ ]{} & P
\end{array}
\qquad (25)
$$

is the pushout in the category of groups, then α and β are monomorphisms. The group P is referred to as the free product of X and Y with amalgamated subgroup S.

Let **B** be the category of sets with a single partially defined binary operation. The diagram $X \longleftarrow S \to Y$ of groups can be regarded as a diagram in **B** and can be completed to a diagram

$$
\begin{array}{ccc}
S & \xrightarrow{\ \subseteq\ } & X \\
{\scriptstyle\subseteq}\downarrow & & \downarrow{\scriptstyle\gamma} \\
Y & \xrightarrow[\ \delta\ ]{} & Z
\end{array}
\qquad (26)
$$

commuting in **B** where Z is the disjoint union of X and Y with common subset S identified and ab is defined if both $a, b \in X$ or if both $a, b \in Y$, otherwise it is undefined. Clearly (26) is a pushout in **Set** and in **B**. The morphisms γ and δ are the obvious monomorphisms. The next Lemma and Proposition describe how this approach yields the Schreier Theorem.

Lemma. *The Schreier Theorem holds if in* (26) *the pushout codomain Z in **B** is embeddable in a group.*

Proof. Let $\iota : Z \to P'$ be a monomorphism in **B** with P' a group. Then

$$
\begin{array}{ccc}
S & \xrightarrow{\ \subseteq\ } & X \\
{\scriptstyle\subseteq}\downarrow & & \downarrow{\scriptstyle\iota\gamma} \\
Y & \xrightarrow[\ \iota\delta\ ]{} & P'
\end{array}
$$

commutes in groups. Thus for some group homomorphism ϕ we have $\iota\gamma = \phi\alpha$ and $\iota\delta = \phi\beta$ since (25) is a pushout in groups. Thus α, β are monic since ι, γ and δ are. $\qquad\square$

Proposition. *Let Z be as in* (26). *Then Z is embeddable in a group.*

Proof. We embed Z in a particular semigroup which turns out to be a group. Begin with the diagram

$$\mathbf{A} \overset{U}{\underset{L}{\rightleftarrows}} \mathbf{B} \overset{V}{\longrightarrow} \mathbf{Set}$$

where $\mathbf{A}$ is the category of semigroups (not necessarily with 1), U forgetful, V forgetful and $(L, VU, \eta', \varepsilon')$ an adjunction. We then have a preorder $\mathbf{C}_Z$ which is a quotient of the free category $\mathbf{F}_Z$ generated by $C_V(Z)$. Proposition 3.4 (which also holds for the category $\mathbf{B}$ of partial algebras, see [24]) shows that an adjunction $(F, U, \eta, \varepsilon) : \mathbf{B} \rightharpoonup \mathbf{A}$ exists and describes it. It is sufficient to show that the unit $\eta_Z : Z \to UFZ$ of the adjunction is a monomorphism. By Theorem 3.5 it is sufficient to verify conditions (a) through (d). These conditions are trivial except for (c), which requires that the categories $(X/\mathbf{C})_{\mathcal{P}}$ be connected for each $\mathbf{G}$-reducible object X of $\mathbf{C}$. The objects of $\mathbf{C}_Z$ are elements of the free semigroup LVZ on VZ. An object X may be written as a string $(a_1, \cdots, a_n)$ of length $n \geq 1$ where $a_i \in VZ$ for $i = 1, \cdots, n$. It is sufficient to show that C_V arrows

$$(\cdots, a_i a_{i+1}, \cdots) \overset{\alpha}{\longleftarrow} (a_1, \cdots, a_n) \overset{\beta}{\longrightarrow} (\cdots, a_j a_{j+1}, \cdots)$$

regarded as $(X/\mathbf{C}_Z)_{\mathcal{P}}$ objects can be connected by a finite sequence of morphisms in the same category. This requires a detailed argument when $i = j - 1$ or $i = j + 1$, otherwise it is trivial (cf. Baer [3], MacDonald [24]). $\qquad\square$

4. The Kleisli triples of Manes and their generated monads

E. Moggi [33] wrote that "There is an alternative description of a monad which is easier to justify computationally". He was referring to the Kleisli triples introduced by E. Manes in his 1976 book called Algebraic Theories [30].

4.1. Kleisli triples. A Kleisli triple over a category $\mathbf{C}$ is a system consisting of functions

    (i) $T : \mathrm{Obj}\mathbf{C} \to \mathrm{Obj}\mathbf{C}$,
    (ii) $\eta : \mathrm{Obj}\mathbf{C} \to \mathrm{Arr}\mathbf{C}$ with value $\eta_A : A \to TA$ on object A and
    (iii) $-^*$ a function on arrows $f : A \to TB$ with value $f^* : TA \to TB$.

This data is subject to the following equations

    (K1) $\eta_A^* = \mathrm{id}_{TA}$.
    (K2) $f^*\eta_A = f$
    (K3) $g^* f^* = (g^* f)^*$ for $g : B \to TC$.

Proposition. *Every Kleisli triple* $(T, \eta, -^*)$ *over* **C** *has a unique extension to a monad* (T, η, μ) *on* **C** *with* $Th = (\eta_B h)^*$ *for each morphism* $h : A \to B$ *of* **C** *and with* $\mu_A = \mathrm{id}_{TA}{}^*$. *Conversely, every monad on* **C** *restricts to a Kleisli triple with* $f^* = \mu_B \cdot Tf$ *for each* $f : A \to TB$ *in* **C**. *These processes are inverse to each other.*

Proof. Starting with the Kleisli triple $(T, \eta, -^*)$ we show that the function $T : \mathrm{Obj}\mathbf{C} \to \mathrm{Obj}\mathbf{C}$ can be extended to a functor T with value $Th = (\eta_B h)^*$ on morphisms $h : A \to B$. We have that $T(\mathrm{id}_A) = (\eta_A \mathrm{id}_A)^* = \mathrm{id}_{TA}$ by (K1). Let $k : B \to C$. Then

$$TkTh = (\eta_C k)^*(\eta_B h)^* = ((\eta_C k)^* \eta_B h)^* = ((\eta_C k)h)^* = (\eta_C(kh))^* = T(kh),$$

by (K3) and (K2). Thus T is a functor.

For naturality of η we suppose $\eta_A : A \to TA$ and $h : A \to B$. Then,

$$Th\eta_A = (\eta_B h)^* \eta_A = \eta_B h : A \to TB,$$

by (K2).

By definition $\mu_A = (\mathrm{id}_{TA})^* : T^2 A \to TA$. For naturality, by (K2) and (K3), we have that

$$
\begin{aligned}
Th\mu_A &= Th(\mathrm{id}_{TA})^* = (\eta_B h)^*(\mathrm{id}_{TA})^* = ((\eta_B h)^* \mathrm{id}_{TA})^* \\
&= ((\eta_B h)^*)^* = (Th)^* = (\mathrm{id}_{TB} Th)^* = ((\mathrm{id}_{TB})^* \eta_{TB} Th)^* \\
&= (\mathrm{id}_{TB})^*(\eta_{TB} Th)^* = (\mathrm{id}_{TB})^* T^2 h = \mu_B T^2 h.
\end{aligned}
$$

Also

$$\mu_A \eta_{TA} = (\mathrm{id}_{TA})^* \eta_{TA} = \mathrm{id}_{TA}.$$

Thus $\mu \cdot \eta T = 1$. Furthermore

$$\mu_A \cdot T\eta_A = (\mathrm{id}_{TA})^*(\eta_{TA}\eta_A)^* = ((\mathrm{id}_{TA})^* \eta_{TA}\eta_A)^* = (\mathrm{id}_{TA}\eta_A)^* = \eta_A^* = \mathrm{id}_{TA},$$

and so $\mu \cdot T\eta = 1$.

Finally we verify that $\mu \cdot \mu_T = \mu \cdot T\mu$.

$$
\begin{aligned}
\mu_A \mu_{TA} &= (\mathrm{id}_{TA})^*(\mathrm{id}_{T^2 A})^* = ((\mathrm{id}_{TA})^*(\mathrm{id}_{T^2 A}))^* = ((\mathrm{id}_{TA})^*)^* = \mu_A^* \\
&= (\mu_A \eta_{TA} \mu_A)^* = ((\mathrm{id}_{TA})^*(\eta_{TA}\mu_A))^* \\
&= (\mathrm{id}_{TA})^*(\eta_{TA}\mu_A)^* = \mu_A T\mu_A.
\end{aligned}
$$

Conversely, start with a monad (T, η, μ) on **C**. Functions $T : \mathrm{Obj}\mathbf{C} \to \mathrm{Obj}\mathbf{C}$ and $\eta : \mathrm{Obj}\mathbf{C} \to \mathrm{Arr}\mathbf{C}$ with value $\eta_A : A \to TA$, of the required type, are obtained by restriction.

We define $-^*$ on arrows $f : A \to TB$ by $f^* = \mu_B \cdot Tf : TA \to TB$. Then

- $\eta_A^* = \mu_A T\eta_A = \mathrm{id}_{TA}$
- $f^* \eta_A = \mu_B Tf\eta_A = \mu_B \eta_{TB} f = f$ and
- $g^* f^* = \mu_C Tg\mu_B Tf = \mu_C \mu_{TC} T^2 gTf = \mu_C T\mu_C T^2 gTf = \mu_C Tg^* Tf$
  $\mu_C T(g^* f) = (g^* f)^*.$ $\qquad\qquad\qquad\qquad\qquad\qquad\qquad\qquad\square$

4.2. Properties of the Kleisli category. The *Kleisli category* $\mathbf{C}_T$ *for a Kleisli triple* $(T, \eta, -^*)$ can be defined as the Kleisli category of the generated monad as in Proposition 4.1. However, it can be specified directly in terms of the triple in the following way. Let the objects of $\mathbf{C}_T$ be the same as those of $\mathbf{C}$ and let $\mathbf{C}_T(X, Y) = \mathbf{C}(X, TY)$. The composite $g \circ f : X \to Y \to Z$ in $\mathbf{C}_T$ is the $\mathbf{C}$ composite $g^* f : X \to TY \to TZ$ and the identities of $\mathbf{C}_T$ are the unit morphisms $\eta_X : X \to TX$ of $\mathbf{C}$.

Proposition. *The morphism f is an isomorphism in $\mathbf{C}_T$ if and only if f^* is an isomorphism in $\mathbf{C}$.*

Proof. Suppose $f : A \to B$ is in $\mathbf{C}_T$ with inverse g. Then, in $\mathbf{C}$, $g^* f = \eta_A : A \to TA$ and $f^* g = \eta_B : B \to TB$. Then

$$f^* g^* = (f^* g)^* = \eta_B^* = \mathrm{id}_{TB} \text{ and}$$

$$g^* f^* = (g^* f)^* = \eta_A^* = \mathrm{id}_{TA} \text{ by (K3) and (K1).}$$

Conversely, suppose $f^* : TA \to TB$ is an isomorphism in $\mathbf{C}$ with inverse k. Then we show that the $\mathbf{C}$ morphism $h = k\eta_B : B \to TA$ in $\mathbf{C}$ is inverse as a $\mathbf{C}_T$ morphism $B \to A$ to f. Clearly $f \circ h = f^* h = f^* k\eta_B = \eta_B$ which is the identity in $\mathbf{C}_T$. Furthermore $h \circ f = h^* f = \mu_A Thf = \mu_A TkT(\eta_B)f = k\mu_A T(\eta_B)f = kf = kf^* \eta_A = \eta_A$, by the description given at beginning and (K2), which is the identity in $\mathbf{C}_T$. $\qquad \square$

Theorem. *Given diagrams*

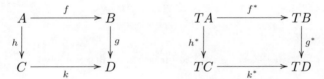

it then follows that

(1) *the left square commutes in $\mathbf{C}_T$ if and only if the right square commutes in $\mathbf{C}$.*

(2) *the left square is a pullback in $\mathbf{C}_T$ if and only if the right square is a pullback in $\mathbf{C}$.*

4.3. Computations with side effects. Given a set S we define a Kleisli triple $(T, \eta, -^*)$ over the category **Set** of sets. T is defined on objects A by

$$TA = (A \times S)^S.$$

An element of TA is called a *computation*. It is a function $S \to A \times S$ that takes a store and returns a value together with a modified store. Similarly $\eta_A : A \to TA$ is the function which assigns to on element a of A the function

$$\eta_A(a) : S \to A \times S$$

given by $\eta_A(a)s = (a, s)$.

Given $f : A \to TB = (B \times S)^S$ we define $f^* : TA \to TB$ as follows:
For $c : S \to A \times S$ in TA let $f^*(c) : S \to B \times S$ in TB be the composite

$$
\begin{array}{ccc}
S & & B \times S \qquad\qquad (27)\\
\downarrow{\scriptstyle c} & & \uparrow{\scriptstyle \text{eval}} \\
A \times S & \xrightarrow{\ f \times 1\ } & TB \times S
\end{array}
$$

That is, $f^*(c)s = \text{eval}(f \times 1)c(s)$ in $B \times S$.

Given $c : S \to A \times S$ in TA it follows that $\eta_A^*(c) = \text{eval}(\eta_A \times 1)c$. Let $c(s) = (a, s')$.
Then $(\eta_A^*(c))s = \text{eval}(\eta_A \times 1)c(s) = \text{eval}(\eta_A \times 1)(a, s') = \text{eval} \cdot (\eta_A a, s')$. But
$(\eta_A a)s' = (a, s') = c(s)$. Thus $\eta_A^*(c) = c$ and $\eta_A^* = \text{id}_{TA}$.

Furthermore

$$f^*(\eta_A(a))(s) = \text{eval}(f \times 1)(\eta_A(a))s = \text{eval}(f \times 1)(a, s) = \text{eval}(fa, s) = (fa)s,$$

thus $f^*(\eta_A(a)) = fa$ and $f^* \eta_A = f$, as required in (K2).

Let $f : A \to TB$ and $g : B \to TC$, then $g^* f^*$ is the composite

$$
\begin{array}{ccccc}
S & & B \times S & \xrightarrow{\ g \times 1\ } & TC \times S \qquad (28)\\
\downarrow{\scriptstyle c} & & \uparrow{\scriptstyle \text{eval}} & & \downarrow{\scriptstyle \text{eval}} \\
A \times S & \xrightarrow{\ f \times 1\ } & TB \times S & & C \times S
\end{array}
$$

by (27). Hence

$$g^* f^*(c)(s) = g^*(\text{eval}(f \times 1)c(s)) = \text{eval}(g \times 1)\text{eval}(f \times 1)c(s)$$

and for, c(s) = (a,s'), this is equal to

$$\text{eval}(g \times 1)\text{eval}(fa, s') = \text{eval}(g \times 1)((fa)s' = g^*(fa)s' = \text{eval}(g^*(fa), s') =$$

$$\text{eval}(g^* \times 1)(f \times 1)(a, s') = \text{eval}(g^* \times 1)(f \times 1)cs = \text{eval}(g^* f \times 1)cs = (g^* f)^*(c)s.$$

Thus $g^* f^* = (g^* f)^*$ as required in (K3).

4.4. The monad of state transformers. We next present the monad of state transformers for imperative programming in a pure functional setting (cf. [31]). This is presented as a Kleisli triple $(T, \eta, -^*)$ on **Set** which in turn has the monad extension given by Proposition 4.1.

We begin with the following data. Let Q be a fixed set (of "states"). The functor $T :$ Obj(**Set**) $\to$ Obj(**Set**) is given by $TA = Q^Q \times A^Q$. The function $\eta :$ Obj(**Set**) $\to$ Arr(**Set**) takes value $\eta_A : A \to TA$ on set A. The arrow η_A is itself a function $\eta_A : A \to Q^Q \times A^Q$ defined on element a of A as the pair $(\text{id}_Q, \hat{a})$ where $\hat{a} : Q \to A$ is the function with constant value a.

Suppose $f : A \to TB = Q^Q \times B^Q$. Let the value of f on an element a be written $f(a) = (h_a, g_a) = (h_a : Q \to Q, g_a : Q \to B)$. The function $-^* : \mathbf{C}(A, TB) \to \mathbf{C}(TA, TB)$ is given on f as follows: $f^* : Q^Q \times A^Q \to Q^Q \times B^Q$ is the function defined by

$$f^*(t : Q \to Q, u : Q \to A) = (\lambda_q h_{uq}(tq), \lambda_q g_{uq}(tq))$$

in $Q^Q \times B^Q$. In more detail, if we let $(\tilde{t}, \tilde{u}) = (\lambda_q h_{uq}(tq), \lambda_q g_{uq}(tq))$, then $\tilde{t} : Q \to Q$ is defined by $\tilde{t}(q) = h_{uq}(tq)$ and $\tilde{u} : Q \to B$ is defined by $\tilde{u}(q) = g_{uq}(tq)$.

Proposition. *The functions T, η and $-^*$ given above determine a Kleisli triple $(T, \eta, -^*)$.*

Proof. Given $f : A \to TB$ and $g : B \to TC$,

(K1) $\eta_A(a) = (\mathrm{id}_Q, \hat{a}) = (h_a, g_a)$. But $\eta_A^*(t, u) = (\tilde{t}, \tilde{u})$ where $\tilde{t}(q) = h_{uq}(tq) = \mathrm{id}_Q(tq) = tq$. Thus $t = \tilde{t}$. Similarly $\tilde{u}(q) = g_{uq}(tq) = \widehat{uq}(tq) = uq$ and $\tilde{u} = u$. Thus $\eta_A^* = \mathrm{id}_{TA}$.

(K2) $f(a) = (h_a, g_a)$ and $f^* \eta_A(a) = f^*(\mathrm{id}_Q, \hat{a}) = (\widetilde{\mathrm{id}_Q}, \tilde{\hat{a}})$. But

$$\widetilde{\mathrm{id}_Q}(q) = h_{\hat{a}q}(\mathrm{id}_Q q) = h_{\hat{a}q}(q) = h_a(q).$$

Thus $\widetilde{\mathrm{id}_Q} = h_a$. Furthermore

$$\tilde{\hat{a}}(q) = g_{\hat{a}q}(\mathrm{id}_Q q) = g_{\hat{a}q}(q) = g_a(q).$$

Thus $f^* \eta_A = f$.

The verification of (K3) is longer but straightforward. $\qquad \square$

4.5. The filter monad. The Kleisli triple $(F, \eta, -^*)$ generating the filter monad on **Set** is easily described. It has $FX = \{\mathcal{F} \subset 2^X : \mathcal{F} \text{ is a filter on } X\}$ for each set X, $\eta_X : X \to FX$ defined by $\eta_X(x) = \mathrm{prin}(x) = \{A \subset X : x \in A\}$ and $\alpha^* : FX \to FY$ defined by $\alpha^*(\mathcal{F}) = \{B \subset Y : \{x \in X : B \in \alpha(x)\} \in \mathcal{F}\}$, for $\alpha : X \to FY$.

5. Eilenberg-Moore and Kleisli objects in a 2-category

In this section we show how the notions of algebra and coalgebra as well as that of the Eilenberg-Moore and Kleisli object can be described in a way which makes it possible to identify such objects in 2-categories other than **Cat**, e.g. in a slice category, without requiring a separate definition in each case.

Kelly and Street [18] define an Eilenberg-Moore object to be the representing object, when it exists, for a certain abstractly defined contravariant algebra functor from **K** to **Cat**. Such an algebra functor is determined by each monad defined on an object of **K**.

There are in fact two different algebra functors, one covariant and the other contravariant, whose representability lead us to the Kleisli and Eilenberg-Moore objects, respectively. For comonads (i.e. for monads in the dual category), there are likewise two, possibly representable, coalgebra functors.

We first define a 2-functor $\mathbf{Alg}t : \mathbf{K} \to \mathbf{CAT}$ of *algebras of an endomorphism t* for arbitrary fixed endomorphism t of a 2-category $\mathbf{K}$. When t is part of a monad $T = (t, \eta, \mu)$ then we examine a certain subfunctor $\mathbf{Alg}T$ of $\mathbf{Alg}t$ to see if it has a representing object. Such an object for $\mathbf{Alg}T$ will be called a *Kleisli object*.

5.1. Algebras of an endomorphism. Let $t : X \to X$ be a 1-cell in a 2-category $\mathbf{K}$.

Outgoing endomorphism algebras. The functor $\mathbf{Alg}t : \mathbf{K} \to \mathbf{CAT}$ is defined in the following way. For object D of $\mathbf{K}$, the category $(\mathbf{Alg}t)D = D^t$ has as objects those pairs (y, σ) such that y is an *outgoing* arrow from X, that is, $y : X \to D$, and $\sigma : yt \Longrightarrow y$ is a 2-cell of $\mathbf{K}$. The morphisms $a : (y, \sigma) \to (y', \sigma')$ of D^t are those 2-cells $a : y \Longrightarrow y'$ of $\mathbf{K}$ such that $a \cdot \sigma = \sigma' \cdot at$.

Given $f : D \to E$ of $\mathbf{K}$, the functor $(\mathbf{Alg}t)f = f^t : D^t \to E^t$ is defined by $f^t(y, \sigma) = (fy, f\sigma)$ and $f^t(a) = fa$. Given $\rho : f \Longrightarrow g$ of $\mathbf{K}$, the natural transformation $(\mathbf{Alg}t)\rho = \rho^t : f^t \Longrightarrow g^t : D^t \to E^t$ is defined on object (y, σ) of D^t by $\rho^t(y, \sigma) : f^t(y, \sigma) = (fy, f\sigma) \to (gy, g\sigma) = g^t(y, \sigma)$ in E^t which is just $\rho y : fy \Longrightarrow gy$ as a 2-cell in $\mathbf{K}$.

Outgoing endomorphism coalgebras. The category $(\mathbf{Coalg}t)D = D_t$ is the category with objects (y, ν) such that $y : X \to D$ and $\nu : y \Longrightarrow yt$ in $\mathbf{K}$ and with morphisms $a : (y, \nu) \to (y', \nu')$ those 2-cells $a : y \Longrightarrow y'$ of $\mathbf{K}$ such that $\nu' \cdot a = at \cdot \nu$. Given $f : D \to E$ of $\mathbf{K}$, the functor $(\mathbf{Coalg}t)f = f_t : D_t \to E_t$ is defined by $f_t(y, \nu) = (fy, f\nu')$, $f_t(a) = fa$ and for 2-cell $\rho : f \Longrightarrow g$ of $\mathbf{K}$ $(\mathbf{Coalg}t)\rho = \rho_t$ is defined by $\rho_t(y, \nu) = \rho_y$.

Incoming structures and the classical connection. For t, as before, we consider arrows $y : D \to X$ incoming to X with $\sigma : ty \Longrightarrow y$ for algebras or $\nu : y \Longrightarrow ty$ for coalgebras. The pairs (y, σ) are the objects of a category $(Alg_i t)D$. If $\mathbf{K} = \mathbf{CAT}$ and $D = \mathbf{1}$ then we have essentially the classical notions of Eilenberg-Moore algebras and coalgebras, but without laws.

Contravariance of incoming structures. In describing the incoming functor $\mathbf{Alg}_i t$ we see that the contravariance arises naturally since for $f : D \to E$ of $\mathbf{K}$ and $(y : E \to X, \sigma : ty \Longrightarrow y)$ of E_i^t, the functor $f_i^t : E_i^t \to D_i^t$ is defined on objects by $f_i^t(y, \sigma) = (yf, \sigma f : tyf \Longrightarrow yf)$. Thus $\mathbf{Alg}_i t : \mathbf{K}^{\mathrm{op}} \to \mathbf{CAT}$ and similarly $\mathbf{Coalg}_i t : \mathbf{K}^{\mathrm{op}} \to \mathbf{CAT}$ where, as usual, $\mathbf{K}^{\mathrm{op}}$ turns around the 1-cells of $\mathbf{K}$.

5.2. Algebras for a monad defined on an object of a 2-category. Let $t : X \to X$ be a 1-cell which is part of a monad $T = (t, \eta, \mu)$, that is, $\eta : 1 \Longrightarrow t$ and $\mu : t^2 \Longrightarrow t$

are 2-cells subject to the usual laws

$$\mu \cdot t\mu = \mu \cdot \mu t \text{ and } \mu \cdot \eta t = 1 = \mu \cdot t\eta.$$

This introduction of 2-cells, subject to equations, leads to an algebraic substructure which takes the additional 2-cells and equations into account. Accordingly, we define the *outgoing algebra* functor $\mathbf{Alg}T : \mathbf{K} \to \mathbf{CAT}$ for the monad $T = (t, \eta, \mu)$ to be the one obtained by restricting $\mathbf{Alg}t : \mathbf{K} \to \mathbf{CAT}$ by letting $(\mathbf{Alg}T)D = D^T$ be the full subcategory of D^t consisting of objects (y, σ) such that $\sigma \cdot y\eta = 1_y$ and $\sigma \cdot y\mu = \sigma \cdot \sigma t : yt^2 \Longrightarrow y$. The values of $\mathbf{Alg}T$ on 1- and 2-cells are given by the equations defining $\mathbf{Alg}t$ on such cells, only restricted in this case to categories of the form D^T.

For T and incoming algebras $\mathbf{Alg}_i T : \mathbf{K}^{\mathrm{op}} \to \mathbf{CAT}$ is obtained by making a similar restriction.

For the dual cases, we consider a comonad $G = (t, \varepsilon, \delta)$ and let $(\mathbf{Coalg}G)D = D_G$ be the full subcategory of D_t consisting of objects (y, ν) such that $y\varepsilon \cdot \nu = 1_y$ and $y\delta \cdot \nu = \nu t \cdot \nu$. We obtain $\mathbf{Coalg}G : \mathbf{K} \to \mathbf{CAT}$. Similarly $\mathbf{Coalg}_i G : \mathbf{K}^{\mathrm{op}} \to \mathbf{CAT}$.

5.3. Eilenberg-Moore and Kleisli objects.

Definitions. An object X_T of $\mathbf{K}$ is a *Kleisli object* for a monad $T = (t, \eta, \mu)$ on X if there is a natural isomorphism $\mathbf{Alg}T \Longrightarrow \mathbf{K}(X_T, -) : \mathbf{K} \to \mathbf{Cat}$. Similarly, an *Eilenberg-Moore object* X^T is one with an isomorphism $\mathbf{Alg}_i T \Longrightarrow \mathbf{K}(-, X^T)$. Finally, if $\mathbf{Coalg}G \Longrightarrow \mathbf{K}(X_G, -)$ or $\mathbf{Coalg}_i G \Longrightarrow \mathbf{K}(-, X^G)$ are natural isomorphisms, then X_G is a *co-Kleisli* or X^G is a *co-Eilenberg-Moore* object, respectively.

If $\mathbf{K} = \mathbf{Cat}$ then the Eilenberg-Moore object X^T is just the usual category of Eilenberg-Moore algebras for the monad. To prove this claim it is sufficient to show that there is an isomorphism

$$\mathbf{Alg}_i T(C) \to \mathbf{Cat}(C, X^T)$$

natural in C for C a category and X^T the category of Eilenberg-Moore algebras. We give the value of this isomorphism on an object of $\mathbf{Alg}_i T(C)$ and leave the rest of the details for the reader. An object of $\mathbf{Alg}_i T(C)$ is a pair (y, σ) where $y : C \to X$ is a functor and $\sigma : ty \Rightarrow y$ is a natural transformation for t the functor of the monad $T = (t, \eta, \mu)$ on X. Given (y, σ), a functor $B : C \to X^T$ is defined on an object c by letting $B(c)$ be the pair $(yc, \sigma_c : tyc \to yc)$ determined by

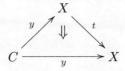

In particular, the correspondence between these notions of algebra becomes most transparent in the case where $C = 1$.

With the definition just given, the problem of identification, for example, of finding a co-Kleisli object, is that of finding a suitable candidate object Y and then finding a natural isomorphism $\mathbf{Coalg}G \Longrightarrow \mathbf{K}(Y, -)$ as we illustrate next.

Co-Kleisli in the slice. Given a 2-category $\mathbf{K}$, there is a slice 2-category defined for each object A. Each comonad on object $D : X \to A$ of $\mathbf{K}/A$ becomes a comonad on X when the projection functor $\mathbf{K}/A \to \mathbf{K}$ is applied.

Proposition. *Let $G = (t, \varepsilon, \delta)$ be a comonad on object $D : X \to A$ of $\mathbf{K}/A$. Suppose further that the co-Kleisli object X_G of G, as comonad on X, exists in $\mathbf{K}$. Then the co-Kleisli object D_G of G on D exists in $\mathbf{K}/A$ and is a $\mathbf{K}$-composite of the form $D \cdot L_G : X_G \to X \to A$ for suitable morphism $L_G : X_G \to X$.*

Proof. Let $\varphi : \mathbf{Coalg}G \to \mathbf{K}(X_G, -) : \mathbf{K} \to \mathbf{Cat}$ be the co-Kleisli isomorphism with inverse ψ. The pair (t, δ) is in $(\mathbf{Coalg}G)D$ since $D \cdot t = D$ and $D \cdot \delta = D$ which implies $t : D \to D$ in $\mathbf{K}/A$ and $\delta : t \to t^2 : D \to D$ and, secondly, since the coalgebra laws $t\varepsilon \cdot \delta = 1_t$ and $\delta t \cdot \delta = t\delta \cdot \delta$ hold because they are part of the comonad rules. After projecting $\mathbf{K}/A \to \mathbf{K}$ the pair (t, δ) is in $(\mathbf{Coalg}G)X$ and $\varphi(t, \delta) : X_G \to X$. We let $L_G = \varphi(t, \delta)$ and let $D_G = D \cdot L_G : X_G \to X \to A$.

The object $D_G : X_G \to A$ is the co-Kleisli object for the comonad defined on the object $D : X \to A$ of $\mathbf{K}/A$. To prove this requires providing a natural isomorphism

$$\Theta : \mathbf{Coalg}G \to (\mathbf{K}/A)(D_G, -) : \mathbf{K}/A \to \mathbf{Cat}.$$

This (rather lengthy) argument is presented next.

We do this by finding natural transformations Θ and Φ such that for each $f : E \to F$ in $\mathbf{K}/A$, with $E : Y \to A$ and $F : Z \to A$, the diagram

$$
\begin{array}{ccccc}
(\mathbf{Coalg}G)E & \xrightarrow{\Theta_E} & \mathbf{K}/A(D_G, E) & \xrightarrow{\Phi_E} & (\mathbf{Coalg}G)E \\
\downarrow{\scriptstyle f^\#} & & \downarrow{\scriptstyle f^*} & & \downarrow{\scriptstyle f^\#} \\
(\mathbf{Coalg}G)F & \xrightarrow{\Theta_F} & \mathbf{K}/A(D_G, F) & \xrightarrow{\Phi_F} & (\mathbf{Coalg}G)F
\end{array}
$$

commutes with Θ and Φ inverse. The vertical maps are clearly defined.

The initial step is to define $\Phi_{D_G}(1_{D_G})$. The co-Kleisli isomorphism

$$\varphi : \mathbf{Coalg}G \to \mathbf{K}(X_G, -)$$

with inverse ψ provides us with a commutative diagram for each $\mathbf{K}$ morphism $f : Y \to Z$, namely,

$$
\begin{array}{ccccc}
(\mathbf{Coalg}G)Y & \xrightarrow{\varphi_Y} & \mathbf{K}(X_G, Y) & \xrightarrow{\psi_Y} & ((\mathbf{Coalg}G)Y \\
\downarrow{\scriptstyle f^\#} & & \downarrow{\scriptstyle f^*} & & \downarrow{\scriptstyle f^\#} \\
(\mathbf{Coalg}G)Z & \xrightarrow{\varphi_Z} & \mathbf{K}(X_G, Z) & \xrightarrow{\psi_Z} & (\mathbf{Coalg}G)Z
\end{array}
$$

with ψ_Y inverse to φ_Y for each object Y of $\mathbf{K}$.

Now 1_{D_G} as a $\mathbf{K}$-morphism is 1_{X_G} and $\psi_{X_G}(1_{X_G}) = (R_G, \nu_R)$ in $(\mathbf{Coalg}G)X_G$, where $R_G : X \to X_G$ and $\nu_R : R_G \to R_G t$ satisfy $R_G\varepsilon \cdot \nu_R = 1_{R_G}$ and $R_G\delta \cdot \nu_R = \nu_R t \cdot \nu_R : R_G \to R_G t^2$. Naturality of ψ yields

$$\psi_Y(f : X_G \to Y) = \psi_Y(f * (1_{X_G})) = f^{\#}\psi_{X_G}(1_{X_G}) = f^{\#}(R_G, \nu_R) = (fR_G, f\nu_R).$$

Thus, starting with (t, δ) in $(\mathbf{Coalg}G)D$ we have

$$(t, \delta) = \psi_D\varphi_D(t, \delta) = \psi_D L_G = (L_G R_G, L_G \nu_R).$$

Hence $t = L_G R_G$ and $\delta = L_G \nu_R$.

Thus we may define $\Phi_{D_G}(1_{D_G})$ to be (R_G, ν_R), because

$$D_G R_G = D \cdot L_G \cdot R_G = Dt = D \text{ and } D_G \nu_R = D \cdot L_G \nu_R = D\delta = D,$$

and so $\Phi_{D_G}(1_{D_G}) = (R_G, \nu_R)$ is in $(\mathbf{Coalg}G)D_G$, since the coalgebra equations for (R_G, ν_R) in $(\mathbf{Coalg}G)D_G$ follow from the equations for (R_G, ν_R) in $(\mathbf{Coalg}G)X_G$. It is now easy to define $\Phi_E(f)$ in general by

$$\Phi_E(f) = \Phi_E(f * (1_{D_G})) = f^{\#}\Phi_{D_G}(1_{D_G}) = f^{\#}(R_G, \nu_R) = (fR_G, f\nu_R).$$

Naturality of Φ follows since

$$\Phi_F f^*(g : D_G \to E) = \Phi_F f * g * (1_{D_G}) = \Phi_F(fg) * (1) = \Phi_F(fg)$$

$$= (fg)^{\#}\Phi_{D_G}(1_{D_G}) = f^{\#}g^{\#}\Phi_{D_G}(1_{D_G}) = f^{\#}\Phi_E(g).$$

Conversely, starting with (R_G, ν_R) in $(\mathbf{Coalg}G)D_G$, let $\Theta_{D_G}(R_G, \nu_R) = 1_{D_G}$. Then for arbitrary (y, ν) in $(\mathbf{Coalg}G)E$ we show that the corresponding $\varphi_Y(y, \nu) : X_G \to Y$ may be regarded as a morphism $f : D_G \to E$. Then let

$$\Theta_E(y, \nu) = \Theta_E f^*(R_G, \nu_R) = f^*\Theta_{D_G}(R_G, \nu_R)$$

define Θ_E. The values of Θ_E and Φ_E on morphisms are just those of φ_Y and ψ_Y, respectively. $\qquad\square$

6. Monads, idempotent monads, and commutative monads as algebras

In [29] Mac Lane gives a presentation of the simplicial category Δ by generators and relations and shows that Δ contains a monoid universal for monoids in strict monoidal categories. It is then only a few short steps to the characterization of the category of 1-monads in a 2-category $\mathbf{A}$ with tensor products as the algebras for a 2-monad $T = (T, \eta, \mu)$ on $\mathbf{A}$ with $T = \Delta \otimes -$. Using results contained in the thesis of Peter Stone [37] we show that there is a similar characterization of the category of commutative monads in terms of Δ_C and of the category of idempotent monads in terms of Δ_Q, respectively, where Δ_C is the category of finite ordinals and all functions and Δ_Q is the quotient category of Δ obtained by identifying the morphisms in each given non-empty hom set. More details of the canonical form for Δ_C morphisms appear in [37].

6.1. Monoidal background. A *strict monoidal category* $(D, +, 0)$ consists of a category D, a functor $+ : D \times D \to D$ and an object 0 of D such that $+$ is associative and 0 is the identity.

A *monoid* (c, η, μ) in $(D, +, 0)$ consists of an object c of D together with morphisms $\eta : 0 \to c$ and $\mu : c + c \to c$ such that $\mu(\eta + c) = 1 = \mu(c + \eta)$ and $\mu(\mu + c) = \mu(c + \mu)$.

Such a $(D, +, 0)$ can be regarded in an obvious way as a 2-category $\mathbf{D}$ with a single object $*$ and a monoid (c, η, μ) as a monad on the single object $*$ of $\mathbf{D}$. A monoidal functor $(D, +, 0) \to (D', +', 0')$ can be identified with a 2-functor $\mathbf{D} \to \mathbf{D}'$ between the corresponding 2-categories.

If $\mathbf{A}$ is a 2-category, then for each object B of $\mathbf{A}$ the hom category $\mathbf{A}(B, B)$ is strict monoidal and a monad (B, t, η, μ) is a monoid in $\mathbf{A}(B, B)$ (see 1.1).

Let A be an object of a 2-category $\mathbf{A}$ and let D be a small category. A tensor product $D \otimes A$ is an object of $\mathbf{A}$ for which there is an isomorphism of categories

$$\mathbf{A}(D \otimes A, B) \to \mathbf{Cat}(D, \mathbf{A}(A, B)) \tag{29}$$

which is natural in B. We say that $\mathbf{A}$ has tensor products if $D \otimes A$ exists in $\mathbf{A}$ for each A in $\mathbf{A}$ and each small category D. If $\mathbf{A}$ has tensor products, then $\otimes$ extends to a 2-functor $\mathbf{Cat} \times \mathbf{A} \to \mathbf{A}$.

Proposition. *Let $(D, +, 0)$ be a strict monoidal category and $\mathbf{A}$ be a 2-category with tensor products. Let $T = D \otimes - : \mathbf{A} \to \mathbf{A}$ and let $\eta : 1 \to T$, $\mu : T^2 \to T$ be the strictly natural transformations given by*

- $\eta_A = \sigma_{(D, A)}(0) : A \to D \otimes A$
- $\mu_A : D \otimes (D \otimes A) \to (D \times D) \otimes A \to D \otimes A,$

where, under the isomorphism (29)*, $\sigma_{(D, A)} : D \to \mathbf{A}(A, D \otimes A)$ corresponds to the identity 1-cell on $D \otimes A$ and η_A is just the value of $\sigma_{(D, A)}$ on the object 0. Then (T, η, μ) is a strict 2-monad.*

We note that the isomorphism $D \otimes (E \otimes A) \to (D \times E) \otimes A$ arises from a repeated use of (29), namely

$$\mathbf{A}(D \otimes (E \otimes A), B) \to \mathbf{Cat}(D, \mathbf{A}(E \otimes A, B)) \to \mathbf{Cat}(D, \mathbf{Cat}(E, \mathbf{A}(A, B))) \to$$
$$\to \mathbf{Cat}(D \times E, \mathbf{A}(A, B)) \to \mathbf{A}((D \times E) \otimes A), B).$$

6.2. The algebras of a 2-monad. Let (T, η, μ) be a 2-monad defined on a 2-category $\mathbf{A}$. A (strict) T-algebra (B, p) is an object B of $\mathbf{A}$ together with a 1-cell $p : TB \to B$ such that $p\eta_B = 1$ and $p\mu_B = pTp$. A morphism of T-algebras $(B, p) \to (B', p')$ is a pair (h, α) where $h : B \to B'$ is a 1-cell and $\alpha : p'Th \to hp$ is a 2-cell satisfying $\alpha \cdot \eta_B = 1$ and $\alpha \cdot \mu_B = \alpha Tp \cdot p'T\alpha$, where a central dot represents 2-cell composition. A 2-cell $\varphi : (h, \alpha) \to (g, \beta)$ is a 2-cell $\varphi : h \to g$ such that $\beta \cdot p'T\varphi = \varphi p \cdot \alpha$. For more information see Blackwell, Kelly, Power [9].

Proposition. *Let $(D, +, 0)$ be a strict monoidal category and let (T, η, μ) be the 2-monad of Proposition 6.1 with $T = D \otimes -$. Then $p : D \otimes B \to B$ is a T-algebra structure map if and only if the corresponding functor $\Phi : D \to \mathbf{A}(B, B)$ is a monoidal functor, where $\mathbf{A}(B, B)$ is given the obvious strict monoidal category structure.*

Corollary. *Let $(D \otimes -, \eta, \mu)$ be a strict 2-monad on $\mathbf{A}$ and let $\mathbf{D}$ be a one object 2-category both of which correspond to a strict monoidal category $(D, +, 0)$. Then the 2-category of D-algebras is a 2-category whose objects correspond to strict 2-functors $\mathbf{D} \to \mathbf{A}$ under (29).*

Proof. By the previous proposition a D-algebra corresponds to a monoidal functor $\Phi : D \to \mathbf{A}(B, B)$. This corresponds to a 2-functor $\mathbf{D} \to \mathbf{A}$. □

6.3. Monoidal Categories of Ordinals. Let **Mnd** be the 2-category freely generated by an object $*$, one non-identity 1-cell t and two 2-cells $\eta : 1 \to t$ and $\mu : t^2 \to t$ subject to equations

$$\mu \cdot t\mu = \mu \cdot \mu t \text{ and } \mu \cdot \eta t = 1 = \mu \cdot t\eta. \tag{30}$$

Suppose that D is a category whose objects are all finite ordinal numbers $n = [0, 1, \cdots, n-1]$. These objects may be regarded as the free monoid on 1 under the usual ordinal addition.

Now let Δ be such a category D with morphisms all weakly monotone functions $f : n \to n'$, that is, $0 \leq i \leq j \leq (n-1)$ implies $f(i) \leq f(j)$ (cf. Mac Lane [29]). Thus $(\Delta, +, 0)$ is a strict monoidal category if we define $+$ on morphisms $f : n \to n'$, $g : m \to m'$ by $(f + g)(i) = f(i)$ for $0 \leq i \leq n-1$ and by $(f+g)(i) = n' + g(i-n)$ for $n \leq i \leq n+m-1$.

Let $(1, \eta, \mu)$ be the monoid in $(\Delta, +, 0)$ determined by the unique morphisms $\mu : 2 \to 1$ and $\eta : 0 \to 1$ to the terminal object 1. Thus the analogues of the equations of (30) hold, namely,

$$\mu \cdot (1 + \mu) = \mu \cdot (\mu + 1) : 3 \to 1 \text{ and } \mu \cdot (\eta + 1) = \mathrm{id}_1 = \mu \cdot (1 + \eta). \tag{31}$$

Lemma. *There is a unique strict 2-functor $\mathbf{Mnd} \to \Delta$ taking η to η, μ to μ and t to 1 where Δ is just $(\Delta, +, 0)$ considered as a 2-category.*

Observe that this functor takes t^2 to $1+1$ and μt to $\mu + 1$.

We consider two further examples of categories whose objects are the finite ordinals. One of them Δ_Q is the quotient category of Δ defined by introducing into each non-empty hom set of Δ the equivalence relation identifying all the morphisms in that hom set. The operations of $+$ and composition in Δ induce well defined operations in Δ_Q so that $(\Delta_Q, +, 0)$ becomes a strict monoidal category. Let $\boldsymbol{\Delta_Q}$ be the associated 2-category.

Let Δ_C be the category whose morphisms are all functions $f : n \to n'$. We can define $+$ on such functions in the same way as in Δ. Let $\mathbf{\Delta}_C$ be the 2-category associated to the monoidal category $(\Delta_C, +, 0)$.

As 2-categories we have a diagram

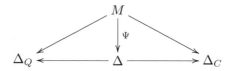

6.4. Δ and monads. Mac Lane in [29] shows that the monoid $(1, \eta, \mu)$ in $(\Delta, +, 0)$ is universal in the sense that for any other monoid (c, η, μ) in a strict monoidal category $(B, \times, e)$ there is a unique monoidal functor $(\Delta, +, 0) \to (B, \times, e)$ taking $(1, \eta, \mu)$ to (c, η, μ). The proof relies on "showing that the arrows of Δ are exactly the iterated formal products (for the binary product μ)."

Proposition. *Given a monad* $\mathbf{B} = (B, t, \eta, \mu)$ *defined on the object B of a 2-category* $\mathbf{A}$, *then there exists a unique monoidal functor,* $\Phi_{\mathbf{B}} : \Delta \to \mathbf{A}(B, B)$ *such that* $\Phi_{\mathbf{B}}(1) = t$, $\Phi_{\mathbf{B}}(\eta) = \eta$ *and* $\Phi_{\mathbf{B}}(\mu) = \mu$. *The functor $\Phi_{\mathbf{B}}$ may also be regarded as a 2-functor* $\Delta \to \mathbf{A}(B, B)$ *between one object 2-categories.*

Proof. A monad (B, t, η, μ) in $\mathbf{A}$ may be regarded as a monoid in the strict monoidal category $\mathbf{A}(B, B)$. The Proposition then follows from Mac Lane's result stated above. $\qquad\square$

Corollary A. *Let* (B, t, η, μ) *be a 1-monad in a 2-category* $\mathbf{A}$ *with tensor products. Then there is a unique Δ-algebra structure $\Delta \otimes B \to B$ for the 2-monad $(\Delta \otimes -, \eta, \mu)$ whose image under* (29) *is $\Phi_{\mathbf{B}}$.*

Proof. It follows from Propositions 6.4 and 6.2. $\qquad\square$

The correspondence of this corollary can be extended to an isomorphism

$$\mathbf{Mon}(\mathbf{A}) \to \Delta\text{-}\mathbf{Alg}(\mathbf{A}).$$

We use the definition of 1- and 2-cells in the 2-category Δ-$\mathbf{Alg}(\mathbf{A})$ of Δ-algebras for the 2-monad $(\Delta \otimes -, \eta, \mu)$, as defined in Section 6.1. For $\mathbf{Mon}(\mathbf{A})$ a 1-cell

$$(h, \alpha) : (B, t, \eta, \mu) \to (B', t', \eta', \mu')$$

consists of a 1-cell $h : B \to B'$ and a 2-cell $\alpha : t'h \to ht$ such that $\alpha \cdot \eta'h = h\eta$ and $\alpha\mu'h = \mu \cdot (\alpha\alpha)$. The 2-cells $\phi : (h, \alpha) \to (g, \beta)$ are 2-cells $\phi : h \to g$ such that $\beta \cdot t'\phi = \phi t \cdot \alpha$.

Corollary B. *The functor* $\Psi : \mathbf{Mnd} \to \Delta$ *has an inverse.*

Proof. Clearly $(*, t, \eta, \mu)$ is a monad in $\mathbf{Mnd}$. Thus by Proposition 6.4 there is a unique monoidal functor $\Phi : \Delta \to \mathbf{Mnd}(*, *)$ taking $(1, \eta, \mu)$ to (t, η, μ). Thus we have a 2-functor $\Delta \to \mathbf{Mnd}$. The universal properties indicated in 6.3 and 6.4 combine to show that Φ is inverse to Ψ. $\qquad\qquad\qquad\qquad\qquad\qquad\square$

A well known Lemma from [29] gives a unique representation of each morphism of Δ into canonical form involving morphisms of the form

$$\delta_i^n = 1_i + \eta + 1_{n-i} : n \to (n+1), \quad i = 0, \cdots, n$$

and

$$\sigma_i^n = 1_i + \mu + 1_{n-i-1} : (n+1) \to n, \quad i = 0, \cdots, n-1.$$

Putting a morphism into canonical form can be regarded as a formal set of moves involving some laws holding between particular composites of morphisms involving δ and μ. These laws are included in a larger set of rules involving commutative monads in the next section.

6.5. Δ_C and commutative monads. A *commutative monad* $(B, t, \eta, \mu, \lambda)$ in a 2-category $\mathbf{A}$ consists of an object B of $\mathbf{A}$, a 1-cell $t : B \to B$ and 2-cells $\eta : 1 \to t$, $\mu : t^2 \to t$ and $\lambda : t^2 \to t^2$ such that the following relations hold.

(CM1) $\mu \cdot t\eta = \mu \cdot \eta t = \mathrm{id}$
(CM2) $\mu \cdot t\mu = \mu \cdot \mu t$
(CM3) $\lambda \cdot \lambda = \mathrm{id}$
(CM4) $t\lambda \cdot \lambda t \cdot t\lambda = \lambda t \cdot t\lambda \cdot \lambda t$
(CM5) $\lambda \cdot t\eta = \eta t$ and $\lambda \cdot \eta t = t\eta$
(CM6) $\mu \cdot \lambda = \mu$
(CM7) $\mu t \cdot t\lambda \cdot \lambda t = \lambda \cdot t\mu$
(CM8) $t\mu \cdot \lambda t \cdot t\lambda = \lambda \cdot \mu t$.

There are some redundancies in these relations. In view of (CM3) either part of (CM5) implies the other; in view of (CM5) and (CM6), either part of (CM1) implies the other.

Example. Let G be an abelian group and let the following commutative monad be defined on a (large) category. The endofunctor $t : \mathbf{Set} \to \mathbf{Set}$ is given by $tS = G \times S$. Furthermore, $\eta_S : S \to G \times S$ and $\mu_S : G \times G \times S \to G \times S$ are defined by $(\eta_S)(x) = (e, x)$ and $(\mu_S)(g, g', x) = (gg', x)$ (see the monad considered in 1.1 taking a group G instead of a monoid M).

Let $\lambda : t^2 \to t^2$ be defined by

$$(\lambda_S)(g, g', x) = (g', g, x).$$

Then $(\mathbf{Set}, t, \eta, \mu, \lambda)$ is a commutative monad.

There is clearly an inclusion functor

$$(\Delta, +, 0) \to (\Delta_C, +, 0)$$

of monoids. Along with the morphisms $\tilde{\eta} : 0 \to 1$ and $\tilde{\mu} : 2 \to 1$ which generate Δ, we single out the isomorphism $\tilde{\lambda} : 2 \to 2$ given by $\tilde{\lambda}(0) = 1$ and $\tilde{\lambda}(1) = 0$.

A *commutative monoid* (c, η, μ, λ) in a strict monoidal category $(D, +, 0)$ consists of an object c of D together with morphisms $\eta : 0 \to c$, $\mu : c + c \to c$ and $\lambda : c + c \to c + c$ satisfying the axioms

(cm1) $\mu \cdot (c + \eta) = \mu \cdot (\eta + c) = \mathrm{id}$
(cm2) $\mu \cdot (c + \mu) = \mu \cdot (\mu + c)$
(cm3) $\lambda \cdot \lambda = \mathrm{id}$
(cm4) $(c + \lambda) \cdot (\lambda + c) \cdot (c + \lambda) = (\lambda + c) \cdot (c + \lambda) \cdot (\lambda + c)$
(cm5) $\lambda \cdot (c + \eta) = \eta + c$ and $\lambda \cdot (\eta + c) = c + \eta$
(cm6) $\mu \cdot \lambda = \mu$
(cm7) $(\mu + c) \cdot (c + \lambda) \cdot (\lambda + c) = \lambda \cdot (c + \mu)$
(cm8) $(c + \mu) \cdot (\lambda + c) \cdot (c + \lambda) = \lambda \cdot (\mu + c)$.

Lemma A. $(1, \tilde{\eta}, \tilde{\mu}, \tilde{\lambda})$ *is a commutative monoid in* $(\Delta_C, +, 0)$.

Lemma B. *A commutative monoid* (c, η, μ, λ) *in a strict monoidal category* $(D, +, 0)$ *is a commutative monad in* D *considered as a one object 2-category* $\mathbf{D}$. *Conversely, a commutative monad* $(B, t, \eta, \mu, \lambda)$ *defined on an object* B *of the 2-category* $\mathbf{D}$ *is a commutative monoid in* $\mathbf{D}(B, B)$.

Proposition. *Any morphism* $\xi : m \to n$ *of* Δ_C *has a unique factorization*

$$\xi = \xi_\Delta \cdot \xi_P \tag{32}$$

where $\xi_\Delta : m \to n$ *is a morphism of* Δ *and* $\xi_P : m \to m$ *is a permutation with the additional property that if* $i \leq (j - 1)$ *and* $\xi(i) = \xi(j)$, *then* $\xi_P(i) \leq \xi_P(j) - 1$. *That is,* ξ_P *does not change the order of any points which have the same image under* ξ *or, more simply,* ξ_P *has no "unnecessary switching".*

We call ξ_P the *switching part* of ξ and ξ_Δ the *monotonic part* of ξ. We call (32) the *canonical factorization* of the morphism ξ.

6.6. Canonical factorizations. We next give background information and definitions needed to describe a canonical form for permutations of Δ_C.

Let $\lambda^{(h,k)} : (h + k) \to (h + k)$ for $h, k \geq 1$ be given by

$$\lambda^{(h,k)}(i) = (i + k) \text{ if } 0 \leq i \leq (h - 1)$$

and

$$\lambda^{(h,k)}(i) = (i - h) \text{ if } h \leq i \leq (h + k - 1).$$

In particular,

$$\lambda^{(h,1)}(i) = \begin{cases} i + 1 & \text{if } 0 \leq i \leq (h - 1). \\ 0 & \text{if } i = h. \end{cases}$$

Then

$$\lambda^{(2,1)} = (\tilde{\lambda} + 1)(1 + \tilde{\lambda}) \text{ and } \lambda^{(1,2)} = (1 + \tilde{\lambda})(\tilde{\lambda} + 1).$$

Similarly it follows that

$$\lambda^{(h+1,1)} = (\tilde{\lambda} + h)(1 + \lambda^{(h,1)}) \text{ and } \lambda^{(1,k+1)} = (k + \tilde{\lambda})(\lambda^{(1,k)} + 1),$$

which can be taken as the recursive definition of $\lambda^{(h,1)}$ and $\lambda^{(1,k)}$ in terms of $\tilde{\lambda}$.
A permutation $\sigma : (h+k+m+1) \to (h+k+m+1)$ of the form $\sigma = h + \lambda^{(k,1)} + m$ with $k \geq 1$ is called an *elementary permutation*. It is a special type of cycle. We define the *height* of such an elementary permutation to be $\mathbf{H}(C) = (h+k+1)$.

Lemma. *For any non-identity permutation $\xi : n \to n$ where $n \geq 2$, there are unique elementary permutations $\sigma_p, \sigma_{p-1}, \cdots, \sigma_1$ such that*

$$\xi = \sigma_p \sigma_{p-1} \cdots \sigma_1$$

and $\mathbf{H}(\sigma_i) \leq \mathbf{H}(\sigma_{i+1}) - 1$ for $i = 1, \cdots, p-1$. We call such a representation the canonical factorization of the permutation ξ.

Furthermore note that if $\xi : n \to n$ and $\zeta : n \to n$ are two permutations given in canonical form, then the composite permutation can be obtained in canonical form using the relations (cm3) and (cm4).

Canonical factorizations of functions. We can now describe a canonical form for a general morphism $\xi : m \to n$ of Δ_C namely: $\xi = \xi_\Delta \cdot \xi_P$ where ξ_P is the switching part of ξ and ξ_Δ is the monotonic part of ξ with each of ξ_P and ξ_Δ given in their respective canonical forms, that is,

$$\xi_\Delta = \mu^{m_0} + \mu^{m_1} + \cdots + \mu^{m_{n-1}}$$

where m_i is the number of elements in $\xi_\Delta^{-1}(i)$ or $\xi^{-1}(i)$ for $i = 1, \cdots, n$ and

$$\xi_P = \sigma_p \sigma_{p-1} \cdots \sigma_1$$

where each σ_j is an elementary permutation and $\mathbf{H}(\sigma_j) \leq \mathbf{H}(\sigma_{j+1}) - 1$ for $j = 1, \cdots, p-1$.

We next state two Propositions regarding the canonical form.

Proposition A. *Let $\xi_\Delta : m \to n$ be a morphism of Δ given in canonical form and let $\zeta_P : m \to m$ be a permutation also given in the canonical form*

$$\zeta_P = \sigma_q \sigma_{q-1} \cdots \sigma_1$$

where σ_i is an elementary permutation for each i and $\mathbf{H}(\sigma_i) \leq \mathbf{H}(\sigma_{i+1}) - 1$ for $i \leq q-1$. Then there are indices $i_1, \cdots, i_k$ with $0 \leq i_1 \leq i_2 \cdots \leq i_k$ such that if

$$\xi_P = \sigma_{i_k} \sigma_{i_{k-1}} \cdots \sigma_{i_1}$$

then $\xi_\Delta \cdot \xi_P$ is the canonical factorization of $\xi = \xi_\Delta \cdot \zeta_P$ with monotonic part ξ_Δ and switching part ξ_P (which now contains no unnecessary switching).

Furthermore this can be achieved by purely formal algebraic means using the relations (cm2), (cm6) and (cm8).

Proposition B. *Let $\xi : m \to n$ and $\zeta : n \to k$ be morphisms of Δ_C each given in canonical form. Then the composite morphism $\zeta \cdot \xi : m \to k$ can be obtained in canonical form by purely formal algebraic means using the relations (cm1) to (cm8).*

6.7. Commutative monads and Δ_C-algebras. In this section we show that the 2-category of commutative monads in the 2-category **A** is isomorphic to the 2-category of Δ_C-algebras in **A**.

Proposition. *Given a commutative monoid (c, η, μ, λ) in a strict monoidal category $(D, +, 0)$ there is a unique monoidal functor $F : (\Delta_C, +, 0) \to (D, +, 0)$ such that $F1 = c$, $F\tilde{\mu} = \mu$, $F\tilde{\eta} = \eta$, and $F\tilde{\lambda} = \lambda$.*

Corollary. *Let $(B, t, \eta, \mu, \lambda)$ be a commutative monad in a 2-category **A**. Then there is a unique 2-functor $\Phi : \Delta_C \to \mathbf{A}$ such that*

$$\Phi 1 = t, \Phi\tilde{\mu} = \mu, \Phi\tilde{\eta} = \eta \text{ and } \Phi\tilde{\lambda} = \lambda.$$

Theorem. *The 2-category $\mathbf{Mon}_C(\mathbf{A})$ of commutative monads in the 2-category **A** is isomorphic to the 2-category of Δ_C-algebras in **A**.*

Proof. D-algebras were identified with monoidal functors $D \to \mathbf{A}(B, B)$ in 6.2. For $D = \Delta_C$, these correspond by proposition above to commutative monads. The 1- and 2-cells of $\mathbf{Mon}_C(\mathbf{A})$ are defined in exactly the same fashion as those of $\mathbf{Mon}(\mathbf{A})$. Remaining details follow the same pattern as those for Δ.   □

6.8. Generators and relations for the category Δ_C. We now consider the problem of finding generators and relations for Δ_C using only the operation of composition. In order to do this we extend ideas already available for the corresponding problem for Δ.

A morphism of Δ_C of the form

$$i + \tilde{\lambda} + n - i - 2 : n \to n$$

where $n \geq 2$ and $0 \leq i \leq (n - 2)$ is called a *primary permutation*. The formula

$$\lambda^{(k,1)} = (\tilde{\lambda} + k - 1)(1 + \tilde{\lambda} + k - 2) \cdots (k - 1 + \tilde{\lambda})$$

gives a canonical factorization of an elementary permutation $h + \lambda^{(k,1)} + m$ as a composition of primary permutations.

Theorem. *The category Δ_C, with objects all finite ordinals, is generated by the monotonic and injective morphisms $n \to (n + 1)$, the monotonic and surjective morphisms $(n + 1) \to n$ and the primary permutations $n \to n$ subject to certain relations (see [37]).*

6.9. Δ_Q and idempotent monads. In this section we show that the 2-category of idempotent monads in the 2-category **A** is isomorphic to the 2-category of Δ_Q-algebras in **A**.

Proposition. *Let (B, t, η, μ) be a monad in a 2-category **A**. Then the following conditions are equivalent:*

(i) $t\eta = \eta t$
(ii) $t\eta = \eta t$
(iii) μ *is an isomorphism.*

A monad (B, t, η, μ) in a 2-category $\mathbf{A}$ which satisfies any one of these three equivalent conditions is called an *idempotent monad.*

Recall that for a monad (B, t, η, μ) in a 2-category $\mathbf{A}$, $\mu^k : t^2 \to t$ is defined recursively by

- $\mu^0 = \eta : 1 \to t$
- $\mu^1 = 1 : t \to t$
- $\mu^2 = \mu : t^2 \to t$
- $\mu^k = \mu \cdot \mu^{(k-1)} t : t^2 \to t$ for $k = 3, 4, \cdots$

or equivalently, if $\Phi : (\Delta, +, 0) \to \mathbf{A}(B, B)$ is the monoidal functor corresponding to (B, t, η, μ), then $\mu^k = \Phi(\mu^k)$.

Let $k\tilde\eta : 0 \to k$ be the "k summand" morphism of Δ given by

$$k\tilde\eta = \tilde\eta + \cdots + \tilde\eta : 0 \to k$$

and write $\eta^k = \eta \cdots \eta : 1 \to t^k$ for the "k factors" morphism so that

$$\Phi(k\tilde\eta) = \eta^k.$$

Recall that Δ_Q is the quotient category of Δ defined by introducing into each nonempty hom-set of Δ the equivalence relation which identifies all the morphisms in that hom-set. In particular, for $n, k \geq 0$, every morphism $n \to (n + k)$ is identified with $n + k\tilde\eta$ and every morphism $n + k \to n + 1$ is identified with $n + \mu^k$. Then, denoting the equivalence class containing a morphism ξ by $[\xi]$, we have $[n + k\tilde\eta] = [k\tilde\eta + n]$ and $[n + \mu^k] = [\mu^k + n]$.

The operations of $+$ and composition in Δ induce well defined operations of $+$ and composition in Δ_Q, so that $(\Delta_Q, +, 0)$ becomes a strict monoidal category. We shall show that idempotent monads can be regarded as Δ_Q-algebras.

Lemma. *In* $(\Delta_Q, +, 0)$ *the following relations hold for* $h, k \geq 0$:

(Q1) $(h + k\tilde\eta)h\tilde\eta = (h + k)\eta$
(Q2) $\mu^{(h+1)}(h + \mu^{(k+1)}) = \mu^{(h+k+1)}$
(Q3) $\mu^{(h+1)}(1 + k\tilde\eta) = 1$
(Q4) $h + k\tilde\eta = k\tilde\eta + h$
(Q5) $h + \mu^k = \mu^k + h$
(Q6) $(1 + k\tilde\eta)\mu^{(k+1)} = k + 1$
(Q7) $h\tilde\eta + k\tilde\eta = (h + k)\tilde\eta$
(Q8) $\mu^h + \mu^k = 1 + \mu^{(h+k-1)}$
(Q9) $\mu^{(h+1)} + k\tilde\eta = k\tilde\eta + \mu^{(h+1)}$ *which equals* $h + \mu^{(h-k+1)}$ *if* $h \geq k$ *and equals* $(k - h)\eta$ *if* $h \leq (k - 1)$.

Proposition. *Let (B, t, η, μ) be an idempotent monad in a 2-category $\mathbf{A}$. Then there is a unique monoidal functor*

$$\Phi : (\Delta, +, 0) \to \mathbf{A}(B, B)$$

such that $\Phi(1) = t$, $\Phi(\tilde{\eta}) = \eta$ and $\Phi(\tilde{\mu}) = \mu$.

Proof. In order that Φ becomes a monoidal functor we must set $\Phi(n) = t^n$, $\Phi(n + k\tilde{\eta}) = t^n\eta^k$ and $\Phi(n + \mu^k) = t^n\mu^k$ for $n, k \geq 0$. With this definition, all the relations (Q1) to (Q9) are preserved under Φ. Since these relations determine the structure of $(\Delta_C, +, 0)$ as a monoidal category, it follows that Φ is indeed a monoidal functor. $\qquad\qquad\square$

Let $\mathbf{A}$ be a 2-category with tensor products. Then, given an idempotent monad (B, t, η, μ) in $\mathbf{A}$, we have a corresponding monoidal functor

$$\Phi : (\Delta_Q, +, 0) \to \mathbf{A}(B, B).$$

Let $p : \Delta_Q \otimes B \to B$ be the corresponding 1-cell. Then p is the structure map for a Δ_Q-algebra (B, p). Thus we see that idempotent monads can be identified with the objects of the 2-category Δ_Q-**Alg**$(\mathbf{A})$. Identifying Mon$(\mathbf{A})$ with Δ-**Alg**$(\mathbf{A})$ in the manner described in Section 3 we can regard Δ_Q-**Alg**$(\mathbf{A})$ as a 2-subcategory of Δ-**Alg**$(\mathbf{A})$. The inclusion 2-functor is induced by the quotient functor $Q : \Delta \to \Delta_Q$.

If (B, p) is an idempotent monad (Δ_Q-algebra) with structure map $p : \Delta_Q \otimes B \to B$, then its structure map as a monad (Δ-algebra) is given by the composition

$$p \cdot (Q \otimes 1) : \Delta \otimes B \to \Delta_Q \otimes B \to B.$$

References

[1] J. Adámek, Colimits of algebras revisited, *Bull. Austral. Math. Soc.* **17** (1977) 433-450.

[2] J. Adámek and G. M. Kelly, $\mathcal{M}$-completeness is seldom monadic over graphs, *Theory and Appl. of Categories*, Vol. **7**, No. 8 (2000) 171-205.

[3] R. Baer, Free sums of groups and their generalizations, *Amer. J. Math.* **71** (1949) 708-742.

[4] M. Barr and J. Beck, Homology and standard constructions, in: *Seminar on triples and categorical homology*, Lecture Notes in Mathematics **80** (Springer-Verlag 1969), pp. 245-335.

[5] M. Barr and C. Wells, *Toposes, Triples and Theories*, Springer-Verlag, 1985.

[6] M. Barr and C. Wells, *Category Theory for Computing Science*, Prentice Hall, 1990.

[7] J. Bénabou, *Treillis locaux et paratopologies*, Séminaire C. Ehresmann(1957/58) Fac. Sciences de Paris (1959).

[8] J. Beck, *Triples, Algebras and Cohomology*, Dissertation, Columbia University, 1967.

[9] R. Blackwell, G. M. Kelly and J. Power, *Two-dimensional monad theory*, Sydney Category Seminar Reports 1987.

[10] G. Birkhoff, Representability of Lie algebras and Lie groups by matrices, *Ann. of Math. II* **38** (1937) 526-532.

[11] A. Burroni, Algèbres graphiques, *Cahiers de Top. at Géom. Différentielle* **22** (1981) 249-265.

[12] E. J. Dubuc and G. M. Kelly, A presentation of topoi as algebraic relative to categories or graphs, *J. of Algebra* **81** (1983) 420-433.

[13] J. Duskin, Variations on Beck's tripleability criterion, in: *Reports of the Midwest Category Seminar III*, Lecture Notes in Mathematics **106** (1969) pp. 74-129.

[14] S. Eilenberg and J. C. Moore, Adjoint functors and triples, *Illinois J. Math.* **9** (1965) 231-244.

[15] R. Godement, *Topologie algébrique et théorie des faisceaux*, Hermann Paris, 1958.

[16] P. Higgins, Baer invariants and the Birkhoff-Witt theorem, *J. Algebra* **11** (1969) 469-482.

[17] P. Huber, Homotopy theory in general categories, *Math. Ann.* **144** (1961) 361-385.

[18] G. M. Kelly and R. H. Street, Review of the elements of 2-categories, in: *Category Seminar*, Lecture Notes in Mathematics **420**, Springer-Verlag 1974, pp. 75-109.

[19] H. Kleisli, Every standard construction is induced by a pair of adjoint functors, *Proc. Amer. Math. Soc.* **16** (1965) 544-546.

[20] J. Lambek, Toposes are monadic over categories, in: *Category Theory*, Lecture Notes in Mathematics 962, Springer-Verlag 1982, pp. 153-166.

[21] F. W. Lawvere, *Functorial semantics of algebraic theories*, Dissertation, Columbia University (1963).

[22] I. J. Le Creurer, F. Marmolejo and E. M. Vitale, Beck's theorem for pseudo-monads, *J. of Pure and Applied Algebra* (to appear).

[23] F. E. Linton, Some aspects of equational categories, in: Proc. of the Conf. on Categorical Algebra at La Jolla, Springer-Verlag (1966), pp. 84-94.

[24] J. MacDonald, Coherence and embedding of algebras, *Math. Zeitschrift* **135** (1974) 185-220.

[25] J. MacDonald, Conditions for a universal mapping of algebras to be a monomorphism, *Bull. A.M.S.* **80** (1974) 888-892.

[26] J. MacDonald and A. Stone, Soft adjunctions between 2-categories, *J. of Pure and Applied Algebra* **60** (1989) 155-203.

[27] S. Mac Lane, Homologie des anneaux et des modules, in: *Colloque de topologie algebrique*, Louvain 1956, pp. 55-80.

[28] S. Mac Lane, Natural associativity and commutativity, *Rice Univ. Studies* **49** (1963) 28-46.

[29] S. Mac Lane, *Categories for the Working Mathematician*, Springer-Verlag, 1971.

[30] E. Manes, *Algebraic Theories*, Springer-Verlag, 1976.

[31] E. Manes, *Predicate Transformer Semantics*, 1992.

[32] B. Mesabishvili, Pure morphisms of commutative rings are effective descent morphims for modules – a new proof, *Theory and Appl. of Categories*, Vol. **7**, No. 3 (2000) 38-42.

[33] E. Moggi, Computational lambda-calculus and monads, *Fourth Symposium on Logic in Computer Science*, IEEE, 1989.

[34] B. Pierce, *Basic Category Theory for Computer Scientists*, MIT Press, 1991

[35] J. Serre, *Lie algebras and Lie groups*, W.A. Benjamin, 1965.

[36] M. Sobral, **CABool** is monadic over almost all categories, *Journal of Pure and Applied Algebra* **7** (1992) 97-106.

[37] P. Stone, *Monad-like structures in 2-categories and soft adjunctions*, Ph. D. Thesis, University of British Columbia, Vancouver 1989.
[38] R.H.Street, The formal theory of monads, *Journal of Pure and Applied Algebra* **2** (1972) 149-168.
[39] O. Wyler, **CABool**'s many guises, preprint (2001).

Department of Mathematics,
University of British Columbia,
1984 Mathematics Road
Vancouver, Canada V6T 1Z2

E-mail: johnm@math.ubc.ca

Departamento de Matemática
Universidade de Coimbra
Apartado 3008
3001-454 Coimbra, Portugal

E-mail: sobral@mat.uc.pt

VI

Algebraic Categories

Maria Cristina Pedicchio and Fabrizio Rovatti

Introduction

The aim of this note is to make the reader familiar with the notion of algebraic category.

The approach we use is based on Lawvere's pioneering work [20], where the notions of algebraic theory and algebraic category are introduced as invariant formulations of Birkhoff's universal algebra. The new and innovative idea is that an algebraic theory is a definite mathematical object ("the perfect idea of - for example - a group"), namely a small category with finite products; an algebra or model is a finite-product-preserving functor ("a possible meaning") from such a category to the category of sets. Hence an algebraic category will be the category $Mod_{\mathcal{T}}$ of finite-product-preserving functors on a small category $\mathcal{T}$ with finite products.

To make things simpler we will start with one-sorted algebraic categories, i.e. with models of the form $F : \mathcal{T} \longrightarrow \mathbf{Set}$ where $\mathcal{T}$ is a category having as objects a set $\{T^0, T, T^2, ..., T^n, ...\}$ of finite products of a single object T; then we move to the general multi-sorted case.

The beauty of the subject comes from the interplay between properties of the theory (the description, the syntax) and properties of the objects described (the models or semantics).

In Section 1 we present these basic notions and prove the main properties of algebraic categories. One novelty in our approach in comparison to classical treatments on the subject (see for example [6]), is the use of reflexive graphs and of their coequalizers.

The fact that this kind of colimit is formed in an algebraic category as in the underlying category of sets is of basic importance and will appear many times in the note.

In Section 2 we discuss intensively Lawvere's characterization of one-sorted algebraic categories (see Theorem 2.2.4) and show that an algebraic category is a definite mathematical object, more precisely it is equivalent to an exact category with a set of regular projective, finitely presentable, regular generators.

As far as a regular generator P is concerned, observe that we include also the existence of copowers of P itself as part of the definition.

This kind of characterization theorem in terms of exactness of the category together with some extra properties, is very interesting and appears in the literature in other basic theorems like in Giraud's Theorem for sheaves or in Tierney's Theorem for abelian categories (see Chapters IV and VII).

In Section 3 we show how Lawvere's Characterization Theorem is extremely useful when discussing applications to different important categorical situations.

In particular, regular-epireflective subcategories of algebraic ones and quasi-algebraic categories (i.e. regular-epireflective subcategories with a finitary inclusion) will be completely characterized by weakening the hypothesis of the main Theorem 2.2.4.

Then, localizations of algebraic and quasi-algebraic categories will be investigated. To do that we will need sophisticated techniques based on the exact completion construction.

Finally, the case of Mal'cev categories will also be discussed and characterized in a similar way.

In the last section, the idea of a model as a functor preserving a certain property will be generalized to the case of a locally finitely presentable category $\mathcal{K} = \text{Lex}(\mathcal{C}^{op}, \textbf{Set})$ where $\mathcal{K}$ is determined by finite-limit-preserving functors on a small category $\mathcal{C}^{op}$ with finite limits. $\mathcal{C}^{op}$ will be called the essentially algebraic theory of $\mathcal{K}$.

Any algebraic category (or quasi-algebraic category) is locally finitely presentable with essentially algebraic theory the free equalizer completion of its algebraic theory. Conversely, there is no reason for a locally finitely presentable category to be algebraic (or quasi-algebraic). We will characterize these cases syntactically, i.e. in terms of conditions on the corresponding essentially algebraic theories.

It is important to notice that once more the results we give will be obtained as suitable application of Lawvere's Characterization Theorem. In this approach the new notion of effective projective object will play a basic role.

1. Algebraic categories

1.1. Lawvere theories

1.1.1. Definition. By a *Lawvere* or *algebraic theory* we mean a category $\mathcal{T}$ with a denumerable set of distinct objects $\{T^0, T, T^2, ..., T^n, ...\}$, where n is a natural number and each T^n is the n-th power of the object $T = T^1$.

1.1.2. Definition. Given a Lawvere theory $\mathcal{T}$, a *model* of $\mathcal{T}$ is a functor

$$F : \mathcal{T} \longrightarrow \textbf{Set}$$

which preserves finite products.

We will write $\mathcal{M}od_{\mathcal{T}}$ for the category of $\mathcal{T}$-*Models* and natural transformations between them. Categories of this form will be called *algebraic categories*.

We recall the following:

1.1.3. Definition. An object G in a category $\mathcal{C}$ is called *strong (regular) generator* if and only if

- G admits all copowers;
- the canonical morphism

$$\gamma_C : \coprod_{\mathcal{C}(G,C)} G \longrightarrow C$$

  is a strong (regular) epimorphism for any $C \in \mathcal{C}$.

1.1.4. Lemma. *Let $\mathcal{C}$ be a finitely complete category, and consider G in $\mathcal{C}$ such that G admits all copowers; then the following are equivalent:*

- *G is a strong generator;*
- *$\mathcal{C}(G,-) : \mathcal{C} \longrightarrow \mathbf{Set}$ reflects isomorphisms.*

Proof. Let G be a strong generator, and consider the following diagram

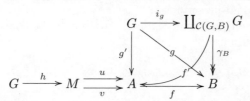

Take $f : A \longrightarrow B$ such that $\mathcal{C}(G,f)$ is an isomorphism. For every $g : G \longrightarrow B$, $g = \gamma_B \circ i_g \in \mathcal{C}(G,B)$, where i_g is the inclusion morphism corresponding to g. Since $\mathcal{C}(G,f) : \mathcal{C}(G,A) \longrightarrow \mathcal{C}(G,B)$ is an isomorphism, there is a unique $g' \in \mathcal{C}(G,A)$ such that $g = \gamma_B \circ i_g = f \circ g'$. So we can get the factorization f' through the coproduct, with $f' \circ i_g = g'$. Now we have that $f \circ f' \circ i_g = \gamma_B \circ i_g$, so we get $f \circ f' = \gamma_B$ by definition of coproduct. Since γ_B is a strong epimorphism, f is a strong epimorphism as well.

To prove that f is a monomorphism, consider $u, v : M \rightrightarrows A$ such that $f \circ u = f \circ v$. For every $h : G \longrightarrow M$, we have $f \circ u \circ h = f \circ v \circ h$, that means $\mathcal{C}(G,f)(u \circ h) = \mathcal{C}(G,f)(v \circ h)$. Since $\mathcal{C}(G,f)$ is an isomorphism, $u \circ h = v \circ h$ and since G is a strong generator $u = v$. So, being f both a monomorphism and a strong epimorphism, it is an isomorphism.

Conversely suppose that $\mathcal{C}(G,-)$ reflects isomorphisms. If $\gamma_C = j \circ p$ with j monomorphism, $\mathcal{C}(G,j)$ is injective. Moreover, given $g : G \longrightarrow C$ we have $g = \gamma_C \circ i_g = j \circ p \circ i_g = \mathcal{C}(G,j)(p \circ i_g)$, proving that $\mathcal{C}(G,j)$ is surjective as well. Therefore $\mathcal{C}(G,j)$ is bijective, and j is an isomorphism. So we get that γ_C is an extremal (and thus strong) epimorphism. $\square$

Consider now the canonical functor

$$\mathbb{U} : \mathcal{M}od_T \longrightarrow \mathbf{Set}$$

of evaluation in T. We get the following:

1.1.5. Proposition. *The functor* $\mathbb{U} : \mathcal{M}od_\mathcal{T} \longrightarrow \mathbf{Set}$ *satisfies the following proper-*
ties:

(1) $\mathbb{U}$ *is representable, represented by* $\mathcal{T}(T, -)$;
(2) $\mathbb{U}$ *is faithful;*
(3) $\mathbb{U}$ *reflects isomorphisms;*
(4) $\mathcal{M}od_\mathcal{T}$ *is complete, and* $\mathbb{U}$ *preserves and reflects limits;*
(5) $\mathbb{U}$ *preserves and reflects monomorphisms.*

Proof.

(1) is trivial by Yoneda Lemma.
(2) take two morphisms α, β of $\mathcal{T}$-Models, and $\mathbb{U}\alpha = \mathbb{U}\beta$. This implies $\alpha_T = \beta_T$, hence $\alpha_{T^n} = \beta_{T^n}$, and so $\alpha = \beta$.
(3) follows from $\alpha_{T^n} = (\alpha_T)^n$.
(4) $\mathcal{M}od_\mathcal{T}$ is complete and limits are computed pointwise; this follows from the fact that limits are pointwise in the presheaf category $\mathbf{Set}^\mathcal{T}$, and that the limit of a diagram of finite-product-preserving functors is again finite-product-preserving.
(5) Trivial by (4), since any monomorphism can be characterized by a pull-back condition. $\qquad\square$

1.1.6. Proposition. *Let* $\mathcal{T}$ *be an algebraic theory; then* $\mathcal{M}od_\mathcal{T}$ *is reflective in the presheaf category* $\mathbf{Set}^\mathcal{T}$.

Proof. $\mathcal{M}od_\mathcal{T}$ is complete and limits are computed pointwise, by Proposition 1.1.5, (4); consider then the inclusion of $\mathcal{M}od_\mathcal{T}$ in $\mathbf{Set}^\mathcal{T}$: the existence of a left adjoint to the inclusion follows by the Adjoint Functor Theorem (see [6] for further details). $\qquad\square$

1.1.7. Proposition. *Let* $\mathcal{T}$ *be an algebraic theory, then*

(1) $\mathcal{M}od_\mathcal{T}$ *is cocomplete;*
(2) $\mathcal{T}(T, -)$ *is a strong generator for* $\mathcal{M}od_\mathcal{T}$.

Proof. $\mathcal{M}od_\mathcal{T}$ is cocomplete by Proposition 1.1.6, then $\mathcal{T}(T, -) \in \mathcal{M}od_\mathcal{T}$ will admit all copowers. Hence (2) follows from 1.1.4 and from the third property of 1.1.5. $\qquad\square$

We recall that a category $\mathcal{D}$ is called filtered if:

(1) $\mathcal{D}$ is non empty;
(2) for each pair D_1, D_2 of objects there exists an object D and morphisms in $\mathcal{D}$

$$f : D_1 \longrightarrow D$$
$$g : D_2 \longrightarrow D.$$

(3) for each pair $f, g : D_1 \rightrightarrows D_2$ of morphisms in $\mathcal{D}$ there exists a morphism $h : D_2 \longrightarrow D$ with $h \circ f = h \circ g$.

Colimits of filtered diagrams are called filtered colimits.

1.1.8. Proposition. *Let T be an algebraic theory. The category Mod_T has filtered colimits, and they are computed pointwise.*
In particular filtered colimits commute with finite limits in Mod_T, and $\mathbb{U} : Mod_T \longrightarrow \mathbf{Set}$ preserves and reflects filtered colimits.

Proof. Since colimits in $\mathbf{Set}^T$ are computed pointwise and filtered colimits commute with finite limits, we get that Mod_T admits filtered colimits which are pointwise. In particular $\mathbb{U}$ preserves filtered colimits, and since it reflects isomorphisms, $\mathbb{U}$ reflects filtered colimits as well. $\qquad\square$

1.1.9. Proposition. *Let T be an algebraic theory; then the functor $\mathbb{U} : Mod_T \longrightarrow \mathbf{Set}$ admits a left adjoint $\mathbb{F} : \mathbf{Set} \longrightarrow Mod_T$.*

Proof. $T(T, -) \in Mod_T$ admits all copowers.
We can then define $\mathbb{F}$ by

$$\mathbb{F}(X) = \coprod_{x \in X} T(T, -),$$

for any set X.
Given $f : X \longrightarrow Y$ in $\mathbf{Set}$, $\mathbb{F}(f)$ will be the canonical morphism defined by the following diagram:

$$
\begin{array}{ccc}
T(T, -) & & \\
\Big\downarrow{\scriptstyle i_x} & \searrow^{\delta_{f(x)}} & \\
\coprod_{x \in X} T(T, -) & \xrightarrow{\mathbb{F}(f)} & \coprod_{y \in Y} T(T, -)
\end{array}
$$

where δ_y and i_x are the inclusions.
For any finite set X, with $card(X) = n$, we get

$$\mathbb{F}(X) = \coprod_X T(T, -) \cong T(T^n, -).$$

To check that $\mathbb{F} \dashv \mathbb{U}$ it suffices to produce the counit $\varepsilon : \mathbb{F} \circ \mathbb{U} \longrightarrow 1_{Mod_T}$ of the adjoint pair.
For any model $G \in Mod_T$, ε_G is canonically determined by the following diagram:

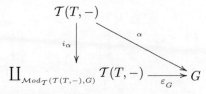

where i_α is the sum inclusion corresponding to $\alpha \in Mod_T(T(T, -), G)$. Routine verifications show that ε is natural and universal. $\qquad\square$

1.2. Reflexive coequalizers

1.2.1. Definition. A *graph* X in a category $\mathcal{C}$ is given by two parallel arrows in $\mathcal{C}$, usually denoted by $s_X, t_X : X_1 \rightrightarrows X_0$, where X_0 and X_1 are two objects in $\mathcal{C}$. A graph is *reflexive* if there exists a morphism $r_X : X_0 \longrightarrow X_1$ such that

$$s_X \circ r_X = 1_{X_0} = t_X \circ r_X.$$

By a coequalizer of a reflexive graph we mean the coequalizer of the pair of parallel maps.

We recall the following well known lemma ([18, Lemma 0.17]):

1.2.2. Lemma. *Let*

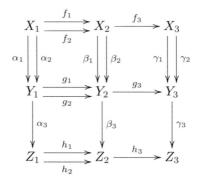

be a diagram in any category satisfying the "obvious" commutativity conditions (i.e. $\beta_j \circ f_i = g_j \circ \alpha_i$ for $i = 1, 2$ and $j = 1, 2$ etc.), in which the rows and columns are coequalizers and the pairs (f_1, f_2) and (α_1, α_2) are reflexive.
Then the diagonal

$$X_1 \xrightarrow[\beta_2 \circ f_2]{\beta_1 \circ f_1} Y_2 \xrightarrow{\gamma_3 \circ g_3} Z_3$$

is a coequalizer.

1.2.3. Lemma. *In* **Set** *coequalizers of reflexive graphs commute with finite products.*

Proof. Take two coequalizers of reflexive graphs $A_1 \rightrightarrows A_0 \longrightarrow A$ and $B_1 \rightrightarrows B_0 \longrightarrow B$ in **Set**, and consider the following diagram:

$$
\begin{array}{ccccc}
A_1 \times B_1 & \rightrightarrows & A_0 \times B_1 & \longrightarrow & A \times B_1 \\
\downdownarrows & & \downdownarrows & & \downdownarrows \\
A_1 \times B_0 & \rightrightarrows & A_0 \times B_0 & \longrightarrow & A_1 \times B_0 \\
\downarrow & & \downarrow & & \downarrow \\
A_1 \times B & \rightrightarrows & A_0 \times B & \longrightarrow & A \times B
\end{array}
$$

Now, $- \times X$ preserves colimits in **Set**, being a left adjoint to $hom(X, -)$, then the rows and columns are coequalizers.

Hence, applying Lemma 1.2.2 we get that the diagonal is a coequalizer. $\square$

1.2.4. Corollary. *In $\mathcal{M}od_{\mathcal{T}}$ coequalizers of reflexive graphs are formed pointwise, and*

$$\mathbb{U} : \mathcal{M}od_{\mathcal{T}} \longrightarrow \textbf{Set}$$

preserves and reflects coequalizers of reflexive graphs.

1.2.5. Lemma. *Let $\mathcal{C}$ be a category with finite sums, then any coequalizer can be seen as coequalizer of a reflexive graph.*

Proof. Take f to be the coequalizer of (u, v). Then, if we consider the sum $B \coprod X$ we get a canonically induced reflexive graph $(B \coprod X, u', v', i_X)$.

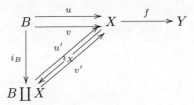

It is easy to check that f coequalizes (u', v'). $\square$

1.2.6. Corollary. $\mathbb{U} : \mathcal{M}od_{\mathcal{T}} \longrightarrow \textbf{Set}$ *preserves and reflects regular epimorphisms.*

Proof. Let $f : X \twoheadrightarrow Y$ be a regular epimorphism in $\mathcal{M}od_{\mathcal{T}}$.

Since $\mathcal{M}od_{\mathcal{T}}$ has finite sums, then by the previous lemma we know that f is a coequalizer of a reflexive graph. Apply $\mathbb{U}$, then, since $\mathbb{U}$ preserves coequalizer of reflexive graph by 1.2.4, $\mathbb{U}(f)$ is a coequalizer, hence it is a regular epimorphism. $\square$

We will see (Theorem 2.1.10) how this corollary will play a basic role to prove the exactness of $\mathcal{M}od_{\mathcal{T}}$.

1.3. $\mathcal{M}od_{\mathcal{T}}$ and $\textbf{Set}^{\mathbb{T}}$

In this section we want to compare categories of models for a Lawvere theory to classical categories of $\mathbb{T}$-*algebras* for a monad $\mathbb{T}$.

For what concerns the theory of $\mathbb{T}$-algebras we mainly refer to Chapter V and to [21] and [6].

1.3.1. Definition. A monad $\mathbb{T} = (\textbf{T}, \eta, \mu)$ on **Set** has finite rank when the functor $\textbf{T} : \textbf{Set} \longrightarrow \textbf{Set}$ preserves filtered colimits. We call such a monad *finitary*.

Denote by $\textbf{Set}^{\mathbb{T}}$ the corresponding category of $\mathbb{T}$-algebras over **Set**, and by $\mathbb{F}^{\mathbb{T}} \dashv \mathbb{U}^{\mathbb{T}}$ the canonical adjunction between **Set** and $\textbf{Set}^{\mathbb{T}}$.

We recall the following:

1.3.2. Theorem. *Consider the diagram*

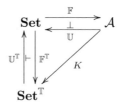

where $\mathbb{T}$ *is the monad defined by the adjunction* $\mathbb{F} \dashv \mathbb{U}$. *Then there is a unique comparison functor* $K : A \longrightarrow \mathbf{Set}^{\mathbb{T}}$ *with* $\mathbb{U}^{\mathbb{T}} \circ K = \mathbb{U}$ *and* $K \circ \mathbb{F} = \mathbb{F}^{\mathbb{T}}$.

Proof. See [21].   □

Then we can prove the main result:

1.3.3. Theorem. *Given a Lawvere theory* $\mathcal{T}$ *there exists a monad* $\mathbb{T}$ *on* $\mathbf{Set}$ *with finite rank such that* $\mathbf{Set}^{\mathbb{T}} \cong \mathcal{M}od_{\mathcal{T}}$.
Conversely, given a monad $\mathbb{T}$ *on* $\mathbf{Set}$ *with finite rank we can find a Lawvere theory* $\mathcal{T}$ *such that* $\mathcal{M}od_{\mathcal{T}} \cong \mathbf{Set}^{\mathbb{T}}$.

Proof. Given a Lawvere theory $\mathcal{T}$, let us consider the adjunction between $\mathbf{Set}$ and $\mathcal{M}od_{\mathcal{T}}$, that exists by 1.1.9. Then, consider the induced monad on $\mathbf{Set}$, $\mathbb{T} = \mathbb{U} \circ \mathbb{F}$, and the corresponding adjoint pair $\mathbb{U}^{\mathbb{T}}$ and $\mathbb{F}^{\mathbb{T}}$.

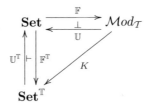

Let K be the comparison functor.
We apply Beck's Characterization Theorem (see [21] and Chapter V) to show that the comparison functor K is an isomorphism:
Let $u, v : G \longrightarrow H$ be two morphisms in $\mathcal{M}od_{\mathcal{T}}$ such that $(\mathbb{U}(u), \mathbb{U}(v)) = (u_T, v_T)$ has a split coequalizer Q in $\mathbf{Set}$, and consider the following diagram for any morphism $\gamma : T^n \longrightarrow T^m$ in $\mathcal{T}$

$$
\begin{array}{ccccc}
G(T^n) & \underset{v_{T^n}}{\overset{u_{T^n}}{\rightrightarrows}} & H(T^n) & \overset{q^n}{\longrightarrow} & Q^n \\
{\scriptstyle G(\gamma)}\big\downarrow & & {\scriptstyle H(\gamma)}\big\downarrow & & \big\downarrow{\scriptstyle \exists! Q(\gamma)} \\
G(T^m) & \underset{v_{T^m}}{\overset{u_{T^m}}{\rightrightarrows}} & H(T^m) & \overset{q^m}{\longrightarrow} & Q^m
\end{array}
$$

Horizontal rows are coequalizers, and $q^m \circ H(\gamma)$ coequalizes (u_{T^n}, v_{T^n}), then there exists a unique $Q(\gamma) : Q^n \longrightarrow Q^m$ such that $q^m \circ H(\gamma) = Q(\gamma) \circ q^n$.

We can define a functor $Q : \mathcal{T} \longrightarrow \mathbf{Set}$ such that $Q(T^n) = Q^n$, acting on morphisms as above.

Q is a $\mathcal{T}$-model, and moreover (Q, q), where $q : H \longrightarrow Q$ is defined by $q_{T^n} = q^n$, is a coequalizer of (u, v) in $\mathcal{M}od_{\mathcal{T}}$.

Being a left adjoint, $\mathbb{F}$ preserves all the colimits, $\mathbb{U}$ preserves filtered colimits by Proposition 1.1.8, hence the monad $\mathbb{T}$ has finite rank.

Conversely, given a monad $\mathbb{T}$ on $\mathbf{Set}$ with finite rank, consider the canonical adjunction between $\mathbf{Set}$ and $\mathbf{Set}^{\mathbb{T}}$ (see [21]).

$$\mathbf{Set} \underset{\mathbb{U}^{\mathbb{T}}}{\overset{\mathbb{F}^{\mathbb{T}}}{\underset{\perp}{\rightleftarrows}}} \mathbf{Set}^{\mathbb{T}} \ .$$

Take now the subcategory $\mathcal{F}$ of $\mathbf{Set}^{\mathbb{T}}$, spanned by the free finitely generated algebras $\mathbb{F}^{\mathbb{T}}(X)$, where X is a finite set, and define $\mathcal{T} = \mathcal{F}^{op}$.

We get that $\mathcal{T}$ is a Lawvere theory with $T = \mathbb{F}^{\mathbb{T}}(1)$, since $\mathbb{F}^{\mathbb{T}}(X) = \coprod_X \mathbb{F}^{\mathbb{T}}(1)$.

Consider the adjunction between $\mathbf{Set}$ and $\mathcal{M}od_{\mathcal{T}}$, and let $\mathbb{T}' = \mathbb{U} \circ \mathbb{F}$ be the induced monad on $\mathbf{Set}$, then $\mathcal{M}od_{\mathcal{T}} \cong \mathbf{Set}^{\mathbb{T}'}$. Now, we show that the two monads $\mathbb{T}$ and $\mathbb{T}'$ are isomorphic.

We check it on objects: for any finite set X, with $card(X) = n$, we get:

$\mathbb{T}'(X) = \mathbb{U}\mathbb{F}(X) = \mathbb{U}(\mathcal{T}(T^n, -)) = \mathcal{T}(T^n, T) = \mathcal{F}(T, T^n) = \mathcal{F}(\mathbb{F}^{\mathbb{T}}(1), \mathbb{F}^{\mathbb{T}}(X)) \cong Set(1, \mathbb{U}^{\mathbb{T}}\mathbb{F}^{\mathbb{T}}(X)) \cong \mathbb{U}^{\mathbb{T}}\mathbb{F}^{\mathbb{T}}(X) = \mathbb{T}(X)$.

Now, any arbitrary set X' can be written as filtered colimit of its finite subsets. Then, since $\mathbb{T}$ and $\mathbb{T}'$ have finite rank, we get:

$\mathbb{T}'(X') = \mathbb{T}'(colim_{Y \subseteq X'}Y) \cong colim_{Y \subseteq X'}\mathbb{T}'(Y) \cong colim_{Y \subseteq X'}\mathbb{T}(Y) \cong \mathbb{T}(colim_{Y \subseteq X'}Y) = \mathbb{T}(X')$, where X' is an arbitrary set and Y is a finite subset.

Then the categories $\mathbf{Set}^{\mathbb{T}}$ and $\mathbf{Set}^{\mathbb{T}'}$ are equivalent. $\qquad\square$

Observe that the two constructions of Theorem 1.3.3 are (up to equivalence of categories) inverse to each other.

So, from now on when considering an algebraic category we will be able to use, according to the context and to the results we need, the Lawvere theory approach as well as the $\mathbb{T}$-algebra approach.

1.3.4. Remark. It is quite natural to develop results similar to those of the previous sections in a multi-sorted context.

For any set S, denote by $\mathbf{Set}^S$ the S-th power of $\mathbf{Set}$ and consider a finitary monad $\mathbb{T}$ on $\mathbf{Set}^S$ (i.e. a filtered-colimit-preserving monad on $\mathbf{Set}^S$).

It is possible to show that $(\mathbf{Set}^S)^{\mathbb{T}}$ is equivalent to the category $\mathcal{M}od_{\mathcal{T}}$ of finite-product-preserving functors from a small category (the corresponding Lawvere theory) $\mathcal{T}$ with finite products to $\mathbf{Set}$ (see [3]).

Categories of the form $\mathcal{M}od_{\mathcal{T}}$ with $\mathcal{T}$ a small category with finite products will be called *multi-sorted* or *S-sorted algebraic categories* .

Now, the theory $\mathcal{T}$ is given by the dual category of the subcategory of $(\mathbf{Set}^S)^{\mathbb{T}}$ spanned by the free finitely generated $\mathbb{T}$-algebras $\mathbb{F}^{\mathbb{T}}(X)$, where $X = (X_s)_{s \in \mathbf{Set}} \in$

Set^S and the power $\sharp X = \sum card X_s$ is finite.

For any fixed s_0 in S, consider the family $_{s_0}X \in$ **Set**^{S} defined by

$$_{s_0}X = (\,_{s_0}X_s)_{s \in S}$$

where

$$_{s_0}X_s = 1 \text{ if } s = s_0,$$

$$_{s_0}X_s = \emptyset \text{ otherwise.}$$

If we denote by T_s, $s \in S$ the free algebras corresponding to each $_sX$, $s \in S$, then any element of $\mathcal{T}$ can be expressed as a finite product of T_s.

Hence, by defining $\mathbb{U} : \mathcal{M}od_{\mathcal{T}} \longrightarrow$ **Set**^{S} as $\mathbb{U}(G) = (G(T_s))_{s \in S}$, all the results of the previous sections apply.

In this multi-sorted context the single strong generator T (of 1.1.7) has been replaced by a set of strong generators T_s.

We recall that:

1.3.5. Definition. A set $\mathcal{G}$ of objects of a category $\mathcal{C}$ is called a *strong (regular) generator* if and only if all (small) coproducts of elements of $\mathcal{G}$ exist and the canonical morphism

$$\coprod_{G \in \mathcal{G}, f \in \mathcal{C}(G,C)} (\text{domain of } f) \longrightarrow C$$

is a strong (regular) epimorphism for any C in the category $\mathcal{C}$.

1.3.6. Examples.

(1) Any finitary variety in the sense of universal algebra is an algebraic category (groups, Abelian groups, monoids, Boolean algebras,etc.).

(2) **Set** is an algebraic category with the dual of the category of finite sets as its algebraic theory.

(3) The category of complete ∨-lattices is not an algebraic category (it is not finitary).

(4) Torsion free Abelian groups do not form an algebraic category.

(5) The category of compact Hausdorff spaces is a category of $\mathbb{T}$-algebras over **Set** but $\mathbb{T}$ does not have finite rank (in fact does not admit any rank)([21]).

(6) Modules over a ring form a two-sorted algebraic category (observe that they could be also presented as a one-sorted algebraic category).

(7) Any presheaf category **Set**$^{\mathcal{A}}$ with $\mathcal{A}$ small can be considered as an $\mathcal{A}$-sorted algebraic category. It has only unary operations.

(8) If $\mathcal{T}$ is the opposite of the category of finite rectangular matrices with entries in $\mathbb{Z}$ and matrix multiplication as composition, then $\mathcal{M}od_{\mathcal{T}} \cong Ab$ the category of abelian groups.

2. The Lawvere Theorem for algebraic categories

2.1. Regular and exact categories

We start by recalling some basic facts on regular and exact categories (see Chapter IV and [6], [5] for more details).

2.1.1. Definition. A category $\mathcal{C}$ is *regular* if it satisfies the following conditions:

(1) $\mathcal{C}$ has finite limits;
(2) $\mathcal{C}$ has coequalizers of kernel pairs;
(3) the pullback of a regular epimorphism along any morphism is a regular epimorphism.

2.1.2. Theorem. *In a regular category, every morphism factors as a regular epimorphism followed by a monomorphism, and the factorization is unique up to isomorphism.*

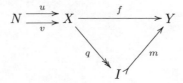

Proof. Given a morphism $f : X \longrightarrow Y$ the factorization $f = m \circ q$ is obtained by taking the kernel pair (u, v) of f and its coequalizer q. Since $f \circ u = f \circ v$ we get a unique factorization m. By using the third regularity axiom we can show that m is a monomorphism. $\square$

The object I obtained in the factorization is called the image of f.

2.1.3. Proposition. *In any regular category the following hold:*

(1) *strong epimorphisms coincide with regular epimorphisms;*
(2) *the composite of regular epimorphisms is a regular epimorphism;*
(3) *if a composite $g \circ f$ is a regular epimorphism, then g is a regular epimorphism;*
(4) *a morphism that is both a regular epimorphism and a monomorphism is an isomorphism.*

2.1.4. Definition. A *relation* R from an object X to an object Y is a subobject $r : R \rightarrowtail X \times Y$.

Equivalently we can give the following composites

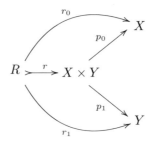

$r_0 = p_0 \circ r$ and $r_1 = p_1 \circ r$ where p_0, p_1 are the product projections and (r_0, r_1) is a monomorphic pair. This second formulation will be used all the time.

2.1.5. Definition. A relation $R \rightarrowtail X \times X$ on X is called an *equivalence relation* if it is:

(1) reflexive: there exists a morphism $\delta : X \longrightarrow R$ with $r_0 \circ \delta = 1_X = r_1 \circ \delta$;
(2) symmetric: there exists a morphism $\sigma : R \longrightarrow R$ with $r_0 \circ \sigma = r_1$ and $r_1 \circ \sigma = r_0$;
(3) transitive: there exists a morphism $\xi : R \times_X R \longrightarrow R$ such that $r_0 \circ \xi = r_0 \circ \rho_0$ and $r_1 \circ \xi = r_1 \circ \rho_1$, where $R \times_X R, \rho_0, \rho_1$ are defined by the following pullback

$$
\begin{array}{ccc}
R \times_X R & \xrightarrow{\ \rho_1\ } & R \\
{\scriptstyle \rho_0}\downarrow & & \downarrow{\scriptstyle r_0} \\
R & \xrightarrow{\ r_1\ } & X
\end{array}
$$

Observe that we could speak of equivalence relations also in a category without products and pullbacks simply by taking a monomorphic pair $(r_0, r_1) : R \rightrightarrows X$ of arrows in $\mathcal{C}$, such that for any $C \in \mathcal{C}$ the corresponding relation $\mathcal{C}(C, R)$ on $\mathcal{C}(C, X)$ is an equivalence relation.

2.1.6. Definition. In a regular category $\mathcal{C}$, given a relation R from X to Y and S from Y to Z, the composite $S \circ R$ is a relation from X to Z defined as the image of the morphism

$$(r_1 \circ u_1, s_0 \circ u_0) : R \times_Y S \longrightarrow X \times Z.$$

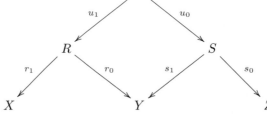

Regularity of $\mathcal{C}$ will imply that such a composition is associative.

2.1.7. Remark. Regular categories are the precise context where it is possible to develop a "good calculus" of relations; see [13] for more details.

2.1.8. Definition. An equivalence relation (R, r_0, r_1) on an object X of $\mathcal{C}$ is *effective* when the coequalizer q of (r_0, r_1) exists and (r_0, r_1) is the kernel pair of q.

It is trivial to see that any kernel pair is an equivalence relation.

2.1.9. Definition. A category $\mathcal{C}$ is *exact* if:

(1) $\mathcal{C}$ is regular;
(2) any equivalence relation in $\mathcal{C}$ is effective.

Now, we can prove the main result of this section:

2.1.10. Theorem. *Let $\mathcal{T}$ be an algebraic theory; then $\mathcal{M}od_{\mathcal{T}}$ is an exact category.*

Proof. First, we show that $\mathcal{M}od_{\mathcal{T}}$ is regular: let $f : X \twoheadrightarrow Y$ be a regular epimorphism in $\mathcal{M}od_{\mathcal{T}}$, and take its pullback

$$
\begin{array}{ccc}
X' & \longrightarrow & X \\
\scriptstyle g \downarrow & & \downarrow \scriptstyle f \\
Y' & \longrightarrow & Y
\end{array}
$$

Applying $\mathbb{U}$, by 1.1.5 we get another pullback in **Set**, where $\mathbb{U}(f)$ is a regular epimorphism by 1.2.6. Now, **Set** is regular, so the pullback $\mathbb{U}(g)$ is a regular epimorphism. Then, since $\mathbb{U}$ reflects regular epimorphism by 1.2.6, g is a regular epimorphism in $\mathcal{M}od_{\mathcal{T}}$.

To show that $\mathcal{M}od_{\mathcal{T}}$ is exact, take an equivalence relation R on X in $\mathcal{M}od_{\mathcal{T}}$, let q be the coequalizer and N the kernel pair of q.

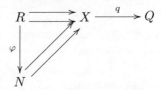

Then apply the functor $\mathbb{U}$,

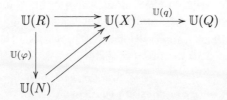

$\mathbb{U}(R)$ is an equivalence relation, since $\mathbb{U}$ preserves limits by 1.1.5; for the same reason $\mathbb{U}(N)$ is the kernel pair of $\mathbb{U}(q)$. q is a coequalizer of an equivalence relation, so it is a coequalizer of a reflexive graph, then $\mathbb{U}(q)$ is a coequalizer of $\mathbb{U}(R)$ by

1.2.4.
Now, since $\mathbb{U}(R)$ is effective in **Set**, $\mathbb{U}(\varphi)$ is an isomorphism hence φ is an isomorphism by 1.1.5, and this means that R is effective.   □

Further examples follow:

2.1.11. Examples.
(1) **Set** is exact (and coexact).
(2) Any presheaves category $\mathbf{Set}^{\mathcal{C}}$ is exact (and coexact).
(3) Any topos is exact (and coexact, [6]).
(4) Any category of $\mathbb{T}$-algebras over **Set** is exact.
(5) The category of compact Hausdorff spaces is exact (and coexact).
(6) The category of torsion free abelian group is regular but not exact.
(7) The category of compact Hausdorff 0-dimensional spaces is regular but not exact.
(8) The category Top of topological spaces and continuous maps is not regular.
(9) The category Cat of small categories and functors is not regular. Indeed strong and regular epimorphisms do not coincide in Cat.
(10) The category of topological groups is regular, but not exact.

2.2. The Characterization Theorem

We recall the following definitions:

2.2.1. Definition. An object P in a category $\mathcal{C}$ is *regular projective* if and only if $\mathcal{C}(P, -) : \mathcal{C} \longrightarrow \mathbf{Set}$ preserves regular epimorphisms.

2.2.2. Definition. An object P in a category $\mathcal{C}$ is *finitely presentable* if and only if $\mathcal{C}(P, -) : \mathcal{C} \longrightarrow \mathbf{Set}$ preserves filtered colimits.

Observe that finite presentability of P means exactly that, for any filtered colimit $L = colim_{i \in I} M_i$ the following two conditions are satisfied:
(1) every morphism $h : P \longrightarrow L$ factors through one of the canonical morphisms $s_i : M_i \longrightarrow L$ of the colimit;
(2) when two morphisms $h_1, h_2 : P \longrightarrow M_i$ are such that $s_i \circ h_1 = s_i \circ h_2$ for some $s_i : M_i \longrightarrow L$ of the colimit, there exists a morphism $s_{ij} : M_i \longrightarrow M_j$ in the original diagram such that $s_{ij} \circ h_1 = s_{ij} \circ h_2$.

Before proving the main characterization theorem, we recall the following technical lemma that we will need in the proof of 2.2.4:

2.2.3. Lemma. *Let* $\mathbb{U} : \mathcal{C} \longrightarrow \mathcal{X}$ *be a functor with left adjoint* $\mathbb{F}$ *and counit* $\varepsilon : \mathbb{F}\mathbb{U} \longrightarrow 1_{\mathcal{C}}$*. The following properties are equivalent:*
- ε *is an epimorphism if and only if* $\mathbb{U}$ *is faithful;*
- ε *is an extremal epimorphism if and only if* $\mathbb{U}$ *is faithful and reflects isomorphisms;*

- ε is a regular epimorphism if and only if $\mathbb{U}$ has a full and faithful comparison functor K into its corresponding category of $\mathbb{T}$-algebras.
- ε is an isomorphism if and only if $\mathbb{U}$ is full and faithful.

Proof. See [6] for the details of the proof. $\qquad\square$

Now, we have all the notions we need to discuss the main result, the *Lawvere Characterization Theorem* for algebraic categories (see [20]).

2.2.4. Theorem.[Lawvere] *A category C is equivalent to a one-sorted algebraic category if and only if:*

(1) C *is exact;*
(2) *there exists a regular generator* $P \in C$;
(3) P *is regular projective;*
(4) P *is finitely presentable.*

Proof. Let C be equivalent to an algebraic category, then $C \cong Mod_\mathcal{T}$. By Theorem 2.1.10, C is exact.
Consider the canonical adjunction of Proposition 1.1.9

$$\mathbf{Set} \underset{\mathbb{U}}{\overset{\mathbb{F}}{\rightleftarrows}} \bot\ Mod_\mathcal{T}$$

where $\mathbb{F}(1) = \mathcal{T}(T, -)$ and $\mathbb{U}$ is representable, represented by $\mathcal{T}(T, -)$.
Define $P = \mathcal{T}(T, -)$. Then, by 1.1.7 and 2.1.3, P is a regular generator; by 1.2.6, P is regular projective, and by 1.1.8, P is finitely presentable.
To prove the converse, we will strongly use Theorem 1.3.3 by comparing C with a category of $\mathbb{T}$-algebras for a monad $\mathbb{T}$ with finite rank.
Let C satisfy the conditions from (1) to (4).
Define the following pair of functors

$$\mathbf{Set} \underset{\mathbb{U}}{\overset{\mathbb{F}}{\rightleftarrows}} \bot\ C$$

by: $\mathbb{U} = C(P, -)$, and $\mathbb{F}(X) = \coprod_{x \in X} P$ for any $X \in \mathbf{Set}$.
$\mathbb{F}$ is clearly a functor; for any $f : X \longrightarrow Y$ in $\mathbf{Set}$ and $x \in X$, $\mathbb{F}(f)$ is in fact defined by:

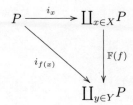

As already seen in Proposition 1.1.9, $\mathbb{F} \dashv \mathbb{U}$, with counit $\varepsilon : \mathbb{F} \circ \mathbb{U} \longrightarrow 1_{\mathcal{C}}$ defined by $\varepsilon_C \circ i_g = g$ for any $C \in \mathcal{C}$ and $g : P \longrightarrow C$.

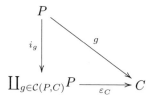

Now consider the monad $\mathbb{T} = \mathbb{U} \circ \mathbb{F}$ on **Set** and the corresponding category $\mathbf{Set}^{\mathbb{T}}$ of $\mathbb{T}$-algebras.

We will show that the comparison functor $K : \mathcal{C} \longrightarrow \mathbf{Set}^{\mathbb{T}}$ defined by $K(C) = (\mathbb{U}(C), \mathbb{U}(\varepsilon_C))$ and $K(f) = \mathbb{U}(f)$ is an equivalence of categories.

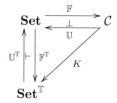

By construction, $\mathbb{U} = \mathcal{C}(P, -)$ preserves regular epimorphisms, since P is projective, and it also preserves finite limits.

Then, since $\mathbb{U}^{\mathbb{T}} \circ K = \mathbb{U}$ and $\mathbb{U}^{\mathbb{T}}$ reflects limits and regular epimorphisms, K preserves limits and regular epimorphisms.

Moreover, since P is a regular generator, the counit of the adjunction is a regular epimorphism, so K is full and faithful by Lemma 2.2.3. Then we can suppose, up to equivalences, that $\mathcal{C}$ is a full subcategory of $\mathbf{Set}^{\mathbb{T}}$.

It remains to verify that K is essentially surjective on objects, i.e. that for any $Z \in \mathbf{Set}^{\mathbb{T}}$ there exists $X \in \mathcal{C}$ with $K(X) \cong Z$. Consider the following presentation as $\mathbb{T}$-algebra for Z:

$$\mathbb{F}^{\mathbb{T}} \mathbb{U}^{\mathbb{T}}(M) \xrightarrow{\varepsilon_M^{\mathbb{T}}} M \underset{t}{\overset{s}{\rightrightarrows}} \mathbb{F}^{\mathbb{T}} \mathbb{U}^{\mathbb{T}}(Z) \xrightarrow{\varepsilon_Z^{\mathbb{T}}} Z$$

where $\varepsilon^{\mathbb{T}}$ is the counit of the adjoint pair $\mathbb{F}^{\mathbb{T}} \dashv \mathbb{U}^{\mathbb{T}}$ and (s, t) is the kernel pair of $\varepsilon_Z^{\mathbb{T}}$ in $\mathbf{Set}^{\mathbb{T}}$.

Since $K \circ \mathbb{F} = \mathbb{F}^{\mathbb{T}}$ and K is full, from

$$(K\mathbb{F})\mathbb{U}^{\mathbb{T}}(M) \underset{t \circ \varepsilon_M^{\mathbb{T}}}{\overset{s \circ \varepsilon_M^{\mathbb{T}}}{\rightrightarrows}} (K\mathbb{F})\mathbb{U}^{\mathbb{T}}(Z)$$

we get two morphisms (s', t')

$$\mathbb{F}\mathbb{U}^{\mathbb{T}}(M) \underset{t'}{\overset{s'}{\rightrightarrows}} \mathbb{F}\mathbb{U}^{\mathbb{T}}(Z)$$

in $\mathcal{C}$ such that $K(s') = s \circ \varepsilon_M^{\mathbb{T}}$ and $K(t') = t \circ \varepsilon_M^{\mathbb{T}}$.

Let I be the image in the regular epi-mono factorization in $\mathcal{C}$ of (s', t'). Since K preserves limits and regular epimorphisms, it also preserves factorizations, hence $K(I) \cong M$.

Now let q be the coequalizer in $\mathcal{C}$ of (s', t'); it exists since it coincides with the coequalizer of I, and I is an equivalence relation, from M an equivalence relation.

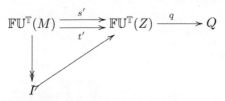

$\mathcal{C}$ is exact, hence I is the kernel pair of q, then, since K preserves limits and regular epimorphisms, $K(q)$ is the coequalizer of $K(I)$ in $\mathbf{Set}^{\mathbb{T}}$. So $K(q) \cong \varepsilon_Z^{\mathbb{T}}$ and $K(Q) \cong Z$.

To complete the proof, observe that P finitely presentable implies that $\mathbb{T}$ has finite rank, hence $\mathbf{Set}^{\mathbb{T}}$ is an algebraic category and $\mathcal{C} \cong \mathbf{Set}^{\mathbb{T}}$. $\qquad\square$

2.2.5. Remark. It is quite easy to generalize Theorem 2.2.4 to the case of a monadic category over $\mathbf{Set}$, i.e. of a category $\mathcal{C} \cong \mathbf{Set}^{\mathbb{T}}$ for an arbitrary monad $\mathbb{T}$ on $\mathbf{Set}$. To get such a characterization theorem it suffices to erase condition (4) in the assumptions of 2.2.4 (see also Chapter V).

2.2.6. Remark. Theorem 2.2.4 applies to the multi-sorted case simply by replacing the single projective, finitely presentable regular generator P by a regular generating set $\mathcal{P} = \{P_s\}_{s \in S}$, ($S$ is the set of sorts) where each P_s is regular projective and finitely presentable. So we get the following:

2.2.7. Corollary. *A category $\mathcal{C}$ is equivalent to an S-sorted algebraic category if and only if:*

    (1) *$\mathcal{C}$ is exact;*
    (2) *$\mathcal{C}$ has a regular generator $\mathcal{P} = \{P_s\}_{s \in S}$;*
    (3) *any P_s is regular projective;*
    (4) *any P_s is finitely presentable.*

2.3. A remark on multi-sorted algebraic categories

As an application of Corollary 2.2.7 we can prove the following Theorem 2.3.1, relating S-sorted and one-sorted algebraic categories.

Recall that an initial object 0 is called *strict initial* if and only if any morphism with codomain 0 is an isomorphism.

2.3.1. Theorem. *An S-sorted algebraic category* $\mathcal{C}$*, with* S *finite, can be presented as a one-sorted algebraic category if and only if the terminal object* 1 *has as subobjects either* 1 *itself or the strict initial.*

Proof. One direction is trivial; conversely let $\mathcal{C}$ be an S-sorted algebraic category satisfying the hypothesis.

By Corollary 2.2.7, there exists a regular generator $\mathcal{P} = \{P_s\}_{s\in S}$, where any P_s is regular projective and finitely presentable as well as an adjoint pair $\mathbb{F} \dashv \mathbb{U}$

$$\mathcal{C} \underset{\mathbb{U}}{\overset{\mathbb{F}}{\rightleftarrows}} \mathbf{Set}^S$$

where

$$\mathbb{U} = (\mathcal{C}(P_s, -))_{s\in S}$$

$$\mathbb{F}(X) = \coprod_{s\in S}(\coprod_{x\in X_s} P_s)$$

for any $X = (X_s)_{s\in S}$ in $\mathbf{Set}^S$.

The objects P_s, $s \in S$, are of the form $P_s = \mathbb{F}(\,_sX)$, where $\,_sX \in \mathbf{Set}^S$ is the S-set having all components empty except the s-component that is equal to 1.

Define $P = \coprod_{s\in S} P_s$. P is regular projective and finitely presentable since the two properties are stable under finite sums.

Then to prove that $\mathcal{C}$ is a one-sorted algebraic category, it suffices to show that P is a regular generator and apply Theorem 2.2.4.

Since $\mathcal{P} = \{P_s\}_{s\in S}$ is a regular generator in $\mathcal{C}$, for any $X \in \mathcal{C}$ there exists a regular epimorphism

$$q : \coprod_{s\in S}(\coprod_{j_s} P_s) \twoheadrightarrow X.$$

We must show that there exists a regular epimorphism

$$p : \coprod_k P \twoheadrightarrow X.$$

To do that, first observe that for any $X \in \mathcal{C}$, where X is not the strict initial object 0, any sort $(\mathbb{U}(X))_s$ of $\mathbb{U}(X)$ is not empty. In fact, suppose that there exists $i \in S$ such that $(\mathbb{U}(X))_i = \emptyset$. Without loss of generality we can take $i = 1$, and consider the regular epimorphism-monomorphism factorization of the unique morphism $f : X \longrightarrow 1$, in $\mathcal{C}$.

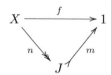

Let J be the image. Applying $\mathbb{U}$ we get the following diagram in $\mathbf{Set}^S$:

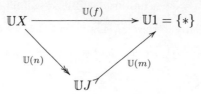

with $\mathbb{U}(m)$ a monomorphism in $\mathbf{Set}^S$. Since $(\mathbb{U}(X))_1 = \emptyset$, then $\mathbb{U}J \neq \{*\}$, so $J \neq 1$. Also $J \neq 0$ since $X \neq 0$, hence we get a contradiction.

So we have proved that every sort $(\mathbb{U}(X))_s$ of $\mathbb{U}(X)$ is not empty for $X \neq 0$; then, since $P_s = \mathbb{F}(\,_sX)$ and $\mathbb{F} \dashv \mathbb{U}$, it follows that for any $s \in S$ and any $X \neq 0$, there exists at least one morphism $\xi_s : P_s \longrightarrow X$.

This remark allows us to construct the morphism p as canonical extension of q

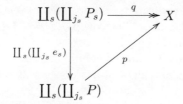

where e_s denotes the sum inclusion.

In fact, for any fixed $s \in S$ and for any morphism $\xi_s : P_s \longrightarrow X$, we can define a corresponding morphism $t : P \longrightarrow X$ such that the following diagram commutes.

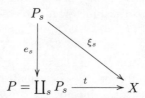

The existence of t, follows from the fact that for any $i \in S$, $i \neq s$, there exists one morphism $\xi_i : P_i \longrightarrow X$.

We conclude by observing that q regular epimorphism and $\mathcal{C}$ regular category imply that p is a regular epimorphism. $\qquad\square$

3. Applications of Lawvere's Theorem

3.1. Quasi-algebraic categories

As an application of the Lawvere Theorem 2.2.4, by using similar techniques, we are now able to describe quasi-algebraic categories.

Recall the following

3.1.1. Definition. A *quasi-algebraic category* is a regular-epireflective subcategory of an algebraic category, with a finitary (=filtered-colimit-preserving) inclusion.

Then, from the main Characterization Theorem 2.2.4, we easily get the following:

3.1.2. Theorem. *A category* $\mathcal{C}$ *is equivalent to a quasi-algebraic category if and only if:*

    (1) $\mathcal{C}$ *is regular;*
    (2) $\mathcal{C}$ *has coequalizers of equivalence relations;*
    (3) *there exists a regular generator* $P \in \mathcal{C}$;
    (4) P *is regular projective;*
    (5) P *is finitely presentable.*

Proof. Let $\mathcal{C}$ be equivalent to a regular-epireflective subcategory of an algebraic category $\mathcal{V}$, with $E : \mathcal{C} \longrightarrow \mathcal{V}$ the inclusion functor, where E preserves filtered colimits, and $H : \mathcal{V} \longrightarrow \mathcal{C}$ with $H \dashv E$.

If P denotes the regular projective, finitely presentable, regular generator of $\mathcal{V}$, then it is easy to see that the reflected object $H(P) \in \mathcal{C}$ is regular projective and finitely presentable in $\mathcal{C}$. This follows from the fact that E preserves regular epimorphisms and E preserves filtered colimits.

Similarly, since P is a regular generator in $\mathcal{V}$, one easily sees that $H(P)$ is a regular generator in $\mathcal{C}$.

Regularity of the category $\mathcal{V}$ together with the fact that E preserves regular epimorphisms implies regularity of $\mathcal{C}$.

Conversely, let $\mathcal{C}$ satisfy the conditions (1) to (5) and apply the same construction as in Theorem 2.2.4. In order to show that $\mathcal{C}$ is regular-epireflective in $\mathbf{Set}^{\mathbb{T}}$, we have to find a left adjoint to the functor $K : \mathcal{C} \longrightarrow \mathbf{Set}^{\mathbb{T}}$ defined by $K(C) = (\mathbb{U}(C), \mathbb{U}(\varepsilon_C))$ and $K(f) = \mathbb{U}(f)$.

Define the functor $H : \mathbf{Set}^{\mathbb{T}} \longrightarrow \mathcal{C}$, by $H(Z) = Q$ for any $Z \in \mathbf{Set}^{\mathbb{T}}$ where (Q, q) is the coequalizer constructed in the proof of Theorem 2.2.4; H is defined on morphisms by the universal property of coequalizers.

Observe that the coequalizer exists since by (2) $\mathcal{C}$ has coequalizers of equivalence relations and I is an equivalence relation.

We now have to verify $H \dashv K$.

Define the unit of the adjunction $\nu_Z : Z \longrightarrow KH(Z)$ by the coequalizer $\varepsilon_Z^{\mathbb{T}}$, as the unique morphism such that $\nu_Z \circ \varepsilon_Z^{\mathbb{T}} = K(q)$.

So $\mathcal{C}$ is, up to equivalence, a reflective subcategory of $\mathbf{Set}^{\mathbb{T}}$; furthermore, since K preserves regular epimorphisms, ν_Z is a regular epimorphism for any $Z \in \mathbf{Set}$.

To complete the proof observe that K preserves filtered colimits since $\mathbb{U}$ preserves filtered colimits and $\mathbb{U}^{\mathbb{T}}$ reflects filtered colimits. $\qquad\qquad \square$

3.1.3. Remark. We conclude by noticing that in Theorem 3.1.2 it suffices to assume as hypothesis conditions (2) to (5), plus the following:

(1a) $\mathcal{C}$ has finite limits.

Another equivalent way to formulate the hypothesis of Theorem 3.1.2 is given by conditions (3) to (5) plus the following:

(2a) $\mathcal{C}$ has coequalizers.

3.1.4. Examples.

(1) Any algebraic category is a quasi-algebraic category.
(2) The category of torsion free abelian groups is a quasi-algebraic category.
(3) The category of sets endowed with a binary relation can be presented as a two-sorted quasi-algebraic category.

3.1.5. Remark. Theorem 3.1.2 applies to the multi-sorted case simply by replacing the single generator P by a regular generating set $\mathcal{P}$ where any $P \in \mathcal{P}$ is regular projective and finitely presentable.

3.1.6. Remark. Observe that in the proof of Theorem 2.3.1 we only use the regularity of the algebraic category. So we can conclude that the Theorem holds for quasi-algebraic categories as well.

It is interesting to notice that Lawvere [20] in his characterization theorem of algebraic categories was not working with the notion of a finitely presentable object P, but he was using the following:

3.1.7. Definition. An object P is called *abstractly finite* if for any set S the copower $\coprod_S P$ exists and, moreover, for any arrow $f : P \longrightarrow \coprod_S P$ there exists a finite subset F of S and an arrow $g : F \longrightarrow S$ such that f factors through

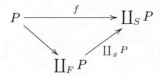

Consider the canonical adjoint pair

$$\mathbf{Set} \underset{\mathcal{C}(P,-)}{\overset{\coprod P}{\underset{\perp}{\rightleftarrows}}} \mathcal{C}$$

then P is abstractly finite in $\mathcal{C}$ if and only if the monad $\mathbb{T} = \mathcal{C}(P, \coprod_{(-)} P)$ induced by P preserves filtered colimits.

Clearly finite presentability of P implies its abstract finiteness, and in the presence of the hypothesis (1), (2), (3), (4) of Theorem 2.2.4 the two properties are in fact equivalent.

This is no longer true if equivalence relations in $\mathcal{C}$ fail to be effective.

This situation has been studied in [17] and [27]; we recall the main result:

3.1.8. Theorem. *The following are equivalent:*

(1) *$\mathcal{C}$ is equivalent to a regular-epireflective subcategory of an algebraic category;*

(2) C *satisfies the following conditions:*
   (a) C *is regular;*
   (b) C *has coequalizer of equivalence relations;*
   (c) *there exists a regular generator* $P \in C$;
   (d) P *is regular projective;*
   (e) P *is abstractly finite.*

Proof. Let C be equivalent to a regular-epireflective subcategory of an algebraic category V, with $E : C \longrightarrow V$ the inclusion functor, and $H : V \longrightarrow C$ with $H \dashv E$.

$$C \xleftarrow[\ E\ ]{\ \ H\ \ }_{\bot} V \, .$$

Then, as in the proof of 3.1.2, consider the reflected object $H(P) \in C$ with P the regular generator, regular projective, finitely presentable object of V. It only remains to show that $H(P)$ is abstractly finite in C. We know that P finitely presentable in V implies P abstractly finite in V.

Then consider a morphism $f : H(P) \longrightarrow \coprod_S H(P)$ in C, and denote by $\eta_P :$ $P \twoheadrightarrow H(P)$ and by $\eta_{\coprod_S P} : \coprod_S P \to H(\coprod_S P) \cong \coprod_S H(P)$ the unit maps of the adjunction $H \dashv E$.

Projectivity of P in V implies that there exists a map $h : P \longrightarrow \coprod_S P$ such that the following diagram commutes

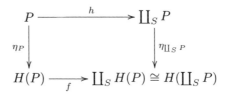

Then, P abstractly finite in V implies $H(P)$ abstractly finite in C.

To show the converse, (2) implies (1), apply the same construction as in the proof of Theorem 3.1.2.

Hypothesis (e) implies that the monad $\mathbb{T}$ has finite rank, hence $\mathbf{Set}^{\mathbb{T}}$ is an algebraic category. $\square$

3.2. Mal'cev categories

The aim of this section is to prove that Theorem 2.2.4 is extremely useful to describe Mal'cev categories ([9], [11]).

3.2.1. Definition. A regular category C is a *Mal'cev category* if and only if for any $X \in C$ and equivalence relations R, S on X

$$R \circ S = S \circ R.$$

If we denote by $EquivX$ the ordered set of equivalence relations on an object X of C, then we get the following characterization theorem:

3.2.2. Theorem. *For a regular category $\mathcal{C}$ the following conditions are equivalent:*

(1) *$\mathcal{C}$ is Mal'cev;*

(2) *for equivalence relations R and S on an object X, $R \circ S$ is an equivalence relation, and therefore the join $R \vee S$ in $\mathrm{Equiv}X$;*

(3) *every relation R on $X \times Y$ is difunctional, that is $R \circ R^{op} \circ R = R$, where R^{op} is the opposite relation of R;*

(4) *every reflexive relation R on X is an equivalence relation;*

(5) *every reflexive relation is transitive;*

(6) *every reflexive relation is symmetric.*

Proof. (1) $\Rightarrow$ (2): We recall that a relation R is symmetric iff $R = R^{op}$, and it is transitive iff $R \circ R = R$. Then, to show that $R \circ S$ is an equivalence it suffices to see that

$$(R \circ S)^{op} = S^{op} \circ R^{op} = S \circ R = R \circ S$$

and that

$$(R \circ S) \circ (R \circ S) = R \circ (S \circ R) \circ S = R \circ (R \circ S) \circ S = (R \circ R) \circ (S \circ S) = R \circ S.$$

$R \vee S$ is the smallest equivalence relation containing both R and S, then since $R \subseteq R \circ S$, $S \subseteq R \circ S$ we get $R \vee S \subseteq R \circ S$; since $R \circ S$ is always contained in $R \vee S$ we get $R \circ S = R \vee S$.

(2) $\Rightarrow$ (1): Trivial since $\vee$ is permutable.

(1) $\Rightarrow$ (3) $\Rightarrow$ (4) $\Rightarrow$ (2):

(1) $\Rightarrow$ (3): Since for any relation R, $R \subseteq R \circ R^{op} \circ R$ it suffices to check that

$$R \circ R^{op} \circ R \subseteq R.$$

Consider the domain and codomain arrows of R, $r_0, r_1 : R \rightrightarrows X$, and the corresponding kernel pairs, $N_0 = Ker\, r_0$ and $N_1 = Ker\, r_1$.

Any kernel pair is an equivalence relation, hence $N_0 \circ N_1 = N_1 \circ N_0$, and this equality exactly means $R \circ R^{op} \circ R = R$.

This last statement can be easily understood by thinking in terms of elements: then, difunctionality means:

$$(x, y) \in R \circ R^{op} \circ R \text{ iff } \exists t, \exists s \text{ with } xRt, sRt, sRy;$$

$N_0 \circ N_1 = N_1 \circ N_0$ implies that:

$$\forall (x, t), (s, t), (s, y) \in R, \text{ if } ((x, t), (s, y)) \in N_1 \circ N_0, \text{ then} ((x, t), (s, y)) \in N_0 \circ N_1,$$

so necessarily $(x, y) \in R$.

(3) $\Rightarrow$ (4): To show that any reflexive R is an equivalence, it suffices to see that

$$R^{op} \subseteq 1 \circ R^{op} \circ 1 \subseteq R \circ R^{op} \circ R \subseteq R$$

and

$$R \circ R \subseteq R \circ 1 \circ R \subseteq R \circ R^{op} \circ R \subseteq R$$

where 1 denotes the identity relation on X.

(4) $\Rightarrow$ (2): Trivial since $R \circ S$ is reflexive.

$(4) \Rightarrow (5) \Rightarrow (6) \Rightarrow (5) \Rightarrow (4)$:
The non-trivial implication $(5) \Rightarrow (6)$ can be seen as follows:
for any reflexive relation R, define a new relation S on R by

$$(x, y) S(x', y') \text{ iff } (x, y') \in R.$$

We use here a description in terms of elements, but this process is clearly a finite limits construction.
Then transitivity of S implies that for $(x, y) \in R$,

$$(y, y) S(x, y) S(x, x),$$

so $(y, y) S(x, x)$ and $(y, x) \in R$, hence R is symmetric.
Similarly for the step $(6) \Rightarrow (5)$.   □

In the case of a classical one-sorted algebraic category, the Mal'cev axiom can be interpreted syntactically as in the following *Mal'cev Theorem* ([22], [10], [30]).
Recall that a *weak limit* (or *colimit*) is defined as in the usual notion of limit (or colimit) except that we only require the existence of a factorization and not the uniqueness.

3.2.3. Theorem. *Let C be a one-sorted algebraic category, then the following are equivalent:*

(1) *C is a Mal'cev category;*
(2) *the Lawvere theory T of C admits a ternary operation*

$$p : T \times T \times T \longrightarrow T$$

satisfying the following two axioms:

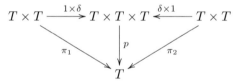

where δ is the diagonal and π_i the product projections;
(3) *for any object X in the Lawvere theory T of C the following coassociativity square determined by the diagonal δ is a weak pushout in T.*

$$
\begin{array}{ccc}
X & \xrightarrow{\;\;\delta\;\;} & X \times X \\
\delta \downarrow & & \downarrow \delta \times 1 \\
X \times X & \xrightarrow[1 \times \delta]{} & X \times X \times X
\end{array}
$$

Proof. We will prove the following implications: $(1) \Rightarrow (3) \Rightarrow (2) \Rightarrow (1)$.
$(1) \Rightarrow (3)$:
Let C be a Mal'cev algebraic category.
Suppose $C = \mathcal{M}od_T$ for a Lawvere theory T. We know (see the proof of Theorem

1.3.3) that $\mathcal{T}^{op}$ is equivalent to the subcategory $\mathcal{F}$ of $\mathcal{C}$ spanned by the free finitely generated algebras and that the inclusion $\mathbb{Y} : \mathcal{F} \cong \mathcal{T}^{op} \hookrightarrow \mathcal{M}od_{\mathcal{T}}$ (it is just the restricted Yoneda embedding) preserves finite sums. Moreover, any $X \in \mathcal{F}$ is regular projective in $\mathcal{C}$ since any sum of regular projective objects is regular projective and the free algebra on 1 is regular projective by 1.2.6.

Then, consider the following diagram in $\mathcal{F}$

$$
\begin{array}{ccc}
X + X + X & \xrightarrow{\Delta + 1} & X + X \\
{\scriptstyle 1 + \Delta} \downarrow & (*) & \downarrow {\scriptstyle \Delta} \\
X + X & \xrightarrow{\Delta} & X
\end{array}
$$

where Δ denotes the codiagonal morphism.

It is easy to check that $(*)$ is a pushout.

Now, if N denotes the image of the following reflexive graph in $\mathcal{C}$

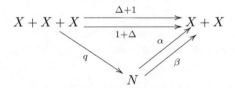

then its reflexivity implies that N is an equivalence relation by 3.2.2.

Moreover, since Δ is the coequalizer of the above graph, from exactness of $\mathcal{C}$ we get the following pullback in $\mathcal{C}$

$$
\begin{array}{ccc}
N & \xrightarrow{\alpha} & X + X \\
{\scriptstyle \beta} \downarrow & & \downarrow {\scriptstyle \Delta} \\
X + X & \xrightarrow{\Delta} & X
\end{array}
$$

Hence $(*)$ is a weak pullback in $\mathcal{F}$, since any object in $\mathcal{F}$ is regular projective in $\mathcal{C}$ and q is a regular epimorphism.

Then the dual result holds in $\mathcal{T}$.

$(3)\Rightarrow(2)$:

p can be easily defined from the universal property of the weak pushout for $X = T$ and from $\pi_1 \circ \delta = \pi_2 \circ \delta$.

$(2)\Rightarrow(1)$:

Observe that a ternary operation p on T, satisfying the two axioms of conditions (2), exactly correspond to a natural transformation

$$
\bar{p} : \mathbb{U}^3 \longrightarrow \mathbb{U}
$$

(where $\mathbb{U} : \mathcal{C} \longrightarrow \mathbf{Set}$ is the usual forgetful functor) such that, for any $X \in \mathcal{C}$ and $x, y \in \mathbb{U}X$ the following Mal'cev equations hold:

$$\bar{p}_X(x, x, y) = y, \qquad \bar{p}_X(x, y, y) = x. \qquad \text{(M)}$$

This fact follows by Yoneda Lemma, since $\mathbb{U} = \mathcal{C}(\mathcal{T}(T, -), -)$ by 1.1.5 (1) and $\mathcal{C}(\mathcal{T}(T, -), -) \times \mathcal{C}(\mathcal{T}(T, -), -) \times \mathcal{C}(\mathcal{T}(T, -), -) \cong \mathcal{C}(\mathcal{T}(T, -) \amalg \mathcal{T}(T, -) \amalg \mathcal{T}(T, -), -)$ $\cong \mathcal{C}(\mathcal{T}(T^3, -), -)$.

Then, for any equivalences R, S on X, to show $R \circ S = S \circ R$ it suffices to show $\mathbb{U}R \circ \mathbb{U}S = \mathbb{U}S \circ \mathbb{U}R$ since $\mathbb{U}$ preserves regular epimorphisms and limits and reflects isomorphisms by 1.1.5 and 1.2.6.

Now if $(x, y) \in \mathbb{U}R \circ \mathbb{U}S$, then there exists $t \in \mathbb{U}X$ such that $(x, t) \in \mathbb{U}R$ and $(t, y) \in \mathbb{U}S$. If we consider $\bar{p}_X(x, t, y) \in \mathbb{U}X$ it is easy to check by naturality of $\bar{p}$ and by the Mal'cev equations (M) for $\bar{p}_X$ that $(x, \bar{p}(x, t, y)) \in \mathbb{U}S$ and $(\bar{p}(x, t, y), y) \in \mathbb{U}R$ hence $(x, y) \in \mathbb{U}S \circ \mathbb{U}R$. $\qquad \square$

3.2.4. Examples. The following are all examples of Mal'cev categories:

(1) The category of groups. Define $p_G(x, y, z) = x - y + z$ for any group G.
(2) The categories of Abelian groups and of R-modules.
(3) The category of Heyting algebras. For any Heyting algebra H, define p_H by

$$p_H(a, b, c) = ((a \to b) \to c) \wedge ((c \to b) \to a)$$

or also by

$$p'_H(a, b, c) = ((b \to (a \wedge c)) \wedge (a \vee c).$$

Note that the Mal'cev operation is not necessarily unique.
(4) The dual category of any topos [23].
(5) The category of topological groups.
(6) The category of localic groups [25].
(7) Any naturally Mal'cev category [19].
(8) Any protomodular category (see Chapter IV and [7]).
(9) Any additive regular category (see Chapter IV and [8]).

Observe now that we could describe Mal'cev algebraic categories simply by adding to Theorem 2.2.4 an extra condition on the regular projective generator.

3.2.5. Theorem. *A category $\mathcal{C}$ is equivalent to a Mal'cev algebraic category if and only if:*

(1) $\mathcal{C}$ *is exact;*
(2) *there exists a regular generator $P \in \mathcal{C}$;*
(3) P *is regular projective;*
(4) P *is finitely presentable;*

(5) *P is an internal Mal'cev coalgebra, i.e. there exists* $\rho : P \longrightarrow P + P + P$
with:

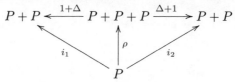

$$P + P \xleftarrow{\;1+\Delta\;} P + P + P \xrightarrow{\;\Delta+1\;} P + P$$

with i_1, i_2 *the sum inclusion and* Δ *the codiagonal.*

Proof. It is Theorem 2.2.4 plus Mal'cev Theorem 3.2.3 together with the remarks that $\mathcal{T}^{op} \cong \mathcal{F}$ where $\mathcal{F}$ is the full subcategory of $\mathcal{C}$ spanned by the free finitely generated algebras, and $\mathcal{F}$ is closed in $\mathcal{C}$ with respect to finite sums. $\qquad\square$

3.2.6. Remark. Observe that Theorem 3.2.5 can be generalized to the quasi-algebraic case, since the exactness condition is not necessary for all the considerations involved in Mal'cev axioms (see [24]).

3.3. The exact completion

In this section we recall some basic facts on exact completion that we will need in 3.4. For more details and complete proofs see [12], [31]; see also Chapter VII for further applications of the exact completion construction.

3.3.1. Definition. Let $\mathcal{C}$ be a category with weak finite limits.
The *exact completion* $\mathcal{C}_{ex}$ of $\mathcal{C}$ is defined as follows:

- objects of $\mathcal{C}_{ex}$ are pseudo equivalence relations in $\mathcal{C}$, i.e. pairs of parallel arrows $r_0, r_1 : R \rightrightarrows X$, not necessarily jointly monic, that are reflexive, symmetric, and transitive (in the definition of transitivity we work now with weak pullbacks);
- a morphism $(f, \overline{f})$ between two objects is an equivalence class of pairs of compatible arrows $((f, \overline{f})$ is said to be compatible if $s_0 \circ \overline{f} = f \circ r_0$ and $s_1 \circ \overline{f} = f \circ r_1)$; two pairs $(f, \overline{f})$ and $(g, \overline{g})$ are considered to be equivalent if there is an arrow $k : X \longrightarrow S$ such that $f = s_o \circ k$ and $g = s_1 \circ k$;

$$
\begin{array}{ccc}
R & \overset{r_0}{\underset{r_1}{\rightrightarrows}} & X \\
{\scriptstyle \overline{f}}\downarrow & & \downarrow{\scriptstyle f} \\
S & \overset{s_0}{\underset{s_1}{\rightrightarrows}} & Y
\end{array}
$$

- composition and identities are the obvious ones.

Then it is possible to show that:

- The functor $\Gamma : \mathcal{C} \longrightarrow \mathcal{C}_{ex}$ sending each object into the pair of identities is full and faithful;

- The image $\Gamma(\mathcal{C})$ generates $\mathcal{C}_{ex}$ via coequalizers; that is, if $(f, \bar{f})$ is an arrow in $\mathcal{C}_{ex}$, then we get the following diagram in $\mathcal{C}_{ex}$:

$$
\begin{array}{ccc}
\Gamma R \rightrightarrows \Gamma X \longrightarrow (R \rightrightarrows X) \\
\Gamma\bar{f} \downarrow \qquad \Gamma f \downarrow \qquad\qquad \downarrow (f,\bar{f}) \\
\Gamma S \rightrightarrows \Gamma Y \longrightarrow (S \rightrightarrows Y)
\end{array}
$$

where the rows are coequalizers.

To formally write the universal property of $\mathcal{C}_{ex}$ we need the following

3.3.2. Definition. A functor $F : \mathcal{C} \longrightarrow \mathcal{A}$, with $\mathcal{C}$ a category with weak finite limits and $\mathcal{A}$ regular, is called *"left covering"* if for all functors $\mathcal{L} : \mathcal{D} \longrightarrow \mathcal{C}$ defined on a finite category $\mathcal{D}$ and for all weak limits L, the canonical factorization $p : FL \longrightarrow \overline{L}$, where $\overline{L} = \lim F\mathcal{L}$, is a regular epimorphism.

Then we can state the main result:

3.3.3. Theorem. *Let $\mathcal{C}$ be a category with weak finite limits and $\mathcal{A}$ an exact one; let $\Gamma : \mathcal{C} \longrightarrow \mathcal{C}_{ex}$ be the exact completion of $\mathcal{C}$. Then Γ induces an equivalence between the category of left covering functors from $\mathcal{C}$ to $\mathcal{A}$ and the category of exact functors (that means finite-limit and regular-epimorphism preserving functors) from $\mathcal{C}_{ex}$ to $\mathcal{A}$.*

Recall that a category $\mathcal{C}$ is said to have enough regular projectives if for any $X \in \mathcal{C}$, there exists a regular projective P and a regular epimorphism $g : P \longrightarrow X$.

We also say that $\mathcal{C}$ has a regular projective cover $\mathcal{S}$ if for any $X \in \mathcal{C}$ there exists $P \in \mathcal{S}$, P regular projective, and a regular epimorphism $g : P \longrightarrow X$. Then we can recall the following theorems characterizing "free exact" categories by means of regular projective covers:

3.3.4. Theorem. *Let $\mathcal{C}$ be a category with weak finite limits, and $\Gamma : \mathcal{C} \longrightarrow \mathcal{C}_{ex}$ the exact completion of $\mathcal{C}$. Then for any $X \in \mathcal{C}$, ΓX is regular projective in $\mathcal{C}_{ex}$. Moreover the image $\Gamma(\mathcal{C})$ is a regular projective cover of $\mathcal{C}_{ex}$.*

3.3.5. Theorem. *Every exact category with enough regular projectives is the exact completion of a category with weak finite limits, given by its projectives.*

$\mathcal{C}_{ex}$ could also be constructed (for $\mathcal{C}$ small) as a full subcategory of $\mathbf{Set}^{\mathcal{C}^{op}}$ (see[16]). Denoting by $\mathbb{Y}$ the Yoneda embedding $\mathcal{C} \longrightarrow \mathbf{Set}^{\mathcal{C}^{op}}$, we get:

3.3.6. Proposition. *$\mathcal{C}_{ex}$ is, up to equivalence, the full subcategory of $\mathbf{Set}^{\mathcal{C}^{op}}$ of functors F such that there is a regular epimorphism e from a representable to F*

$$
e : \mathbb{Y}(C) \longrightarrow F
$$

whose kernel pair admits a cover by a regular epimorphism with domain a representable $\mathbb{Y}(C')$

$$
\mathbb{Y}(C') \longrightarrow N \rightrightarrows \mathbb{Y}(C) \xrightarrow{e} F.
$$

Similar discussions could be done for the regular completion of a category C with weak finite limits as well as for the exact completion associated to a regular category.

3.4. Localizations

In this section we want to show that characterization theorems of the form of 2.2.4 apply to investigate localizations of categories.

In this case it will suffice to erase from Theorem 2.2.4 the assumption (3) of regular projectivity to get a complete characterization of finitary localizations of algebraic categories.

An elementary proof of such a result is not known; in fact the proof given by Vitale in [31] strongly uses techniques of exact completion. For completeness we give a sketch of such a proof without technical details that can be found in [31], [32].

3.4.1. Definition. A *localization* $\mathcal{L}$ of a category C is a reflective subcategory

$$\mathcal{L} \xrightarrow[\substack{F \\ \perp \\ J}]{} C$$

such that F preserves finite limits.

3.4.2. Definition. A localization $\mathcal{L}$ of C is *finitary* if the inclusion $J : \mathcal{L} \longrightarrow C$ preserves filtered colimits.

The idea of Theorem 2.2.4 can now be used to characterize localizations of algebraic categories.

3.4.3. Theorem. *A category C is equivalent to a finitary localization of a one-sorted algebraic category if and only if:*

(1) *C is exact;*
(2) *there exists a regular generator $P \in C$;*
(3) *P is finitely presentable.*

Proof. (Sketch) Given the three conditions, to prove the non-trivial part, consider as in Theorem 2.2.4 the following situation determined by P:

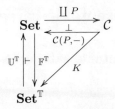

To prove that C is a localization in the algebraic category $\mathbf{Set}^{\mathbb{T}}$, we must show that there is a functor $H : \mathbf{Set}^{\mathbb{T}} \longrightarrow C$ with $H \dashv K$ that preserves finite limits. This step can be obtained by a very elegant use of the universal property of exact completion.

First, define the subcategory $\mathcal{S}$ of $\mathcal{C}$ spanned by copowers of P, and then check that the full inclusion $J : \mathcal{S} \longrightarrow \mathcal{C}$ is a left covering functor in the sense of Definition 3.3.2. Moreover we get $\mathcal{S}_{ex} \cong \mathbf{Set}^{\mathbb{T}}$ by 3.3.5.

Now, by the universal property of the exact completion $\Gamma : \mathcal{S} \longrightarrow \mathcal{S}_{ex}$, since $\mathcal{C}$ is exact, there exists an exact functor $\hat{H} : \mathcal{S}_{ex} \longrightarrow \mathcal{C}$ such that $\Gamma \circ \hat{H} \cong J$.

Then the result follows from $\mathcal{S}_{ex} \cong \mathbf{Set}^{\mathbb{T}}$ by defining $H \cong \hat{H} \dashv K$.   □

We recall the following related result on localizations:

3.4.4. Theorem. *A category $\mathcal{C}$ is equivalent to a localization of a one-sorted algebraic category if and only if:*

    (1) *$\mathcal{C}$ is exact;*
    (2) *there exists a regular generator $P \in \mathcal{C}$;*
    (3) *filtered colimits exist and commute with finite limits in $\mathcal{C}$.*

Proof. (Sketch) To show that any localization of an algebraic category verifies the conditions is simple since in any algebraic category filtered colimits commute with finite limits and the condition is stable under localizations.

Conversely, as in 3.4.3, we can show by the same arguments that $\mathcal{C}$ is equivalent to a localization of a monadic category over $\mathbf{Set}$ for a monad $\mathbb{T}$.

Then the trick consists of constructing a new monad $\mathbb{T}'$, the finitary part of $\mathbb{T}$ and show that $\mathbb{T}'$ has finite rank and $\mathcal{C}$ is still a localization in $\mathbf{Set}^{\mathbb{T}'}$ (see [32]).   □

3.4.5. Remark. The previous results admit a canonical generalization in the multi-sorted case.

Moreover, if we erase condition (3) in 3.4.3 we get a characterization theorem for localizations of monadic categories over $\mathbf{Set}$ (i.e. categories of the form $\mathbf{Set}^{\mathbb{T}}$ with $\mathbb{T}$ a monad on $\mathbf{Set}$).

A discussion of more specialized cases like localizations in Mal'cev algebraic categories, or essential localizations in algebraic categories can be found in recent papers. See for example [15], [4].

4. Locally finitely presentable categories

4.1. Locally finitely presentable categories

We recall the following notion of locally finitely presentable category:

4.1.1. Definition. A category $\mathcal{K}$ is *locally finitely presentable* if and only if:

    (1) $\mathcal{K}$ is cocomplete;
    (2) $\mathcal{K}$ has a set $\mathcal{P}$ of finitely presentable objects such that every object of $\mathcal{K}$ is a filtered colimit of objects of $\mathcal{P}$.

The classical presentation of a locally finitely presentable category as a category of models or more precisely of finite-limit-preserving functors come from the following basic theorem:

4.1.2. Theorem.[14] *Let $\mathcal{K}$ be a locally finitely presentable category, then there exists a small finitely complete category $\mathcal{S}$ such that $\mathcal{K}$ is equivalent to the category* $\mathsf{Lex}(\mathcal{S}, \mathbf{Set})$ *of set-valued left exact functors (i.e. finite-limit-preserving functors). The dual of $\mathcal{S}$ itself is equivalent to the full subcategory of $\mathcal{K}$ generated by the finitely presentable objects.*

If we write $\mathcal{C}$ for $\mathcal{S}^{op}$, where $\mathcal{C}$ is small and finitely cocomplete, then we can always think of $\mathcal{K}$ as $\mathcal{K} \cong \mathsf{Lex}(\mathcal{C}^{op}, \mathbf{Set})$, and consider the following restricted Yoneda embedding:

$$\mathcal{C} \xrightarrow{\quad \mathbb{Y} \quad} \mathcal{K} \cong \mathsf{Lex}(\mathcal{C}^{op}, \mathbf{Set}).$$

$\mathbb{Y}$ preserves finite colimits and all limits that exist in $\mathcal{C}$.

With the above notations, and from the remark that $\mathcal{S}$ is unique up to equivalence of categories, we get the following

4.1.3. Definition. $\mathcal{S} = \mathcal{C}^{op}$ is called the *essentially algebraic theory* of $\mathcal{K}$.

Moreover, by using the approach in terms of models (i.e. lex functors) one can show that:

- $\mathcal{K}$ is equivalent to a full reflective subcategory in $\mathbf{Set}^{\mathcal{C}^{op}}$ closed under filtered colimits;
- $\mathcal{K}$ has small limits and colimits;
- filtered colimits and finite limits commute in $\mathcal{K}$.

For detailed results and complete proofs on locally finitely presentable categories see [3] or [14].

4.1.4. Examples.

(1) **Set** is locally finitely presentable with finite sets as finitely presentables.
(2) The category $\mathbf{Set}_{fin}$ of finite sets is not locally finitely presentable since it is not cocomplete.
(3) Algebraic categories and quasi-algebraic categories are locally finitely presentable (see next section).
(4) The category *Pos* of partial ordered sets and order-preserving functions is locally finitely presentable. The finitely presentable objects are precisely the finite ones. Observe that *Pos* is not regular.
(5) The category *Cat* of small categories and functors is locally finitely presentable. The one arrow category 2 is a strong generator, finitely presentable. Observe that *Cat* is not regular.
(6) The category *Top* of topological spaces and continuous functions is not locally finitely presentable. In this case a topological space is finitely presentable iff it is finite and discrete.

4.2. Algebraic categories and quasi-algebraic categories are locally finitely presentable

Let Q be a one-sorted quasi-algebraic category, with $\mathbb{U}$ and $\mathbb{F}$ the forgetful and free functor respectively. To show that Q is a locally finitely presentable category, we need to understand the role of finitely presentable objects in this case.

Recall that a small full subcategory $\mathcal{D}$ of Q is called a *dense generator* provided that every object Q of Q is a canonical colimit of objects of $\mathcal{D}$; more precisely a colimit for which the colimiting cone consists of all arrows from objects D of $\mathcal{D}$ to Q. Then it is immediate to see that the free finitely generated objects of Q form a dense generator of Q.

4.2.1. Lemma. *Let Q be a one-sorted quasi-algebraic category, and consider the set $\mathcal{P}$ of objects $P \in Q$ that can be presented as coequalizers of the following form:*

$$\mathbb{F}(m) \rightrightarrows \mathbb{F}(n) \longrightarrow P$$

with m, n finite sets.

Then $\mathcal{P}$ determines a dense generator of Q, closed under finite colimits. Moreover any $Q \in Q$ is a filtered colimit of objects of $\mathcal{P}$.

Proof. To prove that $\mathcal{P}$ is closed under finite colimits, we first notice that $\mathcal{P}$ contains the initial object $\mathbb{F}(\emptyset)$, then, given the following coequalizers

$$\mathbb{F}(m) \xrightarrow[\;v\;]{\;u\;} \mathbb{F}(n) \xrightarrow{\;p\;} P$$

$$\mathbb{F}(k) \xrightarrow[\;s\;]{\;r\;} \mathbb{F}(l) \xrightarrow{\;q\;} Q$$

observe that the sequence

$$\mathbb{F}(m+k) \xrightarrow[v \amalg s]{u \amalg r} \mathbb{F}(n+l) \xrightarrow{p \amalg q} P \amalg Q$$

is again a coequalizer.

To verify that $\mathcal{P}$ is closed with respect to coequalizers consider the following diagram:

$$
\begin{array}{ccccc}
\mathbb{F}(m) & \xrightarrow[b]{a} & \mathbb{F}(n) & \xrightarrow{p} & P \\
& & & & \downarrow{\scriptstyle u} \;\; \downarrow{\scriptstyle v} \\
\mathbb{F}(k) & \xrightarrow[d]{c} & \mathbb{F}(l) & \xrightarrow{q} & Q \\
& & & & \downarrow{\scriptstyle r} \\
& & & & R
\end{array}
$$

with $(R, r) = Coeq(u, v)$ and $P, Q \in \mathcal{P}$.

Then since $\mathbb{F}(n)$ is projective we get $x, y : \mathbb{F}(n) \longrightarrow \mathbb{F}(l)$ with $q \circ x = u \circ p$, $q \circ y = v \circ p$.

Routine verifications show that $r \circ q$ is the coequalizer of the pair $((x,c),(y,d))$, where (x,c) and (y,d) are defined by the universal property of coproducts:

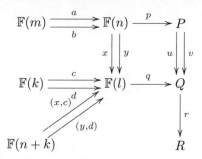

Now since the free objects $\mathbb{F}(n)$, with n finite, form a dense generator and $\mathcal{P}$ is stable under finite colimits, we immediately get the result. □

4.2.2. Proposition. *An object P in a one-sorted quasi-algebraic category $\mathcal{Q}$ is finitely presentable if and only if it can be presented by a coequalizer diagram*

$$\mathbb{F}(m) \rightrightarrows \mathbb{F}(n) \longrightarrow P.$$

Proof. We start by showing that any P presented as coequalizer

$$\mathbb{F}(m) \xrightarrow[\ v\ ]{\ u\ } \mathbb{F}(n) \xrightarrow{\ q\ } P$$

is finitely presentable.

Consider a filtered colimit $N = colim_{i \in I} N_i$ in $\mathcal{Q}$ and a morphism $f : P \longrightarrow N$.

If $f \circ q : \mathbb{F}(n) \longrightarrow N$ is the composite, since N is filtered (filtered colimits in $\mathcal{Q}$ are formed like in **Set**), then there exists an index i and a morphism $g : \mathbb{F}(n) \longrightarrow N_i$ such that $s_i \circ g = f \circ q$ with s_i the colimit inclusion.

Then from $s_i \circ g \circ u = s_i \circ g \circ v$ we get that there exists $s : N_i \longrightarrow N_j$ such that $s \circ g \circ u = s \circ g \circ v$; so, since q is a coequalizer, we obtain a morphism $h : P \longrightarrow N_j$ with $s_j \circ h = f$.

A similar argument shows that h is essentially unique.

Conversely, let $M \in \mathcal{Q}$ be finitely presentable; by 4.2.1 we know that M is a filtered colimit of objects of $\mathcal{P}$:

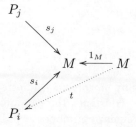

$M = colim_{i \in I} P_i$, I filtered.

Consider the identity $1_M : M \longrightarrow M$, then M finitely presentable implies that there exists $t : M \longrightarrow P_i$ for a certain index i with $s_i \circ t = 1_M$, hence M can be presented as the following coequalizer:

$$P_i \underset{t \circ s_i}{\overset{1_{P_i}}{\rightrightarrows}} P_i \longrightarrow M. \qquad \square$$

Exactly the same argument of 4.2.2 applies to multi-sorted quasi-algebraic categories, as well as to multi-sorted algebraic categories. So we get the following:

4.2.3. Corollary. *An object P in a multi-sorted quasi-algebraic category $\mathcal{Q}$ is finitely presentable if and only if it is the coequalizer of a graph of free finitely generated objects.*

In the multi-sorted case a free finitely generated algebra is clearly of the form $\mathbb{F}(X)$, where $X = (X_i)_{i \in I}$ is a multi-sorted set of finite power, and the power

$$\sharp X = \sum_i card X_i.$$

Then, since any multi-sorted algebraic category is cocomplete, and similarly any quasi-algebraic category is cocomplete, as a reflective subcategory of a cocomplete one, we can easily conclude by 4.2.3 that algebraic categories and quasi-algebraic categories are locally finitely presentable.

At this point, it is quite natural to analyze the converse problem, i.e. to understand when a locally finitely presentable category is algebraic or quasi-algebraic.

We approach the problem in terms of the corresponding essentially algebraic theory $\mathcal{S}$, by finding suitable conditions on $\mathcal{S}$ (or on its dual) that suffices to completely characterize algebraic and quasi-algebraic locally finitely presentable categories.

We conclude this section by giving an answer to the quasi-algebraic case; the algebraic situation will be discussed in 4.3. To do that, consider the following consequence of the multi-sorted version of 3.1.2.

4.2.4. Proposition. *A category $\mathcal{K}$ is equivalent to a multi-sorted quasi-algebraic category if and only if:*

(1) *$\mathcal{K}$ has finite limits;*
(2) *$\mathcal{K}$ has coequalizers of equivalence relations;*
(3) *$\mathcal{K}$ has a strong generator $\mathcal{P}$;*
(4) *any $P \in \mathcal{P}$ is regular projective;*
(5) *any $P \in \mathcal{P}$ is finitely presentable.*

Proof. By applying the multi-sorted version of 3.1.2 together with the remark 3.1.3, it suffices to prove that the strong generator $\mathcal{P}$ is in fact a regular generator. To do that observe that in $\mathcal{K}$ any morphism f admits a regular epi-monomorphism factorization: In fact you can construct the canonical factorization of f in $\mathcal{K}$ as in **Set** and show that it is a regular epi-monomorphism factorization, by applying $\mathbb{U} = \mathcal{K}(P, -)$, for any $P \in \mathcal{P}$ and using that $\mathcal{P}$ is a strong generator and any $P \in \mathcal{P}$ is regular projective. Hence any strong epimorphism in $\mathcal{K}$ is regular. $\qquad \square$

Then we get the following characterization theorem:

4.2.5. Theorem.[2] *Let $\mathcal{K}$ be a locally finitely presentable category with $\mathcal{C}$ the full subcategory of finitely presentables. Then $\mathcal{K}$ is equivalent to a multi-sorted quasi-algebraic category if and only if $\mathcal{C}$ has enough regular projectives.*

Proof. Observe that by 4.1.2, and without any loss of generality, we can suppose $\mathcal{C}$ to be small and $\mathcal{K} \cong \mathsf{Lex}(\mathcal{C}^{op}, \mathbf{Set})$. Then, for any quasi-algebraic category $\mathcal{K}$, since $\mathcal{C}$ is given by finitely presentable objects, any $C \in \mathcal{C}$ is a coequalizer of free finitely generated algebras by 4.2.2. Since any such free algebra is regular projective, then $\mathcal{C}$ has enough regular projectives.

Conversely, to apply Proposition 4.2.4 it suffices to show that, if $\mathcal{P} = \{P \in \mathcal{C} : P$ is regular projective in $\mathcal{C}\}$ is a projective cover of $\mathcal{C}$ (i.e. for any $C \in \mathcal{C}$ there exist $P \in \mathcal{P}$ and $h : P \longrightarrow C$ with h regular epimorphism) then the following conditions are satisfied:

(1) $\mathcal{P}$ is a strong generator in $\mathcal{K}$;
(2) any $P \in \mathcal{P}$ is regular projective in $\mathcal{K}$.

To prove (1), observe that for all $K \in \mathcal{K}$, K is a filtered colimit of objects C_i in $\mathcal{C}$, giving rise to a regular epimorphism $\sum C_i \longrightarrow K$, while each such C_i is a regular quotient of an object P_i in $\mathcal{P}$.
Thus for each K in $\mathcal{K}$ we have

$$\sum P_i \longrightarrow \sum C_i \longrightarrow K$$

and the composite is a strong epimorphism.
To prove condition (2), that P is regular projective in $\mathcal{K}$, use the fact that every regular epimorphism in $\mathsf{Lex}(\mathcal{C}^{op}, \mathbf{Set})$ is a filtered colimit of regular epimorphisms between finitely presentable objects and recall that $\mathbb{Y} : \mathcal{C} \longrightarrow \mathcal{K}$ preserves regular epimorphisms. $\qquad\square$

4.3. Effective projective objects

We intend now to give an answer to the following question: When is a locally finitely presentable category an algebraic category?
As in the quasi-algebraic case we look for syntactic conditions in terms of suitable properties of the corresponding essentially algebraic theories. To do that, consider the following:

4.3.1. Definition. An object P in a category $\mathcal{K}$ is said to be an *effective projective object* if $\mathcal{K}(P, -)$ preserves coequalizers of reflexive graphs.

Observe that Definition 4.3.1 means the following:

4.3.2. Remark. Consider the coequalizer q of a reflexive graph (r_0, r_1) in $\mathcal{K}$.

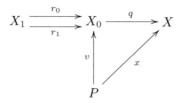

To say that q is preserved by $\mathcal{K}(P, -)$ means that for any arrow $x : P \longrightarrow X$ there exists an arrow $v : P \longrightarrow X_0$ with $q \circ v = x$, such that for any pair $v, w : P \longrightarrow X_0$ with $q \circ v = q \circ w$, (v, w) belongs to the equivalence relation generated by the property of factorizing through the graph X (we write $v \approx_X w$). More precisely this means that there exists a natural number n and n arrows $h_1, ... h_n : P \longrightarrow X_1$ such that

$$
\begin{aligned}
r_0 \circ h_0 &= v, \\
r_1 \circ h_0 &= r_1 \circ h_1, \\
r_0 \circ h_1 &= r_0 \circ h_2, \\
r_1 \circ h_n &= w.
\end{aligned}
$$

Definition 4.3.1 has been introduced in [28] to characterize the coequalizer completion of categories with sums.

We mention now some basic properties of effective projectives:

4.3.3. Lemma.

    (1) *The effective projectives of a category are closed with respect to finite sums and retracts.*

    (2) *In any category with enough effective projectives, any regular projective is effective.*

Proof. (1) It is easy to see that an initial object is effective projective and that a retract of an effective projective is effective projective. A binary sum of effective projectives is effective projective since binary products commute with reflexive coequalizers in **Set** by 1.2.3.

(2) Take $e : E \twoheadrightarrow D$, where e is a regular epimorphism, E an effective projective and take D regular projective. Then lifting the identity $1_D : D \longrightarrow D$ through e exhibits D as a retract of E, and then D is an effective projective by (1).   □

4.3.4. Remark. Unlike the case of (ordinary) projectives, an arbitrary sum of effective projectives is not necessarily effective projective.

For example in **Set** every object is regular projective, while the effective projectives are precisely the finite sets.

To approach the main problem consider the following:

4.3.5. Lemma. *Let $\mathcal{K}$ be an exact, locally finitely presentable category. Then any regular projective and finitely presentable object $P \in \mathcal{K}$ is effective projective.*

Proof. Let $(r_0, r_1) : X_1 \rightrightarrows X_0$ be a reflexive graph in $\mathcal{K}$ and $q : X_0 \longrightarrow Q$ its coequalizer.
Consider the following diagram:

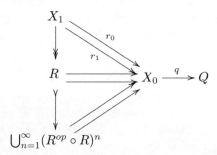

in which the middle fork is the image of the reflexive graph, hence a reflexive relation, $\bigcup_{n=1}^{\infty}(R^{op} \circ R)^n$ is the equivalence relation generated by R, and $(R^{op} \circ R)^n$ denotes the n-fold composition of $R^{op} \circ R$. Observe that, since R is reflexive, the equivalence relation can be generated in this way, and the result is a filtered colimit. Necessarily all three forks are coequalizers and the bottom one is also a kernel, because $\mathcal{K}$ is exact.
Now apply $\mathcal{K}(P, -)$ to the diagram, for P an object satisfying the conditions of the Lemma.
Projectivity of P means of course that $\mathcal{K}(P, -)$ takes regular epimorphisms to surjections so that $\mathcal{K}(P, q)$ is a surjection, with kernel

$$\mathcal{K}(P, \bigcup_{n=1}^{\infty}(R^{op} \circ R)^n) \rightrightarrows \mathcal{K}(P, X_0).$$

It follows that $\mathcal{K}(P, -)$ preserves images, relational composition, and filtered unions, so that its effect on the vertical column of the diagram is isomorphic to

$$\mathcal{K}(P, X_1) \longrightarrow\!\!\!\!\!\rightarrow \mathcal{K}(P, R) \longrightarrow \bigcup_{n=1}^{\infty}(\mathcal{K}(P, R)^{op} \circ \mathcal{K}(P, R))^n.$$

It follows from the way in which coequalizers can be calculated in the category of sets that $\mathcal{K}(P, -)$ preserves all three of the coequalizers above.
So P is an effective projective object. $\qquad\qquad\square$

Now we can prove the following corollary of 2.2.7:

4.3.6. Corollary. *For a category $\mathcal{K}$ the following are equivalent:*

    (1) *$\mathcal{K}$ is equivalent to a multi-sorted algebraic category;*
    (2) *$\mathcal{K}$ satisfies the following conditions:*
        (a) *$\mathcal{K}$ has finite limits;*
        (b) *equivalence relations in $\mathcal{K}$ are effective;*
        (c) *$\mathcal{K}$ has a strong generator $\mathcal{P}$;*

(d) *any $P \in \mathcal{P}$ is regular projective;*
(e) *any $P \in \mathcal{P}$ is finitely presentable;*
(3) $\mathcal{K}$ *satisfies the following conditions:*
    (a) $\mathcal{K}$ *has finite limits;*
    (b) $\mathcal{K}$ *has coequalizers of equivalence relations;*
    (c) $\mathcal{K}$ *has a strong generator $\mathcal{P}$;*
    (d) *any $P \in \mathcal{P}$ is effective projective;*
    (e) *any $P \in \mathcal{P}$ is finitely presentable.*

Proof. The equivalence of (1) and (2) follows from 2.2.7, together with Remark 3.1.3 and the observation on strong generators made in 4.2.4.

For (3) implies (2) we observe that P effective projective implies P regular projective is trivial, and we are left to show the necessity of (2b).

Consider an equivalence relation R on an objecty X in $\mathcal{K}$, its coequalizer $q : X \longrightarrow Q$, and the kernel pair N of q:

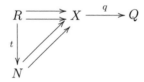

If for any P in $\mathcal{P}$ we apply $\mathcal{K}(P, -)$ to the diagram above we get

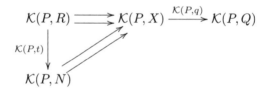

where since **Set** is exact and P is effective projective, $\mathcal{K}(P, t)$ is an isomorphism. Since $\mathcal{P}$ is a set of strong generators it follows that t is an isomorphism hence $\mathcal{K}$ has effective equivalence relations.

For (2) implies (3) note first that (2b) implies (3b) is trivial.

Next, (2) certainly provides that $\mathcal{K}$ is exact and locally finitely presentable from 2.1.10 and 4.2.

Then, it follows from Lemma 4.3.5 that every object in $\mathcal{P}$ is an effective projective. $\qquad\square$

4.3.7. Remark. The reader may have observed that in this proof we did not use the full force of the effective projectivity of P but only that $\mathcal{K}(P, -)$ preserves coequalizers of equivalence relations.

We will return to this point at the end of the section.

Now we can conclude with the main result:

4.3.8. Theorem.[29] *Let $\mathcal{K}$ be a locally finitely presentable category with $\mathcal{C}$ the full subcategory of finitely presentables. Then $\mathcal{K}$ is equivalent to a multi-sorted algebraic category if and only if $\mathcal{C}$ has enough effective projectives.*

Proof. If $\mathcal{K}$ is a multi-sorted algebraic category then the set of finitely presentable regular projectives $\mathcal{P}$ is a cover in $\mathcal{C}$ (meaning that for every C in $\mathcal{C}$ there exists a regular epimorphism $P \longrightarrow C$, with P in $\mathcal{P}$) by the multi-sorted version of 4.2.2. The conditions of Lemma 4.3.5 are satisfied, so the P in $\mathcal{P}$ are effective projectives (we could also use similar arguments as in the 9-lemma 1.2.2 and in 1.2.4). Conversely, suppose $\mathcal{C}$ has enough effective projectives and let $\mathcal{P}$ be the full subcategory of $\mathcal{C}$ determined by the set of effective projectives.

Now (3a), (3b), (3e) in Theorem 4.3.6 are trivially satisfied and $\mathcal{K}$ has all sums, so it suffices to show that $\mathcal{P}$ is a set of effective projective strong generators in $\mathcal{K}$. They are strong generators as already proved in 4.2.5, then it remains to be shown that if an object P is an effective projective in $\mathcal{C}$ then it is also an effective projective in $\mathcal{K}$.

To see this, one has only to express a reflexive graph in $\mathcal{K}$ as a filtered colimit of reflexive graphs in $\mathcal{C}$ and note that coequalizers and filtered colimits commute.

$\square$

4.3.9. Remark. In the last sentence above it is not possible to replace "reflexive graph" by "equivalence relation", so if $\mathcal{C}(P, -)$ preserves coequalizers of equivalence relations it does not follow that $\mathcal{K}(P, -)$ preserves coequalizers of equivalence relations. In fact we can notice that, when we seek the "concentration" of a property of a locally finitely presentable category $\mathcal{K}$ in its theory $\mathcal{C}$, circumlocution is required to deal with anything having to do with limits and monomorphisms, even if $\mathcal{C}$ actually has the required limits.

Theorem 4.3.8 gives a complete understanding of algebraic categories in syntactic terms, i.e. in terms of the corresponding theories.

So if $\mathcal{K} \cong \mathsf{Lex}(\mathcal{C}^{op}, \mathbf{Set})$ is algebraic, with $\mathcal{C} = \{C : C \text{ finitely presentable in } \mathcal{K}\}$ its essentially algebraic theory, it follows that $\mathcal{K}$ corresponds to the closure of $\mathcal{C}$ under filtered colimits.

Then

$$\mathcal{P} \;=\; \{P : P \text{ finitely presentable and regular projective in } \mathcal{K}\}$$
$$=\; \{P : P \text{ finitely presentable and effective projective in } \mathcal{K}\}$$

is a projective cover of $\mathcal{C}$.

Moreover, it is possible to show (see [28]) that $\mathcal{C}$ is the coequalizer completion of $\mathcal{P}$, hence we get

$$\mathsf{Lex}(\mathcal{C}^{op}, \mathbf{Set}) \cong \mathcal{M}od_{\mathcal{P}^{op}}$$

where $\mathcal{M}od_{\mathcal{P}^{op}}$ denotes all finite-product-preserving functors from $\mathcal{P}^{op}$ to $\mathbf{Set}$. Then $\mathcal{P}^{op}$ can be chosen as Lawvere theory of $\mathcal{K}$; it can be proved to be Cauchy complete (i.e. idempotents split), moreover it is invariant in the sense that it does not depend on free algebras (for details and for the corresponding duality theorem see [1]).

4.3.10. Remark. It is also possible to completely describe those locally finitely presentable categories that are exact or regular.

Various approaches can be used, both in terms of properties of the corresponding essentially algebraic theories and in terms of localizations.

We recall the results related to localizations:

- a locally finitely presentable category is regular if and only if it is a finitary localization of a multi-sorted quasi-algebraic category;
- a locally finitely presentable category is exact if and only if it is a finitary localization of a multi-sorted algebraic category.

See [26] for further details.

References

[1] J. Adamek, F.W. Lawvere, J. Rosicky. *On a duality between varieties and algebraic theories.* preprint.

[2] J. Adamek, H.E. Porst. *Algebraic Theories of Quasivarieties.* Journal of Algebra, 208, 379-398, 1998.

[3] J. Adamek, J. Rosicky. *Locally presentable and accessible categories.* LMS Lecture Notes Series 189, Cambridge University Press, 1994.

[4] J. Adamek, J. Rosicky, E.M. Vitale. *On algebraically exact categories and essential localizations of varieties.* Journal of Algebra, 244, 450-477, 2001.

[5] M. Barr. *Exact categories.* Springer Lecture Notes 236. Springer-Verlag, 1971.

[6] F. Borceux. *Handbook of Categorical Algebra.* Cambridge University Press, 1994.

[7] D. Bourn. *Normalization equivalence, kernel equivalence and affine categories.* Springer Lecture Notes Math., 1488, 43-62, 1991.

[8] A. Carboni, G.M. Kelly, M.C. Pedicchio. *Some remarks on Maltsev and Goursat categories.* Appl. Categ. Structures, 1, 385-421, 1993.

[9] A. Carboni, J. Lambek, M.C. Pedicchio. *Diagram chasing in Mal'cev categories.* Journal of Pure and Applied Algebra, 69, 271-284, 1991.

[10] A. Carboni, M.C. Pedicchio. *A new proof of Mal'cev Theorem.* Suppl. Rendiconti Circolo Mat. Palermo, serie II, n.64, 2000.

[11] A. Carboni, M.C. Pedicchio, N. Pirovano. *Internal graphs and internal groupoids in Mal'cev categories.* Canadian Mathematical Society, Conference Proceedings, 13, 97-109, 1992.

[12] A. Carboni, E.M. Vitale. *Regular and exact completions.* Journal of Pure and Applied Algebra, 125, 79-116, 1998.

[13] P. Freyd, A. Scedrov. *Categories-Allegories.* North Holland, 1990.

[14] P. Gabriel, F. Ulmer. *Lokal präsentierbare Kategorien.* Lecture Notes in Mathematics 221. Springer-Verlag, Berlin, 1971.

[15] M. Gran, E.M. Vitale. *Localizations of Maltsev varieties*, volume 5, 12, 281-291. Theory and Applications of Categories, 1999.

[16] H. Hu, W. Tholen. *A note on free regular and exact completions, and their infinitary generalizations.* Theory Appl. Categories 2, 113-132, 1996.

[17] J.R. Isbell. *Subobjects, adequacy, completeness, and categories of algebras.* Rozprawy Matem. 36, Warszawa, 1964.

[18] P.T. Johnstone. *Topos Theory.* London Math. Soc. Monographs, vol. 10, Academic Press, New York, 1977.

[19] P.T. Johnstone. *Affine categories and naturally Mal'cev categories.* J. Pure Appl. Algebra, 61, 251-256, 1989.

[20] F.W. Lawvere. *Functorial semantics of Algebraic Theories.* Dissertation, Columbia University, 1963.

[21] S. Mac Lane. *Categories for the Working Mathematician.* Springer-Verlag, 1971.

[22] A.I. Mal'cev. *On the general theory of algebraic system.* Mat. Sbornik N.S., 35, 3-20, 1954.

[23] M.C. Pedicchio. *Maltsev categories and Maltsev operations.* Journal of Pure and Applied Algebra, 98, 67-71, 1995.

[24] M.C. Pedicchio. *On k-permutability for categories of T-algebras.* Logic and algebra, Lecture Notes in Pure and Applied Mathematics, 180, 637-646, 1996.

[25] M.C. Pedicchio, P.T. Johnstone. *Remarks on continuous Mal'cev algebras.* Rend Ist. Mat. Univ. Trieste, 25, 277-307, 1993.

[26] M.C. Pedicchio, J. Rosicky. *Localization of varieties and quasivarieties.* Journal of Pure and Applied Algebra, 148, 275-284, 2000.

[27] M.C. Pedicchio, E.M. Vitale. *On the abstract characterization of quasi varieties.* Algebra Universalis, 43, 269-278, 2000.

[28] M.C. Pedicchio, R.J. Wood. *A Simple Characterization of Theories of Varieties.* Journal of Algebra, 233, 483–501, 2000.

[29] M.C. Pedicchio, R.J. Wood. *A note on effectively projective objects.* Journal of Pure and Applied Algebra, 158, 83-87, 2001.

[30] J.D.H. Smith. *Mal'cev Varieties.* Springer Lecture Notes in Math., 554, Berlin, 1976.

[31] E.M. Vitale. *Localization of algebraic categories.* Journal of Pure and Applied Algebra, 108, 315-320, 1996.

[32] E.M. Vitale. *Localization of algebraic categories 2.* Journal of Pure and Applied Algebra, 133, 317-326, 1998.

Dipartimento di Matematica,
Università di Trieste,
Piazzale Europa 1,
34100 Trieste, Italia
E-mail: pedicchi@univ.trieste.it
rovatti@mathsun1.univ.trieste.it

VII
Sheaf Theory

Claudia Centazzo and Enrico M. Vitale

1. Introduction

To write a few lines of introduction to a few pages of work on a real corner stone of mathematics like sheaf theory is not an easy task. So, let us try with ... two introductions.

1.1. First introduction: for students (and everybody else). In the study of ordinary differential equations, when you face a Cauchy problem of the form

$$\left\{ y^{(n)} = f(x, y, y', \dots, y^{(n-1)}) , \;\; y^{(i)}(x_0) = y_0^{(i)}, \right.$$

you know that the continuity of f is enough to get a local solution, i.e. a solution defined on an open neighborhood U_{x_0} of x_0. But, to guarantee the existence of a global solution, the stronger Lipschitz condition on f is required.

In complex analysis, we know that a power series $\sum a_n(z - z_0)^n$ uniformly converges on any compact space strictly contained in the interior of the convergence disc. This is equivalent to the local uniform convergence: for any z in the open disc, there is an open neighborhood U_z of z on which the series converges uniformly. But local uniform convergence does not imply uniform convergence on the whole disc. This gap between local uniform convergence and global uniform convergence is the reason why the theory of Weierstrass analytic functions exists.

These are only two simple examples, which are part of everybody's basic knowledge in mathematics, of the passage from local to global. *Sheaf theory is precisely meant to encode and study such a passage.*

Sheaf theory has its origin in complex analysis (see, for example, [18]) and in the study of cohomology of spaces [8] (see also [26] for a historical survey of sheaf theory). Since local-to-global situations are pervasive in mathematics, nowadays sheaf theory deeply interacts also with mathematical logic [3, 24, 38, 41], algebraic geometry [27, 28, 29, 30], algebraic topology [9, 22], algebraic group theory [15], ring theory [23, 48], homological algebra [16, 21, 51] and, of course, category theory [39].

The references mentioned above are not at all exhaustive. Each item is a standard textbook in the corresponding area, and the reader probably has already been in touch with some of them. We have listed them here because, just by having a quick glance at them, one can realize that sheaves play a relevant (sometimes crucial) role. In this way, we have no doubt that the reader will find motivations to attack sheaf theory directly from his favorite mathematical point of view.

In this chapter, we focus our attention on three aspects of sheaf theory.

A presheaf on a topological space X is a variable set indexed by the open subsets of X. More precisely, it is a functor

$$F \colon \mathcal{O}(X)^{op} \to Set,$$

where $\mathcal{O}(X)$ is the ordered set of open subsets of X and Set is the category of sets. Think, as examples, of the presheaf of continuous functions

$$\mathcal{C} \colon \mathcal{O}(X)^{op} \to Set \; ; \quad \mathcal{C}(U) = \{U \to Y \text{ continuous }\}$$

or of the presheaf of constant functions

$$\mathcal{K} \colon \mathcal{O}(X)^{op} \to Set \; ; \quad \mathcal{K}(U) = \{U \to Y \text{ constants }\}$$

for Y a given topological space. Roughly speaking, a presheaf F is a sheaf when we can move from local elements to global elements, i.e. when we can paste together (compatible) elements $\{f_i \in F(U_i)\}_I$ to get a unique element $f \in F(\cup_I U_i)$. The above-mentioned presheaf $\mathcal{C}$ is a sheaf, whereas the presheaf $\mathcal{K}$ is not. The first important result we want to discuss is the fact that the abstract notion of sheaf can be concretely represented by variable sets of the form "continuous functions". More precisely, any sheaf is isomorphic to the sheaf of continuous sections of a suitable étale map (= a local homeomorphism).

A simple but important result (not analyzed in this chapter) is that presheaves with values in the category of abelian groups, that is, functors of the form

$$F \colon \mathcal{O}(X)^{op} \to Ab$$

(where Ab is the category of abelian groups), constitute an abelian category (Chapter IV). In order to apply homological techniques to sheaves, it is then important to observe that the category of sheaves on a topological space is a localization of the corresponding category of presheaves. This means that there is a universal way to turn a presheaf into a sheaf, and this process is an exact functor. The fact that sheaves are localizations of presheaves is true also for set-valued presheaves, and this is the second main point of sheaf theory treated in this chapter. In fact, we show that, up to the necessity of generalizing sufficiently the notion of topological space (here the notion of Grothendieck topology on a small category is needed), sheaf categories are precisely the localizations of presheaf categories.

From the category theorist's point of view, an exciting question in this subject is: Is it possible to give an abstract characterization of sheaf categories? In other words, what assumptions an abstract category has to satisfy in order to prove that it is equivalent to the category of sheaves for a Grothendieck topology? The answer to this question is provided by Giraud's Theorem characterizing Grothendieck toposes. The third scope of this chapter is precisely to discuss such a theorem together with the various conditions involved in its statement and in its proof.

1.2. Second introduction: for teachers (and everybody else). Assume you have to teach an introductory course in category theory for students in mathematics or engineering. Probably, you spend half of the course to establish the basic categorical language and to give a reasonable amount of examples to support the intuition of the students. After this, you have to choose between going deeply into a single topic, proving non-trivial results but completely neglecting other interesting subjects, or to surf on a number of important topics, but hiding their complexity and their mutual relationships because of the lack of time. As the good teacher you are, you feel unhappy with both of these solutions. So, let us try an honorable compromise between them. Choose a single topic, and use it as a kind of *fil rouge* that the students can follow to go far enough in your selected subject (far enough to appreciate the theory), but also to have a first glance at a lot of other topics and their interaction with the development of the main theme.

The present chapter is an example of this approach: sheaf theory is a mathematically relevant skeleton to which to attach several other topics, classical or more recent, which can enter into the picture in a natural way. It is maybe worthwhile to make it clear here and at once that the idea behind this chapter is not to provide a new neither easier treatment of sheaf theory as it already appears in literature. What we are doing is to cruise around quite a lot of different, heterogeneous - sometimes advanced - aspects in category theory to get the reader more and more involved into this interesting part of mathematics. Nevertheless we hope that this *tour* has an internal coherence: from a motivating example as sheaves on a topological space we gradually lead the reader to a rather sophisticated result as Giraud's Theorem, whose proof - although not essentially different from the classical ones - is achieved thanks to the various techniques presented, and aspires to be the aim of the whole chapter.

This chapter does not contain new results. All the results can be found either in one of the standard textbooks in sheaf theory [2, 6, 34, 35, 40, 52] or in some research paper quoted below. For this reason, we include sketches of the proofs only when we think they can be useful to capture the interest of the reader. Some of the proofs we omitted, and some of the exercises we left to the reader, are far to be easy.

1.3. Contents. The chapter is organized as follows:
Section 2 is a short introduction to sheaves on a topological space and serves as basic motivation to the rest of the chapter. The main result here is the equivalence between sheaves and étale maps. Section 3 contains the characterization of localizations of presheaf categories as categories of sheaves. We pass through several categorical formulations of the notion of topology: universal closure operator, pretopology, Grothendieck topology, Lawvere-Tierney topology, and elementary topos. *En passant*, we introduce also categories of fractions, regular categories, and the coproduct completion, and we have a glance at the existence of finite colimits in an elementary topos, which gives a strong link with Chapter V. The last section is devoted to Giraud's characterization of Grothendieck toposes. We put

the accent on lextensive categories (that is, categories with "good" coproducts), seen also as pseudo-algebras for a convenient pseudo-monad, and we touch on Kan extensions, calculus of relations, left covering functors, filtering functors, and on the exact completion.

Apart from the already quoted textbooks, our main references are Carboni-Lack-Walters [11] for extensive and lextensive categories, Carboni-Mantovani [12] for the calculus of relations, and Menni [43, 44] for pretopologies. The latter is the most recent topic we present in this chapter. We have generalized Menni's definition and main result to the case of categories with weak finite limits because of our main example, which is the coproduct completion of a small category. The notion of pretopology is of interest also in the study of realizability toposes (see [44]), but we do not develop this argument here. The reference for the exact completion, sketched in Section 4, is the paper [14] by Carboni and the second author.

We would like to thank R.J. Wood: subsection 4.3 is the result of a stimulating discussion with Richard. We are also grateful to W. Tholen for a number of useful comments and suggestions and, in particular, for the big effort Walter and Jane did to turn the language of our chapter into something more similar to English than to Italian. Grazie!

2. Sheaves on a topological space

Let us start with a slogan:

What is locally true everywhere, is not necessarily globally true.

In other words, a problem could have a lot of interesting local solutions, and fail to have even a single global solution.

2.1. Local conditions. Let us make the previous slogan more precise with some examples.

Example. Consider a map $f \colon Y \to X$ between two topological spaces.

1. The question: is f a continuous map? is a local problem. Indeed, if for each point y of Y there is an open neighborhood U_y containing y and such that the restriction of f to U_y is continuous, then f itself is continuous.

2. The question: is f a constant map? is not a local problem. Assume, for example, that Y is given by the disjoint union of two non-empty open subsets Y_1 and Y_2, and that X contains at least two different points $x_1 \neq x_2$. Define $f(y) = x_1$ if $y \in Y_1$ and $f(y) = x_2$ if $y \in Y_2$. Then f is locally constant, but it is not constant.

3. Now let X be the set of complex numbers $\mathbb{C}$ and let Y be its one-point (or Alexandroff) compactification $\mathbb{C}^*$. The question: is f a holomorphic function? is a local problem. This is a nice example which shows that looking for global solutions to a local problem can trivialize the answer. If U is an open subset of $\mathbb{C}^*$, write $\mathcal{H}(U)$ for the set of holomorphic functions $U \to \mathbb{C}$. If U is strictly included in $\mathbb{C}^*$, then $\mathcal{H}(U)$ separates the points of U (i.e. if x and y are in U

and $x \neq y$, there exists f in $\mathcal{H}(U)$ such that $f(x) \neq f(y)$). But if $U = \mathbb{C}^*$, then $\mathcal{H}(U)$ contains only constant maps.

2.2. Étale maps. For each condition, local or not, we can consider its "localization", which consists in asking locally the condition. Consider again a map $f \colon Y \to X$ between topological spaces. For f to be a homeomorphism is not a local condition. Its localization is known as the condition to be an étale map. This is a crucial notion in sheaf theory.

Definition. Consider two topological spaces X and Y. A map $f \colon Y \to X$ is *étale* if, for each point y in Y, there are open neighborhoods V_y of y and $U_{f(y)}$ of $f(y)$ such that the restriction of f to V_y is a homeomorphism $V_y \simeq U_{f(y)}$. Étale maps are also called *local homeomorphisms*.

A typical example of an étale map which is not a homeomorphism is the projection of the circular helix on the circle, $f(\cos t, \sin t, t) = (\cos t, \sin t)$.

The idea of an étale map is important because, in general, for a map $f \colon Y \to X$ between topological spaces, the best we can discuss is its continuity. But if X has some local structure and f is étale, then we can reconstruct piece-wise this structure on Y. This is the basic idea of variety.

2.3. Local sections. Let $f \colon Y \to X$ be a continuous function. A *continuous section* of f is a continuous map $s \colon X \to Y$ such that $f(s(x)) = x$ for any $x \in X$. A *local section* of f is a continuous map $\sigma \colon U \to Y$ defined on an open subset U of X and such that $f(\sigma(x)) = x$ for any $x \in U$. To have a continuous section is not a local condition for a continuous map $f \colon Y \to X$. Its localization, i.e. to have a local section, is another important ingredient in sheaf theory. Here is a classical example (which leads to the discovery Riemann surfaces).

Example. Let $f \colon \mathbb{C} \to \mathbb{C} \smallsetminus \{0\}$ be the complex exponential, $f(z) = e^z$. (Note that f is an étale map.) For any integer $k \in \mathbb{Z}$, there is a section $g_k \colon \mathbb{C} \smallsetminus \{0\} \to \mathbb{C}$ for f, defined by the complex logarithm $g_k(\rho e^{i\theta}) = \ln\rho + i(\theta + 2k\pi)$ with $\theta \in [0, 2\pi[$. Now, if U is a simply connected open subset of $\mathbb{C} \smallsetminus \mathbb{R}^+$, each of these g_k restricts to a continuous (in fact, holomorphic) section of f. But if U contains a loop around the origin, none of the g_k is continuous. On the other hand, given a map $g \colon \mathbb{C} \smallsetminus \{0\} \to \mathbb{C}$, the equation $f(g(z)) = z$ implies that $g = g_k$ for some k. So, for each $z \in \mathbb{C} \smallsetminus \{0\}$, there is a open neighborhood U_z of z such that f has a continuous section on U_z (if $z \in \mathbb{R}^+$ one has to past together a g_k with g_{k-1}), but f does not have a continuous section on the whole $\mathbb{C} \smallsetminus \{0\}$.

2.4. Presheaves. Let us now formalize the first two items of Example 2.1. Let X and Y be two topological spaces. For each open subset U of X, write $\mathcal{C}(U)$ for the set of continuous maps from U to Y and $\mathcal{K}(U)$ for the set of constant maps from U to Y. If V is an open subset of X contained in U, by restriction we get two maps $\mathcal{C}(U) \to \mathcal{C}(V)$ and $\mathcal{K}(U) \to \mathcal{K}(V)$. Moreover, both of these constructions are functorial, that is they give rise to two presheaves on the topological space X.

Definition.

1. If $\mathbb{C}$ is a small category, a *presheaf* on $\mathbb{C}$ is a functor $F\colon \mathbb{C}^{op} \to Set$ with values in the category Set of sets and mappings. We write $Set^{\mathbb{C}^{op}}$ for the category of presheaves on $\mathbb{C}$ and their natural transformations.
2. If X is a topological space, a presheaf on X is a presheaf on $\mathcal{O}(X)$, the ordered set of open subsets of X, seen as a category with at most one arrow between two objects.

Notation. Having in mind the examples $\mathcal{C}$ and $\mathcal{K}$, for an arbitrary presheaf F on a space X we write $f_{|V}$ for the image of $f \in F(U)$ under $F(V \subseteq U)\colon F(U) \to F(V)$.

2.5. Sheaves. The notion of sheaf will emphasize an additional property of the presheaf $\mathcal{C}$, that the presheaf $\mathcal{K}$ does not share. $\mathcal{C}$ is determined by a local condition, $\mathcal{K}$ is not.

Definitions.

1. Let $F\colon \mathcal{O}(X)^{op} \to Set$ be a presheaf on a topological space X. Consider $U \in \mathcal{O}(X)$ and an open cover $(U_i)_{i \in I}$ of U, that is $U_i \in \mathcal{O}(X)$ for each i and $U = \cup_{i \in I} U_i$. A family of elements $(f_i \in F(U_i))_{i \in I}$ is *compatible* if, for each $i, j \in I$, $f_{i|U_i \cap U_j} = f_{j|U_i \cap U_j}$.
2. A presheaf F on X is a *sheaf* if for each $U \in \mathcal{O}(X)$, for each open cover $(U_i)_I$ of U and for each compatible family $(f_i \in F(U_i))_I$, there is a unique $f \in F(U)$ such that $f_{|U_i} = f_i$ for each $i \in I$. We call f the *glueing* of the family $(f_i)_I$. We write $Sh(X)$ for the full subcategory of $Set^{\mathcal{O}(X)^{op}}$ spanned by sheaves.

Exercise. Show that a presheaf F on X is a sheaf exactly when, for each $U \in \mathcal{O}(X)$ and for each open cover $(U_i)_I$ of U, the following diagram is an equalizer

$$F(U) \to \prod_i F(U_i) \rightrightarrows \prod_{i,j} F(U_i \cap U_j).$$

2.6. Examples of presheaves and sheaves. We give now some basic examples of presheaves and sheaves on a topological space.

Examples.

1. The presheaf of continuous functions $\mathcal{C}\colon \mathcal{O}(X)^{op} \to Set$ is a sheaf. In general, the presheaf of constant functions $\mathcal{K}\colon \mathcal{O}(X)^{op} \to Set$ is not a sheaf.
2. Let X be the complex space $\mathbb{C}$, and, for each $U \in \mathcal{O}(X)$, write $\mathcal{L}(U)$ for the set of bounded holomorphic functions from U to $\mathbb{C}$. Under restriction, $\mathcal{L}\colon \mathcal{O}(\mathbb{C})^{op} \to Set$ is a presheaf, but it is not a sheaf. Consider, for each positive real number r, the open disk D_r centered at the origin and of radius r. Define $f_r \in \mathcal{L}(D_r)$ by the assignment $f_r(z) = z$ for each $z \in D_r$. Clearly, $\mathbb{C} = \cup_r D_r$ and $(f_r)_r$ is a compatible family, but no glueing for this family exists, because a bounded holomorphic function from $\mathbb{C}$ to $\mathbb{C}$ is necessarily constant.

3. The following example of sheaf will turn out to be a generic one (see Theorem 2.8). Fix a continuous map $f\colon Y \to X$ and define, for each $U \in \mathcal{O}(X)$, $\mathcal{S}_f(U)$ to be the set of continuous sections of f defined on U. In other words, an element $\sigma \in \mathcal{S}_f(U)$ is a continuous map $\sigma\colon U \to Y$ such that $f(\sigma(x)) = x$ for all $x \in U$. Once again the action of $\mathcal{S}_f$ on the inclusion $V \subseteq U$ is simply the restriction.

4. The following functor is a sheaf

$$\mathcal{O}\colon \mathcal{O}(X)^{op} \to Set \qquad \mathcal{O}(U) = \{W \in \mathcal{O}(X) \mid W \subseteq U\}$$

with action given by intersection. This simple example will play a special role in Section 3 (see Exercise 3.22).

5. For each $U \in \mathcal{O}(X)$, the representable presheaf $\mathcal{O}(X)(-,U)\colon \mathcal{O}(X)^{op} \to Set$ is a sheaf.

2.7. Internal logic. Roughly speaking, we can say that:

1. A local condition is a condition φ
   - which makes sense in every open subset of a topological space X and
   - which holds in $U \in \mathcal{O}(X)$ exactly when, for any $x \in U$, there is an open neighborhood U_x of x, U_x contained in U, such that the condition φ holds in U_x and in every open $V_x \subseteq U_x$.
2. A local problem is a problem P
   - which makes sense in every open subset of a topological space X and
   - which has a solution in $U \in \mathcal{O}(X)$ exactly when, for any $x \in U$, there is an open neighborhood U_x of x, U_x contained in U, such that the problem P has a solution in U_x and in every open $V_x \subseteq U_x$.

Looking at the previous examples, we have:

1. To be a continuous function is a local condition, to be a constant function is not a local condition.
2. To have a continuous section is not a local problem, to have a local section is a local problem.

The idea of local condition or local problem leads to the notion of local validity of a formula, which is the key ingredient to codify the internal logic of a sheaf. This is another important topic which we do not pursue in this chapter (a full treatment can be found in [6]). Let us only observe that the "definition" of local condition implies that such a condition is inherited by open subset: if φ holds in an open subset U, then it holds in any open subset V contained in U. For example, for a map $f\colon \mathbb{C}^* \to \mathbb{C}$ (see Example 2.1.3), the formula

$$(f \text{ holomorphic} \Rightarrow f \text{ constant})$$

holds for $U = \mathbb{C}^*$ but it does not hold for proper open subsets of $\mathbb{C}^*$.

2.8. The equivalence between sheaves and étale maps. It is a matter of experience that each local problem gives rise to a sheaf, as we have seen for $\mathcal{C}$ and $\mathcal{S}_f$. It would be nice to turn this experience into a theorem, but a more formalized notion of local problem would be needed for this. What we can do is to express the converse statement as a theorem, that is to show that each sheaf is the variable set of answers to some local problem. This will be done in the next theorem, which represents the main achievement of this section, but, before that, some preliminary work is needed.

If X is a topological space, we write Et/X for the category having as objects étale maps $f: Y \to X$. An arrow from $f: Y \to X$ to $f': Y' \to X$ is an étale map $g: Y \to Y'$ such that $f' \cdot g = f$.

Exercises.

1. Show that Et/X is a full subcategory of the comma category Top/X, where Top is the category of topological spaces and continuous functions.
2. Show that, if $\alpha: F \to G$ is an arrow in $Sh(X)$, then the compatibility with the glueing operation is a consequence of the naturality of α. This is why we consider $Sh(X)$ full in $Set^{\mathcal{O}(X)^{op}}$.

We already know how to get a sheaf on a space X from a continuous map $f: Y \to X$, it is the sheaf of local sections S_f defined in Example 2.6.3. Consider now an arrow $g: f \to f'$ in Top/X and an element $\sigma \in S_f(U)$, $U \in \mathcal{O}(X)$. Composition with g gives us an element $g \cdot \sigma \in S_{f'}$. In this way, we obtain a functor

$$S: Top/X \to Sh(X).$$

By composition with the full inclusion $i: Et/X \to Top/X$, we get a functor

$$S \cdot i: Et/X \to Sh(X).$$

The next theorem makes precise our claim that each sheaf is (up to natural isomorphism) the variable set of answers to a local problem.

Theorem. *Let X be a topological space. The functor*

$$S \cdot i: Et/X \to Sh(X)$$

is an equivalence of categories.

2.9. Sketch of the proof, I. The most interesting part of the proof is the construction of the functor $Sh(X) \to Et/X$ quasi-inverse of $S \cdot i$. To discover the construction of $Sh(X) \to Et/X$, we can start with an étale map f, consider the functor S_f and then try to recover f from S_f.

First of all, observe that, since f is étale, for each $y \in Y$ there are open neighborhoods V_y of y and $U_{f(y)}$ of $f(y)$ such that $f_{|V_y}: V_y \to U_{f(y)}$ is a homeomorphism. In this way, we get a local section $s_y = (f_{|V_y})^{-1} \in S_f(U_{f(y)})$ such that $s_y(f(y)) = y$. Such a section is not necessarily unique, that is: it may be possible to find another local element $s'_y \in S_f(U'_{f(y)})$ such that $s'_y(f(y)) = y$. Even if s_y and s'_y are not equal, they are "locally equal". This will be explained in the following lemma.

Lemma. *Let* $f \colon Y \to X$ *be an étale map and consider* $s, s' \in \mathcal{S}_f(U)$. *If there is an* $x \in U$ *such that* $s(x) = s'(x)$, *then there exists an open neighborhood* U_x *of* x *such that* $s_{|U_x} = s'_{|U_x}$.

This is what we need to get a bijection

$$Y \simeq \coprod_{x \in X} (\mathcal{S}_f)_x, \quad \text{where} \quad (\mathcal{S}_f)_x = (\coprod_{U \ni x} \mathcal{S}_f(U))/\approx$$

(here, the symbol $\coprod$ means disjoint union, and $\approx$ is the equivalence relation defined as follows: $s \in \mathcal{S}_f(U) \approx s' \in \mathcal{S}_f(U')$ iff there exists an open neighborhood U_x of x, $U_x \subseteq U \cap U'$, such that $s_{|U_x} = s'_{|U_x}$). Explicitly, the bijection sends $y \in Y$ into the class $[s_y] \in (\mathcal{S}_f)_{f(y)}$. Conversely, given $s \in \mathcal{S}_f(U)$ and $x \in U$, the class $[s] \in (\mathcal{S}_f)_x$ is sent to the point $s(x)$ of Y.

The meaning of the previous bijection is that we can reconstruct the set underlying the space Y by looking at the sheaf of local sections of f. Now, what about the map $f \colon Y \to X$? And what about the topology of Y? The first question is easy. It suffices to compose the map f with the bijection $\coprod (\mathcal{S}_f)_x \simeq Y$ to get a map

$$\pi_f \colon \coprod_{x \in X} (\mathcal{S}_f)_x \to X \qquad [s] \in (\mathcal{S}_f)_x \mapsto f(s(x)) = x \in X.$$

As far as the topology of Y is concerned, the key remark is once again a simple exercise on étale maps.

Exercise. Let $f \colon Y \to X$ be an étale map and consider a subset V of Y. Show that $V \in \mathcal{O}(Y)$ iff $s^{-1}(V) \in \mathcal{O}(U)$ for all $s \in \mathcal{S}_f(U)$ and for all $U \in \mathcal{O}(X)$. In other words, Y has the final topology with respect to all the local sections of f.

Since we want the bijection $Y \simeq \coprod (\mathcal{S}_f)_x$ to be a homeomorphism, we have to put on $\coprod (\mathcal{S}_f)_x$ the topology induced by that of Y, that is the final topology with respect to all the compositions $t \colon U \to Y \simeq \coprod (\mathcal{S}_f)$ for $t \in \mathcal{S}_f(U)$ and $U \in \mathcal{O}(X)$. Once again, these compositions can be expressed without explicit reference to Y. If x is in U, we have $x \mapsto t(x) \mapsto [s_{t(x)}] = [t] \in (\mathcal{S}_f)_{f(t(x))=x}$. Finally, the topology on $\coprod (\mathcal{S}_f)_x$ is the final topology with respect to all the maps

$$\sigma_s^U \colon U \to \coprod_{x \in X} (\mathcal{S}_f)_x \qquad x \mapsto [s] \in (\mathcal{S}_f)_x$$

for $s \in \mathcal{S}_f(U)$ and $U \in \mathcal{O}(X)$.

2.10. The total space. The previous discussion makes evident the construction of a functor $Sh(X) \to Et/X$, or, more generally, of a functor $Set^{\mathcal{O}(X)^{op}} \to Et/X$.

Definition. Let $F \colon \mathcal{O}(X)^{op} \to Set$ be a presheaf on a topological space X. Its *total space* $\mathcal{T}(F)$ is given by

$$\pi_F \colon \coprod_{x \in X} F_x \to X$$

where F_x (the *stalk* of F at the point x) is the quotient set $(\coprod_{U \ni x} F(U))/\approx$, and $\approx$ is the equivalence relation defined as follows: $s \in F(U) \approx s' \in F(U')$ iff there exists an open neighborhood U_x of x, $U_x \subset U \cap U'$, such that $s_{|U_x} = s'_{|U_x}$. The space $\coprod F_x$ has the final topology with respect to all maps

$$\sigma^U_s : U \to \coprod_{x \in X} F_x \qquad x \mapsto [s] \in F_x$$

for $s \in F(U)$ and $U \in \mathcal{O}(X)$. The map π_F is defined by $\pi_F([s] \in F_x) = x$.

Exercises.

1. Show that the map $\pi_F : \coprod F_x \to X$ defined just above is an étale map. [Hint: Recall that if a space Y has the final topology with respect to a family of maps $(g_i : Y_i \to Y)_I$, then a map $f : Y \to X$ is continuous iff all the composites $f \cdot g_i$ are continuous.]

2. Describe the stalk F_x as a filtered colimit.

Consider now two presheaves F and G on X and a natural transformation $\alpha : F \to G$. We get an arrow $\hat{\alpha} : \mathcal{T}(F) \to \mathcal{T}(G)$ in the following way:

$$\hat{\alpha} : \coprod_{x \in X} F_x \to \coprod_{x \in X} G_x \qquad [s \in F(U)] \in F_x \mapsto [\alpha_U(s) \in G(U)] \in G_x.$$

Such $\hat{\alpha}$ is continuous because the composite $\hat{\alpha} \cdot \sigma^U_s$ is nothing but $\sigma^U_{\alpha_U(s)}$. This completes the construction of the total space functor

$$\mathcal{T} : Set^{\mathcal{O}(X)^{op}} \to Et/X.$$

Now we are able to compute

$$Top/X \xrightarrow{\;S\;} Sh(X) \xrightarrow{\;i\;} Set^{\mathcal{O}(X)^{op}} \xrightarrow{\;\mathcal{T}\;} Et/X$$

(here i is again the full inclusion); we get an arrow, natural with respect to $f \in Top/X$,

given by $[s] \in (\mathcal{S}_f)_x \mapsto s(x)$.

So we can summarize the previous discussion saying that ϵ_f is a homeomorphism if and only if f is an étale map. In other words, we have a natural isomorphism $\epsilon : \mathcal{T} \cdot S \Rightarrow Id : Et/X \to Et/X$.

2.11. Sketch of the proof, II. We only sketch what happens when we go the other way round. Let F be in $Set^{\mathcal{O}(X)^{op}}$ and apply

$$Set^{\mathcal{O}(X)^{op}} \xrightarrow{\;\mathcal{T}\;} Et/X \xrightarrow{\;i\;} Top/X \xrightarrow{\;S\;} Sh(X).$$

We want to compare the presheaf F and the resulting sheaf $\mathcal{S}_{\pi_F}$. For each $U \in \mathcal{O}(X)$, there is a map $\eta_F : F(U) \to \mathcal{S}_{\pi_F}(U)$ which sends $s \in F(U)$ into $\sigma^U_s : U \to \coprod F_x : x \mapsto [s] \in F_x$. In fact, these $\eta_F(U)$ collectively give an arrow $\eta_F : F \to$

$\mathcal{S}_{\pi_F}$ in $Set^{\mathcal{O}(X)^{op}}$, which is natural with respect to F. Moreover, each $\eta_F(U)$ is a bijection if and only if F is a sheaf (the surjectivity is given by the existence of the glueing, the injectivity by its uniqueness). In other words, we have a natural isomorphism $\eta \colon Id \Rightarrow \mathcal{S} \cdot \mathcal{T} \colon Sh(X) \to Sh(X)$.

2.12. Surjectivity. Let us point out a simple fact, which will be related to the example of logarithm (Example 2.3). If $\alpha \colon F \to G$ is an arrow in $Sh(X)$ or in $Set^{\mathcal{O}(X)^{op}}$, by definition we have a map $\alpha_U \colon F(U) \to G(U)$ for each $U \in \mathcal{O}(X)$. But, for each x in X, α induces also a map

$$\alpha_x \colon F_x \to G_x \qquad [s] \mapsto [\alpha_U(s)].$$

The difference between α_U and α_x becomes clear if we think of what their surjectivity means. The surjectivity of α_U is the existence of a global solution defined on the open set U, whereas the surjectivity of α_x is the existence of a local solution at x.

Example. Let $X = \mathbb{C}$ (complex numbers), $F = \mathcal{H}$ (the sheaf of holomorphic functions) and $G = \mathcal{H}^*$ (the subsheaf of $\mathcal{H}$ of those $g \colon U \to \mathbb{C}$ such that $g(z) \neq 0$ for all z in U). We can define an arrow $exp \colon \mathcal{H} \to \mathcal{H}^*$ by

$$exp_U \colon \mathcal{H}(U) \to \mathcal{H}^*(U) \qquad (g \colon U \to \mathbb{C}) \mapsto (e^g \colon U \to \mathbb{C}).$$

Now, exp_U is not surjective if U contains a loop around the origin. On the contrary, for each $x \neq 0$, exp_x is surjective because we can find a simply connected open neighborhood U_x of x, not containing 0, where the logarithmic function is well-defined and holomorphic.

It is here that sheaf theory meets homological algebra. In fact, sheaves as $\mathcal{H}$ or $\mathcal{H}^*$ have, for each $U \in \mathcal{O}(X)$, a natural structure of abelian group (and even more), and the restriction operation is a morphism of abelian groups. One says that $\mathcal{H}$ and $\mathcal{H}^*$ are sheaves of abelian groups. The categories of presheaves and sheaves of abelian groups are *abelian categories*, so that all the machinery of homological algebra can be used to study problems like surjectivity and injectivity of arrows. (For example, the exactness of a sequence $F \to G \to H$ between sheaves of abelian groups means that, for each $x \in X$, the sequence of abelian groups and homomorphisms $F_x \to G_x \to H_x$ is exact in the usual sense.) We do not enter into details. Chapter IV gives a glance at abelian categories and the homological techniques therein.

2.13. Sheaves are a localization. To close this section, let us look more carefully at the problem of surjectivity and injectivity. Fix an arrow $\alpha \colon F \to G$ in $Sh(X)$. We have seen, with the example $exp \colon \mathcal{H} \to \mathcal{H}^*$, that the statement

$$(\forall\, x \in X \;\; \alpha_x \text{ surjective }) \;\; \Rightarrow \;\; (\forall\, U \in \mathcal{O}(X) \;\; \alpha_U \text{ surjective })$$

does **not** hold. On the other hand, the injectivity is preserved passing from stalks to local sets:

$$(\forall\, x \in X \;\; \alpha_x \text{ injective }) \;\; \Rightarrow \;\; (\forall\, U \in \mathcal{O}(X) \;\; \alpha_U \text{ injective }).$$

The complete situation is given in the following exercise.

Exercise.

1. Let α be an arrow in $Sh(X)$.
   - Show that α is an epimorphism iff α_x is surjective for all $x \in X$.
   - Show that α is a monomorphism iff α_x is injective for all $x \in X$ iff α_U is injective for all $U \in \mathcal{O}(X)$.
2. Let α be an arrow in $Set^{\mathcal{O}(X)^{op}}$.
   - Show that α is an epimorphism iff α_U is surjective for all $U \in \mathcal{O}(X)$.
   - Show that α is a monomorphism iff α_U is injective for all $U \in \mathcal{O}(X)$.

[Hint:

1. Show that two parallel arrows α, β in $Sh(X)$ are equal iff $\alpha_x = \beta_x$ for all $x \in X$.
2. Recall that in a functor category, limits and colimits are computed point-wise.]

The previous exercise shows that the full inclusion $Sh(X) \to Set^{\mathcal{O}(X)^{op}}$ preserves monomorphisms but not epimorphisms. The ultimate reason for this is a deep one: the full subcategory of sheaves is reflective in the category of presheaves. And even more, it is a *localization*, that is, the left adjoint to the full inclusion preserves finite limits.

Theorem. *Let X be a topological space. The full inclusion $Sh(X) \to Set^{\mathcal{O}(X)^{op}}$ has a left adjoint, given by the composite functor*

$$Set^{\mathcal{O}(X)^{op}} \xrightarrow{\ T\ } Et/X \xrightarrow{\ i\ } Top/X \xrightarrow{\ S\ } Sh(X).$$

Moreover, the left adjoint preserves finite limits.

Proof. This is a particular case of a more general result discussed in the next section. □

3. Topologies, closure operators, and localizations

The final result of the previous section has been that the category $Sh(X)$ is a localization of $Set^{\mathcal{O}(X)^{op}}$, that is a reflective subcategory such that the left adjoint is *left exact* (a functor between categories with finite limits is called left exact if it preserves finite limits). This section is devoted to answer the following question: is any localization of a presheaf category equivalent to some category of sheaves? This question achieves its right level of generality if we consider presheaf categories of the form $Set^{\mathbb{C}^{op}}$ for $\mathbb{C}$ a small category. A way to get a positive answer is to generalize the notion of topological space, considering so-called Grothendieck topologies on the small category $\mathbb{C}$.

3.1. Universal closure operators. The first step is reminiscent of the fact that a topological space can be defined as a set X with a closure operator $\overline{(\ )} : \mathcal{P}(X) \to \mathcal{P}(X)$. This idea can be transposed to an arbitrary category (see, for example

[17]). For an object B of a category $\mathbb{E}$, $Sub(B)$ denotes the partially ordered set of subobjects of B, given by monomorphisms into B.

Definition. Let $\mathbb{E}$ be a category with finite limits. A *universal closure operator* on $\mathbb{E}$ consists of a class of operations $\overline{(\;)}\colon Sub(B) \to Sub(B)$, one for each object B of $\mathbb{E}$, such that

c1. for all $S \in Sub(B)$, $S \subseteq \overline{S}$;
c2. for all $S, T \in Sub(B)$, if $S \subseteq T$ then $\overline{S} \subseteq \overline{T}$;
c3. for all $S \in Sub(B)$, $\overline{\overline{S}} \subseteq \overline{S}$;
c4. for all $f\colon B \dashrightarrow C$ in $\mathbb{E}$, the following diagram commutes (f^* is the pullback operator)

$$
\begin{array}{ccc}
Sub(B) & \xrightarrow{\;\overline{(\;)}\;} & Sub(B) \\
{\scriptstyle f^*}\uparrow & & \uparrow{\scriptstyle f^*} \\
Sub(C) & \xrightarrow[\;\overline{(\;)}\;]{} & Sub(C)
\end{array}
$$

Observe that, in the presence of the other axioms, condition c2 can be equivalently replaced by the following one:

c2'. for each $S, T \in Sub(B)$, $\overline{S \cap T} = \overline{S} \cap \overline{T}$, where $\cap$ is the intersection of subobjects, that is their pullback.

3.2. From localizations to universal closure operators. It is easy to establish a first link between localizations and universal closure operators.

Proposition. *Any localization* $i\colon \mathbb{A} \leftrightarrows \mathbb{E}\colon r$, $r \dashv i$, *with* $\mathbb{E}$ *finitely complete, induces a universal closure operator on* $\mathbb{E}$.

Proof. Let B be an object in $\mathbb{E}$ and consider a subobject $a\colon A \rightarrowtail B$. Since both r and i preserves monos, we get a subobject $i(r(a))\colon i(r(A)) \rightarrowtail i(r(B))$. We define $\overline{(\;)}\colon Sub(B) \to Sub(B)$ by the following pullback, where η_B is the unit of $r \dashv i$,

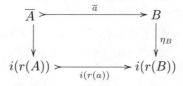

$$
\begin{array}{ccc}
\overline{A} & \xrightarrow{\;\overline{a}\;} & B \\
\downarrow & & \downarrow{\scriptstyle \eta_B} \\
i(r(A)) & \xrightarrow[\;i(r(a))\;]{} & i(r(B))
\end{array}
$$

$\square$

3.3. Bidense morphisms. The previous proposition allows us to associate with any localization of $\mathbb{E}$, a universal closure operator on $\mathbb{E}$. Moreover, we will see that, when $\mathbb{E}$ is a presheaf category, this process is essentially a bijection between localizations of $\mathbb{E}$ and universal closure operators on $\mathbb{E}$. In the more general case where $\mathbb{E}$ is finitely complete and has strong epi-mono factorizations, the mapping from the class of localizations to the class of universal closure operators is only essentially injective. To prove this, we need the notion of bidense morphism and a short digression on categories of fractions.

Definition. Let $\overline{(\ )}$ be a universal closure operator on a category $\mathbb{E}$ with finite limits.

1. A subobject $a \colon A \rightarrowtail B$ is *dense* if $\bar{a} = 1_B$ as subobjects of B (that is, if $\bar{a}$ is an isomorphism).
2. If $\mathbb{E}$ has strong epi-mono factorizations, an arrow is *bidense* if its image is dense and the equalizer of its kernel pair is dense.

Lemma. *Let $i \colon \mathbb{A} \leftrightarrows \mathbb{E} \colon r$, $r \dashv i$ be a localization of a finitely complete category $\mathbb{E}$ with strong epi-mono factorizations. Consider the universal closure operator $\overline{(\ )}$ associated to $r \dashv i$ as in Proposition 3.2. An arrow $f \in \mathbb{E}$ is bidense with respect to $\overline{(\ )}$ if and only if $r(f)$ is an isomorphism in $\mathbb{A}$.*

Proof. Consider the following diagram, where (e, m) is the strong epi-mono factorization of f, (f_0, f_1) is the kernel pair of f and φ is the equalizer of f_0 and f_1,

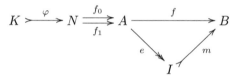

Since r preserves equalizers, kernel pairs, monos (being left exact), and strong epis (being left adjoint), $(r(e), r(m))$ is the factorization of $r(f)$, $(r(f_0), r(f_1))$ is the kernel pair of $r(f)$ and $r(\varphi)$ is the equalizer of $r(f_0)$ and $r(f_1)$. It follows that $r(f)$ is an iso iff $r(\varphi)$ and $r(m)$ are isomorphisms. As a consequence, in order to prove our statement it suffices to prove that, given an arbitrary mono $m \colon I \rightarrowtail B$, $r(m)$ is an iso iff m is dense. For this, consider the following diagram, where the internal square is the pullback defining the closure $\bar{m}$ of m, and j is the unique factorization through such a pullback,

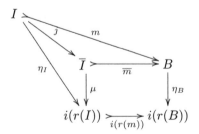

Plainly, if $r(m)$ is an iso, then m is dense. Conversely, for any mono m, $r(j)$ is an iso (because r preserves pullbacks, so that $r(\bar{m})$ is the pullback of $r(m)$ along the identity). If we assume that $\bar{m}$ is an iso, then $r(m) = r(\bar{m}) \cdot r(j)$ is an iso. $\qquad\square$

3.4. Categories of fractions. The previous lemma suggests to pay special attention to the class of arrows inverted by the reflector $r \colon \mathbb{E} \to \mathbb{A}$.

Definition. Let Σ be a class of morphisms in a category $\mathbb{E}$. A *category of fractions* of $\mathbb{E}$ with respect to Σ is a functor $P_\Sigma \colon \mathbb{E} \to \mathbb{E}[\Sigma^{-1}]$ such that $P_\Sigma(s)$ is an iso for any $s \in \Sigma$ and which is universal with respect to this property.

Here universal means that if $F \colon \mathbb{E} \to \mathbb{A}$ is a functor such that $F(s)$ is an iso for any $s \in \Sigma$, then there exists a functor $G \colon \mathbb{E}[\Sigma^{-1}] \to \mathbb{A}$ and a natural isomorphism $\varphi \colon G \cdot P_\Sigma \Rightarrow F$. Moreover, given another functor $G' \colon \mathbb{E}[\Sigma^{-1}] \to \mathbb{A}$ with a natural isomorphism $\varphi' \colon G' \cdot P_\Sigma \Rightarrow F$, there is a unique natural isomorphism $\psi \colon G \to G'$ such that the following diagram commutes

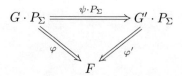

3.5. Calculus of fractions. The category of fractions is characterized, up to equivalence, by its universal property. But its explicit description, and even its existence, is in general a hard problem. Nevertheless, when the class Σ has a calculus of fractions, the description of $\mathbb{E}[\Sigma^{-1}]$ becomes quite easy.

Definition. Let Σ be a class of morphisms in a category $\mathbb{E}$. The class Σ has a *left calculus of fractions* if the following conditions hold:

1. For any object $X \in \mathbb{E}$, the identity $1_X \in \Sigma$;
2. If $s, t \in \Sigma$ and $t \cdot s$ is defined, then $t \cdot s \in \Sigma$;
3. Given $g, t \in \mathbb{E}$ with $t \in \Sigma$, then there are $s, f \in \mathbb{E}$ such that $s \in \Sigma$ and $s \cdot g = f \cdot t$

$$
\begin{array}{ccc}
D & \xrightarrow{\ g\ } & C \\
{\scriptstyle t}\downarrow & & \downarrow{\scriptstyle s} \\
A & \xrightarrow[\ f\ ]{} & B
\end{array}
$$

4. If $t \in \Sigma$ and $f \cdot t = g \cdot t$, then there is $s \in \Sigma$ such that $s \cdot f = g \cdot s$

$$
D \xrightarrow{\ t\ } A \underset{g}{\overset{f}{\rightrightarrows}} B \xrightarrow{\ s\ } C.
$$

Lemma. *Let Σ be a class of morphisms in a category $\mathbb{E}$. If Σ has a left calculus of fractions, then $P_\Sigma \colon \mathbb{E} \to \mathbb{E}[\Sigma^{-1}]$ can be described as follows:*

- *objects of $\mathbb{E}[\Sigma^{-1}]$ are those of $\mathbb{E}$;*

- *a premorphism $A \to B$ in $\mathbb{E}[\Sigma^{-1}]$ is a triple (f, I, s) with $A \xrightarrow{\ f\ } I \xleftarrow{\ s\ } B$ and $s \in \Sigma$;*

- two parallel premorphisms (f, I, s) and (g, J, t) are equivalent if there exist $i\colon I \to X$ and $j\colon J \to X$ such that $i \cdot f = j \cdot g$, $i \cdot s = j \cdot t$ and $i \cdot s \in \Sigma$; a morphism is an equivalence class of premorphisms;
- the composite of $[f, I, s]$ and $[g, J, t]$ is given by $[g' \cdot f, K, s' \cdot t]$, with g', s' any pair of arrows such that $s' \in \Sigma$ and $g' \cdot s = s' \cdot g$

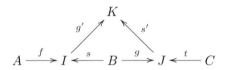

- $P_\Sigma\colon \mathbb{E} \to \mathbb{E}[\Sigma^{-1}]$ sends $f\colon A \to B$ into $[f, B, 1_B]\colon A \to B$. If $f \in \Sigma$, then $P_\Sigma(f)$ is invertible, with $P_\Sigma(s)^{-1} = [1_B, B, f]$.

Proof. We omit the strightforward verification that $\mathbb{E}[\Sigma^{-1}]$ is well-defined and is a category. As far as the universal property is concerned, with the notations of Definition 3.4, we have:
- G sends $[f, I, s]\colon A \to B$ into $F(s)^{-1} \cdot F(f)\colon F(A) \to F(B)$;
- φ is given by $\varphi_A = 1_{F(A)}$ for any $A \in \mathbb{E}$;
- ψ is given by $\psi_A = (\varphi'_A)^{-1}$ for any $A \in \mathbb{E}$; its naturality depends on the fact that $[f, I, s] = P_\Sigma(s)^{-1} \cdot P_\Sigma(f)$.   □

Proposition. *Let* $i\colon \mathbb{A} \leftrightarrows \mathbb{E}\colon r$, $r \dashv i$ *be a reflective subcategory of a category* $\mathbb{E}$ *and let* Σ *be the class of arrows* s *of* $\mathbb{E}$ *such that* $r(s)$ *is an isomorphism. The comparison functor* $r'\colon \mathbb{E}[\Sigma^{-1}] \to \mathbb{A}$ *is an equivalence.*

Proof. First of all, let us check that Σ has a left calculus of fractions. With the notations of Definition 3.5, we have:
1) and 2) are obvious;
3) let $f = i(r(g)) \cdot i(r(t))^{-1} \cdot \eta_A\colon A \to i(r(A)) \to i(r(D)) \to i(r(C))$ and $g = \eta_C\colon C \to i(r(C))$;
4) let $s = \eta_B\colon B \to i(r(B))$.
Now we can use Lemma 3.5 to check that $r'\colon \mathbb{E}[\Sigma^{-1}] \to \mathbb{A}$ is an equivalence:
- essentially surjective: obvious;
- full: given $h\colon r(A) \to r(B)$ in $\mathbb{A}$, then $h = r'[\overline{h}, i(r(B)), \eta_B]$, where $\overline{h}\colon A \to i(r(B))$ corresponds to h via $r \dashv i$;
- faithful: let $[f, I, s], [g, J, j]\colon A \to B$ be two arrows in $\mathbb{E}[\Sigma^{-1}]$ and assume they have the same image under r', that is $r(s)^{-1} \cdot r(f) = r(t)^{-1} \cdot r(g)$. The next diagram, where σ and τ corresponds to $r(s)^{-1}$ and $r(t)^{-1}$ via $r \dashv i$, shows that

$$[f, I, s] = [g, J, j]$$

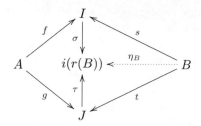

$\square$

Corollary. *Let* $\mathbb{E}$ *be a finitely complete category with strong epi-mono factorization. A localization* $i\colon \mathbb{A} \leftrightarrows \mathbb{E}\colon r,\ r \dashv i$ *is completely determined by the associated universal closure operator (see Proposition 3.2).*

Proof. By the previous proposition, the localization is determined by the class Σ of arrows inverted by r. By Lemma 3.3, Σ is determined by the closure operator.

$\square$

3.6. Localizations as categories of fractions. The next exercise allows us to recognize localizations among reflective subcategories using fractions (see also [4, 5]).

Exercise. Consider a reflective subcategory $i\colon \mathbb{A} \to \mathbb{E}$ of a finitely complete category $\mathbb{E}$. The reflector $r\colon \mathbb{E} \to \mathbb{A}$ is left exact if and only if the class of morphisms inverted by r has a right calculus of fractions (a condition dual to that of Definition 3.5).

3.7. Examples of categories of fractions. To end our discussion on categories of fractions, let us report some examples. They have no relations with the rest of the chapter and we quote them only to give categories of fractions back to their natural context, which is homotopy theory (see [20, 31]).

Examples.

1. The homotopy category of Top is equivalent to the category of fractions of Top with respect to homotopy equivalences.
2. Let R be a commutative ring with unit and let $\mathbb{E} = Ch(R)$ be the category of chain complexes of R-modules. The homotopy category of $\mathbb{E}$ is equivalent to the category of fractions of $\mathbb{E}$ with respect to homotopy equivalences.
3. Let $\mathbb{E}_c^+$ be the subcategory of positive chain complexes which are projective in each degree. The homotopy category of $\mathbb{E}_c^+$ is equivalent to the category of fractions of $\mathbb{E}_c^+$ with respect to arrows inducing an isomorphism in homology.

3.8. Grothendieck topologies. We take now the crucial step indicated at the beginning of this section: passing from sheaves on a topological space to sheaves for a Grothendieck topology. Before giving the formal definition of Grothendieck topology on a small category, let us observe two simple facts about the definition of sheaf on a topological space (Definition 2.5) which will enlight the next notion:

1. The notion of sheaf depends on the fact that we require the glueing condition with respect to *all* open covers $(U_i)_{i \in I}$ of an open subset U of the topological space X. In principle, one could select *some* open covers of U, that is some families $(U_i \to U)_I$ of arrows in $\mathcal{O}(X)$, and require the glueing condition only with respect to the selected open covers. In this way, the notion of sheaf would be meant with respect to the selected system of open covers.

2. On the other hand, there is no restriction in considering only *hereditary* open covers, that is open covers $(U_i)_I$ containing, together with an open subset U_i, all its open subsets. In fact, any open cover $(U_i)_I$ can be made hereditary (by adding to each U_i its open subsets) and compatible families on the original cover are in bijection with compatible families on the new one.

Now we are ready to introduce the notion of Grothendieck topology on a small category. This notion describes the behavior of hereditary open covers in a topological space.

Definition. Let $\mathbb{C}$ be a small category. Write $\mathbb{C}_0$ for the set of objects of $\mathbb{C}$, and $\mathbb{C}_1$ for its set of arrows.

1. If C is an object of $\mathbb{C}$, a *sieve* on C is a subobject $s \colon S \rightarrowtail \mathbb{C}(-, C)$ of the presheaf represented by C. Equivalently, S is a set of arrows with codomain C such that, if $f \colon X \to C$ is in S and $g \colon Y \to X$ is any arrow, then $f \cdot g$ is in S.

2. A *Grothendieck topology* on $\mathbb{C}$ is a map $T \colon \mathbb{C}_0 \to \mathcal{P}(\mathcal{P}(\mathbb{C}_1))$ (that is, for each object C, $T(C)$ is a collection of families of arrows of $\mathbb{C}$) such that:

   g1. for each $C \in \mathbb{C}$ and for each $S \in T(C)$, S is a sieve on C;

   g2. the total sieve $\mathbb{C}(-, C)$ is in $T(C)$;

   g3. if $S \in T(C)$ and $g \colon D \to C$ is any arrow, then $g^*(S) \in T(D)$, where $g^*(S)$ is the following pullback

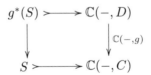

   g4. if $S \in T(C)$ and R is a sieve on C such that $g^*(R)$ is in $T(D)$ for all $g \colon D \to C$ in S, then $R \in T(C)$.

Exercise. Show that, for each $X \in \mathbb{C}$, $g^*(S)(X) = \{x \colon X \to D \mid g \cdot x \colon X \to D \to C$ is in $S(X)\}$.

3.9. From universal closure operators to Grothendieck topologies. By Proposition 3.2, we are able to associate with any localization of a presheaf category $Set^{\mathbb{C}^{op}}$ a universal closure operator on $Set^{\mathbb{C}^{op}}$. The next step will involve Grothendieck topologies on $\mathbb{C}$.

Proposition. *Let $\mathbb{C}$ be a small category. Any universal closure operator on $Set^{\mathbb{C}^{op}}$ induces a Grothendieck topology on $\mathbb{C}$.*

Proof. Let C be an object in $\mathbb{C}$. We get a Grothendieck topology $\mathcal{T}$ on $\mathbb{C}$ in the following way: a sieve S on C is in $\mathcal{T}(C)$ when, regarded as a subobject of $\mathbb{C}(-, C)$, it is dense with respect to the universal closure operator. $\qquad\square$

The previous construction of a Grothendieck topologies from a universal closure operators is, in fact, a bijection. This will be explained in Corollary 3.14.

3.10. Sheaves for a Grothendieck topology. It is time to recall the question addressed at the beginning of this section. Is any localization of a presheaf category equivalent to a sheaf category? To answer this question, we need an appropriate notion of sheaf for a Grothendieck topology. The notion of sheaf for a Grothendieck topology is much like that of sheaf on a topological space (Definition 2.5).

Definition. Let $\mathcal{T}$ be a Grothendieck topology on a small category $\mathbb{C}$ and consider a presheaf $F\colon \mathbb{C}^{op} \to Set$.

1. Consider an object C and a sieve $S \in \mathcal{T}(C)$. An S-compatible family is a family of elements $(f_k \in F(K) \mid k\colon K \to C$ is in $S)$ such that, for each $y\colon K' \to K$ in $\mathbb{C}$, $F(y)(f_k) = f_{k \cdot y}$.
2. The presheaf F is a $\mathcal{T}$-*sheaf* if, for each object $C \in \mathbb{C}$, for each $S \in \mathcal{T}(C)$ and for each S-compatible family $(f_k)_{k \in S}$, there exists a unique $f \in F(C)$ such that $F(k)(f) = f_k$ for all $k \in S$.

We write $Sh(\mathcal{T})$ for the full subcategory of $Set^{\mathbb{C}^{op}}$ given by $\mathcal{T}$-sheaves. Since a sieve on C is a subobject of the representable presheaf $\mathbb{C}(-, C)$, we can express the notion of sheaf in a slightly different way.

Exercise. Show that a presheaf F is a $\mathcal{T}$-sheaf iff for each $C \in \mathbb{C}$ and for each $S \in \mathcal{T}(C)$, the inclusion $S \rightarrowtail \mathbb{C}(-, C)$ induces a bijection $Nat(\mathbb{C}(-, C), F) \simeq Nat(S, F)$ (where $Nat(G, F)$ is the set of natural transformation from G to F).

Theorem. *Let $\mathcal{T}$ be a Grothendieck topology on a small category $\mathbb{C}$. The full subcategory $Sh(\mathcal{T}) \to Set^{\mathbb{C}^{op}}$ is a localization.*

3.11. Cartesian closed categories. Before giving a sketch of the proof of Theorem 3.10, let us point out two important properties of the category $Set^{\mathbb{C}^{op}}$.

Definition. A category $\mathbb{E}$ with binary products is *cartesian closed* if, for any object X, the functor $X \times -\colon \mathbb{E} \to \mathbb{E}$ has a right adjoint. When this is the case, we denote the right adjoint by $(-)^X \colon \mathbb{E} \to \mathbb{E}$.

Proposition. *Let $\mathbb{C}$ be a small category. The category $Set^{\mathbb{C}^{op}}$ is cartesian closed.*

Proof. Consider two presheaves F and G on $\mathbb{C}$. We seek a functor $G^F \colon \mathbb{C}^{op} \to Set$ such that, for any $H \in Set^{\mathbb{C}^{op}}$, there is a natural bijection $Nat(F \times H, G) \simeq Nat(H, G^F)$. In particular, for $H = \mathbb{C}(-, X)$ the previous bijection and the Yoneda Lemma give $G^F(X) \simeq Nat(F \times \mathbb{C}(-, X), G)$. We take this as the definition of G^F and the rest of the proof is routine. (Hint: to check the natural bijection $Nat(F \times H, G) \simeq Nat(H, G^F)$ for an arbitrary presheaf H, express H as a colimit of representable functors as in Proposition 3.27.) $\qquad\square$

3.12. Subobject classifier. The next important fact is the existence of a subobject classifier in any presheaf category.

Definition. Let $\mathbb{E}$ be a category with finite limits. A *subobject classifier* is a mono $t\colon T \rightarrowtail \Omega$ (T being the terminal object) satisfying the following universal property: for each mono $s\colon S \rightarrowtail A$, there is a unique arrow $\varphi_s\colon A \to \Omega$ (the characteristic function of s) such that the following diagram is a pullback

The terminology comes from the case $\mathbb{E} = Set$, where $t\colon \{*\} \to \{0,1\}\colon * \mapsto 1$, and φ_s is the usual characteristic function: $\varphi_s(a) = 1$ iff $a \in S$. As usual, the subobject classifier is uniquely determined (up to isomorphism) by its universal property.

Proposition. *Let $\mathbb{C}$ be a small category. The category $Set^{\mathbb{C}^{op}}$ has a subobject classifier.*

Proof. If $\Omega\colon \mathbb{C}^{op} \to Set$ is the subobject classifier in $Set^{\mathbb{C}^{op}}$, there is a natural bijection $\{s\colon S \rightarrowtail \mathbb{C}(-,X)\} \simeq Nat(\mathbb{C}(-,X),\Omega) \simeq \Omega(X)$. We take this as the definition of Ω on objects. It extends to arrows $f\colon Y \to X$ by pullback along $\mathbb{C}(-,f)$. $\square$

3.13. Elementary toposes. We can summarize the two previous propositions saying that, for each small category $\mathbb{C}$, the category $Set^{\mathbb{C}^{op}}$ is an elementary topos.

Definition. An *elementary topos* is a finitely complete and cartesian closed category with a subobject classifier. (See Section 1 in Chapter I for an equivalent definition.)

Not every elementary topos is of the form $Set^{\mathbb{C}^{op}}$. For example, the category of finite sets and arbitrary maps is an elementary topos.

3.14. Lawvere-Tierney topologies. Let us now explain the interest of the notion of subobject classifier in sheaf theory.

Definition. Let $t\colon T \rightarrowtail \Omega$ be the subobject classifier of an elementary topos $\mathbb{E}$. A *Lawvere-Tierney topology* is an arrow $j\colon \Omega \to \Omega$ such that the following equations hold:

lt1. $j \cdot t = t$;
lt2. $j \cdot j = j$;
lt3. $\wedge \cdot (j \times j) = j \cdot \wedge$ (where $\wedge\colon \Omega \times \Omega \to \Omega$ is the characteristic function of the diagonal $\Omega \rightarrowtail \Omega \times \Omega$).

Example. Let 0 be the initial object of an elementary topos $\mathbb{E}$ (see 3.22 for the existence of finite colimits in an elementary topos) and let $f\colon T \rightarrowtail \Omega$ be the characteristic function of $0 \rightarrowtail T$ (which is a mono because 0 is strict, see Example 4.1.7 and Exercise 4.2.4). Let $\neg\colon \Omega \to \Omega$ be the characteristic function of $f\colon T \rightarrowtail \Omega$. The "double negation" $\neg\neg\colon \Omega \to \Omega$ is a Lawvere-Tierney topology.

Proposition. *Let $\mathbb{E}$ be an elementary topos. There is a bijection between Lawvere-Tierney topologies on $\mathbb{E}$ and universal closure operators on $\mathbb{E}$.*

Proof. Consider a universal closure operator $\overline{(\ )}$ on $\mathbb{E}$ and take the closure $\overline{t}\colon \overline{T} \rightarrowtail \Omega$ of the subobject classifier. The characteristic function $j\colon \Omega \to \Omega$ of $\overline{t}$ is a Lawvere-Tierney topology.

Conversely, consider a Lawvere-Tierney topology $j\colon \Omega \to \Omega$ and a mono $s\colon S \rightarrowtail A$. We take as closure $\overline{s}\colon \overline{S} \rightarrowtail A$ the pullback of $t\colon T \rightarrowtail \Omega$ along $j\cdot\varphi_s\colon A \to \Omega$, where φ_s is the characteristic function of s. $\qquad\square$

Corollary. *Let $\mathbb{C}$ be a small category. There is a bijection between:*

1. *universal closure operators on $Set^{\mathbb{C}^{op}}$*
2. *Lawvere-Tierney topologies on $Set^{\mathbb{C}^{op}}$*
3. *Grothendieck topologies on $\mathbb{C}$.*

Proof. The correspondence between the Grothendieck topologies $\mathcal{T}$ on $\mathbb{C}$ and the Lawvere-Tierney topologies j on $Set^{\mathbb{C}^{op}}$ is described by the following pullback diagram in $Set^{\mathbb{C}^{op}}$

In fact, the Grothendieck topology $\mathcal{T}$ is a special subobject of Ω, and the corresponding Lawvere-Tierney topology j is its characteristic function. $\qquad\square$

3.15. Sheaves for a Lawvere-Tierney topology. Thanks to the previous proposition, we can define the notion of j-sheaf, where j is a Lawvere-Tierney topology on an elementary topos $\mathbb{E}$. We say that a mono is j-*dense* (j-*closed*) if it is dense (closed) with respect to the universal closure operator corresponding to the topology j.

Definition. Let j be a Lawvere-Tierney topology on an elementary topos $\mathbb{E}$. An object $F \in \mathbb{E}$ is a j-*sheaf* if every j-dense mono $s\colon S \rightarrowtail A$ induces, by composition, a bijection $\mathbb{E}(A, F) \simeq \mathbb{E}(S, F)$.

We write $Sh(j)$ for the full subcategory of $\mathbb{E}$ of j-sheaves. When the elementary topos is of the form $Set^{\mathbb{C}^{op}}$, there is no ambiguity in the notion of sheaf. Indeed, using Exercise 3.10, one can check that, under the bijection of Corollary 3.14, $\mathcal{T}$-sheaves correspond exactly to j-sheaves.

3.16. Sheaves are precisely localizations. We finally arrive to a more general form of Theorem 3.10.

Theorem. *Let j be a Lawvere-Tierney topology on an elementary topos $\mathbb{E}$. The full subcategory $Sh(j) \to \mathbb{E}$ is a localization.*

We limit ourselves to the construction of the left adjoint, even though proving that it is left exact is far from being trivial. (The difficult part is the preservation of equalizers. As far as binary products are concerned, see point 4 of the next exercise.) To construct the reflector $r \colon \mathbb{E} \to Sh(j)$, the existence of cokernel pairs in $\mathbb{E}$ is required. In the case of our main interest, that is when $\mathbb{E}$ is a presheaf category, the existence of colimits is obvious (they are computed point-wise in Set). Even in the case of an elementary topos it is possible to prove that finite colimits exist. We will see this in 3.22. Take an object A in $\mathbb{E}$ and the arrow $\pi^A \colon \Omega^A \to (\Omega_j)^A$, where

$$\Omega \xrightarrow{\;\pi\;} \Omega_j \xrightarrowtail{\;\omega\;} \Omega$$

is the factorization of the idempotent $j \colon \Omega \to \Omega$ through the equalizer ω of j and the identity on Ω. Consider the diagonal $\Delta_A \colon A \rightarrowtail A \times A$, its characteristic function $=_A \colon A \times A \to \Omega$ and the arrow $\{-\}_A \colon A \to \Omega^A$ corresponding to $=_A$ by cartesian closedness. Finally, consider the equalizer $i \colon I \rightarrowtail (\Omega_j)^A$ of the cokernel pair of $\pi^A \cdot \{-\}_A \colon A \to (\Omega_j)^A$. We define $r(A) = \overline{I}$, where $\overline{i} \colon \overline{I} \rightarrowtail (\Omega_j)^A$ is the closure of i with respect to the universal closure operator on $\mathbb{E}$ associated to the topology j. The fact that $\overline{I}$ is a j-sheaf follows from the next exercise.

Exercise. Let j be a Lawvere-Tierney topology on an elementary topos $\mathbb{E}$.

1. Ω_j is a j-sheaf.
2. If F is a j-sheaf, then F^X is a j-sheaf for each X.
3. If F is a j-sheaf and $i \colon I \rightarrowtail F$ is a mono, then I is a j-sheaf iff i is j-closed.
4. Let F be a j-sheaf and consider two objects A and B in $\mathbb{E}$. Using point 2 and the cartesian closedness of $\mathbb{E}$, show that there is a natural bijection

$$Sh(j)(r(A) \times r(B), F) \simeq Sh(j)(r(A \times B), F) .$$

Deduce that $r \colon \mathbb{E} \to Sh(j)$ preserves binary products.

3.17. Classification of localizations. The previous theorem allows us to close the circle localizations $\mapsto$ universal closure operators $\mapsto$ topologies. Indeed, going through all the various constructions described in this section, we are now able to state the following theorem.

Theorem.

1. *Let $\mathbb{E}$ be an elementary topos. There is a bijection between localizations of $\mathbb{E}$ and Lawvere-Tierney topologies on $\mathbb{E}$.*
2. *Let $\mathbb{C}$ be a small category. There is a bijection between localizations of $Set^{\mathbb{C}^{op}}$ and Grothendieck topologies on $\mathbb{C}$.*

In other words, any localization of a presheaf category (more generally, of an elementary topos) is a category of sheaves. A category of the form $Sh(\mathcal{T})$, for $\mathcal{T}$

a Grothendieck topology on a small category, is called a *Grothendieck topos*. We can summarize the situation for Grothendieck toposes by saying that:

- *Grothendieck toposes are exactly the localizations of presheaf categories;*
- *Each localization of a Grothendieck topos is still a Grothendieck topos.*

The first statement has no analogue for elementary toposes. On the other hand, the second statement also holds for elementary toposes.

- *Each localization of an elementary topos is still an elementary topos.*

Since we know that any localization of an elementary topos $\mathbb{E}$ is of the form $Sh(j)$ for j a Lawvere-Tierney topology on $\mathbb{E}$, it suffices to prove that $Sh(j)$ is an elementary topos. By Exercise 3.16.2, cartesian closedness passes from $\mathbb{E}$ to $Sh(j)$. As far as the subobject classifier is concerned, observe that, since $j \cdot t = t$, t factors through $\omega \colon \Omega_j \rightarrowtail \Omega$. Again by Exercise 3.16.3, this factorization $t_j \colon T \rightarrowtail \Omega_j$ classifies subobjects in $Sh(j)$.

Exercise. Since any localization of an elementary topos is an elementary topos, and since any presheaf category is an elementary topos, any Grothendieck topos is an elementary topos. In particular, if X is a topological space, the category $Sh(X)$ is an elementary topos. Show that, in $Sh(X)$, the subobject classifier is the sheaf $\mathcal{O} \colon \mathcal{O}(X)^{op} \to Set$ described in Exercise 2.6.4.

3.18. Back to topological spaces. As an exercise, let us specialize some of the notions introduced in this section to the canonical localization

$$Sh(X) \to Set^{\mathcal{O}(X)^{op}}$$

of Section 2 (Theorem 2.13).

If V is an open subset of the topological space X,

$$\downarrow V = \{V' \in \mathcal{O}(X) \mid V' \subseteq V\}$$

is a sieve. A principal sieve is a sieve of the form $\downarrow V$ for some $V \in \mathcal{O}(X)$.

1. We already know that the subobject classifier in $Sh(X)$ is

$$\mathcal{O} \colon \mathcal{O}(X)^{op} \to Set \quad \mathcal{O}(U) = \{V \in \mathcal{O}(X) \mid V \subseteq U\}.$$

   In terms of sieves, $\mathcal{O}$ can be described in the following way:

$$\mathcal{O}(U) = \{\text{ principal sieves on } U\}.$$

2. The subobject classifier in $Set^{\mathcal{O}(X)^{op}}$ is

$$\Omega \colon \mathcal{O}(X)^{op} \to Set \quad \Omega(U) = \{\text{ sieves on } U\}.$$

3. The Lawvere-Tierney topology $j \colon \Omega \to \Omega$ associated to the canonical localization is given by

$$j_U \colon \Omega(U) \to \Omega(U) \quad j_U(S) = \downarrow (\cup\{V \in S\}).$$

   Clearly, the image Ω_j of $j \colon \Omega \to \Omega$ is $\mathcal{O}$.

4. The Grothendieck topology $\mathcal{T}$ on $\mathcal{O}(X)$ associated to the canonical localization is obtained taking as covering sieves on an open subset U those sieves S such that $U = \cup\{V \in S\}$.

3.19. Locales. The category of sheaves on a topological space is called a *spatial topos*. A natural question is if a localization of a spatial topos is again a spatial topos. The answer is negative. Before giving an explicit counterexample, let us recall some basic facts from Chapter II, where the theory of *locales* is developed.

If $\mathcal{L}$ is a locale, we write $Pt(\mathcal{L})$ for its spectrum, which is a topological space (II.1.4). If X is a topological space and $\mathcal{O}(X)$ is the locale of its open subsets, X in general is not homeomorphic to $Pt(\mathcal{O}(X))$ (which is the free sober space on X), but $\mathcal{O}(X)$ and $\mathcal{O}(Pt(\mathcal{O}(X)))$ are isomorphic as locales (II.1.6).

The definition of sheaf on a topological space plainly transposes to the case of a locale. We write $Sh(\mathcal{L})$ for the topos of sheaves on a locale $\mathcal{L}$. In this way, if X is a topological space, $Sh(X) \simeq Sh(\mathcal{O}(X))$. (Basically, the whole Section 2 can be translated in terms of locales, see [6].) As for spaces, if $\mathcal{L}$ is a locale, then the presheaf

$$\Omega \colon \mathcal{L}^{op} \to Set \qquad \Omega(u) = \{w \in \mathcal{L} \mid w \leq u\}$$

is a sheaf, and it is the subobject classifier in $Sh(\mathcal{L})$. Moreover, $\mathcal{L}$ is isomorphic to $Sh(\mathcal{L})(T, \Omega)$ (where T is the terminal object). This implies that two locales having equivalent categories of sheaves are isomorphic.

Finally, an element u of a locale $\mathcal{L}$ is called *regular* if $\neg\neg u = u$ (the negation $\neg u$ is denoted u^c and called also pseudo-complement in Sections I.3 and II.1). The subset of regular elements of a locale $\mathcal{L}$ is a complete Boolean algebra (II.2.13).

3.20. A counterexample. Let us sketch now the announced counterexample about spatial toposes. Let $\mathbb{R}$ be the real line with its usual topology and consider the double negation topology $\neg\neg$ in the topos $Sh(\mathbb{R})$ (Example 3.14). The topos $\mathbb{E}$ of $\neg\neg$-sheaves is equivalent to the category of sheaves on the Boolean algebra $\mathcal{R}$ of regular open subsets of $\mathbb{R}$. Assume that $\mathbb{E}$ is a spatial topos, say $\mathbb{E} = Sh(X)$ for some topological space X. Then $\mathcal{R}$ should be isomorphic to $\mathcal{O}(X)$ and then also to $\mathcal{O}(Pt(\mathcal{R}))$. But this is impossible because $Pt(\mathcal{R})$ is the empty space, whereas all the open intervals are in $\mathcal{R}$ (see II.2.13 in [6] for more details).

3.21. Localic toposes. To end our *détour* through locales, let us give a glance at two further results concerning *localic toposes*, that is toposes of the form $Sh(\mathcal{L})$ for $\mathcal{L}$ a locale.

1. We have just seen that a localization of a spatial topos is not necessarily spatial. This problem disappears using localic toposes, in the sense that
   - *a localization of a localic topos is a localic topos.*
   Moreover, the fact that a localization of a spatial topos is not necessarily spatial can be related to the fact, proved in II.2.14, that a sublocale of a spatial locale (that is, a locale of the form $\mathcal{O}(X)$ for X a topological space) is not necessarily spatial.

2. A *geometric morphism* $F\colon \mathbb{A} \to \mathbb{B}$ is a functor F with a left exact left adjoint. If X and Y are topological spaces, any continuous function $f\colon Y \to X$ induces a geometric morphism $Sh(Y) \to Sh(X)$ (which is a localization if f is the inclusion of an open subset), but the converse is not true. On the contrary, given two localic toposes $Sh(\mathcal{L})$ and $Sh(\mathcal{L}')$, there is a bijection between geometric morphisms $Sh(\mathcal{L}) \to Sh(\mathcal{L}')$ and morphisms of locales $\mathcal{L} \to \mathcal{L}'$.

3.22. Colimits in an elementary topos. In the proof of Theorem 3.16, to construct the reflector $r\colon \mathbb{E} \to Sh(j)$ we have used that the elementary topos $\mathbb{E}$ has finite colimits. The existence of finite colimits in an elementary topos is a problem strictly related to monadic functors studied in Chapter V.

Observe that cartesian closedness for an elementary topos $\mathbb{E}$ induces, for any object Y, a functor $Y^{(-)}\colon \mathbb{E}^{op} \to \mathbb{E}$. Indeed, given $f\colon X \to Z$, we take as $Y^f\colon Y^Z \to Y^X$ the arrow corresponding, by cartesian closedness, to the composite

$$\epsilon_Y \cdot (f \times 1)\colon X \times Y^Z \to Z \times Y^Z \to Y,$$

where ϵ_Y is the counit at Y for the adjunction $Z \times - \dashv (-)^Z$. If we take $Y = \Omega$, we get a functor $\Omega^{(-)}\colon \mathbb{E}^{op} \to \mathbb{E}$. The non-trivial fact is that $\Omega^{(-)}$ is monadic. This implies that $\mathbb{E}^{op}$ has finite limits (because $\mathbb{E}$ has finite limits), that is, $\mathbb{E}$ has finite colimits.

Let us sketch the proof that $\Omega^{(-)}\colon \mathbb{E}^{op} \to \mathbb{E}$ is monadic. We use, for this, Theorem V.2.4. The left adjoint is provided by $\Omega^{(-)}\colon \mathbb{E} \to \mathbb{E}^{op}$. Now, observe that in an elementary topos, every mono is regular (indeed, a mono $s\colon S \rightarrowtail A$ is the equalizer of $\varphi_s\colon A \to \Omega$ and $t \cdot ! \cdot \varphi_s\colon A \to \Omega \to T \to \Omega$). To prove that $\Omega^{(-)}\colon \mathbb{E}^{op} \to \mathbb{E}$ reflects isomorphisms, it suffices now to prove that it reflects epis and monos, and for this we only need to show that it is faithful. But this last fact follows directly from an inspection of the following diagram

$$
\begin{array}{ccccc}
X & \underset{g}{\overset{f}{\rightrightarrows}} & Z & \longrightarrow & T \\
{\scriptstyle <1,f>}\Big\| {\scriptstyle <1,g>} & & \Big\downarrow{\scriptstyle \Delta_Z} & & \Big\downarrow{\scriptstyle t} \\
X \times Z & \underset{g \times 1}{\overset{f \times 1}{\rightrightarrows}} & Z \times Z & \underset{=_Z}{\longrightarrow} & \Omega
\end{array}
$$

To check the last condition of Theorem V.2.4, we use Beck's condition:

- *if the left hand square is a pullback and f is a mono, then the right hand square commutes*

$$
\begin{array}{ccc}
Z \overset{h}{\longrightarrow} X & \qquad & \Omega^Z \overset{\Omega^h}{\longleftarrow} \Omega^X \\
{\scriptstyle k}\Big\downarrow \qquad \Big\downarrow{\scriptstyle f} & & {\scriptstyle \hat{k}}\Big\downarrow \qquad \Big\downarrow{\scriptstyle \hat{f}} \\
V \overset{g}{\longrightarrow} Y & & \Omega^V \overset{\Omega^g}{\longleftarrow} \Omega^Y
\end{array}
$$

(where $\hat{f}$ corresponds, by cartesian closedness, to the characteristic function of

$$(f \times 1) \cdot e_X\colon \in_X \rightarrowtail X \times \Omega^X \rightarrowtail Y \times \Omega^X,$$

with $e_X \colon \in_X \rightarrowtail X \times \Omega^X$ the mono classified by the counit $\epsilon_\Omega \colon X \times \Omega^X \to \Omega$).
Consider now two arrows $f, g \colon X \to Y$ with a common retraction, and consider
also their equalizer $e \colon E \to X$. Then

$$\Omega^Y \mathrel{\mathop{\rightrightarrows}^{\hat{f}}_{\Omega^g}} \Omega^X \mathrel{\mathop{\rightrightarrows}^{\hat{e}}_{\Omega^e}} \Omega^E$$

is a split coequalizer (to check the equation $\Omega^g \cdot \hat{f} = \hat{e} \cdot \Omega^e$, apply Beck's condition
to the pullback of f along g, which is nothing but E).

3.23. Regular projective covers [14]**.** In our analysis of Giraud's Theorem characterizing localizations of presheaf categories (Section 4), we will use that a presheaf
category has enough regular projective objects (see below). With this example in
mind, we study now universal closure operators in the special case of regular categories with enough regular projective objects (even if the most general result on
localizations, which is part 1 of Theorem 3.17, does not involve regular projective
objects). Our aim here is to show that universal closure operators on a regular category with enough regular projective objects are classified by suitable structures
(called *pretopologies*) defined on regular projectives.

From Chapter IV, recall that a category is *regular* if it is finitely complete, has
regular epi-mono factorizations, and regular epis are pullback stable. Recall also
that in any regular category, strong epis coincide with regular epis. An object P
of a category $\mathbb{E}$ is *regular projective* if the functor $\mathbb{E}(P, -) \colon \mathbb{E} \to Set$ preserves
regular epis (which, in Set, are nothing but surjections). We say that a category
has *enough regular projectives* if for any object X there is a regular epi $x \colon P \twoheadrightarrow X$
with P regular projective. In this case, we say that X is a quotient of P and P is
a regular projective cover of X.
Finally, a *regular projective cover* of a category $\mathbb{E}$ is a full subcategory $\mathbb{P}$ such that

- each object of $\mathbb{P}$ is regular projective in $\mathbb{E}$;
- each object of $\mathbb{E}$ has a $\mathbb{P}$-cover, that is a regular projective cover in $\mathbb{P}$.

Clearly, a category $\mathbb{E}$ has enough regular projectives iff it has a regular projective
cover, but a regular projective cover can be strictly smaller than the full subcategory of all regular projectives.

Example. In the context of presheaf categories, the relevant example of regular
projective cover will be described in 3.27 and 3.28. Let us mention here another
example, exploited in Chapter VI to study algebraic categories and their localizations. If $\mathbb{A}$ is an algebraic category (more generally, a monadic category over a
power of Set), its full subcategory of free algebras is a regular projective cover.
Indeed, free algebras are regular projective objects, and each algebra is a quotient
of a free one.

Lemma. *If $\mathbb{P}$ is a regular projective cover of a finitely complete category $\mathbb{E}$, then $\mathbb{P}$
has weak finite limits.*

(A *weak limit* is defined as a limit, except that one requires only the existence of a factorization, not its uniqueness. A functor can have several non-isomorphic weak limits. For example, every non-empty set is a weak terminal object in *Set*.)

Proof. Take the limit in $\mathbb{E}$ and cover it with an object of $\mathbb{P}$. □

3.24. $\mathcal{J}$-closed arrows. We present the axioms which will bring to the definition of $\mathcal{J}$-closed arrow first, and then to the definition of pretopology. Weak limits are required in this context.

Let $\mathbb{P}$ be a category with weak pullbacks. Write $\mathbb{P}_0$ for the class of objects of $\mathbb{P}$, and $\mathbb{P}_1$ for its class of arrows. Consider a map $\mathcal{J}\colon \mathbb{P}_0 \to \mathcal{P}(\mathbb{P}_1)$ (that is, for each object $X \in \mathbb{P}$, $\mathcal{J}(X)$ is a collection of arrows in $\mathbb{P}$) such that:

p0. if $f \in \mathcal{J}(X)$, then X is the codomain of f;

p1. if $f\colon Y \to X$ is a split epi (i.e. there is s such that $f \cdot s = 1_X$), then $f \in \mathcal{J}(X)$;

p2. consider two arrows f and g and a weak pullback (1); if $g \in \mathcal{J}(X)$, then $f^*(g) \in \mathcal{J}(Y)$

$$W \xrightarrow{g^*(f)} Z \qquad (1)$$

with $f^*(g)\colon W \to Y$, $g\colon Z \to X$, $f\colon Y \to X$.

p3. consider two arrows $g\colon Z \to Y$ and $f\colon Y \to X$; if $f \cdot g \in \mathcal{J}(X)$, then $f \in \mathcal{J}(X)$;

p4. consider two arrows $g\colon Z \to Y$ and $f\colon Y \to X$; if $g \in \mathcal{J}(Y)$ and $f \in \mathcal{J}(X)$, then $f \cdot g \in \mathcal{J}(X)$.

The next exercise will be useful to complete the proof of Proposition 3.26.

Exercise. Let $\mathcal{J}\colon \mathbb{P}_0 \to \mathcal{P}(\mathbb{P}_1)$ be as before.

1. Show that condition p2 is equivalent to:
   p2'. consider two arrows f and g as in p2 and assume that $g \in \mathcal{J}(X)$; then there exists a weak pullback (1) such that $f^*(g) \in \mathcal{J}(Y)$.
2. Consider an arrow $g\colon Z \to X$; show that if there is an arrow $f \in \mathcal{J}(X)$ and a weak pullback (1) such that $f^*(g) \in \mathcal{J}(Y)$, then $g \in \mathcal{J}(X)$.

Definition. Let $\mathbb{P}$ be a category with weak pullbacks and $\mathcal{J}\colon \mathbb{P}_0 \to \mathcal{P}(\mathbb{P}_1)$ as before. An arrow h is $\mathcal{J}$-*closed* if, for every commutative square as below

$$\begin{array}{ccc} W & \xrightarrow{k} & X \\ {\scriptstyle g}\downarrow & & \downarrow{\scriptstyle h} \\ Z & \xrightarrow{f} & Y \end{array}$$

if $g \in \mathcal{J}(Z)$ then f factors through h.

Lemma. *$\mathcal{J}$-closed arrows are stable under weak pullbacks.*

3.25. Pretopologies [43, 44]. We are able now to formulate the proper notion of pretopology.

Definition. Let $\mathbb{P}$ be a category with weak pullbacks. A *pretopology* on $\mathbb{P}$ is a map $J\colon \mathbb{P}_0 \to \mathcal{P}(\mathbb{P}_1)$ satisfying conditions p0, p1, p2, p3, and p4, stated in 3.24, and the following condition:

p5. for any arrow $f\colon Y \to X$, there exist an arrow $g\colon A \to B$ in $J(B)$ and a J-closed arrow $h\colon B \to X$ such that f factors through $h \cdot g$ and $h \cdot g$ factors through f

$$
\begin{array}{ccc}
Y & \rightrightarrows & A \\
{\scriptstyle f}\downarrow & & \downarrow{\scriptstyle g} \\
X & \xleftarrow{\ h\ } & B
\end{array}
$$

3.26. Regular projective covers and pretopologies. The interest of regular projective covers and pretopologies in the study of universal closure operators is attested by the next proposition.

Proposition. *Let $\mathbb{E}$ be a regular category and $\mathbb{P}$ a regular projective cover of $\mathbb{E}$. There is a bijection between universal closure operators on $\mathbb{E}$ and pretopologies on $\mathbb{P}$.*

Proof. Let us start with a universal closure operator $\overline{(\ )}$ on $\mathbb{E}$, as in Definition 3.1. We define a pretopology J on $\mathbb{P}$ in the following way: for each $X \in \mathbb{P}$, an arrow $h\colon Y \to X$ of $\mathbb{P}$ is in $J(X)$ when its image (in $\mathbb{E}$) is a dense subobject of X. It is quite easy to check conditions p1, p2, p3, and p4 (see 3.24), so let us focalize on condition p5: we use part 1 of the next exercise (recall that a subobject is closed with respect to $\overline{(\ )}$ if it is equal to its closure). Given an arrow f in $\mathbb{P}$, take its regular epi-mono factorization $f = m \cdot e$. Consider the following diagram, where the mono a is given by condition c1 in Definition 3.1, the two squares are pullbacks, W is a $\mathbb{P}$-cover of $\bar{I}$ and Z is a $\mathbb{P}$-cover of V

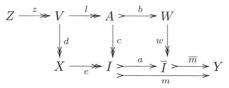

With the notations of condition p5, put $g = b \cdot l \cdot z$ and $h = \overline{m} \cdot w$. Since a is dense (by universality of the operator), also b is dense, so that $g \in J(W)$; moreover (by the next exercise) h is J-closed because $\overline{m}$ is closed. The arrow d shows that $h \cdot g$ factors through f. Conversely, since $d \cdot z$ is a regular epi between regular projective objects, it has a section which shows that f factors through $h \cdot g$.
Consider now a pretopology J on $\mathbb{P}$ and an object $X \in \mathbb{E}$. We need a natural closure operator $\overline{(\ )}\colon Sub_{\mathbb{E}}(X) \to Sub_{\mathbb{E}}(X)$. We assume first that $X \in \mathbb{P}$ and we define the operator in the following way: given a mono $m\colon I \rightarrowtail X$, take a $\mathbb{P}$-cover

$e: Y \twoheadrightarrow I$. Now, condition p5 in Definition 3.25 gives $g: A \to B$ in $\mathcal{J}(B)$ and $h: B \to X$ $\mathcal{J}$-closed such that $f = m \cdot e$ factors through $h \cdot g$ and $h \cdot g$ factors through f. We take as $\overline{m}$ the mono part of the factorization of h. If X is an arbitrary object in $\mathbb{E}$ and $m: I \rightarrowtail X$ is a mono, to define $\overline{m}$ consider a $\mathbb{P}$-cover $x: X' \twoheadrightarrow X$, the pullback $x^*(m)$ of m along x, and its closure $\overline{x^*(m)}$ defined as before (X' is in $\mathbb{P}$). We take as $\overline{m}$ the mono part of the factorization of $x \cdot \overline{x^*(m)}$. To check that this definition does not depend on the chosen $\mathbb{P}$-cover is easy. What is more subtle is to prove that the axioms of universal closure operator still work when we replace a regular projective object by an arbitrary one. In particular, to prove condition c4, one has to use Barr-Kock Theorem for regular categories (see 2.17 in Chapter IV). Finally, to prove that the two constructions just described are one the inverse of the other, one uses part 2 of the next exercise. $\square$

Exercise. Let $\mathbb{P}$ be a regular projective cover of a regular category $\mathbb{E}$. Consider an arrow $h: Y \to X$ in $\mathbb{P}$ and its factorization $h = m \cdot e$ in $\mathbb{E}$.

1. Let $\overline{(\ )}$ be a universal closure operator on $\mathbb{E}$. Show that m is closed with respect to $\overline{(\ )}$ iff h is $\mathcal{J}$-closed (Definition 3.24), where $\mathcal{J}$ is the pretopology on $\mathbb{P}$ induced by $\overline{(\ )}$.
2. Let $\mathcal{J}$ be a pretopology on $\mathbb{P}$. Show that $h \in \mathcal{J}(X)$ iff m is dense with respect to the universal closure operator on $\mathbb{E}$ induced by $\mathcal{J}$.

3.27. Regular projective presheaves. To end this section, we specialize Proposition 3.26 to the case of presheaf categories. Indeed, a presheaf category is regular and has a regular projective cover, as explained in the next proposition.

Proposition. *Let $\mathbb{C}$ be a small category. The category of presheaves $Set^{\mathbb{C}^{op}}$ is regular and its full subcategory of coproducts of representable presheaves is a regular projective cover.*

Proof. The regularity follows from that of *Set*. By Yoneda Lemma, each representable presheaf is regular projective. Moreover, in any category, a coproduct of regular projectives is regular projective. Now, any presheaf F is the colimit of

$$\mathbb{C}/F \xrightarrow{\quad U_F \quad} \mathbb{C} \xrightarrow{\quad Y \quad} Set^{\mathbb{C}^{op}}$$

(where $\mathbb{C}/F$ is the comma category of arrows $\mathbb{C}(-, C) \Rightarrow F$, U_F is the obvious forgetful functor and Y is the Yoneda embedding) and then it is a quotient of a coproduct of representable presheaves. $\square$

3.28. The coproduct completion [11, 14]. The regular projective cover of $Set^{\mathbb{C}^{op}}$ given in the previous proposition can be described directly from $\mathbb{C}$ via a universal property. Consider the following category, which we denote by $Fam\mathbb{C}$:

- An object of $Fam\mathbb{C}$ is a functor $f: I \to \mathbb{C}$, where I is a small set regarded as a discrete category. Sometimes we write (I, f) for such an object.

340 VII. Sheaf Theory

- An arrow $(a, \alpha): (I, f) \to (J, g)$ is a functor $a: I \to J$ together with a natural transformation $\alpha: f \Rightarrow g \cdot a$. Explicitly, α is a family of arrows in $\mathbb{C}$ of the form $\{\alpha_i: f(i) \to g(a(i))\}_{i \in I}$.
- There is a full and faithful functor $\eta: \mathbb{C} \to Fam\mathbb{C}$ which sends an object $X \in \mathbb{C}$ to the functor $X: \{*\} \to \mathbb{C}, \; * \mapsto X$.

Proposition. *Let $\mathbb{C}$ be a (not necessarily small) category. The functor $\eta: \mathbb{C} \to Fam\mathbb{C}$ is the coproduct completion of $\mathbb{C}$. This means:*

1. *The category $Fam\mathbb{C}$ has small coproducts;*
2. *For each category $\mathbb{B}$ with small coproducts, composition with η induces an equivalence from the category of coproduct preserving functors from $Fam\mathbb{C}$ to $\mathbb{B}$, to the category of functors from $\mathbb{C}$ to $\mathbb{B}$.*

Proof. Given $F: \mathbb{C} \to \mathbb{B}$, its extension $F': Fam\mathbb{C} \to \mathbb{B}$ along η sends (I, f) into the coproduct $\coprod_I F(f(i))$ in $\mathbb{B}$, and extends to arrows via the universal property of the coproduct. The essential uniqueness of F' follows from the fact that an object (I, f) of $Fam\mathbb{C}$ is the coproduct of the $\eta(f(i))$'s. $\qquad\square$

The category $Fam\mathbb{C}$ provides the external description of the regular projective cover of $Set^{\mathbb{C}^{op}}$ described in Proposition 3.27. We state this fact in the next lemma, which will be used also in the proof of Giraud's Theorem (Section 4).

Lemma. *Let $\mathbb{C}$ be a small category. Its coproduct completion $Fam\mathbb{C}$ is equivalent to the full subcategory of $Set^{\mathbb{C}^{op}}$ spanned by coproducts of representable presheaves.*

Since $Fam\mathbb{C}$ is a regular projective cover of $Set^{\mathbb{C}^{op}}$, we know that it has weak finite limits (Lemma 3.23). The existence of limits in $Fam\mathbb{C}$ has been studied in [32]. Categories of the form $Fam\mathbb{C}$ are studied also in [7] in connection with categorical Galois theory.

The next corollary represents the last step of this section. In the case of a presheaf category $Set^{\mathbb{C}^{op}}$, the regular projective cover $Fam\mathbb{C}$ is generated by the representable presheaves, and a pretopology on $Fam\mathbb{C}$ can be entirely described looking at the generators. In this way, we rediscover the notion of Grothendieck topology.

Corollary. *Let $\mathbb{C}$ be a small category. There is a bijection between pretopologies on $Fam\mathbb{C}$ and Grothendieck topologies on $\mathbb{C}$.*

Proof. Even if it follows from Corollary 3.14 and Proposition 3.26, it is worthwhile to construct explicitly the bijection between pretopologies and Grothendieck topologies. Before starting, let us point out a simple fact which might help to understand the relation between these two notions. Given a sieve $S \rightarrowtail \mathbb{C}(-, C)$, there is a canonical arrow

$$\sigma: \coprod_{f \in S(X), X \in \mathbb{C}} \mathbb{C}(-, X) \longrightarrow \mathbb{C}(-, C)$$

and we can get S back as the image of σ.

Now, given a pretopology $\mathcal{J}$ on $Fam\mathbb{C}$, consider an object C in $\mathbb{C}$ and its image $\eta(C)$ in $Fam\mathbb{C}$. Then, $\mathcal{J}(\eta(C))$ is a collection of arrows in $Fam\mathbb{C}$ with codomain $\eta(C)$. In fact, an arrow $(b, \beta) \colon (J, g) \to \eta(C)$ in $Fam\mathbb{C}$ is nothing but a family of arrows $\{\beta_j \colon g(j) \to C\}_{j \in J}$ in $\mathbb{C}$. From such a family, we can construct the corresponding arrow in $Set^{\mathbb{C}^{op}}$ and we can consider its regular epi-mono factorization

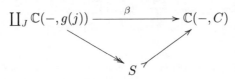

We get a Grothendieck topology taking, as $\mathcal{T}(C)$, all the sieves $S \rightarrowtail \mathbb{C}(-, C)$ arising in this way from an arrow $(b, \beta) \colon (J, g) \to \eta(C)$ in $\mathcal{J}(\eta(C))$.

Conversely, consider a Grothendieck topology $\mathcal{T}$ on $\mathbb{C}$. For every sieve $S \in \mathcal{T}(C)$, we can consider all the $Fam\mathbb{C}$-covers of S. Composing with $S \rightarrowtail \mathbb{C}(-, C)$, each $Fam\mathbb{C}$-cover gives an arrow in $Fam\mathbb{C}$ with codomain $\eta(C)$. We take, as $\mathcal{J}(\eta(C))$, all the arrows in $Fam\mathbb{C}$ arising in this way from the sieves in $\mathcal{T}(C)$. Doing so, we have defined $\mathcal{J}$ on objects of $Fam\mathbb{C}$ coming from $\mathbb{C}$, which means of the form $\eta(C)$, and it remains to extend $\mathcal{J}$ to arbitrary objects of $Fam\mathbb{C}$. To provide such an extension, we will use the fact that any object of $Fam\mathbb{C}$ can be regarded in a unique way as a coproduct of objects coming from $\mathbb{C}$, together with the fact that $Fam\mathbb{C}$ is an extensive category. Extensive categories are the main subject of Section 4, so we suspend the proof at this stage to come back to this problem in 4.4, when the notion of extensivity will have been approached. $\qquad\square$

4. Extensive categories

In Section 2, we have seen that the category of sheaves on a topological space is a localization of the category of presheaves. In Section 3, we have seen how to relax the notion of topological space so that arbitrary localizations of presheaf categories can be interpreted as categories of sheaves. The question we want to answer in this last section is if it is possible to recognize categories of sheaves for a Grothendieck topology from a purely categorical point of view. In other words, we look for a characterization theorem for categories of sheaves (equivalently, for localizations of presheaf categories).

Let us start with some necessary conditions. We have already observed that any category of the form $Set^{\mathbb{C}^{op}}$, for $\mathbb{C}$ a small category, is regular. Even more is true: $Set^{\mathbb{C}^{op}}$ is an *exact* category (see Chapters IV and VI), that is equivalence relations are effective (once again, this follows from the exactness of Set, because limits and colimits of $Set^{\mathbb{C}^{op}}$ are computed pointwisely in Set). Since exactness is preserved by localizations, every Grothendieck topos is an exact category. But this cannot be enough to characterize Grothendieck toposes. For example, any monadic category over a power of Set is exact (and this is one of the main ingredients of the characterization of algebraic categories as given in Chapter VI) and any elementary topos is exact. What make the difference between "algebra" (in the sense of monadic categories over powers of *Set*) and "topology" (in the sense of sheaf categories) is the behavior of coproducts. Coproducts are *disjoint* and *universal* in *Set*, in categories of variable sets and in their localizations (but the same properties do not hold for groups, for instance).

4.1. Extensive and lextensive categories [11]**.** Let us recall the most elegant formulation of disjointness and universality of coproducts.

Definition.

1. A category $\mathbb{A}$ with coproducts is *extensive* when, for each small family of objects $(X_i)_{i \in I}$, the canonical functor $\coprod \colon \prod_I (\mathbb{A}/X_i) \longrightarrow \mathbb{A}/\coprod_I X_i$ is an equivalence.
2. A category is *lextensive* if it is extensive and has finite limits.

A warning: in [11], as well as in III.2.6, only the finitary version of extensive and lextensive categories is considered, that is, the previous condition is required for finite, instead of small, coproducts. With the exception of Example 4.1.7, we always deal with small coproducts.

Examples.

1. Any localization of a lextensive category is lextensive;
2. *Set* and *Top* are lextensive categories;
3. For each small category $\mathbb{C}$, $Set^{\mathbb{C}^{op}}$ is lextensive;
4. If $\mathbb{A}$ is pointed and extensive, then it is trivial;
5. The homotopy category of *Top* is extensive, but not lextensive;
6. For any category $\mathbb{C}$, $Fam\mathbb{C}$ is extensive (but, in general, it is not lextensive);
7. An elementary topos is lextensive (here, replace small coproducts by finite coproducts).

4.2. More on extensive categories. Since the notion of extensivity is crucial in the rest of this chapter, we give now, in the form of exercises, some equivalent formulations or consequences of it. We call injections the canonical arrows $X_i \to \coprod_I X_i$ into the coproduct.

Exercises.

1. A category with coproducts is extensive if and only if it has pullbacks along injections and the following two conditions hold:

   e1. given a family of arrows $(f_i \colon Y_i \to X_i)_I$, for all $i \in I$ the following square is a pullback

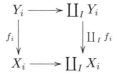

e2. (coproducts are *universal*) given an arrow $f\colon Y \to \coprod_I X_i$ and, for any $i \in I$, the pullback

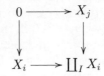

then the comparison morphism $\coprod_I Y_i \to Y$ is an isomorphism.

2. A category with coproducts is extensive if and only if it has pullbacks along injections and the following two conditions hold:

e2. coproducts are universal;

e3. (coproducts are *disjoint*) given a family of objects $(X_i)_I$, for each $i, j \in I$, $i \neq j$, the following square (where 0 is the initial object) is a pullback

$$
\begin{array}{ccc}
0 & \longrightarrow & X_j \\
\downarrow & & \downarrow \\
X_i & \longrightarrow & \coprod_I X_i
\end{array}
$$

3. Consider a category with coproducts and finite limits. The following conditions are equivalent:

e2. coproducts are universal;

e2'. if, for all $i \in I$, the left hand square is a pullback, then the right hand square is also a pullback

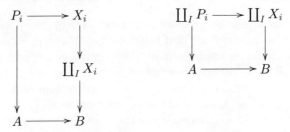

e2". (a) the canonical arrow $\coprod_I (X \times X_i) \to X \times (\coprod_I X_i)$ is an isomorphism (a category with coproducts and binary products which satisfies this conditions is called *distributive*),

   (b) if, for all $i \in I$, $E_i \to X_i \rightrightarrows Y$ is an equalizer, then $\coprod_I E_i \to \coprod_I X_i \rightrightarrows Y$ is an equalizer as well.

4. In an extensive category, injections are monos and the initial object is *strict* (that is, every arrow $X \to 0$ is an isomorphism).

5. In a distributive category, injections are monos and the initial object is strict.

4.3. Lextensive categories as pseudo-algebras [46]. We have already remarked that the coproduct completion $Fam\mathbb{A}$ of any category $\mathbb{A}$ is extensive. Moreover, we can use this completion to give an elegant characterization of extensivity. Assume $\mathbb{A}$

has coproducts. By the universal property of $\eta\colon \mathbb{A} \to Fam\mathbb{A}$, we can extend the identity functor on $\mathbb{A}$ to a coproduct preserving functor $\Sigma\colon Fam\mathbb{A} \to \mathbb{A}$ (which happens to be left adjoint to η).

Proposition. *Let $\mathbb{A}$ be a category with coproducts and finite limits. $\mathbb{A}$ is lextensive if and only if $\Sigma\colon Fam\mathbb{A} \to \mathbb{A}$ preserves finite limits.*

The easy proof uses the explicit description of finite limits in $Fam\mathbb{A}$, which are inherited from those in $\mathbb{A}$. In particular, the statement can be refined saying that $\mathbb{A}$ is distributive iff Σ preserves binary products.

The previous proposition can be incorporated in an elegant "2-dimensional" presentation of lextensive categories. Although it will not be used in the rest of the chapter, we recall it here, because it represents another interesting link with monadic functors, studied in Chapter V. Both the *Fam* construction and the *Lex* (*finite limit completion*) construction give rise to *pseudo-monads Fam*$\colon CAT \to CAT$ and *Lex*$\colon CAT \to CAT$ (where CAT is the 2-category of categories and functors). In fact, *Fam* is a *KZ-doctrine* and *Lex* is a *coKZ-doctrine* over CAT and the two are related by a *distributive law* of *Lex* over *Fam*, so that we can consider the lifting $\widehat{Fam}$ of *Fam* to the 2-category CAT^{Lex} of *Lex*-pseudo-algebras (which are nothing but finitely complete categories and left exact functors). Now, the previous proposition, together with the adjunction $\Sigma \dashv \eta$, essentially means that the 2-category of lextensive categories and functors preserving coproducts and finite limits is bi-equivalent to the 2-category of $\widehat{Fam}$-pseudo-algebras. More on KZ-doctrines and distributive laws for pseudo-monads can be found in [36, 42].

4.4. Back to pretopologies. Let us come back now to the problem pointed out at the end of the proof of Corollary 3.28, that is how to construct a pretopology on $Fam\mathbb{C}$ starting from its values on the objects of $\mathbb{C}$. We consider the problem in terms of the associated universal closure operator on $Set^{\mathbb{C}^{op}}$. Consider a universal closure operator $\overline{(\ )}$ on $\mathbb{A} = Set^{\mathbb{C}^{op}}$ (or on any other lextensive category $\mathbb{A}$). Given a family of objects $(A_i)_I$ in $\mathbb{A}$, we have a canonical bijection $Sub(\coprod_I A_i) \simeq \prod_I Sub(A_i)$. Moreover, for any $i \in I$, the universality of the operator implies that the following diagram commutes

$$
\begin{array}{ccc}
\prod_I Sub(A_i) \simeq Sub(\coprod_I A_i) & \xrightarrow{\ \overline{(\ )}\ } & Sub(\coprod_I A_i) \simeq \prod_I Sub(A_i) \\
\downarrow & & \downarrow \\
Sub(A_i) & \xrightarrow{\ \ \overline{(\ )}\ \ } & Sub(A_i)
\end{array}
$$

In other words, the knowledge of the operator on the A_i's forces its definition on $\coprod_I A_i$. If, moreover, $\mathbb{A}$ is regular, then the factorization of a coproduct of arrows is the coproduct of the various factorizations. The previous argument implies then that a subobject of $\coprod_I A_i$ is dense iff the corresponding subobjects of the A_i's are dense (compare with III.7.5). Translated in terms of the pretopology $\mathcal{J}$ on $Fam\mathbb{C}$, this means that an arrow h is in $\mathcal{J}(\coprod_I \eta(X_i))$ iff each h_i is in $\mathcal{J}(\eta(X_i))$, where h_i is the pullback of h along the injection $\eta(X_i) \to \coprod_I \eta(X_i)$.

4.5. Generators. Now that we have the notion of extensive category, we can come back to the problem of characterizing Grothendieck toposes. We already know that any localization of a presheaf category is exact and extensive. The last ingredient to get such a characterization (the ingredient which essentially makes the difference from elementary toposes) is given by representable presheaves. If $\mathbb{C}$ is a small category, it embeds into $Set^{\mathbb{C}^{op}}$ via the Yoneda embedding. In view of the characterization theorem, the important property of representable presheaves is encoded in the next definition. (A notation: if S is a set and G is an object of a category, $S \circ G$ is the S-indexed copower of G.)

Definition. A set of objects $\mathcal{G}$ of a category $\mathbb{A}$ is a *small generator* if, for each object $A \in \mathbb{A}$, the following conditions hold:

1. the coproduct $\coprod_{G \in \mathcal{G}} \mathbb{A}(G, A) \circ G$ exists;
2. the canonical arrow $a \colon \coprod_{G \in \mathcal{G}} \mathbb{A}(G, A) \circ G \to A$ is an epi.

Observe that the second condition in the previous definition can be equivalently stated saying that two parallel arrows $x, y \colon A \to A'$ are equal iff $x \cdot f = y \cdot f$ for all $f \in \mathbb{A}(G, A)$ and for all $G \in \mathcal{G}$. This formulation does not require the existence of coproducts.

A glance at the proof of Proposition 3.27 shows that, if $\mathbb{C}$ is a small category, the representable presheaves $(\mathbb{C}(-, X))_{X \in \mathbb{C}}$ are a small generator for $Set^{\mathbb{C}^{op}}$. Moreover, if $\mathcal{G}$ is a small generator for a category $\mathbb{E}$ and $\mathbb{A}$ is a reflective subcategory of $\mathbb{E}$, with reflector $r \colon \mathbb{E} \to \mathbb{A}$, then $\{r(G) \mid G \in \mathcal{G}\}$ is a small generator for $\mathbb{A}$.

4.6. Characterization of Grothendieck toposes. We are finally ready to state a semantical characterization of Grothendieck toposes.

Theorem. *Let $\mathbb{A}$ be a category. The following conditions are equivalent:*

1. *$\mathbb{A}$ is equivalent to a localization of a presheaf category;*
2. *$\mathbb{A}$ is exact, extensive, and has a small generator.*

We already know that condition 1 implies condition 2. To prove the converse is less easy. Let us mention in advance our plan. (All notions not yet introduced will be explained at the right moment.)
Step 1: Given $\mathcal{G}$ a small generator for $\mathbb{A}$, we consider the full subcategory $\mathbb{C}$ of $\mathbb{A}$ spanned by the objects of $\mathcal{G}$. We have a full and faithful functor

$$\mathbb{A}(-, -) \colon \mathbb{A} \to Set^{\mathbb{C}^{op}} \qquad A \mapsto \mathbb{A}(-, A) \in Set^{\mathbb{C}^{op}}.$$

Step 2: We get a left adjoint $r \dashv \mathbb{A}(-, -)$ considering the left Kan extension of the full inclusion $i \colon \mathbb{C} \to \mathbb{A}$ along the Yoneda embedding $Y \colon \mathbb{C} \to Set^{\mathbb{C}^{op}}$.
Step 3: We split the construction of r into two steps. First we consider the extension $i' \colon Fam\mathbb{C} \to \mathbb{A}$ of i along the completion $\eta \colon \mathbb{C} \to Fam\mathbb{C}$, and then the extension $i'' \colon Set^{\mathbb{C}^{op}} \to \mathbb{A}$ of i' along the embedding $Fam\mathbb{C} \to Set^{\mathbb{C}^{op}}$.

Step 4: We explain that $r = i''$ is left exact iff i' is left covering iff i is filtering.
Step 5: We check that i is filtering, so that

$$\mathbb{A} \xrightarrow[\mathbb{A}(-,-)]{\quad r \quad} Set^{\mathbb{C}^{op}}$$

is equivalent to a localization.

4.7. Step 1.1: calculus of relations [12, 19]. The next lemma provides the occasion to introduce the *calculus of relations*, an interesting tool available in any regular category.

Lemma. *Let $\mathbb{A}$ be an exact and extensive category.*

1. *In $\mathbb{A}$ cokernel pairs of monos exist and are pullbacks.*
2. *In $\mathbb{A}$ every mono is regular.*
3. *In $\mathbb{A}$ every epi is regular.*

Proof. By point 1, any mono is the coequalizer of its kernel pair. This proves point 2. In any category with regular epi-mono factorization, if every mono is regular, then every epi is regular. It remains to prove point 1. This can be done using the calculus of relations. Let us sketch the argument (see also I.1.5).
The category $Rel(\mathbb{A})$ has the same objects as $\mathbb{A}$. An arrow $R\colon X \relbar\mathrel{\mkern-8mu}\circ Y$ in $Rel(\mathbb{A})$
is a relation $X \xleftarrow{r_0} R \xrightarrow{r_1} Y$, that is a mono $(r_0, r_1)\colon R \rightarrowtail X \times Y$. Given
two relations $X \xleftarrow{r_0} R \xrightarrow{r_1} Y \xleftarrow{s_0} S \xrightarrow{s_1} Z$, their composition is defined
by taking first the pullback

$$
\begin{array}{ccc}
P & \xrightarrow{\;r_1'\;} & S \\
{\scriptstyle s_0'}\downarrow & & \downarrow{\scriptstyle s_0} \\
R & \xrightarrow[\;r_1\;]{} & Y
\end{array}
$$

and then the image of $(r_0 \cdot s_0', s_1 \cdot r_1')\colon P \to X \times Z$. The identity on X is the diagonal
$\Delta_X\colon X \rightarrowtail X \times X$ and the associativity of the composition is equivalent to the
fact that regular epis are pullback stable. If $(r_0, r_1)\colon R \rightarrowtail X \times Y$ is a relation,
the opposite relation is defined to be $(r_0, r_1)^\circ = (r_1, r_0)\colon R \rightarrowtail Y \times X$. This gives
an involution on $Rel(\mathbb{A})$. The category $\mathbb{A}$ can be seen as a non-full subcategory
of $Rel(\mathbb{A})$ identifying an arrow $f\colon X \to Y$ with its graph $(1_X, f)\colon X \rightarrowtail X \times Y$.
Using coproducts in $\mathbb{A}$, we can define the union of two subobjects (and then of
two relations) as the image of their coproduct

$$
\begin{array}{ccc}
S \amalg R & \longleftarrow & R \\
\uparrow & & \downarrow \\
S & \longrightarrow & X
\end{array}
\qquad\qquad
S \amalg R \longrightarrow S \cup R \rightarrowtail X
$$

Moreover, a relation $R\colon X_1 \amalg Y_1 \relbar\mathrel{\mkern-8mu}\circ X_2 \amalg Y_2$ is determined by a matrix $R = (R_{ij})$
of four relations $R_{ij}\colon X_i \relbar\mathrel{\mkern-8mu}\circ Y_j$ (here we are using the distributivity in $\mathbb{A}$). Now,

consider a mono $i\colon A \rightarrowtail X$ and an arbitrary arrow $f\colon A \to Y$. The pushout of i and f is given by the quotient of $X \coprod Y$ with respect to the equivalence relation

$$\begin{pmatrix} \Delta_X \cup R^\circ \cdot R & R \\ R^\circ & \Delta_Y \cup R \cdot R^\circ \end{pmatrix}$$

where $R\colon X \multimap Y$ is $X \xleftarrowtail{\ i\ } A \xrightarrow{\ f\ } Y$. $\qquad\qquad\qquad\square$

4.8. Step 1.2: dense generators. We can now embed $\mathbb{A}$ into $Set^{\mathbb{C}^{op}}$.

Proposition. *Let $\mathbb{A}$ be an exact and extensive category and let $\mathcal{G}$ be a small generator of $\mathbb{A}$. Consider the full subcategory $\mathbb{C}$ of $\mathbb{A}$ spanned by the objects of $\mathcal{G}$. The functor $\mathbb{A}(-,-)\colon \mathbb{A} \to Set^{\mathbb{C}^{op}}$ is full and faithful.*

Proof. By Lemma 4.7, for any $A \in \mathbb{A}$ the canonical arrow $a\colon \coprod_{G \in \mathcal{G}} \mathbb{A}(G, A) \circ G \to A$ is a regular epi. The key of the proof consists of deducing from this fact that $\mathcal{G}$ is a *dense generator*. This means that the functor

$$U_A \colon \mathbb{C}/A \to \mathbb{A} \qquad (f\colon G \to A) \mapsto G$$

(where $\mathbb{C}/A$ is the full subcategory of the comma category $\mathbb{A}/A$) has colimit $\langle A, (f\colon G \to A)_{f,G}\rangle$. (Indeed, this is exactly the condition stated in Proposition 3.27.) We include some details here, to see extensivity at work at least once. Let $\langle C, (g_f\colon G \to A)_{f,G}\rangle$ be another cocone on U_A. By the universal property of the coproduct, we get a unique factorization g such that $g \cdot s_f = g_f$, where the s_f are the injections in the coproduct (see the following diagram, where $N(a)$ is the kernel pair of a)

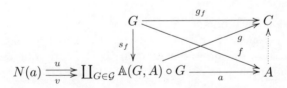

To get the (necessarily unique) factorization $A \to C$, it remains to prove that $g \cdot u = g \cdot v$. Since coproducts in $\mathbb{A}$ are universal, $N(a)$ is the coproduct of the various pullbacks U_f as in the following diagram

$$\begin{array}{ccc}
U_f & \longrightarrow & G \\
\downarrow{\scriptstyle u_f} & & \downarrow{\scriptstyle s_f} \\
N(a) & \xrightarrow{\ u\ } & \coprod_{G \in \mathcal{G}} \mathbb{A}(G, A) \circ G
\end{array}$$

so that to check $g \cdot u = g \cdot v$ simplifies to verify $g \cdot u \cdot u_f = g \cdot v \cdot u_f$ for all f. But $N(a)$ is also the coproduct of the various pullbacks

$$
\begin{array}{ccc}
V_f & \longrightarrow & G \\
\scriptstyle{v_f} \downarrow & & \downarrow \scriptstyle{s_f} \\
N(a) & \underset{v}{\longrightarrow} & \coprod_{G \in \mathcal{G}} \mathbb{A}(G, A) \circ G
\end{array}
$$

Fix now one of the morphisms f, say $f_0 \colon G_0 \to A$. Using once again the universality of the coproducts, U_{f_0} can be described as the coproduct of the various pullbacks

$$
\begin{array}{ccc}
P_{f,f_0} & \longrightarrow & V_f \\
\scriptstyle{p_{f,f_0}} \downarrow & & \downarrow \scriptstyle{v_f} \\
U_{f_0} & \underset{u_{f_0}}{\longrightarrow} & N(a)
\end{array}
$$

so that checking $g \cdot u \cdot u_{f_0} = g \cdot v \cdot u_{f_0}$ simplifies further to verify $g \cdot u \cdot u_{f_0} \cdot p_{f,f_0} = g \cdot v \cdot u_{f_0} \cdot p_{f,f_0}$. Finally, for this it is enough to check that $g \cdot u \cdot u_{f_0} \cdot p_{f,f_0} \cdot l = g \cdot v \cdot u_{f_0} \cdot p_{f,f_0} \cdot l$ for all $l \colon G \to P_{f,f_0}$ and for all $G \in \mathcal{G}$. This follows from a diagram chase, $(g_f \colon G \to A)_{f,G}$ being a cocone on $U_A \colon \mathbb{C}/A \to A$.

Since $\mathcal{G}$ is a dense generator, the functor $\mathbb{A}(-, -)$ is full and faithful (in fact, the converse holds too). Consider two objects $A, B \in \mathbb{A}$ and a natural transformation $\tau \colon \mathbb{A}(-, A) \Rightarrow \mathbb{A}(-, B) \colon \mathbb{C}^{op} \to Set$. For every $G \in \mathcal{G}$ and for every $f \colon G \to A$, we get an arrow $\tau_G(f) \colon G \to B$, and all these arrows form a cocone on U_A. $\mathcal{G}$ being a dense generator, there is a unique factorization $t \colon A \to B$ through the colimit of U_A. This process inverts the obvious construction of a natural transformation $\mathbb{A}(-, A) \Rightarrow \mathbb{A}(-, B)$ from an arrow $A \to B$. $\qquad \square$

4.9. Step 2: Kan extensions. We start by recalling the definition of (left) Kan extension.

Definition. Consider two functors $F \colon \mathbb{C} \to \mathbb{A}$ and $G \colon \mathbb{C} \to \mathbb{B}$. A *Kan extension* of F along G is a functor $K_G(F) \colon \mathbb{B} \to \mathbb{A}$ with a natural transformation $\epsilon_G(F) \colon F \Rightarrow K_G(F) \cdot G$ such that, for any other functor $H \colon \mathbb{B} \to \mathbb{A}$, composing with $\epsilon_G(F)$ gives a natural bijection $Nat(K_G(F), H) \simeq Nat(F, H \cdot G)$.

Being defined by a universal property, a Kan extension, when it exists, is essentially unique. We present also the following easy lemma, which will be useful to split the construction of a Kan extension.

Lemma. *Consider three functors $F \colon \mathbb{C} \to \mathbb{A}$, $G \colon \mathbb{C} \to \mathbb{B}$, $G' \colon \mathbb{B} \to \mathbb{B}'$. Assume the existence of $K_G(F)$ and of $K_{G'}(K_G(F))$. Then, $K_{G' \cdot G}(F) = K_{G'}(K_G(F))$.*

Observe that, with no regards to size conditions, the definition of Kan extension precisely means that the functor

$$
- \cdot G \colon [\mathbb{B}, \mathbb{A}] \to [\mathbb{C}, \mathbb{A}]
$$

between functor categories, has a left adjoint

$$K_G(-)\colon [\mathbb{C}, \mathbb{A}] \to [\mathbb{B}, \mathbb{A}].$$

From this point of view, the previous lemma is a particular instance of the general fact that adjoint functors compose.

Proposition. *Consider a functor $F\colon \mathbb{C} \to \mathbb{A}$ with $\mathbb{C}$ small and $\mathbb{A}$ cocomplete. Consider also the Yoneda embedding $Y\colon \mathbb{C} \to Set^{\mathbb{C}^{op}}$. Then $K_Y(F)$ exists and the natural transformation $\epsilon_Y(F)$ is a natural isomorphism. Moreover, $K_Y(F)$ is left adjoint to the functor $\mathbb{A}(F-,-)\colon \mathbb{A} \to Set^{\mathbb{C}^{op}}$ which sends $A \in \mathbb{A}$ into $\mathbb{A}(F-, A) \in Set^{\mathbb{C}^{op}}$.*

Proof. We know, from Proposition 3.27, that each $E \in Set^{\mathbb{C}^{op}}$ is the colimit of $\mathbb{C}/E \xrightarrow{U_E} \mathbb{C} \xrightarrow{Y} Set^{\mathbb{C}^{op}}$. Since we want $K_Y(F)$ to be a left adjoint and $\epsilon_Y(F)$ to be an isomorphism, we have to define $K_Y(F)(E)$ as the colimit of $\mathbb{C}/E \xrightarrow{U_E} \mathbb{C} \xrightarrow{F} \mathbb{A}$. Let us write $\langle A, (\sigma_f\colon F(C) \to A)_{f \in \mathbb{C}/E} \rangle$ for such a colimit. Given a functor $H\colon Set^{\mathbb{C}^{op}} \to \mathbb{A}$ and a natural transformation $\alpha\colon F \Rightarrow H \cdot Y$, we get a natural transformation $\beta\colon K_Y(F) \to H$ in the following way: the component of β at E is the unique arrow making the diagram commutative

$$
\begin{array}{ccc}
A & \xrightarrow{\ \beta_E\ } & H(E) \\
{\scriptstyle \sigma_f}\big\uparrow & & \big\uparrow{\scriptstyle H(f)} \\
F(C) & \xrightarrow[\ \alpha_C\ ]{} & H(\mathbb{C}(-, C))
\end{array}
$$

That $\epsilon_Y(F)$ is an isomorphism follows from the Yoneda embedding being full and faithful. As far as the adjunction $K_Y(F) \dashv \mathbb{A}(F-,-)$ is concerned, consider a natural transformation $h\colon E \Rightarrow \mathbb{A}(F-, B)$, i.e. a natural family of arrows $\{h_C\colon Nat(\mathbb{C}(-, C), E) \simeq E(C) \to \mathbb{A}(F(C), B)\}$. Now, for every $f \in \mathbb{C}/E$, we get $h_C(f)\colon F(C) \to B$. By the universal property of the colimit, these $h_C(f)$ give rise to a unique $k\colon A \to B$ such that $k \cdot \sigma_f = h_C(f)$. $\qquad\square$

Corollary. *In the situation of Proposition 4.8, the functor $\mathbb{A}(-,-)\colon \mathbb{A} \to Set^{\mathbb{C}^{op}}$ has a left adjoint given by the Kan extension of the full inclusion $\mathbb{C} \to \mathbb{A}$ along the Yoneda embedding $Y\colon \mathbb{C} \to Set^{\mathbb{C}^{op}}$.*

Proof. In order to apply the previous proposition, it remains to prove that $\mathbb{A}$ is cocomplete. For this let us remark only that, because of extensivity and exactness, the coequalizer of two parallel arrows can be constructed as the quotient of the equivalence relation generated by their jointly monic part. $\qquad\square$

4.10. Step 3.1: coproduct extension as a Kan extension. In some special case, the Kan extension of a functor can be described in an easy way. The first case of interest for us is the Kan extension along the coproduct completion $\eta\colon \mathbb{C} \to Fam\mathbb{C}$ of a small category $\mathbb{C}$. Indeed we have the following simple lemma.

Lemma. *Let* $\mathbb{C}$ *be a small category,* $\mathbb{A}$ *a category with coproducts, and* $F\colon \mathbb{C} \to \mathbb{A}$ *an arbitrary functor. The coproduct preserving extension* $F'\colon Fam\mathbb{C} \to \mathbb{A}$ *of* F *(see Proposition 3.28) is the Kan extension of* F *along* $\eta\colon \mathbb{C} \to Fam\mathbb{C}$.

4.11. Step 3.2: left covering functors [14]. We have already observed that $Fam\mathbb{C}$ is a regular projective cover of the exact category $Set^{\mathbb{C}^{op}}$, so that $Fam\mathbb{C}$ has weak finite limits. In this situation, we have an extension property with respect to left covering functors. The notion of left covering functor generalizes the notion of left exact functor to situations in which only weak finite limits exist.

Definition. Let $\mathbb{P}$ be a category with weak finite limits and $\mathbb{A}$ an exact category. A functor $F\colon \mathbb{P} \to \mathbb{A}$ is *left covering* if for any functor $L\colon \mathcal{D} \to \mathbb{P}$ ($\mathcal{D}$ being a finite category) and for any (equivalently, for one) weak limit W of L, the canonical comparison between $F(W)$ and the limit of $F \cdot L$ is a regular epi.

The fact that left covering functors are a good generalization of left exact functors is confirmed by the next exercise.

Exercise. Let $F\colon \mathbb{P} \to \mathbb{A}$ be a left covering functor. Show that F preserves all the finite limits which exist in $\mathbb{P}$.
[Hint: Show that F preserves all finite monomorphic families.]

4.12. Step 3.3: the exact extension of a left covering functor. Let us come back to the situation where $\mathbb{P}$ is a projective cover of an exact category $\mathbb{B}$, and consider another exact category $\mathbb{A}$. In this context, a left covering functor $F\colon \mathbb{P} \to \mathbb{A}$ can be extended to a functor $\overline{F}\colon \mathbb{B} \to \mathbb{A}$. Indeed, given an object $B \in \mathbb{B}$, we can consider a $\mathbb{P}$-cover of B, its kernel pair in $\mathbb{B}$, and again a $\mathbb{P}$-cover of the kernel pair, as in the following diagram

$$P' \overset{n}{\longrightarrow\!\!\!\!\!\rightarrow} N(b) \overset{b_0}{\underset{b_1}{\rightrightarrows}} P \overset{b}{\longrightarrow\!\!\!\!\!\rightarrow} B.$$

Now we can apply the functor F to the $\mathbb{P}$-part of the diagram and factorize its image in $\mathbb{A}$ as a regular epi followed by a jointly monic pair

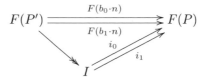

The left covering character of F guarantees that (i_0, i_1) is an equivalence relation. Since $\mathbb{A}$ is an exact category, the coequalizer of (i_0, i_1) exists, and we take this coequalizer as $\overline{F}(B)$.

Exercises.

1. Prove that (i_0, i_1) is an equivalence relation.
2. Prove that $\overline{F}$ is well defined, that is, it does not depend on the choice of P, b, P', and n.

3. Extend the construction of $\overline{F}$ to the arrows of $\mathbb{B}$.

The previous construction gives us our second example of Kan extension.

Lemma. *Let $\mathbb{P}$ be a regular projective cover of an exact category $\mathbb{B}$, $\mathbb{A}$ an exact category and $F: \mathbb{P} \to \mathbb{A}$ a left covering functor. The functor $\overline{F}: \mathbb{B} \to \mathbb{A}$ just described is the Kan extension of F along the full inclusion $\mathbb{P} \to \mathbb{B}$.*

Corollary. *Under the hypothesis of Proposition 4.8, the Kan extension of the full inclusion $i: \mathbb{C} \to \mathbb{A}$ along the Yoneda embedding is given by the Kan extension along $Fam\mathbb{C} \to Set^{\mathbb{C}^{op}}$ of the Kan extension along $\mathbb{C} \to Fam\mathbb{C}$ of i.*

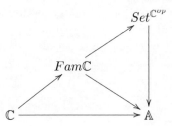

4.13. Step 4.1: exact completion [14, 33]. In Step 3, we have constructed the Kan extension $K_Y(i)$ of $i: \mathbb{C} \to \mathbb{A}$ along $Y: \mathbb{C} \to Set^{\mathbb{C}^{op}}$ in two steps. We can detect the left exactness of $K_Y(i)$ through these two steps.

Lemma. *Let $\mathbb{P}$ be a regular projective cover of an exact category $\mathbb{B}$ and let $F: \mathbb{P} \to \mathbb{A}$ be a left covering functor, with $\mathbb{A}$ exact. The Kan extension $\overline{F}: \mathbb{B} \to \mathbb{A}$ is left exact.*

A complete proof of this lemma requires the *exact completion* of a category with weak finite limits. It amounts to the construction of an exact category $\mathbb{P}_{ex}$ from a category $\mathbb{P}$ with weak finite limits in such a way that:

- the category $\mathbb{P}$ is a regular projective cover of $\mathbb{P}_{ex}$;
- composition with the full inclusion $\gamma: \mathbb{P} \to \mathbb{P}_{ex}$ classifies exact functors (a functor between exact categories is exact if it is left exact and preserves regular epis). This means that, for any exact category $\mathbb{A}$, γ induces an equivalence between the category of exact functors from $\mathbb{P}_{ex}$ to $\mathbb{A}$ and the category of left covering functors from $\mathbb{P}$ to $\mathbb{A}$.

Moreover, if an exact category $\mathbb{B}$ has enough regular projective objects, then $\mathbb{B}$ is equivalent to $\mathbb{P}_{ex}$, where $\mathbb{P}$ is any regular projective cover of $\mathbb{B}$. In other words, an exact category with enough regular projectives is determined by any of its regular projective covers, and an exact functor out from it is determined by its restriction to a regular projective cover. For example, for any small category $\mathbb{C}$, the category $Set^{\mathbb{C}^{op}}$ is equivalent to the exact completion of $Fam\mathbb{C}$.

A complete treatment of the exact completion is off the subject of this chapter. We point out only one of the main ingredients, which is part 2 of next exercise (part 1 is preparatory to part 2).

Exercises.

1. Show that if a category has a weak terminal object, weak binary products, and weak equalizers, then it has all weak finite limits.
2. Let $\mathbb{P}$ be a category with weak finite limits and $\mathbb{A}$ a regular category. Show that a functor $F: \mathbb{P} \to \mathbb{A}$ is left covering iff it is left covering with respect to a weak terminal object, weak binary products and weak equalizers.

Let us mention here that, since an algebraic category is exact and has enough regular projective objects (Example 3.23), the exact completion can be used to study localizations of algebraic categories. This is done in Chapter VI.

The fact that algebraic categories (more generally, monadic categories over a power of Set) are exact and have enough regular projective objects is related to the axiom of choice in Set (see [49] for a detailed discussion). Let us recall two quite different examples.

Examples.

1. Let $\mathbb{E}$ be an elementary topos; the dual category $\mathbb{E}^{op}$ is exact with enough regular projective objects.
2. Let $\mathbb{E}$ be an elementary topos; the category $Sl(\mathbb{E})$ of internal sup-lattices in $\mathbb{E}$ is exact, and the category of relation $Rel(\mathbb{E})$ is a regular projective cover of $Sl(\mathbb{E})$ (see [49]).

To complete the picture, let us also mention that the problem of when the exact completion of a category is extensive, cartesian closed or a topos has been studied respectively in [25, 37], in [13, 47] and in [43, 45].

4.14. Step 4.2: filtering functors. It remains to examine the Kan extension along the coproduct completion.

Definition. Consider a small category $\mathbb{C}$ and a category $\mathbb{A}$ with finite limits. A functor $F: \mathbb{C} \to \mathbb{A}$ is *filtering* when the following conditions hold:

f1. The family of all maps $F(X) \to T$ (with X varying in $\mathbb{C}$ and T the terminal object) is epimorphic;
f2. For any pair of objects $A, B \in \mathbb{C}$, the family of all maps $(F(u), F(v)): F(X) \to F(A) \times F(B)$ (with $u: X \to A, v: X \to B$ in $\mathbb{C}$ and X varying in $\mathbb{C}$) is epimorphic;
f3. For any pair of arrows $u, v: A \to B$ in $\mathbb{C}$, the family of all maps $F(X) \to E_{u,v}$ (induced, via the equalizer $E_{u,v} \to F(A)$ of $F(u)$ and $F(v)$, by maps $w: X \to A$ in $\mathbb{C}$ such that $u \cdot w = v \cdot w$, with X varying in $\mathbb{C}$) is epimorphic.

Note that, when $\mathbb{A} = Set$, this definition means that the category $\mathbb{C}/F$ is filtering, that is F is a filtered colimit of representable functors.

Lemma. *Consider a small category $\mathbb{C}$ and an exact and extensive category $\mathbb{A}$. The Kan extension $F': Fam\mathbb{C} \to \mathbb{A}$ of a functor $F: \mathbb{C} \to \mathbb{A}$ along $\eta: \mathbb{C} \to \mathbb{A}$ is left covering iff F is filtering.*

Proof. Let us describe some weak limits in $Fam\mathbb{C}$, using their canonical presentation as quotients of the corresponding limits in $Set^{\mathbb{C}^{op}}$. A weak terminal object in $Fam\mathbb{C}$ is the coproduct $\coprod \mathbb{C}(-, X)$ of all the representable presheaves. A weak product of two objects $\mathbb{C}(-, A)$ and $\mathbb{C}(-, B)$ in $Fam\mathbb{C}$ is the coproduct $\coprod \mathbb{C}(-, X)$ indexed by all the pairs of arrows $u: X \to A, v: X \to B$ with X varying in $\mathbb{C}$. A weak equalizer of two parallel arrows $u, v: \mathbb{C}(-, A) \to \mathbb{C}(-, B)$ in $Fam\mathbb{C}$ is the coproduct $\coprod \mathbb{C}(-, X)$ indexed over all the arrows $x: X \to A$ such that $u \cdot x = v \cdot x$, with X varying in $\mathbb{C}$.

Since in $\mathbb{A}$ every epi is regular (see Lemma 4.7), the three conditions of the previous definition respectively mean that $F': Fam\mathbb{C} \to \mathbb{A}$ is left covering with respect to weak terminal objects, weak binary products of objects coming from $\mathbb{C}$ and weak equalizers of arrows coming from $\mathbb{C}$. Using the extensivity of $\mathbb{A}$, one can show that such a functor is left covering with respect to weak terminal objects, weak binary products and weak equalizers. We conclude by Exercise 4.13.2. $\qquad\square$

4.15. Step 5: conclusion. Consider now an exact and extensive category $\mathbb{A}$ with a small generator $\mathcal{G}$, as in condition 2 of Theorem 4.6. Let $\mathbb{C}$ be the full subcategory of $\mathbb{A}$ spanned by the objects of $\mathcal{G}$.

Lemma. *The full inclusion* $i: \mathbb{C} \to \mathbb{A}$ *is a filtering functor.*

Proof. The three conditions of Definition 4.14 are satisfied respectively because $\mathcal{G}$ generates $T, A \times B$ and $E_{u,v}$. $\qquad\square$

Corollary. *In the previous situation,* $\mathbb{A}$ *is equivalent to a localization of* $Set^{\mathbb{C}^{op}}$.

Proof. Since $i: \mathbb{C} \to \mathbb{A}$ is filtering, its Kan extension $K_Y(i): Set^{\mathbb{C}^{op}} \to \mathbb{A}$ along the Yoneda embedding $Y: \mathbb{C} \to Set^{\mathbb{C}^{op}}$ is left exact. But $K_Y(i)$ is left adjoint to $\mathbb{A}(-, -): \mathbb{A} \to Set^{\mathbb{C}^{op}}$, which is full and faithful. $\qquad\square$

4.16. Another proof of Giraud's Theorem. The aim of this last section is to sketch a slightly different proof of Giraud's Theorem, a proof which underlines the role of the exact completion. The stream of the proof is simple: first, we characterize categories of the form $Fam\mathbb{C}$ for $\mathbb{C}$ a small category. Second, using that $Set^{\mathbb{C}^{op}}$ is the exact completion of $Fam\mathbb{C}$, we get an abstract characterization of presheaf categories. Third, to prove that $i: \mathbb{A} \to Set^{\mathbb{C}^{op}}$ is a localization, we need an exact functor $r: Set^{\mathbb{C}^{op}} \to \mathbb{A}$, that is, by the universal property of $Set^{\mathbb{C}^{op}} = (Fam\mathbb{C})_{ex}$, a left covering functor $Fam\mathbb{C} \to \mathbb{A}$.

Definition. Let $\mathbb{B}$ be a category with coproducts. An object C of $\mathbb{B}$ is *connected* if the representable functor $\mathbb{B}(C, -): \mathbb{B} \to Set$ preserves coproducts.

(Connected objects are sometimes called *indecomposable*.)

Lemma. [7, 14] *Let* $\mathbb{B}$ *be a category. The following conditions are equivalent:*

1. $\mathbb{B}$ *is equivalent to the coproduct completion of a small category;*
2. $\mathbb{B}$ *has a small subcategory* $\mathbb{C}$ *of connected objects, such that each object of* $\mathbb{B}$ *is a coproduct of objects of* $\mathbb{C}$.

Proof. The implication $1 \Rightarrow 2$ is obvious. Conversely, consider the coproduct-preserving extension $F' \colon Fam\mathbb{C} \to \mathbb{B}$ of the full inclusion $F \colon \mathbb{C} \to \mathbb{B}$. The fact that the objects of $\mathbb{C}$ are connected implies that F' is full and faithful. It is also essentially surjective because each object of $\mathbb{B}$ is a coproduct of objects of $\mathbb{C}$. $\square$

Corollary. [10, 14] *Let $\mathbb{A}$ be a category. The following conditions are equivalent:*

1. $\mathbb{A}$ *is equivalent to a presheaf category;*
2. $\mathbb{A}$ *is exact, extensive, and has a small (regular) generator $\mathcal{G}$ such that each object in $\mathcal{G}$ is regular projective and connected.*

Proof. The implication $1 \Rightarrow 2$ is obvious. Conversely, consider the full subcategory $\mathbb{B}$ of $\mathbb{A}$ spanned by coproducts of objects in $\mathcal{G}$. $\mathbb{B}$ is a regular projective cover of $\mathbb{A}$. Moreover, by the previous lemma, $\mathbb{B} \simeq Fam\mathbb{C}$ for a small category $\mathbb{C}$. Finally, $\mathbb{A} \simeq \mathbb{B}_{ex} \simeq (Fam\mathbb{C})_{ex} \simeq Set^{\mathbb{C}^{op}}$. $\square$

Proposition. *Let $\mathbb{A}$ be a category. The following conditions are equivalent:*

1. $\mathbb{A}$ *is equivalent to a localization of a presheaf category;*
2. $\mathbb{A}$ *is exact, extensive, and has a small (regular) generator $\mathcal{G}$.*

Proof. The implication $1 \Rightarrow 2$ is obvious. Conversely, let $\mathcal{G}$ be the small generator and consider the *not full* subcategory $\mathbb{B}$ of $\mathbb{A}$ spanned by coproducts of objects of $\mathcal{G}$. An arrow $f \colon G \to \coprod_I G_i$ is in $\mathbb{B}$ exactly when it factors through an injection $G_i \to \coprod_I G_i$ (in other words, the objects of $\mathcal{G}$ are not connected in $\mathbb{A}$ because connectedness is not stable under localization, but we force them to be connected in $\mathbb{B}$). Now, by the previous lemma, $\mathbb{B} \simeq Fam\mathbb{C}$ for a small category $\mathbb{C}$, so that $\mathbb{B}_{ex}$ is a presheaf category. The main point to prove that $\mathbb{A}$ is a localization of $\mathbb{B}_{ex}$ is to check that the inclusion $\mathbb{B} \to \mathbb{A}$ is left covering, so that it extends to an exact functor $\mathbb{B}_{ex} \to \mathbb{A}$ (which plays the role of the left exact left adjoint). The fact that $\mathbb{B} \to \mathbb{A}$ is left covering comes from the next exercise. More details can be found in [50]. $\square$

Exercise. Consider a small category $\mathbb{C}$, an exact and extensive category $\mathbb{A}$, and a coproduct-preserving functor $F \colon Fam\mathbb{C} \to \mathbb{A}$. Show that if F is left covering with respect to binary weak products and weak equalizers of objects and arrows of $Fam\mathbb{C}$ coming from $\mathbb{C}$, then it is left covering with respect to all binary weak products and weak equalizers.

Remark. Another interesting aspect of the previous proof is that *exactly the same arguments* can be used to characterize (localizations of) algebraic categories and monadic categories over *Set* (compare with Chapters V and VI). One has to replace $Fam\mathbb{C}$ by the Kleisli category of the (finitary) monad, that is the full subcategory of free algebras, the connectedness condition by the condition to be abstractly finite, and extensivity by exactness of filtered colimits. This striking analogy is exploited in [50] to study *essential localizations* (a localization is essential if the reflector has a left adjoint).

References

[1] M. Barr, *On categories with effective unions,* Lecture Notes in Mathematics 1348, 19–35, Springer-Verlag (1988).

[2] M. Barr and C. Wells, *Toposes, triples and theories,* Grundlehren der Mathematischen Wissenschaften 278, Springer-Verlag (1985).

[3] J. L. Bell, *Toposes and local set theories. An introduction,* Oxford Logic Guides 14, Oxford University Press (1988).

[4] J. Bénabou, *Some remarks on 2-categorical algebra,* Bull. Soc. Math. Belg. Sér. A 41 (1989) 127–194.

[5] J. Bénabou, *Some geometric aspects of the calculus of fractions,* Appl. Categ. Structures 4 (1996) 139–165.

[6] F. Borceux, *Handbook of categorical algebra,* Encyclopedia of Mathematics and its Applications 50-51-52, Cambridge University Press (1994).

[7] F. Borceux and G. Janelidze, *Galois theories,* Cambridge Studies in Advanced Mathematics 72, Cambridge University Press (2001).

[8] A. Borel, *Cohomologie des espaces localement compacts d'après J. Leray,* Lecture Notes in Mathematics 2, Springer-Verlag (1964).

[9] D. G. Bourgin, *Modern algebraic topology,* The Macmillan Co. (1963).

[10] M. Bunge, *Categories of set valued functors,* Ph. D. Thesis, Univ. of Pennsylvania (1966).

[11] A. Carboni, S. Lack and R. F. C. Walters, *Introduction to extensive and distributive categories,* J. Pure Appl. Algebra 84 (1993) 145–158.

[12] A. Carboni and S. Mantovani, *An elementary characterization of categories of separated objects,* J. Pure Appl. Algebra 89 (1993) 63–92.

[13] A. Carboni and G. Rosolini, *Locally cartesian closed exact completions,* J. Pure Appl. Algebra 154 (2000) 103–116.

[14] A. Carboni and E. M. Vitale, *Regular and exact completions,* J. Pure Appl. Algebra 125 (1998) 79–116.

[15] M. Demazure and P. Gabriel, *Introduction to algebraic geometry and algebraic groups,* North-Holland Mathematics Studies 39, North-Holland Publishing Co. (1980).

[16] J. Dieudonné, *A history of algebraic and differential topology. 1900–1960,* Birkhauser Inc. (1989).

[17] D. Dikranjan and W. Tholen, *Categorical structure of closure operators. With applications to topology, algebra and discrete mathematics,* Mathematics and its Applications 346, Kluwer Academic Publishers (1995).

[18] G. Fischer, *Complex analytic geometry,* Lecture Notes in Mathematics 538, Springer-Verlag (1976).

[19] P. J. Freyd and A. Scedrov, *Categories, allegories,* North-Holland Mathematical Library 39, North-Holland (1990).

[20] P. Gabriel and M. Zisman, *Calculus of fractions and homotopy theory,* Ergebnisse der Mathematik und ihrer Grenzgebiete 35, Springer-Verlag (1967).

[21] J. Giraud, *Cohomologie non abélienne,* Die Grundlehren der mathematischen Wissenschaften 179, Springer-Verlag (1971).

[22] R. Godement, *Topologie algébrique et théorie des faisceaux,* Actualités Scientifiques et Industrielles 1252, Hermann (1973).

[23] J. S. Golan, *Structure sheaves over a noncommutative ring,* Lecture Notes in Pure and Applied Mathematics 56, Marcel Dekker Inc. (1980).

[24] R. Goldblatt, *Topoi. The categorial analysis of logic,* Studies in Logic and the Foundations of Mathematics 98, North-Holland Publishing Co. (1984).

[25] M. Gran and E. M. Vitale, *On the exact completion of the homotopy category,* Cah. Topologie Géom. Différ. Catégoriques 39 (1998) 287–297.

[26] J. W. Gray, *Fragments of the history of sheaf theory,* Lecture Notes in Mathematics 753, 1–79, Springer-Verlag (1979).

[27] A. Grothendieck, *Eléments de géométrie algébrique,* Inst. Hautes Etudes Sci. Publ. Math. 4 (1960) 8 (1961) 11 (1961) 28 (1966).

[28] M. Hakim, *Topos annelés et schémas relatifs,* Ergebnisse der Mathematik und ihrer Grenzgebiete 64, Springer-Verlag (1972).

[29] R. Hartshorne, *Algebraic geometry.,* Graduate Texts in Mathematics 52, Springer-Verlag (1977).

[30] F. Hirzebruch, *Topological methods in algebraic geometry,* Die Grundlehren der Mathematischen Wissenschaften 131, Springer-Verlag (1966)

[31] M. Hovey, *Model categories,* Mathematical Surveys and Monographs 63, American Mathematical Society (1999).

[32] H. Hu and W. Tholen, *Limits in free coproduct completions,* J. Pure Appl. Algebra 105 (1995) 277–291.

[33] H. Hu and W. Tholen, *A note on free regular and exact completions and their infinitary generalizations,* Theory and Appl. of Categories 2 (1996) 113–132.

[34] P. T. Johnstone, *Topos theory,* London Mathematical Society Monographs 10, Academic Press (1977).

[35] P. T. Johnstone, *Sketches of an elephant. A topos theory compendium,* Oxford Logic Guides, Oxford: Clarendon Press (2002).

[36] A. Kock, *Monads for which structures are adjoint to units,* J. Pure Appl. Algebra 104 (1995) 41–59.

[37] S. Lack and E. M. Vitale, *When do completion processes give rise to extensive categories?* J. Pure Appl. Algebra 159 (2001) 203–230.

[38] J. Lambek and P. J. Scott, *Introduction to higher order categorical logic,* Cambridge Studies in Advanced Mathematics 7, Cambridge University Press (1986).

[39] F. W. Lawvere and R. Rosebrugh, *Sets for mathematics,* Cambridge University Press (2001).

[40] S. Mac Lane and I. Moerdijk, *Sheaves in geometry and logic: a first introduction to topos theory,* Universitext, Springer-Verlag (1992).

[41] M. Makkai and G. E. Reyes, *First order categorical logic,* Lecture Notes in Mathematics 611, Springer-Verlag (1977).

[42] F. Marmolejo, *Distributive laws for pseudomonads,* Theory and Appl. of Categories 5 (1999) 91–147.

[43] M. Menni, *Exact completions and toposes,* Ph. D. Thesis, Univ. of Edinburgh (2000).

[44] M. Menni, *Closure operators in exact completions,* Theory and Appl. of Categories 8 (2001) 522–540.

[45] M. Menni, *A characterization of the left exact categories whose exact completions are toposes,* J. Pure Appl. Algebra 177 (2003) 287–301.

[46] R. Rosebrugh and R. J. Wood, *Cofibrations II: Left exact right actions and composition of gamuts,* J. Pure Appl. Algebra 39 (1986) 283–300.

[47] J. Rosický, *Cartesian closed exact completions,* J. Pure Appl. Algebra 142 (1999) 261–270.

[48] F. Van Oystaeyen and A. Verschoren, *Reflectors and localization. Application to sheaf theory,* Lecture Notes in Pure and Applied Mathematics 41, Marcel Dekker Inc. (1979).

[49] E. M. Vitale, *On the characterization of monadic categories over Set,* Cah. Topologie Géom. Différ. Catégoriques 35 (1994) 351–358.

[50] E. M. Vitale, *Essential localizations and infinitary exact completion,* Theory and Appl. of Categories 8 (2001) 465–480.

[51] C. A. Weibel, *An introduction to homological algebra,* Cambridge Studies in Advanced Mathematics 38, Cambridge University Press (1994).

[52] O. Wyler, *Lecture notes on topoi and quasitopoi,* World Scientific Publishing Co. (1991).

Département de Mathématique,
Université catholique de Louvain,
2, ch. du Cyclotron
B-1348 Louvain-la-Neuve, Belgium
E-mail: centazzo@math.ucl.ac.be
           vitale@math.ucl.ac.be

VIII
Beyond Barr Exactness:
Effective Descent Morphisms

George Janelidze, Manuela Sobral, and Walter Tholen

The general purpose of descent theory is to provide a unified treatment for various situations in algebra, geometry, and logic, where a problem on a certain *base* object B is first solved for an *extension* E of B, and then for B itself by "descending" from E along a *projection* $p : E \to B$. Of course, only "good" morphisms p will permit us to toss a "problem" back and forth between the two objects. More specifically, in Grothendieck's descent theory one asks which morphisms p allow for an algebraic description of structures over (the presumably "complicated" object) B in terms of structures over (the presumably "easier" object) E. The meaning of "structure" depends on the context, of course; for example, for a topological space X, a structure over X may be a sheaf over X (or a local homeomorphism with codomain X), and for a ring R a structure over R may be an R-module.

According to Grothendieck, the general setting for descent theory is given by a fibration $\Phi : \mathbf{D} \to \mathbf{C}$, so that the category of structures over the object B in $\mathbf{C}$ is given by the fibre $\Phi^{-1}(B)$, and descent theory then aims at a description of the fibre over B in terms of the fibre over E, with p given (see [8], [7]). In this chapter we restrict ourselves to considering the basic fibration of $\mathbf{C}$ (i.e., the codomain functor $\mathbf{C}^2 \to \mathbf{C}$), so the fibre of B is simply the comma category $(\mathbf{C} \downarrow B)$; hence, here "structure over B" simply means "morphism in $\mathbf{C}$ with codomain B". Our task then is to define and study *effective descent morphisms* $p : E \to B$ in $\mathbf{C}$ which, by definition, are those morphisms which facilitate an algebraic description of $(\mathbf{C} \downarrow B)$ by means of $(\mathbf{C} \downarrow E)$. For many categories $\mathbf{C}$, for example for all Barr-exact categories (see Chapter IV), this is easy: the effective descent morphisms are precisely the regular epimorphisms. For general categories it is our contention that the effective descent morphisms are the "good" regular epimorphisms, which explains the title of this chapter.

We give here a new elementary account of (basic-fibrational) descent theory, describing descent data for p in terms of discrete fibrations over the equivalence relation induced by p (Section 3), and we present a sheaf-theoretic characterization of effective descent morphisms (Section 4). Afterwards we discuss alternative and more general approaches to descent theory (Section 5), before turning to the characterization of effective descent morphisms in specific categories (Section 6).

Space does not allow us to discuss here the standard area of application of descent theory, which is cohomology theory. But we do indicate briefly three types of applications at a somewhat wider and more elementary level, in homological algebra, Galois theory and algebraic topology (Section 7).

Throughout this chapter **C** *denotes a fixed category with pullbacks.*

1. The world of epimorphisms

1.A Some special types of epimorphisms. Consider the following table of restricted notions of epimorphism and some of their examples in familiar algebraic categories:

A morphism $p : E \to B$ is said to be	vector spaces over a field	modules over a ring	groups	any variety of Ω-groups**	any variety of universal algebras
an *epimorph.* if $pu = pv$ always implies $u = v$	a product projection (up to iso)	a surjective homomorphism	a surjective homomorphism		
a *regular* epi if p is a coequalizer of some pair	as above	as above	as above	a surjective homomorphism	a surjective homomorphism
a *normal* epi if p is a cokernel* of some morphism	as above	as above	as above	as above	
a *split* epi if $ps = 1_B$ for some $s : E \to B$	as above	a product projection	a semidirect product projection	a semidirect product** projection	

Table 1

where (*), assuming **C** to be pointed, the cokernel of a morphism $f : F \to B$ is a coequalizer of f and the zero morphism $F \to B$, and (**) semidirect products of Ω-groups (i.e. Ω-algebras with an underlying group structure making the trivial group a subalgebra, as for rings, for instance) are to be defined categorically (see [5]); the empty cells in the table correspond to cases where no reasonable description (better than the categorical definition itself) is possible or known.

Both this table and the fact that all varieties of universal algebras are Barr-exact categories suggest that an extension E of B is to be defined categorically as a regular epimorphism $p : E \to B$. However, beyond the algebraic situations and Barr-exactness there are many other candidates for the notion of extension. For instance, if **C** is the opposite category of a variety of universal algebras, then the epimorphisms in **C**, i.e. the monomorphisms of algebras, are just injective homomorphisms, but the regular epimorphisms have no simple universal-algebraic description. And of course, many algebraists would say that the right notion of (co)extension in algebra is the one of monomorphism. On the other hand, in general the notion of regular epimorphism "splits" into at least four (!) important notions that are equivalent in the exact case.

First, an epimorphism $p : E \to B$ is *extremal* if $p = mq$ with a monomorphism m is possible only if m is an isomorphism. (We note that this extremality property already forces p to be an epimorphism, provided that every morphism in **C** factors into an epi followed by a mono, or if **C** has equalizers.) Next, an epimorphism p is *strong* if p is orthogonal to the class of monomorphisms, so that whenever $fp = mg$ with m mono, then $hp = g$ and $mh = f$ for a (necessarily) uniquely determined morphism h. Let us note immediately the following easy facts, the proofs of which we can leave as an exercise:

1.1. Proposition.

(a) *Every regular epimorphism is strong, and every strong epimorphism is extremal.*

(b) *If **C** has pullbacks, every extremal epimorphism is strong, and if in **C** every morphism factors into a regular epi followed by a mono, then every extremal epimorphism is regular.*

Under the general hypothesis on **C** (existence of pullbacks) there is therefore no further need to distinguish between extremal and strong epimorphisms. In a Barr-exact category, even in a regular category (see Chapter IV) they also coincide with the regular epimorphisms which, moreover, are stable under pullback. However, in general, we must distinguish between strong and regular epimorphisms, and we cannot assume pullback stability for any of these two types of epimorphisms. Hence, one calls a morphism p a pullback-stable strong (or simply a *stably strong*) epimorphism if every pullback of p is strong epi; likewise one has the notion of *stably regular* epimorphism. The following chart illustrates the situation for **C** = **Top** (the category of topological spaces) and for **C** = **Cat** (the category of (small) categories), displaying characterizations of special epimorphisms in both categories[1].

We observe that in many *geometrical* categories that are far from being exact (or even regular), including the category of topological spaces, the extremal epimorphisms coincide with the regular ones. On the other hand, despite the fact

---

[1] Epimorphisms in **Cat** were characterized by Isbell [9]. Their failure to be pullback stable had already been observed by Giraud [7], likewise for regular epimorphisms in **Cat**. Although regular epimorphisms were described only recently in [2], their characterization belongs to the "folklore" knowledge; for example, an early source on congruence relations on categories is [16].

of being "almost" an *algebraic* category, we see that **Cat** nevertheless provides a good environment for examining various notions of epimorphism, an observation made by Giraud [7] more than 35 years ago. Here are two simple examples, for illustration:

class of morphisms in **C**	**C = Top**	**C = Cat**
strong epimorphisms $p : E \to B$	quotient maps, i.e., surjective morphisms p such that V is open in B whenever $p^{-1}(V)$ is open in E	functors p such that every morphism in B is a composite of finitely many morphisms in the image of p
stably strong epimorphisms $p : E \to B$	bi-quotient maps (see [17], [6]), i.e., for every open covering (U_i) of the fibre $p^{-1}b$ for a point b in B, the union of finitely many sets $p(U_i)$ form a neighborhood of b	functors surjective on morphisms
regular epimorphisms $p : E \to B$	quotient maps	those strong epis for which equations of finite composites as above are determined by equations of single morphisms only
stably regular epimorphisms	bi-quotient maps	functors surjective on composable pairs of morphisms

Table 2

1.2. Example. Let $\mathbf{2} = \{0 \to 1\}$ and $\mathbf{3} = \{0 \to 1 \to 2\}$ be the categories arising from the ordered sets 2 and 3, respectively. Then the functor $\mathbf{2} + \mathbf{2} \to \mathbf{3}$ which identifies the object 1 of the first copy of $\mathbf{2}$ with the object 0 of the second copy of $\mathbf{2}$ is a regular epimorphism, but not stably so. The quotient functor of $\mathbf{2}$ which identifies its two objects and makes the only non-identical arrow idempotent, is a stably strong epimorphism, but it is not regular.

1.B Pullback stability. Failure of regular epimorphisms in **Cat** to be pullback-stable goes along with their failure to be closed under composition, as we shall see next.

1.3. Proposition. *A composite pq of regular epimorphisms in* **C** *is regular whenever q is a pullback stable epimorphism. In particular, the class of stably regular epimorphisms in* **C** *is closed under composition.*

Proof. Given regular epimorphisms $p : E \to B$ and $q : F \to E$, consider the diagram

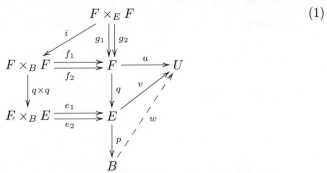 (1)

where

- (e_1, e_2), (f_1, f_2), (g_1, g_2) are the kernel pairs of p, pq, q, respectively;
- i the canonical morphism between pullbacks;
- u any morphism with $uf_1 = uf_2$;
- v the unique morphism with $vq = u$, which does exist since q is a regular epimorphism and $ug_1 = uf_1i = uf_2i = ug_2$;
- w the unique morphism with $wp = v$, whose existence is still to be proved in order to conclude that pq is a regular epimorphism (since pq is an epimorphism, we do not need to prove the uniqueness of w).

Since p is a regular epimorphism, the existence of w will follow from $ve_1 = ve_2$, and since $ve_1(q \times q) = vqf_1 = vqf_2 = ve_2(q \times q)$, we only have to show that $q \times q$ is an epimorphism. However, $q \times q$ is the composite of the morphisms $q \times 1 : F \times_B F \to E \times_B F$ and $1 \times q : E \times_B F \to E \times_B E$ each of which is an epimorphism since q was required to be a pullback-stable epimorphism. □

1.4. Corollary. *For the conditions given below, one always has the implications* (i)$\Rightarrow$ (ii)$\Rightarrow$ (iii), *and all three are equivalent if* **C** *has coequalizers of kernel pairs:*

   (i) *every morphism factors into a regular epi followed by a mono;*
   (ii) *every extremal epimorphism is regular;*
   (iii) *the class of regular epimorphisms is closed under composition.*

Proof. (i)$\Rightarrow$ (ii) is trivial. (ii)$\Rightarrow$ (iii): By Prop. 1.1(a), under hypothesis (ii) the class of regular epimorphisms coincides with the class of strong epimorphisms, which is always closed under composition. (iii)$\Rightarrow$ (i): Consider an extremal epimorphism $p : E \to B$ and, as in 1.3, form its kernel pair (e_1, e_2), but take v to be its coequalizer and $m : U \to B$ the morphism $mv = p$. Furthermore, form the kernel pair (h_1, h_2) of m and its coequalizer w, and let $t : W \to B$ be the morphism with $tw = m$. Then (e_1, e_2) is easily seen to be the kernel pair of wv. Hence, being a regular epi by hypothesis (iii), wv is like v a coequalizer of (e_1, e_2), whence w is an isomorphism. Consequently, $h_1 = h_2$, and m must be a monomorphism. By hypothesis on p then, m is actually an iso, so that regularity of p follows from regularity of v. $\qquad\square$

1.5. Proposition. *If a composite pq of morphisms in $\mathbf{C}$ is a regular epimorphism, then so is the morphism p, provided that any of the following conditions holds:*

(a) *q is an epimorphism;*
(b) *pq is a pullback stable epimorphism.*

In particular the class of stably regular epimorphisms has the strong right cancellation property, i.e., it contains p whenever it contains pq.

Proof. Under condition (a), consider the diagram

$$
\begin{array}{ccc}
F \times_B F & \overset{f_1}{\underset{f_2}{\rightrightarrows}} F \overset{u}{\longrightarrow} U & \\
\Big\downarrow{\scriptstyle q\times q} & \Big\downarrow{\scriptstyle q}\ {\scriptstyle v} & \\
E \times_B E & \overset{e_1}{\underset{e_2}{\rightrightarrows}} E\ \ \ w & \\
& \Big\downarrow{\scriptstyle p} & \\
& B &
\end{array}
\tag{2}
$$

where:

- (e_1, e_2) and (f_1, f_2) are the kernel pairs of p and pq respectively;
- v is any morphism with $ve_1 = ve_2$;
- $u = vq$;
- w is the unique morphism with $wpq = u$, which does exist since pq is a regular epimorphism and $uf_1 = vqf_1 = ve_1(q \times q) = vqf_2 = uf_2$.

Since p is an epimorphism (because pq is one), we only need to prove that $wp = v$, but this follows from $wpq = u$ and the fact that q is an epimorphism.

Suppose now that instead of (a) we have condition (b), and consider the pullback

$$
\begin{array}{ccc}
& \overset{\pi_2}{\longrightarrow} & \\
{\scriptstyle \pi_1}\Big\downarrow & & \Big\downarrow{\scriptstyle pq} \\
& \underset{p}{\longrightarrow} &
\end{array}
\tag{3}
$$

We then have:

- since π_2 can be obtained as a pullback of the pullback of p along p, it is a split epimorphism and therefore a stably regular epimorphism;
- after that, since pq is a regular epimorphism, Proposition 1.3 tells us that $p\pi_1 = pq\pi_2$ is a regular epimorphism;
- since pq is a pullback-stable epimorphism, π_1 is an epimorphism;
- since π_1 is an epimorphism and $p\pi_1$ a regular epimorphism, the first part of the proof tells us that p is an epimorphism — as desired.

$\qquad\square$

We note that if every morphism in **C** has a (regular epi, mono)-factorization, then with pq also p is a regular epimorphism. More refinedly, the validity of the implication

(+) p split epi, q regular epi $\Rightarrow pq$ regular epi

for all composable pairs p, q implies the validity of the implication

(++) pq regular epi $\Rightarrow p$ regular epi

for all p, q (see [3]). Nothing seems to be known about the converse relation between (+) and (++), other than that (++) certainly does not hold true without additional conditions; for a counter-example, see [13].

1.C When pulling back is conservative. In what follows, an important technical tool will be to consider a morphism $p : E \to B$ in **C** as a morphism $p : (E, p) \to (B, 1_B)$ in $(\mathbf{C} \downarrow B)$. Clearly, when p is a (regular) epimorphism in **C**, it is also a (regular) epimorphism in $(\mathbf{C} \downarrow B)$, but generally not conversely. However, if p is an extremal epimorphism in $(\mathbf{C} \downarrow B)$, it is also one in **C**, provided that **C** has equalizers, as one easily verifies; consequently, if furthermore extremal epimorphisms are regular (see Cor. 1.4), there is no need to distinguish between p being regular epi in **C** or in $(\mathbf{C} \downarrow B)$. The question that concerns us most in the general situation is: when is p a stably extremal epimorphism in $(\mathbf{C} \downarrow B)$?

1.6. Proposition. *The following conditions on a morphism $p : E \to B$ in **C** are equivalent:*

(a) *p is a stably strong epimorphism in $(\mathbf{C} \downarrow B)$;*
(b) *p is a stably extremal epimorphism $(\mathbf{C} \downarrow B)$;*
(c) *the pullback functor $p^* : (\mathbf{C} \downarrow B) \to (\mathbf{C} \downarrow E)$ is conservative, i.e. reflects isomorphisms;*
(d) *if*

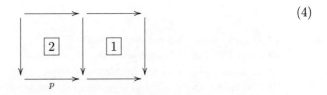

(4)

*is a commutative diagram in **C**, in which the squares $\boxed{2}$ and $\boxed{2} + \boxed{1}$ are pullbacks, then the square $\boxed{1}$ also is a pullback.*

Proof. (a) $\Leftrightarrow$ (b) follows from Proposition 1.1(b) since with **C** also $(\mathbf{C} \downarrow B)$ has pullbacks.

(c) $\Leftrightarrow$ (d): To give a morphism $m : (A, \alpha) \to (A', \alpha')$ in $(\mathbf{C} \downarrow B)$, for which $p^*(m) : p^*(A, \alpha) \to p^*(A', \alpha')$ is an isomorphism, is the same as to give a commutative diagram

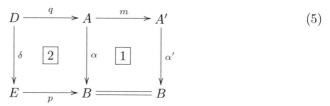

$$(5)$$

in which the squares $\boxed{2}$ and $\boxed{2} + \boxed{1}$ are pullbacks. Since in such a diagram, the square $\boxed{1}$ is a pullback if and only if m is an isomorphism, the implication (d) $\Rightarrow$ (c) is obvious. For the converse, just observe that the situation (4) can be reduced to the situation (5) by pulling back the right hand vertical arrow along the right hand bottom arrow.

(b) $\Rightarrow$ (c): We have to prove that if the squares $\boxed{2}$ and $\boxed{2} + \boxed{1}$ in (5) are pullbacks and p is a stably extremal epimorphism in $(\mathbf{C} \downarrow B)$, then m is an isomorphism. However, since mq, as a pullback of p, is an extremal epimorphism, it suffices to prove that m is a monomorphism (in $(\mathbf{C} \downarrow B)$, or equivalently in $\mathbf{C}$). For that, consider the diagram

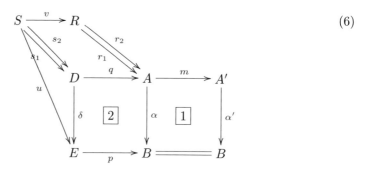

$$(6)$$

where:

- (r_1, r_2) is the kernel pair of m;
- the object S together with the morphisms u and v is constructed as the pullback of p and $\alpha' r_1 = \alpha' m r_1 = \alpha r_2$;
- the morphisms s_1 and s_2 are induced by the pairs (u, r_1) and (u, r_2) respectively.

We have:

- $s_1 = s_2$, since $m r_1 v = m r_2 v$ and the square $\boxed{2} + \boxed{1}$ is a pullback;
- v is an epimorphism, since it is a pullback of p;
- $r_1 v = q s_1 = q s_2 = r_2 v$, and since v is an epimorphism, this implies $r_1 = r_2$.

Hence, m is a monomorphism. (c)$\Rightarrow$ (b): We observe that every time a pullback p' of p factors as mq with m a monomorphism m, we obtain the situation (5), where the square $\boxed{2}$ is a pullback *because* so is $\boxed{2}+\boxed{1}$ and m is a monomorphism. Hence, such an m must be an isomorphism by (c). Since $(\mathbf{C} \downarrow B)$, as a category with pullbacks and a terminal object, has equalizers, the extremality property forces p' to be an epimorphism, in fact an extremal one, in $(\mathbf{C} \downarrow B)$. $\qquad\square$

2. Generalization of the kernel-cokernel correspondence

2.A Regular epimorphisms in $\mathbf{C}^2$. The usual kernel-cokernel correspondence in a pointed category $\mathbf{C}$ can be presented as a category equivalence

$$\text{(normal monomorphisms in } \mathbf{C}) \underset{\text{kernel}}{\overset{\text{cokernel}}{\rightleftarrows}} \text{(normal epimorphisms in } \mathbf{C}), \quad (7)$$

and the familiar non-pointed version of this would then be

$$\text{(effective equivalence relations in } \mathbf{C}) \underset{\text{kernel pair}}{\overset{\text{coequalizer}}{\rightleftarrows}} \text{(regular epimorphisms in } \mathbf{C})$$
$$(8)$$

Since we are not assuming the existence of coequalizers in $\mathbf{C}$, the *effective equivalence relations* are to be defined as those pairs, which admit coequalizers and are kernel pairs of their coequalizers.

The question we are going to consider in this section can be briefly displayed by:

$$\text{(effective equivalence relations in } \mathbf{C}^2) \underset{\text{kernel pair}}{\overset{\text{coequalizer}}{\rightleftarrows}} \text{(regular epimorphisms in } \mathbf{C}^2)$$

$$? \qquad\qquad \rightleftarrows \qquad \text{(regular pullback squares in } \mathbf{C})$$
$$(9)$$

where $\mathbf{C}^2$ denotes the arrow category of $\mathbf{C}$, and where by *regular pullback squares* we mean regular epimorphisms in $\mathbf{C}^2$ that are pullback squares in $\mathbf{C}$. That is, starting from a diagram of the form

$$\begin{array}{ccccc}
R' & \underset{r'_2}{\overset{r'_1}{\rightrightarrows}} & E' & \overset{p'}{\longrightarrow} & B' \\
\rho \downarrow & \boxed{2} & \varepsilon \downarrow & \boxed{1} & \downarrow \beta \\
R & \underset{r_2}{\overset{r_1}{\rightrightarrows}} & E & \underset{p}{\longrightarrow} & B
\end{array} \qquad (10)$$

that is *exact* (in the sense that the squares $\boxed{2}$ and $\boxed{1}$ correspond to each other under the top equivalence of (9)), we are interested in finding the conditions on the square $\boxed{2}$, under which the square $\boxed{1}$ is a pullback. Any such square $\boxed{2}$ is a discrete fibration of equivalence relations (see Chapter 4), but it turns out that not every discrete fibration of equivalence relations is of this form — even if we require its rows to be effective equivalence relations and the category **C** to admit all coequalizers (as easily follows from Remark 6.3(d) below). Accordingly we will use the expression *effective discrete fibration* (of equivalence relations), and rewrite the bottom category equivalence in (9) as

$$(\text{effective discrete fibrations in } \mathbf{C}) \rightleftarrows (\text{regular pullback squares in } \mathbf{C}) \tag{11}$$

2.1. Lemma. *Suppose (10) is a (not necessarily exact) diagram in $\mathbf{C}^2$, which is a fork in $\mathbf{C}^2$. Then it is a coequalizer diagram in $\mathbf{C}^2$ if and only if the top fork is a coequalizer diagram in $(\mathbf{C} \downarrow B)$ (regarding R', E', and B' as objects in $(\mathbf{C} \downarrow B)$ via $pr_1\rho = pr_2\rho = p\varepsilon r'_1 = p\varepsilon r'_2 = \beta p'r'_1 = \beta p'r'_2$, $p\varepsilon = \beta p'$, and β respectively) and the bottom one is a coequalizer diagram in $\mathbf{C}$.*

Proof. "if": We have to show the existence and uniqueness of the dotted arrows $B' \to C'$ and $B \to C$ in the diagram

$$\tag{12}$$

where the "new" solid arrows q, q', γ form a fork in $\mathbf{C}^2$ together with the left hand square, and the dotted arrows are required to make the right hand side of the diagram commute. Since p is the coequalizer of (r_1, r_2), the bottom dotted arrow making the bottom triangle commute is uniquely determined, and we can pull back γ along it. This reduces our problem to the case $(C,q)=(B,p)$, in which it becomes trivial since the top fork is a coequalizer diagram in $(\mathbf{C} \downarrow B)$.

For the two assertions of "only if" just use

$$\tag{13}$$

and

$$
\begin{array}{ccc}
R' \underset{r'_2}{\overset{r'_1}{\rightrightarrows}} E' & \xrightarrow{\ p'\ } B' \\
\end{array}
\tag{14}
$$

respectively. □

2.2. Corollary. *A commutative square*

$$
\begin{array}{ccc}
E' & \xrightarrow{\ p'\ } & B' \\
{\scriptstyle \varepsilon}\downarrow & & \downarrow{\scriptstyle \beta} \\
E & \xrightarrow{\ p\ } & B
\end{array}
\tag{15}
$$

is a regular epimorphism in $\mathbf{C}^2$ if and only if p' is a regular epimorphism in $(\mathbf{C} \downarrow B)$ and p is a regular epimorphism in $\mathbf{C}$. In particular, a pullback square (15) is regular if and only if p' and p satisfy these conditions. □

That is, the regular pullbacks have a simple description "inside $\mathbf{C}$", i.e. without referring to coequalizers in $\mathbf{C}^2$. The situation with effective discrete fibrations is more complicated, as we shall see next.

2.B Stably effective equivalence relations.

2.3. Definition. An equivalence relation is said to be *stably effective*, if it is effective and its coequalizer is a stably regular epimorphism.

2.4. Lemma. *Let*

$$
\begin{array}{ccc}
R' & \underset{r'_2}{\overset{r'_1}{\rightrightarrows}} & E' \\
{\scriptstyle \rho}\downarrow & & \downarrow{\scriptstyle \varepsilon} \\
R & \underset{r_2}{\overset{r_1}{\rightrightarrows}} & E
\end{array}
\tag{16}
$$

be a discrete fibration of equivalence relations (r_1, r_2) and (r'_1, r'_2), in which (r_1, r_2) is effective, and let $p : E \to B$ be the coequalizer of (r_1, r_2). If (r'_1, r'_2) is stably effective in $(\mathbf{C} \downarrow B)$ (regarding R' and E' as objects in $(\mathbf{C} \downarrow B)$ as in Lemma 2.1), then (16) is an effective discrete fibration.

Proof. Consider the diagram

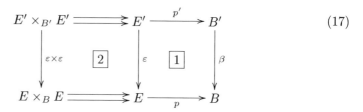

$$(17)$$

in which:

- $p : E \to B$ is the coequalizer of (r_1, r_2), as above;
- $E \times_B E \rightrightarrows E$ is the kernel pair of p, which we identify with the bottom row in (16);
- $\varepsilon : E' \to E$ is as in (16);
- $p' : (E', p\varepsilon) \to (B', \beta)$ is the coequalizer of (r'_1, r'_2) in $(\mathbf{C} \downarrow B)$;
- $E' \times_{B'} E' \rightrightarrows E'$ is the kernel pair of p'.

Since kernel pairs in $(\mathbf{C} \downarrow B)$ are formed as in $\mathbf{C}$, the square $\boxed{2}$ in (17) can be identify with diagram (16). On the other hand, since kernel pairs in $\mathbf{C}^2$ are formed "componentwise", the square $\boxed{2}$ in (17) can be considered as the kernel pair of the morphism (p', p) in $\mathbf{C}^2$. Therefore Corollary 2.2 tells us that we only need to show that the square $\boxed{1}$ in (17) is a pullback. For that, consider the diagram

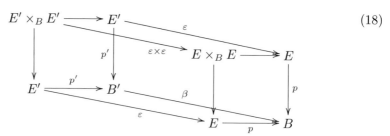

$$(18)$$

where the unnamed arrows are pullback projections. Since three out of its four vertical faces are pullbacks and p' is a stably regular epimorphism, the fourth vertical face, which is the square $\boxed{1}$ in (17), is a pullback by Proposition 1.1(a) and Proposition 1.6(a)⇔(d). (In fact, p' is a stably regular epimorphism in $(\mathbf{C} \downarrow B)$ — not in $\mathbf{C}$, but this does not cause any problem since the diagram above can also be considered as a diagram in $(\mathbf{C} \downarrow B)$.) $\square$

2.5. Corollary.

(a) *Suppose the discrete fibration* (16) *of equivalence relations has the property that* (r_1, r_2) *is effective in* $\mathbf{C}$ *and stably effective in* $(\mathbf{C} \downarrow B)$, *where B is as in Lemma 2.4. Then* (16) *is an effective discrete fibration if and only if* (r'_1, r'_2) *is stably effective in* $(\mathbf{C} \downarrow B)$.

(b) *If a discrete fibration* (16) *of equivalence relations has the property that* (r_1, r_2) *is stably effective in* **C**, *then it is an effective discrete fibration if and only if* (r_1', r_2') *is stably effective in* **C**. □

2.C The generalized correspondence. The Corollary suggests a surprising conclusion: although we have no simple description of the equivalence (11), we would have one (with a restricted equivalence) if we were using all the time stably regular epimorphisms instead of all regular epimorphisms. Indeed, from the previous arguments we easily obtain:

2.6. Theorem. *Let* **StEffDiscFib** *be the category of stably effective equivalence relations in* $\mathbf{C}^2$ *that are discrete fibrations of equivalence relations in* **C**, *and* **StRegPb** *be the category of stably regular epimorphisms in* $\mathbf{C}^2$ *that are pullback diagrams in* **C**. *Then:*

(a) *There are category equivalences*

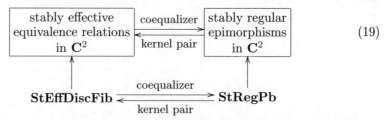

$\hspace{7cm}$ (19)

 i.e., the (obvious) top equivalence induces the equivalence between **StEffDiscFib** *and* **StRegPb**.

(b) *A discrete fibration of equivalence relations* (r_1, r_2) *and* (r_1', r_2') *is in* **StEffDiscFib** *if and only if* (r_1, r_2) *and* (r_1', r_2') *are stably effective in* **C**.

(c) *A pullback diagram* (2.6) *is in* **StRegPb** *if and only if* p *is a stably regular epimorphism in* **C** *or, equivalently, both* p *and* p' *are stably regular. In other words the category* **StRegPb** *can equivalently be described as the category of diagrams in* **C** *of the form* $\to^{\downarrow}$, *where the horizontal arrow is a stably regular epimorphism.*

Proof. We need to prove that for what we called an exact diagram (10), the following conditions are equivalent:

(i) the square $\boxed{2}$ is at the same time a stably effective equivalence relation in $\mathbf{C}^2$ and a discrete fibration in **C** ;

(ii) the square $\boxed{2}$ is a discrete fibration of stably effective equivalence relations in **C**;

(iii) the square $\boxed{1}$ is at the same time a stably regular epimorphism in $\mathbf{C}^2$ and a pullback diagram in **C**;

(iv) the square $\boxed{1}$ is a pullback diagram, and p is a stably regular epimorphism in **C**.

Then (a) follows from (i)⇔(iii), (b) from (i)⇔(ii), and (c) from (iii)⇔(iv). Since all pullbacks and coequalizers in $\mathbf{C}^2$ that exist componentwise actually *are* formed componentwise, the implications (ii)⇒(i) and (iv)⇒(iii) are obvious. The implications (iii)⇒(i) and (iv)⇒(ii) follow from the fact that whenever $\boxed{1}$ is a pullback, $\boxed{2}$ is a discrete fibration (although for (iv)⇒(ii) we also need to observe again that kernel pairs in $\mathbf{C}^2$ are formed componentwise). Therefore it suffices to prove (i)⇒(iv):

Starting from an arbitrary commutative diagram of the form

$$
\begin{array}{ccc}
C' & \xrightarrow{\ f'\ } & B' \\
{\scriptstyle \gamma}\downarrow & & \downarrow{\scriptstyle \beta} \\
C & \xrightarrow[\ f\ ]{} & B
\end{array}
\tag{20}
$$

consider the pullback diagram

$$
\begin{array}{ccc}
F' & \xrightarrow{\ q'\ } & C' \\
{\scriptstyle \varphi}\downarrow & & \downarrow{\scriptstyle \gamma} \\
F & \xrightarrow[\ q\ ]{} & C
\end{array}
\tag{21}
$$

in which the morphism (q', q) from ϕ to γ in $\mathbf{C}^2$ is obtained by pulling back (p', p) along (f', f) (so that q and q' are the pullbacks in $\mathbf{C}$ of p and p' along f and f', respectively). Since the square $\boxed{2}$ in (10) is now required to be a stably effective equivalence relation in $\mathbf{C}^2$, (q', q) is a stably regular epimorphism in $\mathbf{C}^2$, hence q' is a regular epimorphism in $(\mathbf{C} \downarrow B)$ and q is a regular epimorphism in $\mathbf{C}$, by the first assertion of Corollary 2.2. We need this observation twice:

- when f is an arbitrary morphism to B and diagram (20) is a pullback — in order to conclude that p is a pullback stable regular epimorphism in $\mathbf{C}$;
- when $f = 1_B$, f' is an arbitrary morphism to B', and $\gamma = \beta f'$ — in order to conclude that p' is a pullback stable regular epimorphism in $(\mathbf{C} \downarrow B)$.

Now (iv) follows from Lemma 2.4. □

2.7. Remark. If our category $\mathbf{C}$ is regular (or, more generally, if all regular epimorphisms in $\mathbf{C}$ are pullback stable), then diagrams (9) and (19) of course coincide, and the category **StEffDiscFib** becomes the category of discrete fibrations of effective equivalences relations in $\mathbf{C}$. □

3. Elementary descent theory

3.A Descent via discrete fibrations. For a morphism $p : E \to B$ in **C**, consider the diagram

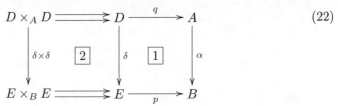

$$ \quad (22) $$

where:

- (A, α) is an arbitrary object in $(\mathbf{C} \downarrow B)$ (equivalently, $\alpha : A \to B$ is an arbitrary morphism with codomain B);
- the square $\boxed{1}$ is a pullback;
- the square $\boxed{2}$ is the kernel pair of (q, p) in the category $\mathbf{C}^2$.

Let

$$ K^p : (\mathbf{C} \downarrow B) \to \mathbf{DiscFib}(Eq(p)) \qquad (23) $$

be the functor determined by the assignment $(A, \alpha) \mapsto \boxed{2}$ (see (22)), where $Eq(p)$ is the equivalence relation $E \times_B E \rightrightarrows E$ (i.e., the kernel pair of of p), and $\mathbf{DiscFib}(Eq(p))$ is the category of discrete fibrations over $Eq(p)$. Since the pullback square $\boxed{1}$ is determined by α (up to an isomorphism, for a fixed p), and since the squares $\boxed{2}$ and $\boxed{1}$ correspond to each other under the equivalence described in Theorem 2.6 when p is a stably regular epimorphism, we obtain:

3.1. Lemma. *Suppose that $p : E \to B$ is a stably regular epimorphism. Then:*

(a) *the functor (23) induces an equivalence between $(\mathbf{C} \downarrow B)$ and the category of all discrete fibrations (F, φ) over $Eq(p)$, in which the equivalence relation F is stably effective; therefore*

(b) *the functor (23) is itself an equivalence if and only if for every discrete fibration (F, ϕ) over $Eq(p)$, the equivalence relation F is stably effective.* $\qquad \square$

3.2. Remark. We could also choose any morphism $\gamma : B \to C$ and apply Lemma 3.1 to p regarded as a morphism from $(E, \gamma p)$ to (B, γ) in $(\mathbf{C} \downarrow C)$. Since $((\mathbf{C} \downarrow C) \downarrow (B, \gamma))$ can be identified with $(\mathbf{C} \downarrow B)$, and since all constructions above remain unchanged (in the same manner), we can make our requirement on p slightly weaker: p must be a stably regular epimorphism just in $(\mathbf{C} \downarrow C)$. Moreover, it is easy to see that this does not depend on **C** and γ. Indeed, if p is a stably regular epimorphism in $(\mathbf{C} \downarrow C)$, then p clearly has the same property in $(\mathbf{C} \downarrow B)$ (regarded as a morphism from (E, γ) to $(B, 1_B)$), and the converse can easily be shown by pulling back along γ. That is, we should simply take $C = B, \gamma = 1_B$, and require p to be a stably regular epimorphism in $(\mathbf{C} \downarrow B)$. $\qquad \square$

On the other hand, we have:

3.3. Lemma. *If the functor* (23) *is full and faithful, then p is a stably regular epimorphism in* $(\mathbf{C} \downarrow B)$.

Proof. We consider an arbitrary object (A, α) in $(\mathbf{C} \downarrow B)$, construct diagram (22), and we have to prove that whenever the functor (23) is full and faithful, for every object (A', α') in $(\mathbf{C} \downarrow B)$ and every morphism $f : (D, \delta) \to (A', \alpha')$ which has the same composites with the two projections $D \times_A D \to D$, there exists a unique morphism $g : (A, \alpha) \to (A', \alpha')$ with $f = gq$. The morphism f determines a morphism from (D, δ) to the pullback (D', δ') of (A', α') along p, and since that morphism also has the same composites with the two projections $D \times_A D \to D$, this gives a morphism $K^p(A, \alpha) \to K^p(A', \alpha')$ and hence the desired morphism $(A, \alpha) \to (A', \alpha')$. □

From Lemma 3.1, Remark 3.2, and Lemma 3.3, we obtain:

3.4. Theorem. *Let $p : E \to B$ be a morphism in $\mathbf{C}$. Then the functor K^p : $(\mathbf{C} \downarrow B) \to \mathbf{DiscFib}(Eq(p))$ is*

(a) *full and faithful if and only if p is a stably regular epimorphism in $(\mathbf{C} \downarrow B)$;*

(b) *a category equivalence if and only if p is a regular epimorphism in $(\mathbf{C} \downarrow B)$ and, for every discrete fibration (F, ϕ) over $Eq(p)$, the equivalence relation F is stably effective in $(\mathbf{C} \downarrow B)$.* □

3.5. Remark. Since the functor U^p in (24) obviously reflects isomorphisms, the functor K^p has the same property if and only if p^* has it. According to Proposition 1.6 this happens if and only if p is a stably extremal (=strong) epimorphism in $(\mathbf{C} \downarrow B)$. Therefore Table 2 of Section 1 shows the difference between the reflection of isomorphisms and being full and faithful for K^p in case $\mathbf{C} = \mathbf{Cat}$: these two conditions correspond to surjectivity of p on morphisms, and on composable pairs (of morphisms), respectively. However, the most interesting observation concerning that case is yet to come: it turns out that K^p is a category equivalence if and only if p is surjective on composable triples — see Proposition 6.2 below. □

We are now able to describe what we call *elementary approach to ("global")* *descent theory.* To this end consider the commutative diagram

$$(\mathbf{C} \downarrow B) \xrightarrow{\quad K^p \quad} \mathbf{DiscFib}(Eq(p)) \qquad (24)$$

with U^p the forgetful functor. The diagram suggests to

- look at objects of $\mathbf{DiscFib}(Eq(p))$ as objects of $(\mathbf{C} \downarrow E)$ equipped with a certain structure, and to
- require K^p to be an equivalence, or at least full and faithful, and then to look at objects of $(\mathbf{C} \downarrow B)$ in the same way.

Thus, according to this suggestion, which is (an elementary reformulation of) one of the great mathematical discoveries of Alexander Grothendieck, the objects (A, α) of $(\mathbf{C} \downarrow B)$ are to be considered as structures on their pullbacks $p^*(A, \alpha)$, which are "intuitively bigger"! In other words, considering the morphism $p : E \to B$ as an extension of B, we wish to study the objects of $(\mathbf{C} \downarrow B)$ by "extending" them from B to E, and then by "descending along p".

The expression "global" used above refers to the fact that we deal with the categories $(\mathbf{C} \downarrow B)$ and $(\mathbf{C} \downarrow E)$ themselves, rather than with their subcategories formed by a specified class of "good" morphisms in $\mathbf{C}$. In fact, Grothendieck's original idea was to use what he called a fibration of categories, which reduces to our present context when the fibration $\Phi : \mathbf{D} \to \mathbf{C}$ in question is the codomain functor $\mathbf{C}^2 \to \mathbf{C}$. Since the fibration $\mathbf{C}^2 \to \mathbf{C}$ is in fact quite "visible" in Section 2, the reader familiar with Grothendieck's fibrations will have no difficulty to extend our arguments to the general case.

Using again our present context, we reformulate Grothendieck's definitions as follows:

3.6. Definition. Let $p : E \to B$ be a morphism in $\mathbf{C}$. The category $\mathbf{DiscFib}(Eq(p))$ is called the category of *descent data* for p, and is denoted by $\mathbf{Des}(p)$. The morphism p is said to be:

(a) a *descent morphism* if the functor $K^p : (\mathbf{C} \downarrow B) \to \mathbf{Des}(p)$ is full and faithful;

(b) an *effective descent morphism* if the functor $K^p : (\mathbf{C} \downarrow B) \to \mathbf{Des}(p)$ is a category equivalence. $\qquad\qquad\square$

Theorem 3.4 gives a general characterization of descent morphisms and of effective descent morphisms. Let us repeat the second part of it together with some obvious simplified reformulations in some important special cases:

3.B Effective descent in special types of categories.

3.7. Theorem. *Each of the following conditions is necessary and sufficient for a morphism* $p : E \to B$ *in* **C** *to be an effective descent morphism, under the additional assumptions given on* **C** *in each case:*

(a) p *is a regular epimorphism in* $(\mathbf{C} \downarrow B)$, *and for every discrete fibration* (F, φ) *over* $Eq(p)$, *the equivalence relation* F *is stably effective in* $(\mathbf{C} \downarrow B)$ *– in the general case;*

(b) p *is a regular epimorphism, and for every discrete fibration* (F, φ) *over* $Eq(p)$, *the equivalence relation* F *is stably effective – if* **C** *has a terminal object;*

(c) p *is a regular epimorphism, and for every discrete fibration* (F, φ) *over* $Eq(p)$, *the equivalence relation* F *is effective – if* **C** *is a regular category;*

(d) p *is a regular epimorphism – if* **C** *is a Barr-exact category;*

(e) p *is a normal epimorphism – if* **C** *is a semi-abelian category (in the sense of* [10]*);*

(f) p *is an epimorphism – if* **C** *is either an abelian category, or a (pre)topos.* ☐

In addition to the Theorem the following obvious corollaries (the first of which also uses Remark 3.2) will be useful:

3.8. Corollary. *For any morphism* $\gamma : B \to C$, *a morphism* $p : E \to B$ *is an (effective) descent morphism in* **C** *if and only if it is an (effective) descent morphism from* $(E, \gamma p)$ *to* (B, γ) *in* $(\mathbf{C} \downarrow C)$. ☐

3.9. Corollary. *Let* **D** *be a category with pullbacks, and* **C** *a full subcategory in* **D** *closed under pullbacks. Then for a morphism* $p : E \to B$ *in* **C**, *which is an effective descent morphism in* **D**, *the following conditions are equivalent:*

(a) $p : E \to B$ *an effective descent morphism in* **C**;

(b) *for every pullback diagram of the form*

(25)

in **D**, *if* D *is in* **C**, *then so is* A. ☐

3.10. Remark. As already indicated by the examples given in Table 2 of the Introduction, in many categories of interest for descent theory the convenient additional hypotheses given in 3.7 are not satisfied. A prominent classical example in this regard is the opposite of the category of commutative unital rings. From V.2.9 one can derive an important result of A. Joyal and M. Tierney, namely that the effective descent morphisms in that category are precisely the pure monomorphisms.

4. Sheaf-theoretic characterization of effective descent

4.A The characterization theorem. Let $\mathbf{P}$ be the class of stably regular epimorphisms in $\mathbf{C}$, and $\check{\mathbf{C}} = Shv(\mathbf{C}, \mathbf{P})$ the category of sheaves with respect to $\mathbf{P}$ regarded as (a base of) a Grothendieck topology on $\mathbf{C}$ (see Chapter VII). We recall that $\mathbf{P}$ is closed under composition by Proposition 1.3, and trivially pullback stable. That is, $\check{\mathbf{C}}$ is the category of all functors $F : \mathbf{C}^{\mathrm{op}} \to \mathbf{Set}$ such that for every $p : E \to B$ in $\mathbf{P}$,

$$F(B) \xrightarrow{\;F(p)\;} F(E) \underset{F(pr_2)}{\overset{F(pr_1)}{\rightrightarrows}} F(E \times_B E) \tag{26}$$

is an equalizer diagram in the category of sets. Since the topology we introduced is clearly subcanonical (=every representable functor is a sheaf), the Yoneda embedding yields a fully finite-limit preserving functor

$$\begin{aligned} Y : \mathbf{C} &\to \check{\mathbf{C}} \\ C &\mapsto \hom(-, C). \end{aligned} \tag{27}$$

4.1. Lemma. *The functor* (27) *carries stably regular epimorphisms in* $\mathbf{C}$ *to effective descent morphisms in* $\check{\mathbf{C}}$.

Proof. For a $p : E \to B$ in $\mathbf{P}$, an object U in $\mathbf{C}$, and $b \in \hom(U, B)$, we construct the pullback of b and p, and consider the diagram

$$
\begin{array}{ccc}
\hom(U \times_B E, E) & \xrightarrow{\;\hom(U \times_B E, p)\;} & \hom(U \times_B E, B) \\[2pt]
{\scriptstyle\hom(pr_1, E)}\big\uparrow & & \big\uparrow{\scriptstyle\hom(pr_1, B)} \\[2pt]
\hom(U, E) & \xrightarrow[\;\hom(U, p)\;]{} & \hom(U, B)
\end{array}
\tag{28}
$$

Although b itself might be outside the image of $\hom(U, p)$, its image under $\hom(pr_1, B)$, being equal to $b(pr_1) = p(pr_2) = \hom(U \times_B E, p)(pr_2)$, certainly belongs to the image of $\hom(U \times_B E, p)$. Since the projection pr_1 belongs to $\mathbf{P}$, for b arbitrary, we conclude that $Y(p) = \hom(-, E) \to \hom(-, B)$ is an epimorphism in $\check{\mathbf{C}}$; since $\check{\mathbf{C}}$ is a topos, this implies that $Y(p)$ is an effective descent morphism by Theorem 3.7(f). $\qquad\square$

From Lemma 4.1 and Corollary 3.9, we obtain:

4.2. Theorem. *A stably regular epimorphism* $p : E \to B$ *in* $\mathbf{C}$ *is an effective descent morphism if and only if for every pullback diagram in* $\check{\mathbf{C}}$ *of the form*

$$
\begin{array}{ccc}
\hom(-, D) & \longrightarrow & F \\[2pt]
\big\downarrow & & \big\downarrow \\[2pt]
\hom(-, E) & \xrightarrow[\;\hom(-, p)\;]{} & \hom(-, B)
\end{array}
\tag{29}
$$

the functor F is representable. If $\mathbf{C}$ has a terminal object, then this describes the class of all effective descent morphisms in $\mathbf{C}$. $\qquad\square$

4.B Stability properties of effective descent morphisms. This theorem (used for $(\mathbf{C} \downarrow B)$ instead of $\mathbf{C}$, if necessary) on the one hand provides a sheaf-theoretic description of effective descent morphisms, and on the other hand (together with Propositions 1.3 and 1.5 and Corollary 3.8) immediately gives:

4.3. Corollary. *The class of effective descent morphisms in $\mathbf{C}$ is closed under composition, pullback stable, and has the strong right cancellation property (see Proposition (1.5). In particular, it contains all split epimorphisms.* $\qquad\square$

The Reader will probably agree with us that pullback stability follows equally easily from Theorem 3.7, but this is not the case for closedness under composition (see [18]).

Corollary 4.3, together with many applications of effective descent morphisms, some of which are briefly described in the next section, suggests that this new class of morphisms is to be added to the World of Epimorphisms (Section 1), and to be considered as a very important one.

5. Links with other categorical constructions

5.A Monadic description of effective descent. In order to establish our first link — with *monadicity* — we are going to show that diagram (24) is canonically equivalent to

$$(\mathbf{C} \downarrow B) \xrightarrow{\quad K^{T^p} \quad} (\mathbf{C} \downarrow E)^{T^p} \qquad (30)$$

$$p^* \searrow \qquad \swarrow U^{T^p}$$

$$(\mathbf{C} \downarrow E)$$

where $T^p = (T^p, \eta^p, \mu^p)$ is the monad on $(\mathbf{C} \downarrow E)$ corresponding to the adjunction

$$(p_!, p^*, \eta^p, \varepsilon^p) : (\mathbf{C} \downarrow E) \to (\mathbf{C} \downarrow B), \qquad (31)$$

in which

- $p_! : (\mathbf{C} \downarrow E) \to (\mathbf{C} \downarrow B)$ is the composition with p (i.e. the functor defined by $p_! : (D, \delta) \mapsto (D, p\delta)$),
- the unit $\eta^p : 1_{(\mathbf{C} \downarrow E)} \to p^* p_!$ is defined by $(\eta^p)_{(D,\delta)} = \langle \delta, 1_D \rangle : D \to E \times_B D$,
- the counit $\varepsilon^p : p_! p^* \to 1_{(\mathbf{C} \downarrow B)}$ is defined by: $(\varepsilon^p)_{(A,\alpha)} =$ the pullback projection $E \times_B A \to A$,

and where U^{T^p} is the forgetful functor and K^{T^p} the comparison functor.

What we mean by a canonical equivalence between (24) and (30) is of course a canonical category equivalence

$$\Theta : \mathbf{DiscFib}(Eq(p)) \to (\mathbf{C} \downarrow E)^{T^p} \tag{32}$$

with $U^{T^p}\Theta \approx U^p$ and $\Theta K^p \approx K^{T^p}$; it turns out that the first of these two isomorphisms will in fact be an identity.

The objects of $(\mathbf{C} \downarrow E)^{T^p}$ are pairs $((C, \gamma), \xi)$, in which (C, γ) is an object in $(\mathbf{C} \downarrow E)$, and $\xi : E \times_B C \to C$ is a morphism in $\mathbf{C}$ making the diagram

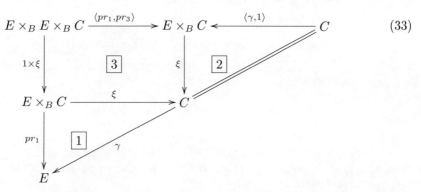

$$(33)$$

commute. Here $\boxed{1}$ says that ξ is to be considered as a morphism in $(\mathbf{C} \downarrow E)$ from $T^p(C, \gamma) = p^*p_!(C, \gamma) = (E \times_B C, pr_1)$ to (C, γ), and $\boxed{2}$ and $\boxed{3}$ express the usual conditions defining an algebra over a monad. The commutativity of (33), rewritten in terms of generalized elements, becomes

$$\gamma(e \cdot c) = e, \gamma(c) \cdot c = c, e \cdot (e' \cdot c) = e \cdot c, \tag{34}$$

where $e \cdot c$ denotes the composite $\xi\langle e, c\rangle$, well defined for an arbitrary object U in $\mathbf{C}$ and morphisms $e : U \to E$ and $c : U \to C$ in $\mathbf{C}$ with $p = p\gamma c$.

Given now an object (X, f) in $\mathbf{DiscFib}(Eq(p))$ displayed as

$$
\begin{array}{ccc}
X_1 & \underset{\varphi_2}{\overset{\varphi_1}{\rightrightarrows}} & X_0 \\
{\scriptstyle f_1}\big\downarrow & & \big\downarrow{\scriptstyle f_0} \\
E \times_B E & \underset{pr_2}{\overset{pr_1}{\rightrightarrows}} & E
\end{array}
\tag{35}
$$

we define $\Theta(X, f)$ as

$$\Theta(X, f) = ((X_0, f_0), \varphi_1\langle(pr_1)f_1, \varphi_2\rangle^{-1}) = ((X_0, f_0), \varphi_1\langle f_0\varphi_1, \varphi_2\rangle^{-1}), \tag{36}$$

where:

- we begin by observing that since (35) is a discrete fibration, the morphism $\langle f_1, \varphi_2\rangle : X_1 \to (E \times_B E) \times_E X_0$ is an isomorphism; therefore

- its composite $\langle (pr_1)f_1, \varphi_2 \rangle : X_1 \to E \times_B X_0$ with the canonical isomorphism $(E \times_B E) \times_E X_0 \approx E \times_B X_0$ also is an isomorphism, and we use its inverse in (36).

It is a routine calculation to show that (36) indeed gives an equivalence of diagrams (24) and (30); however to see this more clearly, let us use generalized elements again. To this end, for a fixed object U in $\mathbf{C}$, let us write just $\sim$ for the equivalence relations on $\hom(U, E)$ and $\hom(U, X_0)$ determined by the two rows of (35). To say that (35) is a discrete fibration, i.e., that $\langle f_1, \varphi_2 \rangle : X_1 \to (E \times_B E) \times_E X_0$ is an isomorphism (or, equivalently, that $\langle (pr_1), f_1, \varphi_2 \rangle : X_1 \to E \times_B X_0$ is an isomorphism), is to say that

$$\boxed{\begin{array}{c} \text{for all } U \text{ in } \mathbf{C} \text{ and } e' \sim e \text{ in } \hom(U, E) \text{ and } x \text{ in } \hom(U, X_0) \\ \text{with } f_0 x = e, \text{ there exists a unique } x' \text{ in } \hom(U, X_0) \\ \text{with } x' \sim x \text{ and } f_0 x' = e' \end{array}} \quad (37)$$

which we could display as

$$\begin{array}{ccc} \boxed{x'} & \sim & x \\ | & & | \\ e' & \sim & e \end{array} \quad (38)$$

According to the notation used in (34) the new generalized element x' here is nothing but $e' \cdot x$, and the three equalities in (34) correspond to the following three obvious situations

$$\begin{array}{ccccccccccc} e \cdot c & \sim & c & & (\gamma c) \cdot c & \sim & c & & e \cdot (e' \cdot c) & \sim & e' \cdot c & \sim & c \\ | & & | & & | & & | & & | & & | & & | \\ e & \sim & \gamma c & & \gamma c & = & \gamma c & & e & \sim & e' & \sim & \gamma c, \end{array} \quad (39)$$

where $(C, \gamma) = (X, f_0)$.

It is also useful to mention that starting from a T^p-algebra $((C, \gamma), \xi)$, we can describe the corresponding discrete fibration as

$$\begin{array}{ccc} E \times_B C & \xrightarrow{\ \xi\ }_{pr_2} & C \\ {\scriptstyle 1 \times \gamma}\Big\downarrow & & \Big\downarrow{\scriptstyle \gamma} \\ E \times_B E & \xrightarrow[pr_2]{\ pr_1\ } & E, \end{array} \quad (40)$$

which gives the following description of the top equivalence for the generalized elements of C:

$$\begin{array}{rl} c' \sim c & \Leftrightarrow \ (\text{there exists } e \text{ with } c' = e \cdot c) \\ & \Leftrightarrow \ (\gamma c' \sim \gamma c \text{ and } c' = (\gamma c') \cdot c) \\ & \Leftrightarrow \ (\gamma c \sim \gamma c' \text{ and } c = (\gamma c) \cdot c') \\ & \Leftrightarrow \ (\text{there exists } e \text{ with } c = e \cdot c'). \end{array} \quad (41)$$

The equivalence of diagrams (24) and (30) brings about an important conclusion:

5.1. Theorem. *A morphism $p : E \to B$ in $\mathbf{C}$ is an effective descent morphism if and only if the functor $p^* : (\mathbf{C} \downarrow B) \to (\mathbf{C} \downarrow E)$ is monadic.* $\qquad\square$

5.2. Remark. An alternative approach to what we called *elementary descent theory* (see Section 3) would be *monadic descent theory*, in which Theorem 5.1 would become the definition of effective descent morphism, and in which various characterizations and other properties of this class of morphisms are deduced from various forms of the Beck monadicity theorem (see [12], [21]). For instance, if **C** is locally cartesian closed and has coequalizers, this immediately tells us that p is an effective descent morphism if and only if p^* reflects isomorphisms. $\square$

5.B Grothendieck's descent theory. Let us now make some calculations with generalized elements, which will lead us to another link — essentially with the original approach to descent theory developed by Grothendieck.

For a fixed object (C, γ) in $(\mathbf{C} \downarrow E)$, we observe that $\hom(T^p(C, \gamma), (C, \gamma))$ can be considered as a subset of

$$
\begin{aligned}
\hom(p_! T^p(C, \gamma), p_!(C, \gamma)) &= \hom((E \times_B C, p(pr_1)), p_!(C, \gamma)) \\
&= \hom(p_!(E \times_B C, \gamma(pr_2)), p_!(C, \gamma)).
\end{aligned}
\tag{42}
$$

This leads us to considering the following two sets:

$$
\Xi(C, \gamma) = \{\xi \in \hom(p_!(E \times_B C, \gamma(pr_2)), p_!(C, \gamma)) \mid ((C, \gamma), \xi) \text{ is in } (\mathbf{C} \downarrow E)^{T^p}\},
\tag{43}
$$

$$
\begin{aligned}
Z(C, \gamma) = \quad &\{\zeta \in \hom((E \times_B C, \gamma(pr_2)), p^* p_!(C, \gamma)) \mid \zeta \text{ is an isomorphism} \\
&\text{and } (pr_2)\, \zeta \text{ satisfies the first and the third equality in } (34)\},
\end{aligned}
\tag{44}
$$

and to establishing a bijection between them:

5.3. Lemma.

(a) *Every ζ in $Z(C, \gamma)$, regarded as an automorphism of $p_!((E \times_B C, \gamma(pr_2)) = (E \times_B C, p\gamma(pr_2)) = (E \times_B C, p(pr_1)) = p_! p^*(C, \gamma)$ in $(\mathbf{C} \downarrow B)$, becomes an involution, i.e., satisfies $(p_!(\zeta))^2 = 1$ in $(\mathbf{C} \downarrow B)$.*

(b) *The canonical bijection*

$$
\hom(p_!(E \times_B C, \gamma(pr_2)), p_!(C, \gamma)) \approx \hom((E \times_B C, \gamma(pr_2)), p^* p_!(C, \gamma))
\tag{45}
$$

(defined by $(\xi \mapsto \langle \gamma(pr_2), \xi \rangle$, with the inverse defined by $\zeta \mapsto (pr_2)\zeta$) induces a bijection $\Xi(C, \gamma) \approx Z(C, \gamma)$.

Proof. To give ξ and ζ corresponding to each other under the bijection (45) is the same as to give a commutative diagram of the form

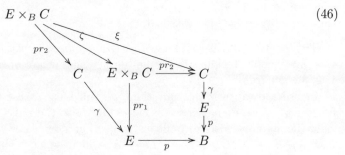

$$
\tag{46}
$$

and what we need to prove can be formulated as follows:
  (i) if ξ is in $\Xi(C, \gamma)$, then $\zeta^2 = 1$ in $(\mathbf{C} \downarrow B)$ (or just in $\mathbf{C}$);
  (ii) if ζ is an isomorphism (say, in $\mathbf{C}$) and ζ satisfies the first and the third equality in (34), then the second equality in (34) is also satisfied.

Using generalized elements as in (34), it is convenient to write

$$\xi\langle e, c\rangle = e \cdot c \text{ and accordingly } \zeta\langle e, c\rangle = \langle \gamma c, e \cdot c\rangle \qquad (47)$$

(even if ζ does not satisfy all identities in (34)).
  (i) Using (47) and (34) we have:

$$\zeta^2\langle e, c\rangle = \zeta\langle \gamma c, e \cdot c\rangle = \langle \gamma(e \cdot c), (\gamma c) \cdot (e \cdot c)\rangle = \langle e, (\gamma c) \cdot c\rangle = \langle e, c\rangle,$$

i.e., $\zeta^2 = 1$, as desired.
  (ii) Without the second equality of (34) the calculation above still gives $\zeta^2\langle e, c\rangle = \langle e, (\gamma c) \cdot c\rangle$, and then

$$\zeta^3\langle e, c\rangle = \zeta\langle e, (\gamma c) \cdot c\rangle = \langle \gamma((\gamma c) \cdot c), e \cdot ((\gamma c) \cdot c)\rangle = \langle \gamma c, e \cdot c\rangle = \zeta\langle e, c\rangle$$

tells us that $\zeta^3 = \zeta$. Since ζ is an isomorphism, we conclude $\zeta^2 = 1$, and then $\zeta^2\langle e, c\rangle = \langle e, (\gamma c) \cdot c\rangle$ gives the desired equality $(\gamma c) \cdot c = c$. □

Now we can propose an alternative description of $Z(C, \gamma)$:

5.4. Lemma. *Using the expression* $\zeta\langle e, c\rangle = \langle \gamma c, e \cdot c\rangle$ *for an arbitrary* $\zeta \in$ $\hom((E \times_B C, \gamma(pr_2)), p^* p_!(C, \gamma))$, *let* $\zeta_{1,2}, \zeta_{2,3}, \zeta_{1,3}$ *be the endomorphisms of* $E \times_B E \times_B C$ *defined by*

$$\zeta_{1,2}\langle e, e', c\rangle = \langle \gamma c, e', e \cdot c\rangle, \zeta_{2,3}\langle e, e', c\rangle = \langle e, \gamma c, e' \cdot c\rangle, \zeta_{1,3}\langle e, e', c\rangle = \langle e', \gamma c, e \cdot c\rangle;$$
$$(48)$$

then ζ *is in* $Z(C, \gamma)$ *if and only if it is an isomorphism satisfying*

$$\zeta_{1,2}\zeta_{2,3} = \zeta_{1,3}. \qquad (49)$$

Proof. A straightforward calculation gives:

$$\zeta_{1,2}\zeta_{2,3}\langle e, e', c\rangle = \zeta_{1,2}\langle e, \gamma c, e' \cdot c\rangle = \langle \gamma(e' \cdot c), \gamma c, e \cdot (e' \cdot c)\rangle, \qquad (50)$$

which of course is equal to $\zeta_{1,3}\langle e, e', c\rangle = \langle e', \gamma c, e \cdot c\rangle$ if and only if the first and the third equality in (34) are satisfied. □

The identity (49) is well known as *Grothendieck's cocycle condition*; presented as above it looks technically simple but strange: where does the choice of indices 1, 2, 3 come from?

In order to answer this question, we will have to use not just the pullback functor p^* as we managed to do before, but also the pullback functors along the eight (i.e. all non-twisted!) pullback projection morphisms $E \times_B E \times_B E \to E \times_B$ $E, E \times_B E \to E, E \times_B E \times_B E \to E$, which will be denoted by $\pi_{i,j}(i = 1, 2$

and $j = 2, 3$ with $i \leq j$, $\pi_i (i = 1, 2)$, $\pi'_i (i = 1, 2, 3)$ respectively. The canonical bijections

$$\hom(p^* p_!(C, \gamma), (C, \gamma)) \approx \hom(\pi_{1!} \pi_2^*(C, \gamma), (C, \gamma)) \approx \hom(\pi_2^*(C, \gamma), \pi_1^*(C, \gamma))$$
(51)

suggest to introduce, in addition to $\Xi(C, \gamma)$ and $Z(C, \gamma)$ above, the set

$$\Phi(C, \gamma) = \{\varphi \in \hom(\pi_2^*(C, \gamma), \pi_1^*(C, \gamma)) \,|\, \varphi \text{ is an isomorphism and the}$$
$$\zeta : p^* p_!(C, \gamma) \to (C, \gamma) \text{ corresponding to it under the bijection (51)} \quad (52)$$
$$\text{satisfies the third equality in (34)}\}.$$

Note that the first equality in (34) holds automatically here since ξ is required to be a morphism from $p^* p_!(C, \gamma)$ to (C, γ). And of course, we have

$$\Phi(C, \gamma) \approx \Xi(C, \gamma) \approx Z(C, \gamma), \tag{53}$$

and $\varphi \in \Phi(C, \gamma)$ is related to the corresponding $\xi \in \Xi(C, \gamma)$ and $\zeta \in \Xi(C, \gamma)$ via

$$\varphi\langle\langle e, e'\rangle, c\rangle = \langle\langle e, \gamma c\rangle, \xi\langle e, c\rangle\rangle = \langle\langle e, \gamma c\rangle, e \cdot c\rangle, \zeta\langle e, c\rangle = \langle \gamma c, e \cdot c\rangle, \tag{54}$$

where:

- $e \cdot c$ is just a convenient abbreviation for $\xi\langle e, c\rangle$ as above, and the second equality just repeats (47);
- $\langle\langle e, e'\rangle, c\rangle$ represents a generalized element of the pullback of γ along the second projection $\pi_2 : E \times_B E \to E$ and therefore in fact has $e' = \gamma c$;
- $\langle\langle e, \gamma c\rangle, \xi\langle e, c\rangle\rangle = \langle\langle e, \gamma c\rangle, e \cdot c\rangle$ must represent a generalized element of the pullback of γ along the first projection $\pi_1 : E \times_B E \to E$ and therefore must satisfy $e = \gamma(e \cdot c)$, which is exactly the first equality in (34).

Now for an arbitrary $\varphi \in \hom(\pi_2^*(C, \gamma), \pi_1^*(C, \gamma))$ consider the diagram:

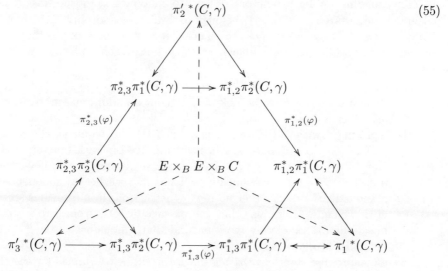

(55)

where:

- the solid arrows are isomorphisms in $(\mathbf{C} \downarrow E \times_B E \times_B E)$;
- the dotted arrows are isomorphisms in $(\mathbf{C} \downarrow B)$ (from $E \times_B E \times_B C$ equipped with the canonical morphism to B to the appropriate underlying object of $\pi'^*_i(C,\gamma), i = 1, 2, 3$);
- all unnamed arrows are canonical isomorphisms.

Let us list the morphisms involved in the hexagon together with their expressions in the language of generalized elements:

the morphism	composed with	where	gives
$\pi^*_{2,3}\pi'^*_2(C,\gamma) \to \pi^*_{2,3}\pi'^*_1(C,\gamma)$	$\langle\langle e,e',e''\rangle,\langle\langle e',e''\rangle,c\rangle\rangle$	$e''=\gamma c$	$\langle\langle e,e',e''\rangle,\langle\langle e',\gamma c\rangle,e'\cdot c\rangle\rangle$
$\pi^*_{2,3}\pi'^*_1(C,\gamma) \to \pi^*_{1,2}\pi'^*_2(C,\gamma)$	$\langle\langle e,e',e''\rangle,\langle\langle e',e''\rangle,c\rangle\rangle$	$e'=\gamma c$	$\langle\langle e,e',e''\rangle,\langle\langle e,e'\rangle,c\rangle\rangle$
$\pi^*_{1,2}\pi'^*_2(C,\gamma) \to \pi^*_{1,2}\pi'^*_1(C,\gamma)$	$\langle\langle e,e',e''\rangle,\langle\langle e,e'\rangle,c\rangle\rangle$	$e'=\gamma c$	$\langle\langle e,e',e''\rangle,\langle\langle e,\gamma c\rangle,e\cdot c\rangle\rangle$
$\pi^*_{2,3}\pi'^*_2(C,\gamma) \to \pi^*_{1,3}\pi'^*_2(C,\gamma)$	$\langle\langle e,e',e''\rangle,\langle\langle e',e''\rangle,c\rangle\rangle$	$e''=\gamma c$	$\langle\langle e,e',e''\rangle,\langle\langle e,e''\rangle,c\rangle\rangle$
$\pi^*_{1,3}\pi'^*_2(C,\gamma) \to \pi^*_{1,3}\pi'^*_1(C,\gamma)$	$\langle\langle e,e',e''\rangle,\langle\langle e,e''\rangle,c\rangle\rangle$	$e''=\gamma c$	$\langle\langle e,e',e''\rangle,\langle\langle e,\gamma c\rangle,e\cdot c\rangle\rangle$
$\pi^*_{1,3}\pi'^*_1(C,\gamma) \to \pi^*_{1,2}\pi'^*_1(C,\gamma)$	$\langle\langle e,e',e''\rangle,\langle\langle e,e''\rangle,c\rangle\rangle$	$e=\gamma c$	$\langle\langle e,e',e''\rangle,\langle\langle e,e'\rangle,c\rangle\rangle$

Table 3

Consequently for the enveloping triangle we have

the morphism	composed with	where	gives
$\pi'^*_3(C,\gamma) \to \pi'^*_2(C,\gamma)$	$\langle\langle e,e',e''\rangle,c\rangle$	$e''=\gamma c$	$\langle\langle e,e',e''\rangle,e'\cdot c\rangle$
$\pi'^*_2(C,\gamma) \to \pi'^*_1(C,\gamma)$	$\langle\langle e,e',e''\rangle,c\rangle$	$e'=\gamma c$	$\langle\langle e,e',e''\rangle,e\cdot c\rangle$
$\pi'^*_3(C,\gamma) \to \pi'^*_1(C,\gamma)$	$\langle\langle e,e',e''\rangle,c\rangle$	$e''=\gamma c$	$\langle\langle e,e',e''\rangle,e\cdot c\rangle$

Table 4

Composed with the dotted arrows (in (55)) and their inverses in the appropriate way, these three morphisms produce three automorphisms of $E \times_B E \times_B C$ that are nothing but $\zeta_{2,3}, \zeta_{1,2}, \zeta_{1,3}$ — which immediately answers our question about the equality (49) in Lemma 5.4, and yields a reformulation of it:

5.5. Conclusion. The morphisms $\zeta_{i,j}$ in Lemma 5.4 are in fact the images of ζ under the corresponding pullback functors $\pi^*_{i,j}$ up to canonical isomorphisms. A morphism $\varphi \in \hom(\pi^*_2(C,\gamma), \pi^*_1(C,\gamma))$ belongs to $\Phi(C,\gamma)$ if and only if the diagram (55) commutes. That is, we propose three possible ways to present Grothendieck's cocycle condition (49), which "coincide up to canonical isomorphisms" with each other:

(a) using $\zeta_{2,3}, \zeta_{1,2}, \zeta_{1,3}$ as in (49), which is technically the simplest way, although it does not really look nice without an additional motivation for introducing $\zeta_{i,j}$;

(b) replacing ζ with φ (related to it as in (54)) and requiring the hexagon part of (55) to commute which, involving five pullback functors and six morphisms in $(\mathbf{C} \downarrow E \times_B E \times_B E)$, looks very complicated; nevertheless it seems to be most convincing as it has its own beauty independently of monadicity or any other notions/constructions from previous sections of this chapter; note also that the passage directly from ξ to φ via the bijections (51) is clearly more natural than the passage from ξ to ζ in Lemma 5.3(b) via (42) and (45);

(c) a compromise solution, writing down the explicit formulas for the three morphisms from Table 4, and requiring the composite of the first two to be equal to the third one; in other words, making the diagram (55) a triangle by "erasing" the three inner (solid arrows and all dotted) arrows.

□

Furthermore, from previous observations it is easy to see that diagrams (24) and (30) are equivalent to the following one:

$$(\mathbf{C} \downarrow B) \xrightarrow{\quad K^p \quad} \widetilde{\mathbf{Des}}(p) \qquad (56)$$

$$p^* \searrow \qquad \swarrow U^p$$

$$(\mathbf{C} \downarrow E)$$

Here $\widetilde{\mathbf{Des}}(p)$ is the category of pairs $((C, \gamma), \varphi)$, where (C, γ) is an object in $(\mathbf{C} \downarrow E)$ and $\varphi : \pi_2^*(C, \gamma) \to \pi_1^*(C, \gamma))$ is an isomorphism making the hexagon part of (55) commute. U^p is the forgetful functor, and K^p is the "Grothendieck comparison functor" $(A, \alpha) \mapsto (p^*(A, \alpha), \varphi_\alpha)$, where φ_α is the canonical isomorphism $\pi_2^* p^*(A, \alpha) \to \pi_1^* p^*(A, \alpha)$. The category $\widetilde{\mathbf{Des}}(p)$ is equivalent to the category $\mathbf{Des}(p)$ of descent data for p, which we decided to define via discrete fibrations (see Definition 3.6). Hence, what is essentially known as Grothendieck's definition of an effective descent morphism becomes a theorem in our presentation:

5.6. Theorem. *A morphism* $p : E \to B$ *in* $\mathbf{C}$ *is an effective descent morphism if and only if the Grothendieck comparison functor involved in* (56) *is a category equivalence.* □

There are more links involving more advanced categorical constructions, which are briefly described or just indicated below for readers familiar with internal categories and 2-dimensional category theory.

5.C Internal category actions and discrete fibrations. Let $C =$

$$C_1 \times_{C_0} C_1 \underset{p_2}{\overset{p_1}{\underset{\longrightarrow}{\overset{\longrightarrow}{\underset{m}{\longrightarrow}}}}} C_1 \underset{c}{\overset{d}{\underset{\longrightarrow}{\overset{\longrightarrow}{\underset{e}{\longleftarrow}}}}} C_0 \qquad (57)$$

be an internal category in $\mathbf{C}$ (see also Section 6 below). Displaying an internal C-action $F = (F_0, \gamma, \xi)$ as

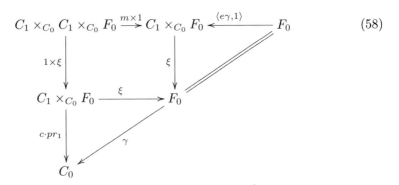

$$\begin{array}{ccc}
C_1 \times_{C_0} C_1 \times_{C_0} F_0 \xrightarrow{m \times 1} & C_1 \times_{C_0} F_0 \xleftarrow{\langle e\gamma, 1 \rangle} & F_0 \\
\end{array} \tag{58}$$

and then taking C to be

$$E \times_B E \times_B E \underset{\substack{\longrightarrow \\ \pi_{1,2}}}{\overset{\substack{\pi_{2,3} \\ \longrightarrow}}{\xrightarrow{\pi_{1,3}}}} E \times_B E \underset{\pi_1}{\overset{\pi_2}{\underset{\longrightarrow}{\xleftarrow{\langle 1,1 \rangle}}}} E, \tag{59}$$

where various π's are as in (55), we observe:

- diagram (58) becomes the same as (33) (where C is to be replaced with F_0), up to the canonical isomorphism $E \times_B F_0 \approx (E \times_B E) \times_E F_0$.
- since (59) is nothing but the equivalence relation $Eq(p)$ considered as an internal category (we see no harm in identifying $Eq(p)$ with $Eq(p)^{\mathrm{op}}$ here), the conclusion is that the category $(C \downarrow E)^{T^p}$ is canonically equivalent to the category $C^{Eq(p)}$ of internal $Eq(p)$-actions.
- moreover, if we identify $(C \downarrow E)^{T^p}$ with $\mathbf{C}^{CEq(p)}$, then the equivalence (32) becomes a special case of the standard equivalence between the category of discrete fibrations over an internal category C and the category of internal C^{op}-actions.
- the morphism $p : E \to B$, when considered as an internal functor between discrete internal categories, has its canonical factorization, which is the internal version of the *(bijective-on-objects, fully faithful)-factorization* (which is different from the *(surjective-on-objects, injective-on-objects-and-fully faithful)-factorization*, but can also be regarded as the 2-dimensional *(essentially surjective-on-objects, fully faithful)-factorization)*, which then gives two more diagrams equivalent to (24), (30), (56):

$$\begin{array}{ccc}
(\mathbf{C} \downarrow B) \approx \mathbf{C}^B & \xrightarrow{\;\;\mathbf{C}^{p''}\;\;} & \mathbf{C}^{Eq(p)} \\
& \mathbf{C}^p \searrow \quad \nearrow \mathbf{C}^{p'} & \\
& (\mathbf{C} \downarrow E) \approx \mathbf{C}^E &
\end{array} \tag{60}$$

$$(\mathbf{C} \downarrow B) \approx \mathbf{DiscFib}(B) \xrightarrow{\quad p''^{\,*} \quad} \mathbf{DiscFib}(Eq(p))$$

$$\searrow p^* \qquad\qquad p'^{\,*} \swarrow$$

$$(\mathbf{C} \downarrow E) \approx \mathbf{DiscFib}(E)$$

$$(61)$$

5.7. Remark. It is useful to consider some of the constructions above once again in the simplest case, namely $\mathbf{C} = \mathbf{Set}$, the category of sets:

(a) In this case a bundle (C, γ) over E can be identified with an E-indexed family $(C_e)_{e \in E}$ of sets $C_e = \gamma^{-1}(e)$, yielding an equivalence

$$(\mathbf{Set} \downarrow E) \sim \mathbf{Set}^E \tag{62}$$

between the categories of "internal" and "external" E-indexed families of sets. Furthermore, the equality $\gamma(e \cdot c) = e$ in (34) tells us that the map ξ involved in (33) splits up into a family of maps

$$(\xi_{e,e'} : C_e \to C_{e'})_{(e,e') \in E \times_B E} \tag{63}$$

defined by

$$\xi_{e,e'}(c) = \xi(e', c) = e' \cdot c, \tag{64}$$

and satisfying the following two equalities corresponding to the second and third identities in (34):

$$\xi_{e,e}(c) = c, \, \xi_{e',e''} \xi_{e,e'}(c) = \xi_{e,e''}(c). \tag{65}$$

In other words, our structure $((C, \gamma), \xi)$ becomes nothing but a functor from the equivalence relation $Eq(p)$ regarded as a category to the category of sets.

(b) Concerning terminology and notation: the equivalence (62) is of course a simplified version of the canonical equivalence between the category of actions of a fixed (small) category and the category of all functors from that category to the category $\mathbf{Set}$, and our passage above from $((C, \gamma), \xi)$ to a functor $Eq(p) \to \mathbf{Set}$ is a passage through the category equivalences

$$(\mathbf{Set} \downarrow E)^{T^p} \sim (Eq(p)\text{-actions in } \mathbf{Set}) \sim (\text{functors } Eq(p) \to \mathbf{Set}). \tag{66}$$

Since in the general case of an abstract category $\mathbf{C}$ there is no notion of a functor from an internal category in $\mathbf{C}$ to $\mathbf{C}$, it is reasonable to look at internal actions *as* such functors, and to ignore the difference between actions and functors in the case of sets. In particular, there is no problem with $\mathbf{C}^{Eq(p)}$ in (60) denoting the category of internal $Eq(p)$-actions, and hence having now two meanings for $\mathbf{Set}^{Eq(p)}$. However we should warn our readers that some authors call internal category actions *internal functors* (which seems to be the right name for morphisms of internal categories rather then for what we called actions) or *internal (covariant) presheaves*.

(c) The assertion (ii) in the proof of Lemma 5.3, which made it possible for us to omit the second equality of (34) in (44), tells us that we can also avoid the first equality in (65). Indeed, if the second equality in (65) holds, then the first one holds if and only if each $\xi_{e,e'}$ is a bijection — which is an instance of the following obvious observation about functors from a groupoid: they can be defined as diagrams that preserve composition and send all morphisms to isomorphisms.

(d) The presentation of descent data as functors, i.e., the presentation of $((C,\gamma),\xi)$ above as the family $(C_e)_{e\in E}$ together with the family

$$(\xi_{e,e'} : C_e \to C_{e'})_{(e,e')\in E\times_B E}$$

of bijections satisfying (65), is very useful not only for $\mathbf{C} = \mathbf{Set}$, but also for various categories of *sets with structure*, and especially for those where every subset is a *substructure*. For example, for $\mathbf{C} = \mathbf{Top}$, the category of topological spaces, a descent datum for $p : E \to B$ can be described as the family $(C_e)_{e\in E}$ of topological spaces equipped with

- a topology on their disjoint union C that induces the given topologies on each C_b, and makes the canonical map from C to B continuous;
- a family $(\xi_{e,e'} : C_e \to C_{e'})_{(e,e')\in E\times_B E}$ of homeomorphisms satisfying (65) and making the map $\xi : E \times_B C \to C$ defined via $\xi_{e,e'}$ by (64) continuous.

5.D 2-Dimensional lax equalizers and pseudo-equalizers. For an arbitrary internal category (57), the category $\mathbf{C}^C$ of internal C-actions can be presented via the following *reflexive lax equalizer diagram*:

$$\mathbf{C}^C \longrightarrow (\mathbf{C} \downarrow C_0) \underset{\underset{c^*}{\xleftarrow{\hspace{1em}e^*\hspace{1em}}}}{\overset{d^*}{\rightrightarrows}} (\mathbf{C} \downarrow C_1) \underset{\underset{p_2^*}{\rightrightarrows}}{\overset{p_1^*}{\underset{m^*}{\rightrightarrows}}} (\mathbf{C} \downarrow C_1 \times_{C_0} C_1)$$

$$(67)$$

(we omit the 2-cells involved from the display), which implies that $\mathbf{C}^C$ can be described as the category of pairs $F = ((F_0,\gamma),\varphi)$, where (F_0,γ) is an object in $(\mathbf{C} \downarrow C_0)$ and $\varphi : d^*(F_0,\gamma) \to c^*(F_0,\gamma)$ a morphism in $(\mathbf{C} \downarrow C_1)$ such that

$$e^*(\varphi) = \text{ canonical isomorphism } e^*d^*(F_0,\gamma) \to e^*c^*(F_0,\gamma) \qquad (68)$$

and the diagram

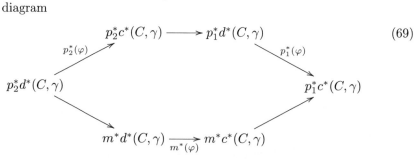

$$(69)$$

commutes (where, again, unnamed arrows are canonical isomorphisms). It is also well known that if C is a groupoid and the diagram (69) commutes, then equality (68) holds if and only if φ is an isomorphism. That is, in the groupoid case diagram (67) is at the same time a *reflexive pseudo-equalizer diagram*, and if we remove the arrow e^* from it, it is still a (non-reflexive) pseudo-equalizer diagram. Since the hexagon in (55) is clearly a special case of the hexagon (69), this yields three new links of descent theory, this time with 2-dimensional limits. We obtain:

5.8. Theorem. *The following conditions on a morphism* $p : E \to B$ *in* **C** *are equivalent:*

(a) *p is an effective descent morphism;*

(b) *the diagram*

$$(\mathbf{C} \downarrow B) \xrightarrow{p^*} (\mathbf{C} \downarrow E) \rightrightarrows (\mathbf{C} \downarrow E \times_B E) \Rrightarrow (\mathbf{C} \downarrow E \times_B E \times_B E) \quad (70)$$

(where the unnamed arrows are the pullback functors along the corresponding arrows in (59), and where the 2-cells involved are omitted) is a reflexive lax equalizer diagram;

(c) *diagram (70) is a reflexive pseudo-equalizer diagram;*

(d) *diagram (70) with the arrow from* $(\mathbf{C} \downarrow E \times_B E)$ *to* $(\mathbf{C} \downarrow E)$ *removed is a pseudo-equalizer diagram.* $\square$

Of course, this theorem comes from the three new ways of looking at the ("comparison") functor $(\mathbf{C} \downarrow B) \to \mathbf{C}^{Eq(p)}$, according to the universal properties of the three kinds of 2-dimensional equalizers above. Note also that the coincidence of the three kinds of 2-dimensional equalizers in our situation follows in fact directly from Lemma 5.3 (see (i) and (ii) in its proof, with the obvious remark that every involution is an isomorphism).

5.E Fibrations, pseudofunctors, cosmoi, and 2-dimensional exponentiation. As we mentioned in Section 3, our elementary approach to descent theory easily extends to the context of general fibrations of categories. The same is essentially true for the other approaches considered in this section; however the connection with monadicity requires the Beck-Chevalley condition, which always holds for the basic fibration (=codomain functor) $\mathbf{C}^2 \to \mathbf{C}$. Note that we have used it explicitly, as the first bijection in (51), but earlier also implicitly. It is not clear from the literature whether that condition was originally meant to be applied like this, but certainly the connection between Grothendieck's definition of an effective descent morphism and monadicity under the Beck-Chevalley condition was clarified already in 1970, by Bénabou and Roubaud in [1].

The idea of replacing fibrations with pseudofunctors and to consider descent theory as a case of 2-dimensional sheaf theory goes back to A. Grothendieck. It was developed further by several authors, and especially by Street, who puts descent theory into the context of *cosmos theory* (which is a 2-dimensional version of topos theory), using also his previous joint work with Walters (see [22], [23]). According to this idea, the diagrams of the form

$$F(B) \xrightarrow{F(p)} F(E) \rightrightarrows F(E \times_B E) \Rrightarrow F(E \times_B E \times_B E) \quad (71)$$

(together with the relevant 2-cells) involving a pseudofunctor $F : \mathbf{C}^{\mathrm{op}} \to \mathbf{Cat}$ play a crucial role, since they are 2-dimensional versions of the diagrams of the form (26), generalizing (70). Whenever this diagram is a reflexive lax (=pseudo-) equalizer diagram (i.e., whenever p is an effective descent morphism with respect to F) for a given "good" class of morphisms p, one also calls F a 2-dimensional sheaf, or a *stack* (see Chapter VII). The following observation of Street makes the cosmos-theoretic approach especially beautiful:

Given a pseudofunctor $F : \mathbf{C}^{\mathrm{op}} \to \mathbf{Cat}$ and an internal category C in $\mathbf{C}$, we can of course define F^C as in the special cases above, i.e., via the reflexive lax equalizer diagram

$$F^C \longrightarrow F(C_0) \rightrightarrows F(C_1) \mathrel{\substack{\textstyle\rightarrow\\[-0.6ex]\textstyle\rightarrow\\[-0.6ex]\textstyle\rightarrow}} F(C_1 \times_{C_0} C_1). \tag{72}$$

This turns out to be the as same as to define it just as the 2-dimensional exponent F^C in the cosmos of pseudofunctors $\mathbf{C}^{\mathrm{op}} \to \mathbf{Cat}$, where C is regarded as such a pseudofunctor via the (2-dimensional) Yoneda embedding. In particular, what looks like exponents in diagram (60) are indeed exponents!

6. Effective descent morphisms in Cat and related remarks

6.A Characterization of effective descent in Cat. Let τ be the category generated by the graph

$$2 \mathrel{\substack{\textstyle\xrightarrow{p_1}\\[-0.3ex]\textstyle\xrightarrow{m}\\[-0.3ex]\textstyle\xrightarrow{p_2}}} 1 \mathrel{\substack{\textstyle\xleftarrow{d}\\[-0.3ex]\textstyle\xleftarrow{e}\\[-0.3ex]\textstyle\xleftarrow{c}}} 0 \tag{73}$$

and the identities

$$de = 1 = ce, \, dm = dp_2, \, cm = cp_1. \tag{74}$$

The functors $\tau \to \mathbf{Set}$, also called *precategories*, play the role of "generalized categories". Indeed, any category C can be regarded as such a functor with

- $C(0) = C_0$ being the set of objects in C,
- $C(1) = C_1$ the set of morphisms,
- $C(2) = C_2$ the set of composable pairs of morphisms,
- $C(d) : C_1 \to C_0$ the domain map,
- $C(c) : C_1 \to C_0$ the codomain map,
- and the C-images of the other arrows in (6.1) defined by

$$e(x) = 1_x, p_1(g, f) = g, p_2(g, f) = f, m(g, f) = gf, \tag{75}$$

and this obviously determines a fully faithful limit-preserving functor

$$\mathbf{Cat} \to \mathbf{Set}^\tau. \tag{76}$$

Although we are not going to use anything from Section 5 here, let us point out that the presentation above of a category C as a functor $\tau \to \mathbf{Set}$ perfectly agrees with the display (57), the only difference is that d, c of (57) are now called $C(d), C(c)$, etc.

The following simple proposition tells us how to distinguish categories among precategories:

6.1. Proposition. *An object A in* $\mathbf{Set}^\tau$ *is (isomorphic to) a category if and only if it satisfies the following conditions* (a) *and* (b):

(a) *the diagram*

$$
\begin{array}{ccc}
A_2 & \xrightarrow{\ A(p_2)\ } & A_1 \\
{\scriptstyle A(p_1)}\downarrow & & \downarrow{\scriptstyle A(c)} \\
A_1 & \xrightarrow[\ A(d)\ ]{} & A_0
\end{array}
\tag{77}
$$

is a pullback, or equivalently, the morphism $A_2 \to A_1 \times_{A_0} A_1$ induced by the pair $(A(p_1), A(p_2))$ is an isomorphism.

Under the condition (a), we will identify A_2 with $A_1 \times_{A_0} A_1$ and write $A(m) : (g, f) \mapsto gf$, and accordingly $A(e) : x \mapsto 1_x$.

(b) m *is associative, and* $1_y f = f = f 1_x$ *whenever* $A(d)(f) = x$ *and* $A(c)(f) = y$. $\qquad\qquad\qquad\square$

Now Corollary 3.9 applied to the embedding (76) helps to prove:

6.2. Proposition. *A functor $p : E \to B$ is an effective descent morphism in* **Cat** *if and only if it is surjective on composable triples of morphisms.*

Proof. We observe:

- Since all stably regular epimorphisms in **Cat** are surjective on composable pairs of morphisms, the effective descent morphism in **Cat** also have this property.
- Since $\mathbf{Set}^\tau$ is a topos, its effective descent morphisms are the same as (regular) epimorphisms. On the other hand the surjectivity of p on composable pairs of morphisms (= surjectivity of $p_2 : E_2 \to B_2$) implies surjectivity of $p_0 : E_0 \to B_0$ and of $p_1 : E_1 \to B_1$. Therefore a functor $p : E \to B$ is an effective descent morphism in $\mathbf{Set}^\tau$ if and only if it is surjective on composable pairs of morphisms.
- By the two previous observations and by Corollary 3.9 and Proposition 6.1 it suffices to prove that the following two conditions on p are equivalent:
  (i) p is surjective on composable triples of morphisms;
  (ii) if $\alpha : A \to B$ is a morphism in $\mathbf{Set}^\tau$ such that $E \times_B A$ satisfies conditions (a) and (b) of Proposition 6.1, then A also satisfies those conditions.

Let us begin with (i)$\Rightarrow$(ii). We will express now condition 6.1(a) on, say A, by saying that every composable pair (a, a') of elements in A_1 has a unique lifting a'' in A_2. Since p is surjective on composable pairs, for every composable (a, a') of elements in A_1 there exists a composable pair (e, e') of elements in E_1 with

$p_1(e) = \alpha(a)$ and $p_1(e') = \alpha(a')$; the existence for a unique lifting for (a, a') in A_2 then follows from the existence for a unique lifting for $((e, a), (e', a'))$ in $(E \times_B A)_2$. That is, whenever p is surjective on composable pairs and $E \times_B A$ satisfies condition 6.1(a), also A satisfies that condition. Furthermore, there is no problem with condition 6.1(b) since it is purely equational (although it really needs the surjectivity of p on composable triples, not just pairs, for the associativity of $A(m)$).

(ii)$\Rightarrow$(i): By a *non-associative category* we mean an object A in $\mathbf{Set}^\tau$ satisfying 6.1(a) and the identity $1_y f = f = f 1_x$ in 6.1(b), but not necessarily the associativity for $A(m)$. In particular, we will use the non-associative category $\mathbf{4}^{n.a.}$ displayed by

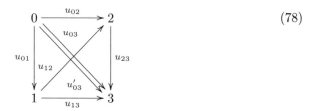

$$\tag{78}$$

where the identity arrows are omitted, and $u_{j,k} u_{i,j} = u_{i,k}$ for all $i < j < k$ with one exception, namely $u_{13} u_{01} = u'_{03}$. For every composable triple (h, g, f) of morphisms in B there exists a unique morphism $\alpha : \mathbf{4}^{n.a.} \to B$ in $\mathbf{Set}^\tau$ with $\alpha(u_{01}) = f$, $\alpha(u_{12}) = g$, $\alpha(u_{23}) = h$, and it is easy to check that it makes $E \times_B \mathbf{4}^{n.a.}$ a category if and only if there is no composable triple of morphisms in E whose p-images form the triple (h, g, f). Therefore, if p is not surjective on composable triples, then there is an $\alpha : A \to B$ in $\mathbf{Set}^\tau$ with $E \times_B A$ but not A in $\mathbf{Cat}$. □

6.B Additional remarks. (a) As follows from Theorem 3.7(a) and Proposition 6.2, whenever p is not surjective on composable triples, there exists a discrete fibration (F, φ) over $Eq(p)$ in which the equivalence relation F is not stably effective. Moreover, the proof of Proposition 6.2 tells us that F can be chosen to be not even effective. For that, consider diagram (22) in $\mathbf{Set}^\tau$, in which we take now A to be the "Nonassociative Four" $\mathbf{4}^{n.a.}$, and p to be a morphism in $\mathbf{Cat}$ that is not surjective on composable triples. Then the top equivalence relation of (22) belongs to $\mathbf{Cat}$, and its coequalizer in $\mathbf{Cat}$ must be the $\mathbf{Cat}$-reflection of its coequalizer in $\mathbf{Set}^\tau$, i.e., the reflection of the Nonassociative Four, which is the Associative Four of course, i.e., the ordered set $\mathbf{4} = \{0, 1, 2, 3\}$ regarded as a category. Since the canonical morphism from $\mathbf{4}^{n.a.}$ to $\mathbf{4}$ is obviously a regular epimorphism in the exact category $\mathbf{Set}^\tau$, we conclude that that equivalence relation is not effective in $\mathbf{Cat}$.

(b) Since we have mentioned the reflection from $\mathbf{Set}^\tau$ into $\mathbf{Cat}$, let us also briefly mention where it comes from. Any functor $T : C \to \mathbf{A}$ from a small category C to a small cocomplete category $\mathbf{A}$ determines a diagram

$$\check{C} \underset{R}{\overset{L}{\rightleftarrows}} A \qquad (79)$$

where:

- Y is the Yoneda embedding (as in Section 4);
- L is the unique (up to an isomorphism) colimit-preserving functor with $LY \approx T$;
- R is the right adjoint of L, explicitly defined by $R(A) = \hom(T(-), A)$.

Among many well-known important examples of adjunctions obtained this way, there are the (fundamental groupoid, nerve)-adjunction between simplicial sets and groupoids, and its standard modification with categories instead of groupoids, in which:

- $C = \Delta$ is the simplicial category ($=$ the category of finite ordinals regarded as categories);
- $A = \mathbf{Cat}$;
- T is the inclusion functor, hence R is what is called the nerve functor.

Clearly, the embedding $\mathbf{Cat} \to \mathbf{Set}^\tau$ used in this section is the similarly *truncated nerve functor* obtained from an appropriate composite $\tau \to \Delta \to \mathbf{Cat}$ instead of the inclusion $\Delta \to \mathbf{Cat}$. Therefore the reflector $\mathbf{Set}^\tau \to \mathbf{Cat}$ is a special case of the functor L in (79). Let us also recall that any of these two nerve functors (and several other good candidates!) yields a presentation of categories as *models of theories over sketches*.

(c) Our observation in (a) leads to the following question: Is it possible to find a discrete fibration (F, φ) over $Eq(p)$ in which the equivalence relation F is effective but not stably effective? If p was an effective descent morphism in $\mathbf{Set}^\tau$, then we at least need to find an A as in (22) (see also (a)) for which the canonical morphism into its reflection is a monomorphism but not an isomorphism. Such an A is like a category, but it does not have "enough composable pairs", i.e. its canonical morphism $A_2 \to A_1 \times_{A_0} A_1$ is injective but not bijective. However, p is an effective descent morphism in $\mathbf{Set}^\tau$ if and only if it is surjective on composable pairs of morphisms, if and only if it is a stably regular epimorphism in $\mathbf{Cat}$. The surjectivity on composable pairs implies that the non-bijectivity of $A_2 \to A_1 \times_{A_0} A_1$ will create similarly non-bijectivity for the pullback $E \times_B A$, and destroy our example. The conclusion is that the category $\mathbf{Cat}$ has the following very special exactness property:

(i) Although there exists a stably regular epimorphism p that is not effective descent morphism, every such p has the property that for every discrete fibration (F, φ) over $Eq(p)$, the effectiveness of F implies its stable effectiveness. In particular, a morphism p is an effective descent morphism if and only if it is a pullback stable regular epimorphism, and for every discrete fibration (F, φ) over $Eq(p)$, the equivalence relation F is effective.

Still, the question we started with has a positive answer: just take any morphism p in **Cat** that is not surjective on composable pairs and use the "Non-transitive Three" instead of the Non-associative Four. That is: we

- define $\mathbf{3}^{n.t.}$ as the subobject of (78) such that $(\mathbf{3}^{n.t.})_0 = \{0, 1, 2\}, (\mathbf{3}^{n.t.})_1$ consists of the identity arrows and u_{01} and u_{12}, and $(\mathbf{3}^{n.t.})_2$ is empty; that is $\mathbf{3}^{n.t.}$ can be briefly displayed as $0 \to 1 \to 2$.
- take $A = \mathbf{3}^{n.t.}$ and define $\alpha : A \to B$ in (22) as the morphism whose image is a composable pair in B that is not an image of a composable pair in E.

Then again (see (a)) the top equivalence relation of (22) belongs to **Cat**, and its coequalizer in **Cat** must be the **Cat**-reflection of its coequalizer in $\mathbf{Set}^\tau$, i.e., the reflection of $\mathbf{3}^{n.t.}$, which is the Transitive Three of course, i.e., the ordered set $\mathbf{3} = \{0, 1, 2\}$ regarded as a category. Since the canonical morphism from $\mathbf{3}^{n.t.}$ to $\mathbf{3}$ is a monomorphism, we conclude that that equivalence relation remains the same when we replace $\mathbf{3}^{n.t.}$ by $\mathbf{3}$; hence, it is effective also in **Cat**. On the other hand, the top equivalence relation of (22) cannot be stably effective because its coequalizer in **Cat** is a morphism to $\mathbf{3}$ that factors through $\mathbf{3}^{n.t.}$ in $\mathbf{Set}^\tau$ and hence is not surjective on composable pairs of morphisms.

(d) Exactly the same arguments apply to the category **Preord** of preordered sets, and they tell us that every effective descent morphism of preorders is surjective on composable pairs. The converse is also true and can be proved like the first part of the proof of Proposition 6.2. (Note that the associativity of composition is not involved here, we do not need composable triples; moreover, the same can be done with the category $1 \rightrightarrows 0$ instead of τ; see [11]). Hence, a morphism $p : E \to B$ in **Preord** is an effective descent morphism if and only if for every $b_0 \leq b_1 \leq b_2$ in B there exists $e_0 \leq e_1 \leq e_2$ in E with $p(e_i) = b_i$ ($i = 0, 1, 2$). On the other hand, it is easy to show that p is a stably regular epimorphism in **Preord** if and only if it is surjective on morphisms (that is: on ordered pairs, i.e., for all $b_0 \leq b_1$ in B there exists $e_0 \leq e_1$ in E with $p(e_i) = b_i$ ($i = 0, 1$)), and that every discrete fibration (F, φ) over an effective equivalence relation in **Preord** has F itself effective. Therefore the category **Preord** has another special exactness property (to be compared with (i) in (c)!):

(ii) Although there exists a stably regular epimorphism p that is not an effective descent morphism, any morphism p has the property that for every discrete fibration (F, φ) over $Eq(p)$, the equivalence relation F is itself effective. In particular, a morphism p is an effective descent morphism if and only if it is a stably regular epimorphism, and for every discrete fibration (F, φ) over $Eq(p)$, the coequalizer of F is a stably regular epimorphism.

So, although **Preord** is a full subcategory in **Cat** closed under limits, their properties, namely (i) in (c) for **Cat** and (ii) here, are "contradictory": if both of them were satisfied in the same category, then every pullback stable regular epimorphism in that category would be an effective descent morphism.

(e) The descent constructions in **Preord** are examined in detail in the above-mentioned paper [11]. It turns out that they provide a really illuminating motivation and explanation for the more complicated topological descent constructions, including the quite difficult characterization of effective descent morphisms in Top in terms of convergence structures as given in the paper [19], which also provides the first example of a non-effective descent morphism. However, the first *finite* example of this type was given in [20], which led to a better understanding of the fact that one should use preordered sets as a guide to topological descent, especially in the finite case where preordered sets are the *same* as topological spaces.

7. Towards applications of descent theory: objects that are "simple" up to effective descent

As explained in Sections 3 and 5, any effective descent morphism $p : E \to B$ in the ground category $\mathbf{C}$ can be used to describe the *bundles* over B, i.e., the objects of $(\mathbf{C} \downarrow B)$, as certain structures that occur in various equivalent forms (as discrete fibrations, or algebras over a monad, or others) in the category $(\mathbf{C} \downarrow E)$. Moreover, a bundle (A, α) (over B) corresponds to such a structure on its *induced bundle* $p^*(A, \alpha)$ (i.e., its pullback along p). In order to be really useful, this obviously requires $p^*(A, \alpha)$ itself to be "simpler" than (A, α). Consequently, it is reasonable to begin by choosing an appropriate notion of a *trivial bundle* (also called a *split bundle*) and then to study those (A, α) for which $p^*(A, \alpha)$ is trivial; one then says that the bundle (A, α) is split (or *splits*, or *trivializes*, or *neutralizes*) over (E, p). We illustrate this idea by a discussion of three fundamental examples.

7.A Extensions in homological algebra. Let R be a ring and $\mathbf{C} = R\text{-}\mathbf{Mod}$ the category of R-modules; the bundles (A, α) in which $\alpha : A \to B$ is an epimorphism are called *extensions* of B. In this case "split" means "split epimorphism", i.e. split (= trivial) bundles are defined as those (A, α) in which α is a split epimorphism. It follows that a bundle (A, α) is an extension if and only if there exists an effective descent morphism $p : E \to B$ such that (A, α) is split over (E, p). There is a beautiful connection between the descent-theoretic description of the category of extensions, and the standard approach in homological algebra:

According to descent theory, as an effective descent morphism every epimorphism $p : E \to B$ of R-modules (see Theorem 3.7(f)) yields a category equivalence $K^p : (R\text{-}\mathbf{Mod} \downarrow B) \to \mathbf{DiscFib}(Eq(p))$. This equivalence restricts to an equivalence between the category $\mathbf{Spl}(E, p)$ of extensions of B split over (E, p) and the category of discrete fibrations (X, f) of equivalence relations over $Eq(p)$ in which f_0 is a split epimorphism (using the notation of (35)). However the category of equivalence relations in $R\text{-}\mathbf{Mod}$ is canonically equivalent to the category of monomorphisms, and under that equivalence the discrete fibrations correspond to the morphisms (u_1, u_0) (of monomorphisms) in which the "domain component"

u_1 is an isomorphism. Therefore, when we fix a kernel $k : K \to E$ of p (say, take $K = \{e \text{ in } E \mid p(e) = 0\}$ and k the inclusion map), we obtain a category equivalence between **DiscFib**$(Eq(p))$ and the category of systems (X_0, f_0, κ), in which $\kappa : K \to X_0$ is a monomorphism, $f_0 : X_0 \to E$ a split epimorphism, and $f_0\kappa = k$ (this equality tells us that we may omit the requirement on κ to be a monomorphism). Since every split epimorphism in R-**Mod** is a product projection, there is a further simplification: the systems (X_0, f_0, κ) can be replaced by pairs (L, λ), where L is an R-module and $\lambda : K \to L$ is an arbitrary homomorphism. Here we should not lose the right notion of morphism: the morphisms between the objects (L, λ) must agree with the morphisms between the corresponding objects (X_0, f_0, κ), and an easy calculation shows that a morphism $(L, \lambda) \to (L', \lambda')$ is to be defined as a pair of homomorphisms $(\varphi, \psi) = (\varphi : E \to L', \psi : L \to L')$ with

$$\varphi k + \psi \lambda = \lambda', \tag{80}$$

and that the composition of morphisms is to be defined as

$$(\varphi', \psi')(\varphi, \varphi) = (\psi' + \psi'\varphi, \psi'\psi). \tag{81}$$

Now, let **Ext**$^{(E,p)}(B, L)$ be the category of short exact sequences

$$0 \longrightarrow L \longrightarrow A \overset{\alpha}{\longrightarrow} B \longrightarrow 0 \tag{82}$$

with fixed B and L, and (A, α) in **Spl**(E, p). The equivalence between **Spl**(E, p) and the category of pairs (L, λ) above induces an equivalence

$$\mathbf{Ext}^{(E,p)}(B, L) \sim \mathbf{Hom}^{(E,k)}(K, L) \tag{83}$$

where **Hom**$^{(E,k)}(K, L)$ is the category whose objects are all homomorphisms $\lambda : K \to L$, and in which a morphism from λ to λ' is a homomorphism $\varphi : E \to L$ satisfying (80) with $\psi = 1_L$, i.e., satisfying

$$\varphi k + \lambda = \lambda'; \tag{84}$$

the composition of morphisms is defined as the addition of homomorphisms. In particular, this implies that **Hom**$^{(E,k)}(K, L)$ is a groupoid (which is a reformulation of the Short Five Lemma; see Chapter IV), and that λ and λ' are isomorphic in **Hom**$^{(E,k)}(K, L)$ if and only if their difference factors through k. Hence there is a bijection

$$Ext^{(E,p)}(B, L) \approx Coker(Hom_R(k, L)) \tag{85}$$

between the set of isomorphism classes $Ext^{(E,p)}(B, L)$ of objects in $Ext^{(E,p)}(B, L)$ and the cokernel of the abelian group homomorphism $Hom_R(k, L) : Hom_R(E, L) \to Hom_R(K, L)$.

On the other hand, as we know from homological algebra, for every R-module L, any short exact sequence

$$0 \longrightarrow K \overset{k}{\longrightarrow} E \overset{p}{\longrightarrow} B \longrightarrow 0 \tag{86}$$

of R-modules determines a long exact sequence of abelian groups starting with

$$0 \to Hom_R(B, L) \to Hom_R(E, L) \to Hom_R(K, L) \to Ext^1_R(B, L) \to Ext^1_R(E, L); \quad (87)$$

in particular, it yields an isomorphism

$$Ker(Ext^1_R(p, L)) \approx Coker(Hom_R(k, L)) \quad (88)$$

between the kernel of the third arrow in (87) and the cokernel of the first one. Since the map $Ext^1_R(p, L)$ is defined via pulling back along p, and since the zero element in $Ext^1_R(E, L)$ is the split exact sequence $0 \to L \to L \oplus E \to E \to 0$, $Ker(Ext^1_R(p, L))$ is the same as $Ext^{(E,p)}(B, L)$. In fact, (85) and (88) provide exactly the same bijections! Let us also make the following remarks:

(a) Our readers should not get the impression that (85) is merely a bijection, in contrast to (88) which is a group isomorphism. Just note that every finite-product preserving functor from an additive category to the category of sets can be regarded as a functor to the category of abelian groups, and the same applies to all natural transformations between such functors. Another thing is that not only the bijection (88) follows from the bijection (85), but even the exact sequence (87) itself naturally appears as a corollary of a descent-theoretic calculation, namely of our description of the category $Ext^{(E,p)}(B, L)$. To this end, let us first observe that any abelian group homomorphism $f : X \to Y$ determines a groupoid $G = G(f : X \to Y)$, because of the standard equivalence between crossed modules and internal categories in the category of groups; the objects of that groupoid are all elements in Y, and a morphism $y \to y'$ is a pair of the form (x, y), where x is an element in X with $f(x) + y = y'$. Next, the set $\Pi_0(G)$ of isomorphism classes of objects of G has a natural group structure determined by the exact sequence

$$0 \longrightarrow \Pi_1(G) \longrightarrow X \longrightarrow Y \longrightarrow \Pi_0(G) \longrightarrow 0. \quad (89)$$

Here $\Pi_1(G) = Aut(0)$, the automorphism group of $0 \in Y$ in G, is sometimes called the fundamental group of G. Then, according to our calculation of $\mathbf{Ext}^{(E,p)}(B, L)$, it is equivalent to the groupoid $G(Hom_R(k, L) : Hom_R(E, L) \to Hom_R(K, L))$, and the sequence (89) written for $f = Hom_R(k, L)$ is nothing but the sequence (87) with its two last members replaced by the kernel of the homomorphism between them.

(b) There is no difficulty in extending all our arguments from R-**Mod** to any abelian category. In particular we could apply them to the opposite of the category of R-modules, and hence establish a connection between (85) and exact sequences

$$0 \to Hom_R(L, K) \to Hom_R(L, E) \to Hom_R(L, B) \to Ext^1_R(L, K) \to Ext^1_R(L, E). \quad (90)$$

(c) It is a basic fact in homological algebra of modules, that each of the exact sequences (87) and (89) provide a method of calculation of the functor Ext^1_R. Indeed, if E is either a projective or an injective R-module, then either $Ext^1_R(E, L) = 0$ or $Ext^1_R(L, E) = 0$ respectively, and accordingly either (87) presents $Ext^1_R(B, L)$ as the cokernel of $Hom_R(E, L) \to Hom_R(K, L)$, or (89) presents $Ext^1_R(L, K)$ as the cokernel of $Hom_R(L, E) \to Hom_R(L, B)$. That is, the calculations of Ext^1_R in homological algebra can be considered as part of descent theory.

(d) There are many important intermediate levels of generality in the "homo-logical-algebraic" theory of extensions between the abelian case and the following two generalizations: the abstract theory of torsors, where (fibrational) descent the-ory leads to cohomological descriptions, and of (generalized, non-abelian) central extensions, where descent theory (in the form presented in this chapter) leads to descriptions in terms of certain internal (pre)category actions. In particular, as soon as non-abelian groups are involved, there is still a connection between the descent constructions and a modified version of (87), with what S. Mac Lane [15] calls *Opext* instead of *Ext*, but nothing like the connection with (88) has ever been studied. The first reason why *Opext* (and many other complications) occurs lies already in the structure of split epimorphisms: product projections become semi-direct product projections, morphisms into semi-direct products are not just pairs of morphisms into components, which makes our passage above from (X_0, f_0, κ) to (L, λ) more complicated and brings cocycles instead of homomorphisms; the κ's must be normal monomorphisms, and so on. Also, in the non-abelian case the monadic approach and Grothendieck's original approach to descent seem to work better than the "discrete-fibrational" one, which again is related to the necessity of working with various cocycles instead of homomorphisms.   □

7.B Separable extensions in classical Galois theory.

A field extension $B \subseteq K$ is a *finite separable extension* if and only if there exists an irreducible *separable polynomial* $u \in B[x]$ with $K \approx B[x]/uB[x]$. For a field E containing the splitting field of u we have $u = (u - a_1) \cdots (u - a_n)$ in $E[x]$ for some $a_1, \cdots, a_n$ in E (recall that separability of u means that $a_i = a_j$ implies $i = j$), which gives

$$
\begin{aligned}
E \otimes_B K &\approx E \otimes_B (B[x]/uB[x]) \approx E[x]/uE[x] \\
&\approx (E[x]/(u - a_1)E[x]) \times \cdots \times (E[x]/(u - a_n)E[x]) \approx E^n
\end{aligned}
\tag{91}
$$

and then suggests to:

- take $p : E \to B$ to be the inclusion map from B to E regarded as a morphism in the opposite category $(\mathbf{CRing})^{\mathrm{op}}$ of commutative rings;
- call a commutative E-algebra *trivial* (or *split*) if it is of the form E^n (=the product of n copies of E in the category of E-algebras) for some natural n;
- consider the pullback functor $p^* : ((\mathbf{CRing})^{\mathrm{op}} \downarrow B) \to ((\mathbf{CRing})^{\mathrm{op}} \downarrow E)$, which is the dual form of the extension-of-scalars functor $E \otimes_B (-)$ from the category of commutative B-algebras to the category of commutative E-algebras (thus, the categorical notion of split will be related to splitting of polynomials!);
- call an object (A, α) in $((\mathbf{CRing})^{\mathrm{op}} \downarrow B)$ (=a commutative B-algebra A, with $\alpha : B \to A$ being the structure map) *separable* if $p^*(A, \alpha) = E \otimes_B A$ is trivial.

According to categorical Galois theory (see [4] and references there), this brings the following conclusions, which we just record here omitting most of the proofs:

(a) A B-algebra is separable if and only if it is (isomorphic to) a finite product of finite separable field extensions of B. Moreover, if $B \subseteq E$ above is a finite Galois extension, then $p^*(A, \alpha) = E \otimes_B A$ is trivial if and only if A a finite product of subextensions of E.

(b) As follows from (a), the Grothendieck's form of the fundamental theorem of Galois theory tells us that there is a category equivalence

$$\mathbf{Spl}(E, p) \sim (\mathbf{FinSet})^G \tag{92}$$

where $\mathbf{Spl}(E, p)$ is the opposite category of all $A = (A, \alpha)$ with trivial $p^*(A, \alpha) = E \otimes_B A$, $\mathbf{FinSet}$ the category of finite sets, and G the Galois group of the Galois extension $B \subseteq E$.

(c) As also follows from (a), the Zariski spectrum $Spec(A)$ of a separable B-algebra A is a finite discrete space, and it can be identified with the set $I(A)$ of minimal idempotents in A. Moreover, this determines a functor $\mathbf{Spl}(E, p) \to \mathbf{FinSet}$, which is the left adjoint of the functor H defined by

$$H(X) = \text{the } B\text{-algebra of all maps } X \to B \tag{93}$$

or, briefly as $H(\{x_1, \cdots, x_n\}) = B^n$.

(d) Applying the calculation (91) to $K = E$, we obtain

$$E \otimes_B E \approx E^n \tag{94}$$

which is used as one of several equivalent definitions of a Galois extension in Galois theory of commutative rings and in Grothendieck's theory of étale coverings in algebraic geometry. However, beyond the case of fields one has also to require $p : E \to B$ to be an effective descent morphism, in $(\mathbf{CRing})^{\mathrm{op}}$, or in the appropriate category of schemes, respectively. Since those categories are far from being exact or regular, the description of descent morphisms there needs a special treatment. For the purposes of Galois theory it is sufficient to know that $p : E \to B$ is an effective descent morphism whenever E is faithfully flat as a B-module, which immediately follows from Beck's monadicity theorem since in the faithfully flat case the functor $p^* = E \otimes_B (-)$ preserves all coequalizers and reflects isomorphisms. In particular p is always an effective descent morphism when B is a field (and E a non-zero ring).

(e) Returning to field extensions, we observe that the number n in (94) must be equal to the dimension of E as a vector space over B, and therefore also to the number $[G : 1]$ of elements in the Galois group. Hence there is a bijection

$$I(E \otimes_B E) \approx G \tag{95}$$

Moreover, there is an amazing fact behind this:

(f) On the one hand the functor $\mathbf{Spl}(E, p) \to \mathbf{FinSet}$ does not preserve the pullback $E \times_B E(= E \otimes_B E$ since the pullback is formed in the opposite category of commutative rings) unless $E = B$, as we see from (94) (or from (95)). On the other hand (94) easily implies that I preserves some other pullbacks, in particular that the canonical maps

$$I((E \times_B E) \times_E (E \times_B E)) \to I(E \times_B E) \times_{I(E)} I(E \times_B E) \tag{96}$$

$$I((E \times_B E) \times_E (E \times_B E) \times_E (E \times_B E)) \to I(E \times_B E) \times_{I(E)} I(E \times_B E)_{I(E)} I(E \times_B E) \tag{97}$$

are bijections. Accordingly, the I-image $I(Eq(p))$ of the equivalence relation $Eq(p)$ $= (E \times_B E \rightrightarrows E)$ (see (59)) is not an equivalence relation, but still a groupoid, in fact a group (since its $I(E)$ of objects has only one element), which turns out to be isomorphic to the Galois group G! Note also, that the fact that $I(Eq(p))$ is a groupoid because (96) and (97) are bijections, can be easily deduced from Proposition 6.1 since $Eq(p)$ itself is a groupoid and the map (96) is what appears in 6.1(a) as $A_2 \to A_1 \times_{A_0} A_1$, which is related to presentations of categories as models of finite limit theories and hence to simplicial sets.

(g) Using the fact that every ring homomorphism preserves idempotents, one easily establishes a category equivalence

$$(\text{trivial } E\text{-algebras})^{\text{op}} \sim \textbf{FinSet} \tag{98}$$

which of course coincides with (92) in the "trivial case" $E = B$. On the other hand we observe that descent theory, formulated via (60), tells us that the category $\textbf{Spl}(E,p)$ is equivalent to the opposite category of trivial E-algebras equipped with $Eq(p)$-actions, and then (after some technical work!) yields an equivalence (92) with G defined as $G = I(Eq(p))$. That is, (92) = (98) + descent theory, or, making the formulation "more impressive",

$$\text{Galois theory} = \text{trivial Galois theory} + \text{descent theory} \tag{99}$$

(h) The critical Reader will ask: does descent theory help to show that the "strange" group $I(Eq(p))$ is isomorphic to the ordinary Galois group G defined as the B-automorphism group of E? The answer is: Yes, it does! Indeed, as soon as we know that the equivalence (92) holds true for some finite group G, we observe:

- Inside the category $\textbf{Spl}(E,p)$, the B-algebra E can be identified categorically as the largest connected object, where connected means indecomposable into a coproduct (of two non-zero objects), and largest means that every morphism from a connected object into it is an isomorphism.
- Since G regarded as a G-set (with the G-action defined as the multiplication of G) can be similarly identified inside the category $(\textbf{FinSet})^G$, we conclude that E and G must correspond to each other under any equivalence $\textbf{Spl}(E,p) \sim (\textbf{FinSet})^G$.
- Previous observation implies that the automorphism group of E in $\textbf{Spl}(E,p)$ must be isomorphic to the automorphism group of G regarded as a G-set, i.e., must be isomorphic to G, as desired. □

7.C Locally trivial fibre bundles and covering spaces in algebraic topology. In 7.A, the trivial bundles were defined as split epimorphisms, and since the ground category was additive with kernels (in fact abelian), they are the same as product projections. This is exactly how algebraic topologists define the trivial bundles in the category **Top** of topological spaces, as the product projections. But since **Top** is far from being additive, they are far from being just split epimorphisms, as one can also see from simple geometrical pictures of course. In fact, the letters we use for $p : E \to B$ arrived from algebraic topology, where B is called the *base space*, E the *extension space*, and p the *projection*, imitating the product projection of a trivial bundle. The same applies to the terminology for several notions used here (and generally in category theory), from induced bundles and bundles themselves to various kinds of fibrations (some explanations are given below). Still, let us ask ourselves, do algebraic topologists ever consider anything like a bundle (A, α) over B split over (E, p), where $p : E \to B$ is an effective descent morphism in **Top**?

A bundle (A, α) over B in **Top** is said to be *locally trivial* if every point b in B has an open neighborhood E_b such that the restriction of α to $\alpha^{-1}(E_b)$ determines a trivial bundle over E_b. That is, (A, α) is locally trivial if and only if for every point b in B there exists a pullback diagram of the form

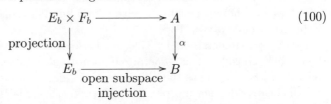

$$(100)$$

Here the space F_b is called the *fibre over b* since it is indeed homeomorphic to the inverse image $\alpha^{-1}(b)$ of b under α. The homeomorphism $F_b \approx \alpha^{-1}(b)$ implies the homeomorphism $F_b \approx F_{b'}'$,

- for $b' \in E_b$,
- therefore also for $E_b \cap E_{b'} \neq \emptyset$,
- therefore whenever B has a connected compact subset containing b and b',
- therefore for arbitrary b and b' whenever B is *path-connected*, i.e., for every two points b_0 and b_1 in B, there exists a path from b to b', i.e. a continuous map f from the closed unit interval $[0, 1]$ to B with $f(0) = b_0$ and $f(1) = b_1$.

That is, all fibers F_b are usually homeomorphic to each other, in which case the index b is to be omitted and "the" space F is called the fibre of α; the bundle (A, α) is then called a *locally trivial fibre bundle with fibre F*. If, in addition, F is discrete, then (A, α) is called a *covering space*. The authors of algebraic topology textbooks should forgive us for writing (A, α) instead of their (E, p); what we would like to take as $p : E \to B$ is the canonical map (to B) from the coproduct $E = \coprod E_b$ of all E_b involved in the pullbacks (100), or just as many of them as necessary to make the canonical map surjective. All the pullbacks (100) (for all

b in B, provided that all F_b are homeomorphic to each other) together carry the same information as the following one:

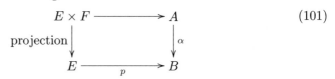

$$(101)$$

Using this, it is not difficult to prove that a bundle (A, α) is a locally trivial fibre bundle (with some fibre F) if and only if it is split over some (E, p), where $p : E \to B$ is a surjective local homeomorphism. Again, the Beck monadicity theorem gives an easy proof of the fact that every surjective local homeomorphism is an effective descent morphisms, and we conclude: algebraic topologists do use what we called bundles split over (E, p) as a very basic notion, but only for a very special class of effective descent morphisms p. Again, let us briefly mention some other related conclusions and remarks:

(a) It is not really important to require all F_b above to be homeomorphic to each other. Without that requirement there is still a reasonable description of locally trivial bundles, namely as those which split over some (E, p), where p is a surjective local homeomorphism, with trivial bundles replaced by their coproducts in the arrow category of **Top**.

(b) As was understood by Grothendieck, descent constructions can be (and should be!) considered as generalized *local-to-global* constructions. This step of generalization is as big as the difference between surjective local homeomorphisms and all effective descent morphisms in **Top**. Yet, there is no evidence that anybody tried to describe effective descent morphisms in **Top** before 1990. After several preliminary attempts by several authors, a characterization was given in [19] (see also the references therein). However these results do not seem to be generally known among algebraic topologists; hence their application still needs to be attended to.

(c) Restricting ourselves from locally trivial bundles to covering spaces (= the case of discrete fibres), we wish to mention that descent theory can be applied in a way very similar to the procedure for separable algebras explained in Example 7.2. In particular, if B is a "good" space and (E, p) a *universal covering space* of B, then the fundamental group(oid) $\Pi_1(B)$ of B can be defined $I(Eq(p))$, where $I = \Pi_0$ is the functor carrying topological spaces to the sets of their connected components. Similarly to (92) there is then a category equivalence

$$(\text{Covering spaces over } B) \sim \mathbf{Set}^{\Pi_1(B)} \qquad (102)$$

which, together with the algebraic-geometrical version of (92) is part of Grothendieck's Galois theory and the reason for his famous conclusion that covering spaces are classified by actions of quotients of equivalence relations.

(d) Since discrete fibrations of equivalence relations play an important role in this chapter, we should say a few words about the corresponding topological notion. There are various kinds of *fibrations* studied in algebraic topology; they are defined via various kinds of *lifting properties* which, roughly speaking, "help to compare fibres". As we already mentioned, the existence of a path from b to b' in B implies the existence of a homeomorphism $\alpha^{-1}(b) = F_b \approx F_{b'} = \alpha^{-1}(b')$ whenever (A, α) is a locally trivial bundle over B (see the explanations given between (100) and (101)). On the other hand, if we start with an arbitrary bundle (A, α), the existence of such homeomorphisms might help to prove (see (e) below) that it is locally trivial. Therefore it is good to have a geometrical method of constructing homeomorphisms $\alpha^{-1}(b) \approx \alpha^{-1}(b')$. Now, an obvious way to try is to fix a path f from b to b', and then to define the image in $\alpha^{-1}(b')$ of a point a in $\alpha^{-1}(b)$ as the end of a path g that is a lifting of f and whose starting point is a:

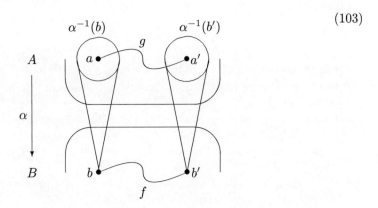

$$(103)$$

However, this procedure can be successful only under further strong geometrical conditions on α, and clearly the whole story would be much simpler if α had the *unique path lifting property*, which says that the g above always does exist and is uniquely determined by f and α. If α is a product projection $B \times F \to B$, then the uniqueness of path liftings simply means that every continuous map $f : [0, 1] \to B$ is constant, and since the spaces studied in algebraic topology usually have open path-connected components, this essentially means that F is discrete. That is, if we allow ourselves to require the unique path lifting property, then our theory will be applicable only to covering spaces, not to general locally trivial bundles. Still, it is useful to observe that the display (103) looks similar to (38), which in fact leads to a connection between topological/geometrical fibrations and fibrations of the corresponding fundamental groupoids. In particular, it turns out that covering spaces correspond *exactly* to the discrete fibrations (a notion closely related to the uniqueness requirement above!), thus yielding a category equivalence

$$\text{(covering spaces over } B) \sim \text{(discrete fibrations over } \Pi_1(B)) \qquad (104)$$

which is a reformulation of (102).

(e) As we have seen above, the connection between locally trivial fibre bundles and fibrations defined via path-lifting properties (with additional conditions) is based on the problem of constructing an isomorphism $(A, \alpha) \approx (B \times F,$ first projection) for some F out of a collection of isomorphisms $\alpha^{-1}(b) \approx \alpha^{-1}(b')$ constructed for all pairs b, b' of elements in B, and/or on doing the same for each E_b instead of B (see above). This is how descent theory arrives once more, this time with $B \to 1$ or $E_b \to 1$ playing the role of $p : E \to B$. Since $B \to 1$ and $E_b \to 1$ as split epimorphisms (provided B and E_b are non-empty) are effective descent morphisms (see Corollary 4.3), descent theory tells us that the conditions on the chosen collection of isomorphisms to be checked are what we described in Remark 5.7(d). □

References

[1] J. Bénabou, J. Roubaud, Monades et descente, C. R. Acad. Sc. 270 (1970) 96-98.

[2] M.A. Bednarczyk, A. Borzyszkowski, W. Pawlowski, Generalized congruences – epimorphisms in **Cat**, Theory Appl. Categories 5 (1999) 266-280.

[3] R. Börger, Making factorizations compositive, Comment. Math. Univ. Carolinae 32 (1991) 749-759.

[4] F. Borceux, G. Janelidze, *Galois Theories*, Cambridge Studies in Advanced Mathematics **72** (Cambridge University Press, Cambridge 2001).

[5] D. Bourn, G. Janelidze, Protomodularity, descent and semidirect products, Theory Appl. Categories 4 (1998) 37-46.

[6] B.J. Day, G.M. Kelly, On topological quotient maps preserved by pullbacks and products, Proc. Cambridge Phil. Soc. 67 (1970) 553-558.

[7] J. Giraud, Methode de la descente, Bull. Soc. Math. France Memoire 2 (1964).

[8] A. Grothendieck, Catégories fibrées et descente, Exposé VI, in: Revêtements Etales et Groupe Fondamental (SGA1), Lecture Notes in Math. 224 (Springer, Berlin 1971), pp. 145-194.

[9] J.R. Isbell, Epimorphisms and dominions III, Amer. J. Math. 90 (1968) 1025-1030.

[10] G. Janelidze, L. Márki, W. Tholen, Semi-abelian categories, J. Pure Appl. Algebra 168 (2002) 367-386.

[11] G. Janelidze, M. Sobral, Finite preorders and topological descent I, J. Pure Appl. Algebra 175 (2002) 187-205.

[12] G. Janelidze, W. Tholen, Facets of descent, I, Appl. Categorical Structures 2 (1994) 1-37.

[13] G.M. Kelly, Monomorphisms, epimorphisms, and pull-backs, J. Austral. Math. Soc. A 9 (1969) 124-142.

[14] I. LeCreurer, Descent of internal categories, thesis (Louvain-la-Neuve 1999).

[15] S. Mac Lane, Homology, Springer (Berlin 1963).

[16] J. Mersch, Le problème du quotient dans les catégories, Mém. Soc. Roy. Sci. Liège (5), vol. 11 (1965), no. 1.

[17] E. Michael, Bi-quotient maps and cartesian products of quotient maps, Ann. Inst. Fourier (Grenoble) 18 (1969) 287-302.

[18] J. Reiterman, M. Sobral, W. Tholen, Composites of effective descent morphisms, Cahiers Topologie Géom. Différentielle Catégoriques 34 (1993) 193-207.

[19] J. Reiterman, W. Tholen, Effective descent morphisms of topological spaces, Topology Appl. 57 (1994) 53-69.

[20] M. Sobral, Another approach to topological descent theory, Appl. Categorical Structures 9 (2001) 505-516.

[21] M. Sobral, W. Tholen, Effective descent morphisms and effective equivalence relations, in: Category Theory 1991, CMS Conference Proceedings (Amer. Math. Soc., Providence, R. I., 1992), pp. 421-432.

[22] R. Street, Cosmoi of internal categories, Trans. Amer. Math. Soc. 258 (1980) 271-318.

[23] R. Street, R.F.C. Walters, Yoneda structures on 2-categories, J. Algebra 50 (1978) 350-379.

Mathematical Institute,
Georgian Academy of Sciences,
M. Alexidze Str. 1
39003 Tbilisi, Georgia
E-mail: george_janelidze@hotmail.com

Departamento de Matemática
Universidade de Coimbra
Apartado 3008
3001-454 Coimbra, Portugal
E-mail: sobral@mat.uc.pt

Department of Mathematics and Statistics
York University
Toronto, Canada M3J 1P3
E-mail: tholen@mathstat.yorku.ca

Index

I.12 indicates chapter I, page 12